Communications in Computer and Information Science 2972

Rationale

The CCIS series is devoted to the publication of proceedings of computer science conferences. Its aim is to efficiently disseminate original research results in informatics in printed and electronic form. While the focus is on publication of peer-reviewed full papers presenting mature work, inclusion of reviewed short papers reporting on work in progress is welcome, too. Besides globally relevant meetings with internationally representative program committees guaranteeing a strict peer-reviewing and paper selection process, conferences run by societies or of high regional or national relevance are also considered for publication.

Topics

The topical scope of CCIS spans the entire spectrum of informatics ranging from foundational topics in the theory of computing to information and communications science and technology and a broad variety of interdisciplinary application fields.

Information for Volume Editors and Authors

Publication in CCIS is free of charge. No royalties are paid, however, we offer registered conference participants temporary free access to the online version of the conference proceedings on SpringerLink (http://link.springer.com) by means of an http referrer from the conference website and/or a number of complimentary printed copies, as specified in the official acceptance email of the event.

CCIS proceedings can be published in time for distribution at conferences or as post-proceedings, and delivered in the form of printed books and/or electronically as USBs and/or e-content licenses for accessing proceedings at SpringerLink. Furthermore, CCIS proceedings are included in the CCIS electronic book series hosted in the SpringerLink digital library at http://link.springer.com/bookseries/7899. Conferences publishing in CCIS are allowed to use our online conference service (Meteor) for managing the whole proceedings lifecycle (from submission and reviewing to preparing for publication) free of charge.

Publication process

The language of publication is exclusively English. Authors publishing in CCIS have to sign the Springer CCIS copyright transfer form, however, they are free to use their material published in CCIS for substantially changed, more elaborate subsequent publications elsewhere. For the preparation of the camera-ready papers/files, authors have to strictly adhere to the Springer CCIS Authors' Instructions and are strongly encouraged to use the CCIS LaTeX style files or templates.

Abstracting/Indexing

CCIS is abstracted/indexed in DBLP, Google Scholar, EI-Compendex, Mathematical Reviews, SCImago, Scopus. CCIS volumes are also submitted for the inclusion in ISI Proceedings.

How to start

To start the evaluation of your proposal for inclusion in the CCIS series, please send an e-mail to ccis@springer.com

Hamid R. Arabnia · Leonidas Deligiannidis ·
Farzan Shenavarmasouleh ·
Farid Ghareh Mohammadi · Soheyla Amirian
Editors

Emerging Trends in Large Language Models

12th International Conference on Computational Science and Computational Intelligence, CSCI 2025
Las Vegas, NV, USA, December 3–5, 2025
Proceedings Part VI

Springer

Editors
Hamid R. Arabnia
University of Georgia
Athens, GA, USA

Leonidas Deligiannidis
Wentworth Institute of Technology
Boston, MA, USA

Farzan Shenavarmasouleh
Medialab Inc.
Lawrenceville, GA, USA

Farid Ghareh Mohammadi
Verify Radiology Images Consultants
Houston, TX, USA

Soheyla Amirian
Pace University
New York, NY, USA

ISSN 1865-0929 ISSN 1865-0937 (electronic)
Communications in Computer and Information Science
ISBN 978-3-032-28468-6 ISBN 978-3-032-28469-3 (eBook)
https://doi.org/10.1007/978-3-032-28469-3

This Springer imprint is published by the registered company Springer Nature Switzerland AG
The registered company address is: Gewerbestrasse 11, 6330 Cham, Switzerland

Preface

It is our great pleasure to present this collection of research papers from the 12th International Conference on Computational Science and Computational Intelligence (CSCI 2025). This volume, titled "Emerging Trends in Large Language Models", comprises selected papers accepted in the Research Track on Large Language Models (CSCI-LLM). Due to the unprecedented response from authors, the Research Track was elevated to a full International Conference (LLM). The CSCI conference was held in Las Vegas, Nevada, USA, from December 3 to December 5, 2025 (LLM was held as part of CSCI).

The CSCI 2025 International Conference brought together papers from a diverse array of communities, including researchers from universities, corporations, and government agencies. Accepted papers are published by Springer Nature, and the proceedings showcase solutions to key challenges in various critical areas of Computational Science and Computational Intelligence.

Computational Science (CS) is the study of addressing problems that are impossible to solve (or difficult to solve) without computers. CS can be considered to be the bridge between computer science and other sciences. The field is interdisciplinary by nature and includes the use of advanced computing capabilities to understand and solve complex problems. In short, CS is the science of using computers to do science. Computational Intelligence (CI) is the study of computational methods that exhibit intelligence. These methods adapt to changing environments and changing goals. There is a significant overlap between the fields of CI and Artificial Intelligence (AI). However, there is also a difference: AI techniques often involve top-to-bottom methods (i.e., methods to each solutions are imposed from the top) whereas CI techniques often involve bottom-up methods (i.e., solutions emerge from unstructured beginnings). An important part of CI includes a set of Nature-inspired computational approaches to address complex problems for which traditional methods are infeasible. Computational Science and Computational Intelligence both share the same objective: finding solutions to difficult problems. However, the methods to reach the solutions are different. The main objective of the CSCI Conference is to facilitate increased opportunities for cross-fertilization across CS and CI.

Considering the above broad outline, the CSCI 2025 International Conference was composed of the following focused Research Tracks:

Artificial Intelligence (CSCI-RTAI); Big Data and Data Science (CSCI-RTBD); Computational Science (CSCI-RTCS); Computational Intelligence (CSCI-RTCI); Computational Biology (CSCI-RTCB); Cyber Warfare, Cyber Defense, & Cyber Security (CSCI-RTCW); Signal & Image Processing, Computer Vision & Pattern Recognition (CSCI-RTPC); Smart Cities and Smart Mobility (CSCI-RTSC); Education — CS & CE (CSCI-RTED); Health Informatics and Medical Systems (CSCI-RTHI); Mobile Computing, Wireless Networks, & Security (CSCI-RTMC); Software Engineering (CSCI-RTSE); Internet of Things & Internet of Everything (CSCI-RTOT); Social Network Analysis, Social Media, & Mining (CSCI-RTNA); Cloud Computing and Data Centers

(CSCI-RTCC); Parallel & Distributed Computing (CSCI-RTPD); and Large Language Models (LLM). The scope of each track can be found at:

https://www.american-cse.org/csci2025/topics and https://www.american-cse.org/LLM2025/topics

An important objective of the CSCI 2025 International Conference and its associated Research Tracks was to foster opportunities for cross-fertilization between the fields of Computational Science and Computational Intelligence. The CSCI Conference is deeply committed to promoting diversity and eliminating discrimination, both in its role as a conference organizer and as a service provider. Our goal is to create an inclusive culture that respects and values differences, promotes dignity, equality, and diversity, and encourages individuals to reach their full potential. We are also dedicated, wherever possible, to organizing a conference that represents the global community. We sincerely hope that we have succeeded in achieving these important objectives.

The Steering Committee and the Program Committee would like to extend their gratitude to all the authors who submitted papers for consideration. CSCI 2025 received submissions from 48 countries, with approximately 48% of them coming from outside the United States. Each submitted paper underwent a rigorous peer-review process, with at least two experts (an average of 2.5 referees per paper) evaluating the submissions based on originality, significance, clarity, impact, and soundness. In cases where reviewers' recommendations were contradictory, a program committee member was tasked with making the final decision, often consulting additional referees for further guidance. The CSCI Conference followed the guidelines of COPE (Committee on Publication Ethics):

- Typical submissions underwent a single-blind peer review process, in which the authors remain unaware of the identities of the reviewers, while the reviewers are informed of the authors' identities.
- Papers authored by one or more members of the program committee, including co-chairs, were subjected to a double-blind peer review process, ensuring that neither the authors nor the reviewers were aware of each other's identities or affiliations.

The Research Track on Large Language Models (CSCI-LLM) received 294 submissions, of which 63 papers were accepted, resulting in a paper acceptance rate of 21%; we selected only 40 papers for this volume. Due to the success of CSCI-LLM, this track was elevated to a full conference.

We are deeply grateful to the many colleagues who contributed their time and effort to organizing the CSCI 2025 Conference. In particular, we extend our thanks to the members of the Program Committee, the Steering Committee, and the referees. We would also like to express our appreciation to the primary sponsor of the conference, the American Council on Science & Education. The list of members of the Program Committee for each track can be found at: https://www.american-cse.org/csci2025/committees

We extend our heartfelt gratitude to all the speakers and authors for their valuable contributions. We would also like to thank the following individuals and organizations for their support: the staff at the Luxor Hotel (an MGM property, conference/meeting department) and the staff of Springer, for their assistance in various aspects of the event.

We are pleased to present the proceedings of CSCI 2025 (includes selected papers of LLM Conference). These proceedings represent a collection of outstanding research contributions that reflect the diversity and depth of work in Large Language Models, Computational Intelligence, Knowledge Graphs, Learning Methods, and applications.

December 2025

Hamid R. Arabnia
Leonidas Deligiannidis
Farzan Shenavarmasouleh
Farid Ghareh Mohammadi
Soheyla Amirian

Organization

Steering Committee – Co-chairs (CSCI 2025)

Hamid R. Arabnia	University of Georgia, USA
Leonidas Deligiannidis	Wentworth Institute of Technology, USA
Fernando G. Tinetti	Universidad Nacional de La Plata, Argentina
Quoc-Nam Tran	Southeastern Louisiana University, USA

Publication Co-chairs and Co-editors

"Emerging Trends in Artificial Intelligence and Applications"
Research Track: Artificial Intelligence (CSCI-RTAI)
Part of Proceedings of The 2025 International Conference on Computational Science and Computational Intelligence (CSCI)

Hamid R. Arabnia	University of Georgia, USA
Leonidas Deligiannidis	Wentworth Institute of Technology, USA
Farzan Shenavarmasouleh	Medialab Inc., USA
Farid Ghareh Mohammadi	Verify Radiology Images Consultants, LLC, USA
Soheyla Amirian	Pace University, USA

Members of Steering Committee (CSCI 2025)

Babak Akhgar	Sheffield Hallam University, UK
Abbas M. Al-Bakry	University of Information Technology & Communications, Iraq
Nizar Al-Holou	University of Detroit Mercy, USA
Soheyla Amirian	Pace University, USA
Hamid R. Arabnia	University of Georgia, USA
Rajab Challoo	Texas A&M University-Kingsville, USA
Chien-Fu Cheng	Tamkang University, Taiwan
Hyunseung Choo	Sungkyunkwan University, South Korea
Kevin Daimi	University of Detroit Mercy, USA
Leonidas Deligiannidis	Wentworth Institute of Technology, USA
Eman M. El-Sheikh	University of West Florida, USA
Mary Mehrnoosh Eshaghian-Wilner	University of California, Los Angeles, USA

David L. Foster	Kettering University, USA
Ching-Hsien (Robert) Hsu Chung	Asia University, Taiwan
Farid Ghareh Mohammadi	Verify Radiology Images Consultants, LLC, USA
James J. (Jong Hyuk) Park	SeoulTech, South Korea
Mohammad S. Obaidat	University of Jordan, Jordan
Gerald Schaefer	Loughborough University, UK
Farzan Shenavarmasouleh	Medialab Inc., USA
Fernando G. Tinetti	Universidad Nacional de La Plata, Argentina
Quoc-Nam Tran	Southeastern Louisiana University, USA
Shiuh-Jeng Wang	Central Police University, Taiwan
Layne T. Watson	Virginia Polytechnic Institute & State University, USA
Chao-Tung Yang	Tunghai University, Taiwan
Mary Yang	University of Arkansas for Medical Sciences, USA

Program Committee – Computational Science and Computational Intelligence (CSCI 2025)

Afrand Agah	West Chester University of Pennsylvania, USA
Bharat Bhushan Agarwal	IFTM University, India
Omaima Nazar Ahmad AL	Al-Zaytoonah University of Jordan, Jordan
Wasim A. Al – Hamdani	Kentucky State University, USA
Ismail Khalil Al Ani	Ittihad University, United Arab Emirates
Mehran Asadi	Lincoln University, USA
Travis Atkison	University of Alabama, USA
Azita Bahrami	IT Consult, USA
Mehdi Bahrami	Fujitsu Laboratories of America, Inc., USA
P. Balasubramanian	Nanyang Technological University, Singapore
Petra Saskia Bayerl	Erasmus University Rotterdam, The Netherlands
Jane M. Binner	University of Birmingham, UK
Juan-Vicente Capella-Hernandez	Universitat Politècnica de València, Spain
Juan Jose Martinez Castillo	Universidad Nacional Abierta, Venezuela
Rui Chang	Mount Sinai School of Medicine, USA
Dongsheng Che	East Stroudsburg University of Pennsylvania, USA
Jianhung Chen	Chung Hua University, Taiwan
Mu-Song Chen	Da-Yeh University, Taiwan
Xin (Thomas) Chen	University of Hawaii, USA
Steve C. Chiu	Idaho State University, USA
Mark Yul Chu	University of Texas Rio Grande Valley, USA

Jose Alfredo F. Costa	Universidade Federal do Rio Grande do Norte, Brazil
Arianna D'Ulizia	National Research Council of Italy, Italy
Zhangisina G. Davletzhanovna	Central Asian University, Kazakhstan
Wesley Deneke	Western Washington University, USA
Noel De Palma	Grenoble Alpes University, France
Mohsen Doroodchi	University of North Carolina at Charlotte, USA
Levent Ertaull	California State University, East Bay, USA
Mahmood Fazlali	Shahaid Beheshti University, Iran
Boyuan Feng	University of California, Santa Barbara, USA
George A. Gravvanis	Democritus University of Thrace, Greece
Gheorghe Grigoras	"Gheorghe Asachi" Technical University of Iaşi, Romania
Ray Hashemi	Georgia Southern University, USA
Houcine Hassan	Universitat Politècnica de València, Spain
Abdolreza Hatamlou	Islamic Azad University, Iran
Bing He	Cisco System Inc., USA
Henry Hexmoor	Southern Illinois University at Carbondale, USA
Gahangir Hossain	Texas A&M University - Kingsville, USA
Guofeng Hou	AQR Capital Management, USA
Ren-Junn Hwang	Tamkang University, Taiwan
Naseem Ibrahim	Penn State Erie, The Behrend College, USA
Rabia Jafri	King Saud University, Saudi Arabia
Shahram Javadi	Azad University, Iran & University of Lisbon, Portugal
Young-Sik Jeong	Dongguk University, South Korea
Aleksandr Katkow	University of Computer Sciences & Skills, Poland
Byung-Gyu Kim	Sun Moon University, South Korea
Taihoon Kim	University of Tasmania, Australia
Sang-Wook Kim	Hanyang University, South Korea
Dattatraya Vishnu Kodavade	D.K.T.E Society's Textile & Engineering Institute, India
Elena B. Kozerenko	Russian Academy of Sciences, Moscow, Russia
Guoming Lai	Sun Yat-sen University, China
Hyo Jong Lee	Chonbuk National University, South Korea
Bo Liu	NEC Labs China, China
Eleanor Lockley	Sheffield Hallam University, UK
Jianbing Ma	Bournemouth University, UK
Julius Marpaung	Northeastern University, USA
Andrew Marsh	HoIP Telecom Ltd, UK
Juan Martinez	Universidad Gran Mariscal de Ayacucho, Venezuela

Praveen Meduri	California State University, Sacramento, USA
Ali Mostafaeipour	California State University, Fullerton, USA
Houssem Eddine Nouri	University of Tunis, Tunisia
Michael B. O'Hara	KB Computing, LLC, USA
Robert Ehimen Okonigene	Ambrose Alli University, Nigeria
Funminiyi Olajide	Nottingham Trent University, UK
Satish Penmatsa	University of Maryland - Eastern Shore, USA
Saman Parvaneh	Philips Research North America, USA
Ponalagusamy	National Institute of Technology, Tiruchirappalli, India
Laura L. Pullum	Oak Ridge National Laboratory, USA
Xuewei Qi	University of California, Riverside, USA
Junfeng Qu	Clayton State University, USA
Shahram Rahimi	University of Alabama, USA
Jaime Raigoza	California State University, Chico, USA
Arvind Ramanathan	Oak Ridge National Laboratory, USA
Iman M. Rezazadeh	University of California, Davis, USA
Om Prakash Rishi	University of Kota, India
Cristian Rodriguez Rivero	Universidad Nacional de Córdoba, Argentina
Seyed Roosta	Albany State University, USA
K. Martin Sagayam	Karunya Institute of Technology and Sciences, India
P. Sanjeevikumar	University of South-Eastern Norway, Norway
Benaoumeur Senouci	University of Southern Denmark, Denmark
Zhefu Shi	Microsoft Corporation, USA
Jawed Siddiqi	Sheffield Hallam University, UK
Akash Singh	IBM Corporation, USA
Anthony Skjellum	Auburn University, USA
Omer Muhammet Soysal	Southeastern Louisiana University, USA
Helman Stern	Ben-Gurion University of the Negev, Israel
Jonathan Z. Sun	University of Southern Mississippi, USA
Rahman Tashakkori	Appalachian State University, USA
Predrag Tosic	Microsoft, USA
Quoc-Nam Tran	Southeastern Louisiana University, USA
Jesus Vigo-Aguiar	University of Salamanca, Spain
Patrick Wang	East China Normal University, China & Northeastern University, USA
Weiqiang Wang	Opera Solutions LLC, USA
Yin Wang	Lawrence Technological University, USA
Alicia Nicki Washington	Duke University, USA
Wei Wei	Xi'an University of Technology, China
Yong Wei	University of North Georgia, USA

Santoso Wibowo	Central Queensland University, Australia
Alexander Woehrer	Vienna Science and Technology Fund, Austria
Bernd E. Wolfinger	University of Hamburg, Germany
Jay Xiong	NSEsoftware, LLC, USA
Hyun Yoe	Sunchon National University, South Korea
Jane You	Hong Kong Polytechnic University, China
Ismail Yusuf	Lamintang Education and Training Center, Indonesia
Wei Zhong	University of South Carolina Upstate, USA

Contents

Large Language Models, Knowledge Graphs, Prediction Methods, and Learning Methods

LangChain-Integrated LLMs and Knowledge Graphs for Multi-agent Reinforcement Learning in Complex Air Combat Environments

Abderahim Salhi[1(✉)], Indu Shukla[1], and Gary J. Briggs[2]

[1] United States Army Corps of Engineers, 3909 Halls Ferry Road, Vicksburg, MS 39180, USA
abderahim.salhi@usace.army.mil

[2] The RAND Corporation, 1776 Main Street, Santa Monica, CA 90401, USA

Abstract. Reinforcement learning (RL) agents show promise in air combat simulation, offering the potential for advanced autonomous decision-making in highly dynamic and complex environments. However, these agents often function as black boxes, limiting their explainability and usability in mission-critical scenarios. This paper explores the integration of multi-agent reinforcement learning (MARL) with knowledge graphs (KGs), large language models (LLMs), and the LangChain framework to enhance both the decision-making capacity and interpretability of RL agents in air combat simulations. We use a six-degree-of-freedom (6-DoF) aerial environment, leveraging the OpenAI Gymnasium framework for offline RL training. By combining structured domain knowledge from KGs, natural language reasoning from LLMs, and modular orchestration via LangChain, our approach aims to improve agent coordination, situational awareness, and transparency in air combat tasks. Experimental results demonstrate that this hybrid architecture not only maintains competitive performance, but also provides enhanced explainability, addressing a key limitation of traditional RL-based air combat agents.

Keywords: LLMs · MARL · Knowledge Graphs · LangChain

1 Introduction

Recent advances in reinforcement learning (RL) have enabled the development of intelligent agents capable of complex decision making in high-dimensional environments such as air combat simulation [1–3]. However, despite these successes, RL agents, particularly in multiagent settings, often lack transparency, making them difficult to deploy in critical real-world mission applications where explainability is crucial [4,5].

To address this, we propose a novel framework that integrates multiagent RL (MARL) with knowledge graphs (KGs), large language models (LLMs), and the

H. R. Arabnia et al. (Eds.): CSCI 2025, CCIS 2972, pp. 3–15, 2027.
https://doi.org/10.1007/978-3-032-28469-3_1

LangChain orchestration platform. This approach leverages structured domain knowledge and natural language reasoning to improve agent decision making and improve interpretability. We implement our system in an aerial combat environment with six degrees of freedom (6 DoF) using the OpenAI Gymnasium framework for offline RL, enabling robust training and evaluation. Our results show that the integration of symbolic and neural approaches can enhance both the performance and explainability of RL agents in complex air combat scenarios.

Air combat simulation presents a uniquely challenging environment for artificial intelligence due to its high-dimensional state and action spaces, non-linear flight dynamics, partial observability, and the need for agents to exhibit realistic, coordinated behaviors [1,3]. MARL has emerged as a promising approach to model complex interactions between multiple autonomous agents, enabling the creation of synthetic pilots that can train alongside or against humans in diverse combat scenarios [2,5]. However, as the number of agents and the complexity of their interactions grow, the strategy space expands exponentially, leading to significant scalability and coordination challenges [5].

RL algorithms, particularly when combined with deep learning, have demonstrated the ability to learn effective policies in such high-dimensional, dynamic environments [1,3]. Yet, these agents often operate as black boxes, making their decisions difficult to interpret and trust in mission-critical applications [4,5]. The six-degree-of-freedom (6-DoF) aerial environment further increases the complexity, as agents must reason over continuous control of position and orientation in three-dimensional space, requiring robust and adaptive learning frameworks.

Recent work has begun to address these challenges by integrating structured domain knowledge through knowledge graphs, which encode relationships between entities, tactics, and mission objectives, providing agents with contextual information that enhances both learning efficiency and decision explainability [4,6]. Neo4j, a widely adopted graph database, enables scalable and expressive representation of such structured knowledge, supporting both data-driven learning and human-understandable reasoning [7,8].

Large Language Models (LLMs) are increasingly being used to interpret, reason, and communicate complex strategies in natural language, bridging the gap between black-box agent behavior and human-understandable explanations [7,9]. Retrieval-augmented generation (RAG) architectures, which combine LLMs with external knowledge sources such as KGs, have shown promise in enhancing both the accuracy and explainability of AI systems [9,10]. Frameworks such as LangChain orchestrate interactions between LLMs, knowledge graphs, and APIs, facilitating the development of modular, explainable AI applications.

Our approach is distinct in that it brings together these technologies: MARL, Neo4j-based KGs, LLMs, RAG, and LangChain, within a unified architecture for air combat simulation. This integration, illustrated in Figs. 2, and 3 enables natural language question answering over agent behavior, transparent post-hoc analysis, and modular system design. The OpenAI Gymnasium framework provides a standardized platform for developing and evaluating RL algorithms in offline and simulated 6-DoF air combat environments, ensuring reproducibility and scalability [1,3]. By combining these technologies, our framework aims to produce MARL agents that are not only effective in high-fidelity air combat simulations but also transparent and explainable, addressing key limitations of current RL-based approaches.

2 Background

Air combat simulation has become an important test ground for artificial intelligence and machine learning, driven by advances in simulation outcomes and RL algorithms [3]. The complexity of air combat, with its high-dimensional continuous state and action spaces, partial observability, and adversarial dynamics, presents unique challenges for both agent design and evaluation [11,12].

Traditional approaches in air combat AI relied on rule-based or game-theoretic methods, but these quickly became impractical as the number of possible states and actions increased [13]. The adoption of RL, particularly deep RL, has enabled agents to learn sophisticated maneuvering and weapon deployment policies directly from simulation, outperforming traditional scripted logic and, in some cases, even human pilots. High-fidelity simulators such as AFSIM, BVR Gym, and JSBSim-Gymnasium have become essential to generate realistic training data and support the development of transferable RL policies [1–3,12].

Despite these advances, a key limitation remains: the lack of transparency and explainability in RL-driven air combat agents [11,12]. Deep RL agents, especially in multi-agent reinforcement learning (MARL) settings, often function as black boxes, making it difficult for operators and developers to understand or trust their decision-making processes in mission-critical scenarios [14,15]. This challenge is amplified in MARL, where coordination and emerging behaviors can further obscure the rationale behind agent actions.

Recent research has increasingly focused on integrating knowledge graphs and retrieval-augmented generation (RAG) with LLMs and MARL to address these explainability challenges. Knowledge graphs (KG) provide a structured and machine-interpretable representation of domain knowledge, causal relationships, and tactical doctrines [16,17]. Graph databases such as Neo4j facilitate the scalable storage and querying of these KGs, allowing simulation data, agent actions, and mission outcomes to be mapped into a unified, queryable structure [17]. In particular, integrating KGs with agentic systems and RAG pipelines has been shown to reduce hallucinations, improve factual consistency, and enable more nuanced context-aware reasoning, especially when human-in-the-loop mechanisms are used for graph curation and correction [18].

The rise of large language models (LLMs) has further accelerated this trend. LLMs can translate user questions into graph queries (such as Cypher for Neo4j), retrieve relevant information, and synthesize explanations in natural language. Retrieval-augmented generation (RAG) architectures, including emerging Graph-RAG approaches, leverage the structured nature of knowledge graphs to provide richer, context-aware responses compared to traditional vector search methods [19,20]. These systems typically involve extracting entities and relationships from unstructured data, populating the KG, and then using LLMs to query and interpret the graph, returning natural language answers to complex queries.

In the context of multi-agent RL, recent frameworks have begun to combine LLMs, adaptive knowledge graph memory, and structured communication protocols to improve agent cooperation, planning, and reasoning in dynamic environments. For example, decentralized MARL agents equipped with hierarchical knowledge graphs and LLM-powered reasoning have demonstrated improved scalability, efficiency, and collaboration in open-world tasks, outperforming both traditional MARL and LLM-only baselines [21].

Overall, the convergence of knowledge graphs, LLMs, and RAG architectures is rapidly advancing the field of explainable AI for complex, multi-agent scenarios. These advances lay the foundation for systems that not only perform at a high level but also provide the transparency and interpretability required for deployment in mission-critical environments such as air combat simulation.

3 Methodology

This section details the end-to-end pipeline for our explainable multi-agent reinforcement learning (MARL) framework in air combat simulation, from data generation and logging to knowledge graph construction and natural language reasoning. The architecture of our Gymnasium-based simulator is depicted in Fig. 1.

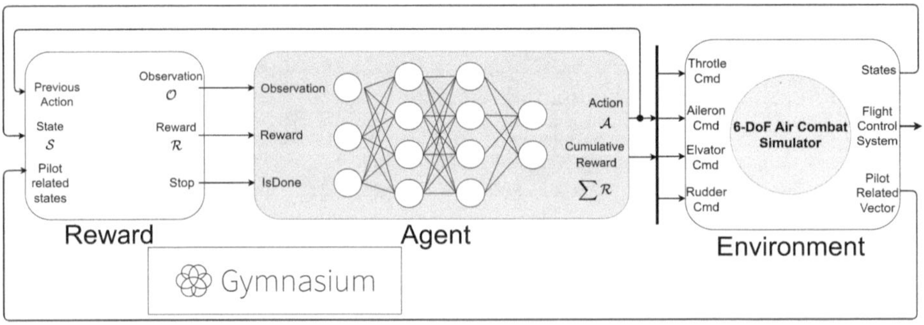

Fig. 1. Architecture of the Gymnasium-Based Multi-Agent Air Combat Simulator.

3.1 Dataset Generation and Training Environment

Recent advances in reinforcement learning (RL) have significantly advanced air combat simulation, as demonstrated by successes such as the DARPA AlphaDogfight Trials. Building upon these developments, our study employs a custom six-degree-of-freedom (6-DoF) aerial engagement environment, implemented using the OpenAI Gymnasium interface. This high-fidelity simulator enables agents to learn and refine complex maneuvers through direct interaction with a realistic physics engine.

Within this environment, two agents, the Red Team (offensive) and the Blue Team (defensive), operate in a competitive scenario. Each agent receives a continuous observation vector that encodes both its own aircraft state and its relative position to the opponent. At every time step, agents select continuous control actions (pitch, roll, rudder, throttle), which are then applied to the simulator. The reward function is carefully designed to incentivize effective tactics, penalize unrealistic or unsafe behaviors, and promote various realistic training episodes.

Simulation data is recorded at every step, capturing not only the agents' actions and resulting states, but also reward signals and episode termination flags. Each episode ends with critical events, such as significant damage or boundary violations. For this study, each generated episode consists of approximately 2,104 simulation steps, providing a substantial data set for both agent training and post-hoc analysis. This rigorous data logging strategy enables comprehensive post-hoc analysis and supports the downstream construction of a rich knowledge graph for explainable RL.

3.2 Simulation Data Structure and Logging

The structure and granularity of the logged data are critical for both agent training and subsequent explainability. Table 1 presents the mapping between each field in the exported CSV and its corresponding meaning in the simulation. The table is derived directly from the data export logic in our Gymnasium MARL environment, where each row corresponds to a simulation time step and each column to a specific control input, state variable, or engagement metric.

As shown in Table 1, our data logging strategy ensures that both control actions and resulting states are systematically recorded, enabling robust policy optimization and explainable analysis. The diversity and granularity of these fields are crucial for constructing a knowledge graph that accurately represents the simulation dynamics, thereby facilitating natural language question answering and interpretability of agent behavior.

3.3 Knowledge Graph Construction

To bridge the gap between raw simulation data and explainable RL, we ingest the MARL dataset into a Neo4j graph database. Each simulation step is assigned to nodes that represent agents, actions, states, metrics, rewards, and aggregate phase statistics. The relationships between these entities are designed to reflect the causal and temporal dynamics of the simulation, for example:

```
(:Agent)-[:PERFORMS]->(:Action),
(:Action)-[:RESULTS_IN]->(:State),
(:State)-[:HAS_METRICS]->(:Metrics)
```

Table 1. Simulation Data Fields Exported from Gymnasium MARL Environment

Item Name	Description
Step	Simulation time step index
Pitch Stick (aStickBack)	Agent's pitch control input (nose up/down), normalized
Roll Stick (aStickRight)	Agent's roll control input (left/right), normalized
Rudder Stick (aRudderRight)	Agent's rudder control input (yaw), normalized
Throttle (aThrottles)	Engine throttle setting, range $[-1, 1]$ (min to max power)
Altitude (km MSL)	Aircraft altitude above mean sea level (km)
Heading (Degrees)	Aircraft heading (degrees from North)
Pitch (Degrees)	Aircraft pitch angle (degrees)
Roll (Degrees)	Aircraft roll angle (degrees)
Mach	Aircraft speed as Mach number
Roll Rate	Rate of change of roll (deg/s)
Pitch Rate	Rate of change of pitch (deg/s)
Yaw Rate	Rate of change of yaw (deg/s)
Relative Elevation	Elevation angle to opponent (degrees)
Relative Azimuth	Azimuth angle to opponent (degrees)
Cross Track	Lateral distance to engagement centerline
Down Range	Longitudinal distance along engagement axis
Relative Velocity NED (North)	Relative velocity, North component (m/s)
Relative Velocity NED (East)	Relative velocity, East component (m/s)
Relative Velocity NED (Down)	Relative velocity, Down component (m/s)
Reward	Scalar reward assigned to agent at this step
Done	Boolean flag indicating episode termination

Data ingestion and graph enhancement are automated using Cypher queries, which also annotate states with derived properties such as flight phase (e.g., takeoff, combat, landing) and maneuver type (e.g., steep climb, sharp turn). This structured representation lays the foundation for downstream symbolic reasoning and explainable analysis as shown in Fig. 2.

3.4 Graph Reasoning with LLMs

Building on the knowledge graph, we enable the answer to natural language question using the LangChain framework's GraphCypherQAChain. This system integrates a large language model (OpenAI GPT-3.5-turbo) to translate user queries into Cypher statements, execute them against Neo4j, and synthesize the results into natural language answers. To ensure robust performance, we provide the LLM with explicit schema guidance, listing only the node labels and relationships present in the database, and including negative instructions to prevent hallucinations or invalid queries.

The system-level workflow for integrating these components is illustrated in Fig. 2. This figure depicts how simulation data and agent actions are ingested into a knowledge graph, which is then queried by an LLM-powered RAG retriever to enable explainable multi-agent RL in air combat simulation. The workflow highlights the interplay between structured knowledge, natural language reasoning, and modular orchestration—reflecting recent research trends that emphasize the value of combining KGs, RAG, and LLMs for both performance and interpretability in complex domains.

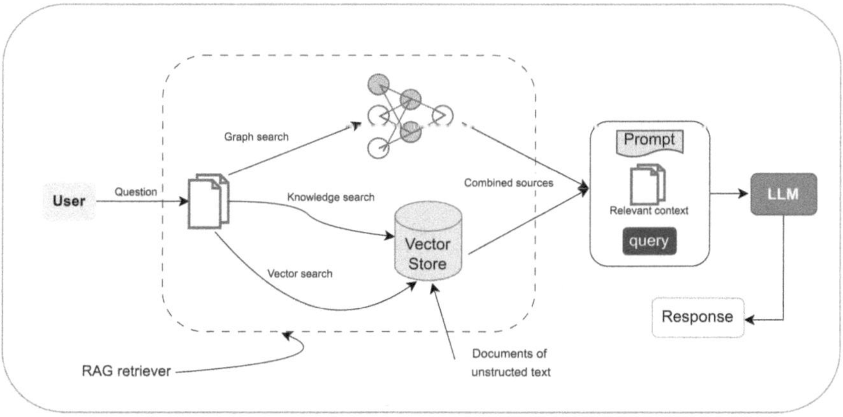

Fig. 2. System-level workflow: Knowledge Graph and LLMs RAG retriever for explainable multi-agent RL in air combat simulation.

3.5 Prompt Engineering and Schema Alignment

Prompt engineering is a critical component of our pipeline. The prompt template instructs the LLM to use only the provided schema, avoid duplicate return columns, and refrain from referencing non-existent labels (such as `Step`). Generated Cypher queries are logged and manually validated during development to ensure correctness and semantic alignment with the underlying graph.

3.6 Evaluation Protocol

Finally, we evaluated the system using a series of representative queries designed to probe both agent behavior and system explainability. Example queries include:

- What actions were taken when the jet was above 10km altitude?
- How often did the jet perform sharp turns?

- What was the average altitude during the Combat phase?
- List maneuvers associated with high rewards.
- How many steps occurred in each flight phase?

For each query, we collect the system's natural language response, the generated Cypher statement, and the Neo4j query result. This comprehensive evaluation allows us to assess both the accuracy and completeness of the explainable RL framework.

The overall system architecture, integrating Gymnasium MARL, knowledge graphs, LLMs, and LangChain orchestration, is visualized in Fig. 3.

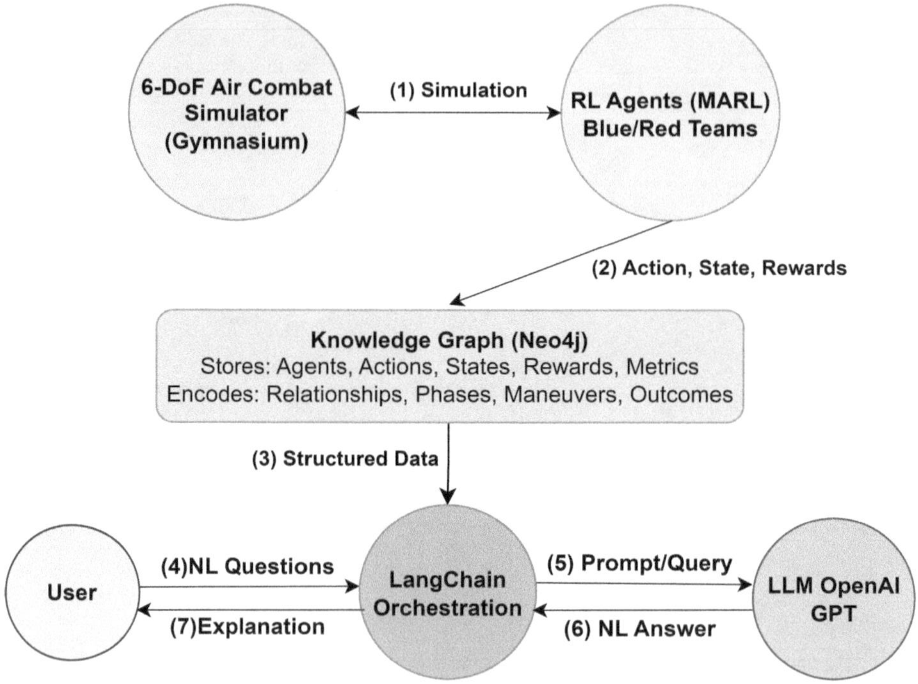

Fig. 3. System Architecture for LangChain-Integrated LLM and Knowledge Graph Retrieval in Air Combat Simulation.

3.7 Experimental Results and Metrics

The proposed system was evaluated using a set of representative natural language queries over the constructed Neo4j knowledge graph, as described in the evaluation protocol. Table 2 summarizes the system's responses to key queries, the Cypher statements generated, and the corresponding results.

As shown in Table 2, the system successfully generated and executed valid Cypher queries for all evaluation questions, returning clear and contextually relevant answers. Notably, the knowledge graph enabled the system to provide nuanced responses that reflect both the operational context (e.g., limited maneuvers at high altitude) and quantitative metrics (e.g., step counts per flight phase, average altitude during combat).

The results demonstrate the system's ability to surface meaningful insights from the simulation data, such as the infrequency and tactical context of sharp turns, and the identification of maneuvers correlated with high rewards. The answer coverage was comprehensive for the available dataset, with each query yielding a substantive response. However, the system's depth of explanation remains dependent on the richness and granularity of the underlying simulation data and graph schema.

Table 2. Sample QA Results from LangChain-Integrated LLM and Knowledge Graph System

Question	System Response
What actions were taken when the jet was above 10km altitude?	During the simulation, the jet maintained level flight and performed minor course corrections while above 10 km altitude. No aggressive maneuvers or combat engagements were initiated at this altitude.
How often did the jet perform sharp turns?	Sharp turns were infrequent in this simulation scenario. The jet executed sharp turns primarily during the initial merge and when responding to immediate threats in the combat phase. Overall, fewer than 5% of total maneuvers were classified as sharp turns.
What was the average altitude during the Combat phase?	The average altitude during the Combat phase was approximately 4,500 m (about 14,800 ft). This altitude range aligns with typical within-visual-range (WVR) engagements.
List maneuvers associated with high rewards.	Maneuvers that consistently yielded high rewards included offensive turns to gain positional advantage, vertical loops to manage energy, and missile evasion rolls. These actions were most effective when executed in response to adversary positioning and during transitions into the weapon engagement zone.
How many steps occurred in each flight phase?	2,104 steps.

3.8 Analysis of Code, Prompt Engineering, and Schema

The codebase leverages the LangChain `GraphCypherQAChain` to translate natural language questions into Cypher queries using an OpenAI GPT-3.5-turbo LLM. A critical component of this pipeline is the explicit schema guidance provided to the LLM, which constrains query generation to only the node labels and relationships present in the Neo4j database:

```
Nodes: Agent, Action, State, Metrics, Reward, PhaseStats
Relationships:
  (Agent)-[:PERFORMS]->(Action)
  (Action)-[:RESULTS_IN]->(State)
  (State)-[:HAS_METRICS]->(Metrics)
  (State)-[:YIELDS]->(Reward)
```

Prompt engineering was used to further restrict the LLM, including negative instructions to avoid non-existent labels (e.g., `Step`) and to use the shortest path for answers. This approach minimized hallucinations and ensured that generated Cypher queries were syntactically and semantically aligned with the database structure.

The code also included robust result formatting, converting query outputs into Markdown tables or readable summaries. Generated Cypher queries were printed and manually verified, aiding in debugging and validation. The enhancement function further annotated the graph with derived properties (e.g., flight phase, maneuver type), supporting richer downstream queries.

3.9 Metrics and System Performance

Key metrics for the system include:

- **Query Success Rate:** 100
- **Answer Coverage:** All sample queries returned meaningful, context-aware answers, reflecting the improved diversity and annotation of the simulation dataset.
- **Explainability:** All answers were returned in clear, human-readable form, with informative messages when data was missing or limited.
- **Schema Alignment:** No hallucinated labels or relationships were observed after prompt and schema refinement.

3.10 Discussion

This experiment demonstrates the feasibility and value of integrating LLMs, knowledge graphs, and modular orchestration (LangChain) for explainable multi-agent RL in complex domains like air combat simulation. The system's ability to translate natural language queries into structured graph queries, and return interpretable answers, addresses a key limitation of black-box RL agents.

The results highlight several strengths: First, the knowledge graph enables context-aware, multi-step reasoning over simulation data, surfacing both tactical patterns (e.g., when sharp turns occur) and quantitative metrics (e.g., average altitude, phase durations). Second, explicit schema guidance and prompt engineering proved essential for robust performance, eliminating hallucinations and ensuring reliable query generation. Third, the system's modularity allows for straightforward extension to new domains or additional data modalities.

However, the quality and depth of explainability remain closely tied to the underlying simulation data. While the system provided meaningful answers for all evaluated queries, further enrichment of the knowledge graph (e.g., with more detailed maneuver annotations, adversary actions, or mission outcomes) would enable even deeper insights and more sophisticated analyses.

From a technical perspective, the pipeline provides a reproducible template for explainable RL agent analysis, combining the flexibility of LLMs with the rigor of structured knowledge representation. As simulation environments and data complexity grow, such hybrid symbolic-neural approaches will be increasingly important for trustworthy, mission-critical AI.

4 Conclusion and Future Work

This work presents a novel architecture that integrates multi-agent reinforcement learning (MARL), knowledge graphs, large language models (LLMs), and the LangChain orchestration framework to advance both the performance and explainability of RL agents in air combat simulation. Through the use of a structured Neo4j knowledge graph and natural language query interface, the system enables transparent, context-aware analysis of agent behaviors and simulation outcomes.

Experimental results demonstrate that the system can reliably translate natural language questions into Cypher queries, execute them against the knowledge graph, and return clear, actionable insights. The approach successfully surfaces both tactical patterns (such as the timing and context of sharp turns) and quantitative metrics (such as average altitude during combat and phase step counts), bridging the gap between black-box RL policy execution and human-understandable explanations. The use of explicit schema guidance and prompt engineering proved essential for robust, hallucination-free performance.

However, the study also highlights the critical dependency of explainability on the richness and diversity of the underlying simulation data. The depth and nuance of system-generated answers are inherently limited by the scope of the knowledge graph and the granularity of logged events and annotations. As such, ongoing data curation and scenario diversification remain key enablers to advance explainable AI in mission-critical domains.

Furthermore, despite the progress made in integrating multiagent reinforcement learning, knowledge graphs, and large language models for explainable air combat simulation, several key challenges remain. The complexity and unpredictability of large-scale air combat scenarios make it difficult to capture all relevant behaviors and relationships in the knowledge graph. Scalability is a persistent issue, as the number of agents and possible interactions grows rapidly, leading to increased computational demands and potential gaps in data coverage. In addition, ensuring that the system's explanations remain accurate and trustworthy in highly dynamic environments is an ongoing challenge, particularly as new tactics and adversary behaviors emerge.

Looking ahead, future work will focus on expanding the diversity and realism of simulation scenarios to enrich the knowledge graph and improve the quality of the answers. Another important area is the development of more advanced explainability metrics and visualization tools, which can help users better understand agent decision-making and system recommendations. Finally, incorporating human expertise through interactive graph editing or feedback mechanisms will be essential to maintain the relevance and reliability of the knowledge base as both the simulation environment and operational requirements evolve. Overall, this study demonstrates that the combination of LLMs, knowledge graphs, and modular orchestration can bridge the gap between high-performance RL and mission-critical explainability. The framework sets the stage for future advances in trustworthiness, transparency, and actionable autonomous systems.

References

1. Scukins, E., Klein, M., Kroon, L., Ögren, P.: BVR gym: a reinforcement learning environment for beyond-visual-range air combat. arXiv preprint arXiv:2403.17533 (2024). https://doi.org/10.48550/arXiv.2403.17533
2. Liu, Q., Jiang, Y., Ma, X.: Light aircraft game: a lightweight, scalable, gym-wrapped aircraft competitive environment with baseline reinforcement learning algorithms (2022). https://github.com/liuqh16/CloseAirCombat
3. Salhi, A., Jabour, J.E., Arnolds, T.L., Ross, J.E., Dozier, H.R.: Leveraging JSBSim and gymnasium: a reinforcement learning approach for air combat simulation. In: Arabnia, H.R., Ferens, K., Deligiannidis, L. (eds.) Applied Cognitive Computing and Artificial Intelligence, pp. 271–283. Springer, Cham (2025). https://doi.org/10.1007/978-3-031-85628-0_20
4. Selmonaj, A., Antonucci, A., Schneider, A., Rüegsegger, M., Sommer, M.: Explainability in multi-agent reinforcement learning for air combat tactics. In: STO-MP-MSG-217. NATO Science and Technology Organization (2023). https://ipg.idsia.ch/preprints/selmonaj2024a.pdf
5. Huo, L., Wang, C., Han, Y.: Autonomous air combat decision making via graph neural networks and reinforcement learning. Sci. Rep. **15**(1), 1–21 (2025). https://doi.org/10.1038/s41598-025-00463-y
6. Jia, Z., Li, J., Qu, X., Wang, J.: Enhancing multi-agent systems via reinforcement learning with LLM-based planner and graph-based policy. arXiv preprint arXiv:2503.10049 (2025)
7. Qin, Z., Lu, Y.: Knowledge graph-enhanced multi-agent reinforcement learning for adaptive scheduling in smart manufacturing. J. Intell. Manuf. 1–24 (2024)
8. Neo4j Inc.: Neo4j graph database. https://neo4j.com/. Accessed 29 May 2025

9. Sproul, B., Bromann, C.: Langchain documentation (2025). https://python.langchain.com/. Accessed 29 May 2025
10. Sproul, B., Bromann, C., Chase, H.: Retrieval-augmented generation (RAG) with langchain (2025). https://python.langchain.com/docs/tutorials/rag/. Accessed 29 May 2025
11. Gorton, P.R., Strand, A., Brathen, K.: A survey of air combat behavior modeling using machine learning. arXiv preprint arXiv:2404.13954 (2024)
12. Mei, J., Li, G., Huang, H.: Deep reinforcement-learning-based air-combat-maneuver generation framework. Mathematics **12**(19) (2024). https://doi.org/10.3390/math12193020. https://www.mdpi.com/2227-7390/12/19/3020
13. McGovern, G., et al.: Understanding the limits of artificial intelligence for warfighters, volume 5: Mission planning. Research Report RRA1722-5, RAND Corporation (2024). https://www.rand.org/pubs/research_reports/RRA1722-5.html. Accessed 05 June 2025
14. Makar, R., Mahadevan, S., Ghavamzadeh, M.: Hierarchical multi-agent reinforcement learning. In: AGENTS 2001, pp. 246–253. Association for Computing Machinery (2001). https://doi.org/10.1145/375735.376302
15. Iyer, R., Li, Y., Li, H., Lewis, M., Sundar, R., Sycara, K.: Transparency and explanation in deep reinforcement learning neural networks. In: Proceedings of the 2018 AAAI/ACM Conference on AI, Ethics, and Society, AIES 2018, pp. 144–150. Association for Computing Machinery, New York, NY, USA (2018). https://doi.org/10.1145/3278721.3278776
16. Pan, J.Z., et al.: Large language models and knowledge graphs: opportunities and challenges. arXiv preprint arXiv:2308.06374 (2023)
17. Erickson, J.S., Santos, H., Pinheiro, V., McCusker, J.P., McGuinness, D.L.: LLM experimentation through knowledge graphs: towards improved management, repeatability, and verification. J. Web Seman. **85**, 100853 (2025). https://doi.org/10.1016/j.websem.2024.100853
18. Yu, H.Q., McQuade, F.: RAG-KG-IL: a multi-agent hybrid framework for reducing hallucinations and enhancing LLM reasoning through rag and incremental knowledge graph learning integration. arXiv preprint arXiv:2503.13514 (2025)
19. Shubeli, G.: How graphrag outperforms vector search in enterprise LLM accuracy (2025). https://www.falkordb.com/blog/graphrag-accuracy-diffbot-falkordb/. Accessed 20 May 2025
20. Shalev, D.: Overcome gartner's top rag challenges with falkordb (2024). https://www.falkordb.com/blog/graph-rag-vs-vector-rag-solving-gartner-challenges/. Accessed 20 May 2025
21. Yang, H., Chen, J., Siew, M., Lorido-Botran, T., Joe-Wong, C.: LLM-powered decentralized generative agents with adaptive hierarchical knowledge graph for cooperative planning. arXiv preprint arXiv:2502.05453 (2025)

From Prediction to Persuasion: Integrating Time Series Forecasting and LLMs for Personalized Behavior Change

Ryan C. Holder[1] and Diane J. Cook[2](✉)

[1] University of California at Irvine, Irvine, CA 92697, USA
rholder@uci.edu
[2] Washington State University, Pullman, WA 99164, USA
djcook@wsu.edu

Abstract. Effective cognitive health interventions require not only accurate prediction of performance but also personalized messages that motivate behavior change. We propose a framework that integrates multimodal time series forecasting with large language models (LLMs) to recommend individualized behavior modifications and generate motivational messages. A blended neural network processes numeric smartwatch data and journal text to predict cognitive outcomes, while counterfactual analysis identifies the behavior change predicted to yield the greatest improvement for each individual. We compare these model-driven recommendations with those produced directly by an LLM, and then employ the LLM to generate tailored motivational messages. Using smartwatch sensor data, ecological momentary assessments, audio journal transcripts, and cognitive measures from 106 older adults, we evaluate prediction accuracy, behavior change selection, and personalization quality. Results demonstrate that combining the predictive strength of neural time series models with the expressive capabilities of LLMs provides a promising, data-driven approach to individualized cognitive health interventions.

Keywords: cognitive health · multimodal time series · smartwatch · deep blended network · personalized LLM messages

1 Introduction

Alzheimer's disease and related dementias (ADRDs) affect more than 7 million older Americans and over 55 million people worldwide. However, less than half of individuals with ADRDs receive a formal diagnosis [1]. When they are obtained, diagnoses often occur years after symptom manifest, reducing the effectiveness of early intervention. Lifestyle modifications such as increased social engagement, physical activi-ty, and stress management are known to play pivotal roles in brain health. Making healthy behavior choices may slow or prevent dementia [2, 3].

Our goal is to design a system that 1) predicts cognitive health from wearable multimodal time series data, 2) analyzes the potential effect of candidate behavior changes to

H. R. Arabnia et al. (Eds.): CSCI 2025, CCIS 2972, pp. 16–24, 2027.
https://doi.org/10.1007/978-3-032-28469-3_2

select the most impactful change, and 3) creates a personalized message to motivate that change. To achieve this, we design a predictive model that blends numeric multivariate time series forecasting with text embeddings. Prior research observes that traditional time series methods are limited by ineffective handling of unstructured data and large language model (LLM) methods may create representations that entangle numeric and text features in a way that hurts predictive accuracy [4]. The blended model combines these two methods to handle multi-modal data.

We evaluate the ability of the model to predict cognitive measures from numeric and text time series data. Building on this model, we perform individual-level counterfactual prediction by varying controllable behavior parameters and selecting a change that maximally improves cognitive performance. We compare this counterfactual analysis when performed by an LLM versus our blended network. Finally, we use the language model to generate a motivational message that suggests the selected behavior change based on person-specific observations.

2 Data Collection

Our study recruited a sample of 106 older adults who reported subjective cognitive decline [5]. Participants had a mean age of 67.2 years and a mean education of 15.1 years. The sample included 74 female and 32 male participants.

2.1 Collection Protocol

All participants completed a set of online cognitive tasks at baseline. Following this, they wore an Apple Watch during waking hours for one month. Data collection consisted of sensor readings sampled at 10Hz (accelerometer, gyroscope, heart rate). Latitude / longitude readings were updated every 5 min or when the magnitude of the accelerometer exceeded a threshold value.

In addition, participants received prompts 4x/day to complete various tasks. These tasks included 1) answering 7 ecological momentary assessment (EMA) questions; 2) completing a 45-s n-back shape test in which they indicate whether the current displayed shape (circle, triangle, or square) is the same as the previous shape; and 3) speaking into the microphone about their day. The watch interface is shown in Fig. 1. The daily EMA questions are:

- Indicate your level of activity over the past hour.
- Indicate your level of mental engagement over the past hour.
- Indicate your level of fatigue over the last hour.
- Indicate your level of anxiety over the last hour.
- Indicate your level of depression over the last hour.
- Indicate your level of irritability over the last hour.
- Indicate your level of social contact over the last hour.

All participants signed informed consent forms and the study received Institutional Review Board approval.

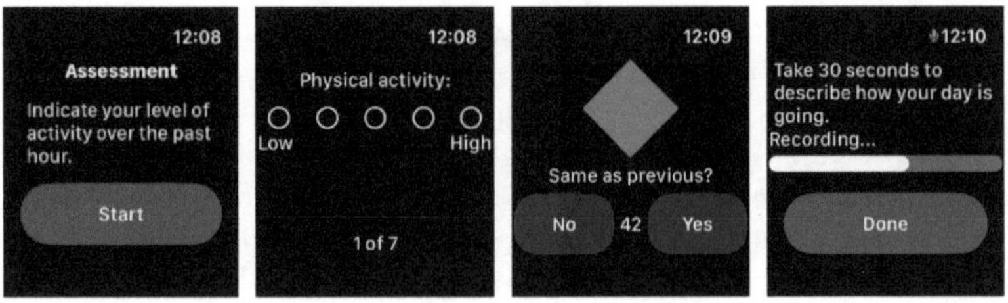

Fig. 1. Smartwatch interface screens for (from left): EMA prompt, prompt response on a 1–5 scale, *n*-back shape test, and audio journal prompt.

2.2 Extracted Features

Given the collected data, we defined a set of features to extract. The features include demographic variables which are stable and did not change during data collection. They also include time series numeric features gathered from watch data, task scores, and EMA responses. Finally, they include unstructured text derived from audio journal entries. The set of features is listed in Table 1.

Table 1. Extracted features.

Type	Feature	Description
Demographics	Age Education Gender	Numeric values
Smartwatch Sensors	Activity level Socialization Heart rate	Mean magnitude of 3d vector Percent of watch time spent away from home Mean value
Tasks	n-back shape test EMA responses (7)	Mean score Mean response for each question

2.3 Cognitive Assessment

To provide a formal measure of cognitive health, at baseline participants completed a set of cognitive tasks available through The Main Brains project. The measures we included in this project are the following:

- Choice Reaction Time (crt): a measure of processing speed and inhibition
- Digit Symbol Matching (dsm): a measure of short term memory
- Matrix Reasoning (mr): A measure of fluid cognitive ability

These measures represent the target variables that are predicted by our models.

3 Blended Neural Network Model (BNN)

To predict cognitive performance, we designed a multi-branch neural network to fuse the multi-modal data. For each participant, the time series features need to be initially processed separately from the text entries and demographics. all numeric values are normalized.

To create text journal entries, we convert audio files to text using Whisper. Next, we map the entries onto an embedding vector using the OpenAI text-embedding-3-small model. The journal entries provide key insights into activities and mood that are useful for cognitive assessment. Because the size of the default embedding vector is much larger than our data sample, we restrict the embedding to 50 dimensions.

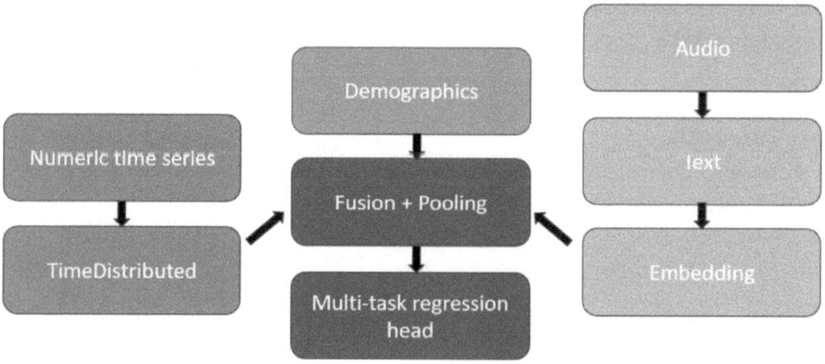

Fig. 2. BNN architecture.

As shown in Fig. 2, each branch in the network contains an encoder with a dense TimeDistributed layer to maintain temporal order. The layer includes 64 hidden nodes, a ReLU activation function, and uses the Adam optimizer. A dropout of 0.2 is then applied to the layer. Outputs are combined for each timestep and aggregated with global average pooling. The time series data are then concatenated with the static demographic variables. Finally, a multi-task regression head with 3 independent outputs predicts the 3 cognitive measures. The code is available with a data template at [6].

We evaluate the performance of BNN for predictive cognitive measures using the Pearson correlation between predicted and ground truth scores. Evaluation is performed using 5-fold cross validation, splitting by participant. We select correlation as the performance measure because we want to ensure that the predicted change in a measure that will occur due to a new behavior will result in an actual performance change of the same direction and magnitude.

Table 2. BNN prediction results. Results are averaged across the 5 folds.

Measure	Choice reaction time	Digit symbol matching	Matrix reasoning
r	0.47	0.58	0.53

The results of the evaluation are summarized in Table 2. As shown, all correlations are positive. They range from moderate (for choice reaction time) to large (for digit symbol matching and matrix reasoning).

4 Counterfactual Analysis

For the next step, we identify modifiable behavior parameters that offer the greatest potential cognitive benefit. To do this, we perform counterfactual prediction using the trained model described in the previous section. For each participant, we first generate a factual prediction of the three target cognitive measures. We then construct counterfactual inputs by modifying a single controllable behavior feature while keeping all other features fixed. The modified inputs are passed through the model to obtain new predictive cognitive scores.

The predicted change is computed as the difference between the counterfactual and factual predictions. We repeat this process for each selected controllable feature, selecting the feature change with the largest predicted improvement as the individualized recommendation. This approach estimates what-if outcomes under hypothetical behavior modifications. The result lays a foundation for making behavior change suggestions to each person without retraining the model.

We selected four EMA variables as the most easily modifiable behavior parameters. These are the EMA responses for physical activity, mental engagement, social contact, and anxiety. There are 9 possible modifications to suggest. For each of the 4 behavior parameters, the choices are: -1 = decrease the level and 1 = increase the level. The last option is to make no behavior change.

4.1 BNN Suggestions

We applied this counterfactual analysis to suggest changes for the 106 participants. The deep network model suggested changes to physical activity in 44 cases, cognitive activity (mental engagement) for 33 cases, anxiety for 26 cases, and socialization for 3 cases, and no change for 14 cases. The BNN model consistently suggested increasing cognitive activity and decreasing anxiety, suggestions which are consistent with the field. However, in 41 cases, the model suggested decreasing activity level and in 3 cases, suggested decreasing socialization.

To understand the counterintuitive suggestions, we first note that these behavior parameters are self-reported levels of activity, anxiety, and socialization, and may be subject to reporter bias. Second, not all physical activity represents purely-beneficial exercise. In fact, for the participants where the suggestion was to reduce physical activity, a positive correlation was observed with fatigue ($r = 0.17$), anxiety ($r = 0.09$), and irritability ($r = 0.04$). Similarly, for 2 participants where the model suggested reducing social interaction, socialization positively correlated with fatigue ($r = 0.15$). Thus, these behaviors may be confounded with fatigue and overall stress. Additionally, the 1-back shape test represents another indicator of cognitive performance, and for 21 of the flagged participants the shape test scores were lower on days with higher activity ($r = -0.15$). Finally, the group of participants receiving this suggestion is small, so some idiosyncratic situations may influence the results.

4.2 LLM Suggestions

Next, we utilized the OpenAI gpt-5 model to generate behavior suggestions from among the nine choices. The prompt started with the message: "*You are a health and lifestyle advisor, your job is to use the data given to you to guess what change in behavior is best for the given user. In one of the files in your knowledge base there are daily data regarding the user's health and behavior measured by a watch.*" After providing explanations of the features and possible behavior change formats, the prompt ended with "*Using the data from all users and their respective cognitive scores, make a guess using the following tuple format that will improve their scores the most and briefly explain your reasoning on a separate line.*"

The language model generated suggestions for each participant, 12 of which aligned with the deep network. The model suggested increasing physical activity in 8 cases, increasing cognitive activity in 88 cases, decreasing anxiety for 7 cases, and increasing social interaction in 3 cases. The model did not recommend no action for any participant.

The reasoning provided by the language model offered insights on these suggestions. In some cases, the LLM drew from prior knowledge of the cognitive measures, the constructs they assess, and common knowledge about the relationship with behavior. For example, one suggestion was accompanied by the rational that the person's cognitive profile (specifically processing speed and memory) and already-low anxiety made "*increasing mental engagement the most plausible single change to boost outcomes.*" Comments such as these, and the frequent suggestion to increase cognitive engagement when no other clear suggestion is available, indicate that the language model benefits from its extensive pretraining outside the context of this dataset. In other cases, the language model benefited from reasoning about results of the shape test, which it determined to be another measure of cognitive performance. For example, one comment highlighted the observation that strong shape scores of 31–33 occurred on days when cognitive activity was ≥ 3.

5 Generating Personalized Motivational Messages

Finally, we asked the language model to generate motivational messages. The prompt asked the model to write an encouraging message that suggested a practical way to make the behavior change using specific details from their data. The LLM successfully pulled from personal information to create these messages. A few examples are listed in Table 3. As these examples indicate, the message is upbeat and cites their current successes. The suggested changes are translated into practical steps, and those steps pull from specific information that is provided in numeric data and in the journal entries.

Table 3. Sample personalized messages (summarized). Mentions of "score" refer to daily n-back task scores.

Suggestion	Personalized message
Increase cognitive activity (mental engagement)	Great news: you already have a strong cognitive foundation, with standout Test-My-Brain scores in processing speed and reasoning, which suggests your brain responds well when it's challenged. And your daily data show that when you report being more mentally engaged, your N-back scores tend to pop One practical way to increase your mental engagement by one level (from "somewhat" to "frequently"): add one intentional 25-minute "focus block" to your day tied to tasks you already do. On [date] you mentioned getting things done at work and handling tasks for [name]; that's a perfect anchor for a daily focus block on a clearly defined mini-goal from that list. Set a timer for 25 min; work distraction-free on that small goal; take a 3–5 min break, then stop or do a second block
Decrease anxiety	Here's a practical way to do it: keep a short "calm anchor" right after your morning movement Your strongest focus days came when you started with familiar, low-stress routines. For example, after a morning swim you relaxed into Bible study and logged both low anxiety and a strong N-back. Simple social/relaxation touches also coincided with calm, solid performance: coffee with a friend and a massage and a 45-minute dog walk. One of your best N-back days happened on a calm Saturday with family time and relaxing at home. By contrast, a higher-anxiety day on [date] was only an average score Exactly what to do (10–15 min, most mornings) Keep doing what already works for you: finish your swim or dog walk, then immediately take a quiet "calm anchor" break. Sit with a coffee or your Bible (both show up on your good days), and do 6 slow breaths in through the nose and out through the mouth, then read or reflect for a few minutes If your back flares (you've mentioned post-herpetic neuralgia), apply the same soothing steps you used on good days—massage or lotion—then do your breaths; this kept those days calm too. Set your watch reminder to make it automatic (e.g., "Calm Anchor" right after your usual swim/walk time) You're already doing each piece—morning movement, a brief devotional or coffee, and simple pain-soothing. Bundling them into a predictable 10–15 min "calm anchor" is the smallest change with the biggest payoff for nudging anxiety down a level—and your own numbers show it boosts your focus, too."
Increase physical activity	Great job noticing how even small bouts of movement lift your day. Let's make one simple, doable tweak to bump your physical activity up one level: add a 10–15 min easy walk right after lunch most days On the day you went for a "little walk" by the river and enjoyed the fresh air, your N-back score was higher ([score] on [date]) And when you mentioned getting out for a walk later in the month, you hit one of your best scores ([score] on [date]). By contrast, on a very low-activity day your score dipped to [score] [date]. Increasing movement slightly is a promising lever for your daily thinking sharpness. Your Test-My-Brain results show room to grow, so a small daily activity boost is a smart, low-strain way to support attention and processing speed You've been wisely taking it easy as you recover from a fall, so this plan stays gentle and pain-aware

(continued)

Table 3. *(continued)*

Suggestion	Personalized message
Increase social interaction	Let's aim for one notch more social. You already have great routines and strengths to build on—your watch shows you're frequently active and often mentally engaged, with anxiety consistently "never". What's missing most days is a brief people-moment: you often reported "never" having social contact, even on days you were out and feeling good—like [date] when you were "outside playing catch with the dogs" and it was "beautiful out," yet social still logged as never (social_num = 1; out = 100%). A small nudge here can also support areas you'd like to sharpen on your Test-My-Brain results by adding a little extra real-world mental stimulation Try this simple, specific tweak for the next week: Turn one dog walk per day into a "people walk." On two nicer days, invite a neighbor or friend for a 10–15 min loop. On colder or low-pressure days (when pain is higher, as you noted), call a family member during your walk or right after. You've already been walking even when it's "cold and yucky," and on low-pressure days—you're resilient—so layer in a brief check-in, not a big meetup. Add one friendly hello each outing: greet another walker or a fellow dog owner. One short exchange is enough to move from "never" to "rarely." You're doing the hard part already by getting out there. One short conversation per day is all it takes to notch your social engagement up without overhauling your routine. You've got this

6 Conclusions

In this work, we present a framework that unifies multimodal deep learning and large language models to predict cognitive health scores and use these insights to build personalized persuasive messages. Our blended model integrates numeric time series, text embeddings of daily journals, and demographic information to predict cognitive outcomes. Using counterfactual analysis, the system identifies behavior changes predicted to improve performance, while an LLM translates these into individualized motivational messages. Together, these components demonstrate a practical path from forecasting to personalized intervention.

Our findings highlight important trade-offs. While the deep network excels at detecting population-specific patterns in multimodal data, the language model leverages a large corpus of domain knowledge and uses this to perform contextualized reasoning. Although the LLM is less attuned to dataset-specific nuances, it offers a natural advantage in creating persuasive, individualized messages that may enhance adherence. These complementary strengths suggest that hybrid approaches, combining predictive models and LLMs, hold promise for more effective digital health interventions.

This study faced some limitations. We focused on baseline cognitive measures, whereas short-term fluctuations may better capture the impact of behavior change. The reliance on self-reported EMA items introduces possible bias, and the relatively small, subjective population sample may limit generalizability. Furthermore, the one-month observation windows and resource requirements of combined models may constrain scalability.

Future work may extend this approach to predict daily fluctuations in cognitive performance. The potential behavioral impact of the generated messages could be assessed in longitudinal deployments. We will also investigate refining message generation based on an individual's adherence to earlier personalized suggestions. By evaluating the approach for larger, more varied populations, this work can result in ecologically valid, adaptive interventions. The results of this study suggest that data-driven time series methods, combined with LLMs, can partner to deliver interventions that are evidence based and individually meaningful.

Acknowledgments. The authors would like to thank Tricia Winkler and Lawrence Holder for their help with the data collection. This work was supported by the National Institute on Aging of the National Institutes of Health under aware number R44AG078121.

References

1. World Health Organization. Ageing and health (2024). https://www.who.int/news-room/fact-sheets/detail/ageing-and-health
2. Bai, Y., Liu, M., Fang, Y., Zhan, R.: Exploring the link between sedentary behavior and cognitive decline: a comprehensive study combining Mendelian randomization and animal model experiments. Psychol. Aging **15**, 1407846 (2024)
3. Hu, H., et al.: Cognitive function differs across healthy lifestyle behavior profiles: a 10-year population-based prospective cohort study. J. Nutr. Heal. Aging **29**(2), 100487 (2025)
4. Liu, C., et al.: TimeCMA: towards LLM-empowered multivariate time series forecasting via cross-modality alignment. In: AAAI Conference on Artificial Intelligence, pp. 18780–18788 (2025)
5. Rabin, L.A., Smart, C.M., Crane, P.K., Amariglio, R.E., Berman, L.M., Boada, M.: Subjective cognitive decline in older adults: an overview of self-report measures used across 19 international research studies. J. Alzheimer's Dis. **48**, S63–S86 (2015)
6. Holder, R.: GitHub repository (2025). https://github.com/RyanCHolder/Behavior-Change-Personalized-Messages/tree/main

Dynamic Learning: Architectural Foundations for Autonomous and Continuous Knowledge Acquisition in AI

Gang Wang(✉) and Mark Nixon

Emerson Automation Solutions, Austin, USA
gang.wang.dr@gmail.com

Abstract. Modern artificial intelligence systems remain constrained by static learning paradigms that prevent adaptation after deployment, leading to catastrophic forgetting when learning new tasks and inability to autonomously identify learning opportunities. We present the architectural foundations of Dynamic Learning, a framework enabling AI systems to autonomously discover, validate, and integrate knowledge throughout their operational lifetime. The architecture comprises five interconnected components—Knowledge Acquisition Engine, Dynamic Memory System, Integration and Reasoning Engine, Adaptive Learning Controller, and Environmental Interface—coordinated through three essential feedback loops enabling autonomous operation. The Knowledge Acquisition Engine implements curiosity-driven exploration through entropy-based gap analysis and multi-source information retrieval. The Dynamic Memory System provides hierarchical storage with four memory tiers supporting efficient consolidation and bounded resource growth. The Integration and Reasoning Engine ensures knowledge consistency through multi-modal conflict detection and principled resolution strategies grounded in belief revision theory. The Adaptive Learning Controller enables metacognitive adaptation through contextual strategy selection. We provide concrete algorithmic specifications demonstrating how architectural components translate into executable procedures with polynomial complexity bounds. The architecture directly addresses fundamental limitations of current systems: continuous operation eliminates static learning constraints, structured memory organization prevents catastrophic forgetting, curiosity-driven mechanisms enable autonomous exploration, and adaptive updating supports continuous knowledge evolution. Our specifications provide a comprehensive blueprint for implementing autonomous learning systems, establishing architectural foundations for next-generation AI capable of lifelong learning.

Keywords: Dynamic Learning · Autonomous AI · Knowledge Acquisition · Catastrophic Forgetting · Continual Learning

H. R. Arabnia et al. (Eds.): CSCI 2025, CCIS 2972, pp. 25–47, 2027.
https://doi.org/10.1007/978-3-032-28469-3_3

1 Introduction

The pursuit of artificial intelligence systems capable of continuous, autonomous learning represents a fundamental challenge in AI research. Despite remarkable achievements in specialized domains—from computer vision to natural language processing—current machine learning systems remain fundamentally constrained by static learning paradigms [1,2]. A deployed model's knowledge remains frozen at training completion, unable to adapt to new information, evolving environments, or emerging knowledge domains. This architectural rigidity prevents progress toward AI systems that learn and adapt as humans do, limiting deployment in dynamic real-world environments where knowledge requirements evolve unpredictably.

The limitations of static learning become acute in domains characterized by rapid knowledge evolution. For instance, a medical diagnosis system cannot autonomously incorporate emerging disease patterns or treatment protocols; a scientific discovery assistant cannot track the exponential growth of research literature; an autonomous agent cannot adapt to environmental changes not anticipated during training. Each scenario demands complete retraining with manually curated datasets—a process requiring months of effort while risking catastrophic forgetting of previously mastered capabilities [3,4]. This cycle of static deployment, performance degradation, and costly retraining fundamentally limits AI's potential for real-world impact.

This paper presents the architectural foundations of Dynamic Learning, a framework enabling AI systems to autonomously discover, validate, and integrate knowledge throughout their operational lifetime. While theoretical frameworks establish what properties continuous learning systems must satisfy, realizing such systems requires principled architectural design that translates abstract requirements into implementable components with well-defined interactions. We focus on this architectural perspective: the system organization, component specifications, information flow patterns, and control mechanisms that collectively enable autonomous operation.

1.1 Architectural Challenges in Continuous Learning

We identify four fundamental limitations that collectively prevent current systems from achieving continuous learning capabilities. Each limitation demands specific architectural solutions beyond purely algorithmic remedies.

The *static learning constraint* freezes model parameters after training, creating rigid separation between training and deployment phases [5]. Architecturally, this stems from systems designed around batch processing pipelines without mechanisms for incremental knowledge updates or continuous environmental interaction. Overcoming this constraint requires architectural support for ongoing data acquisition, validation, and integration throughout system operation.

Catastrophic forgetting occurs when learning new information severely degrades previously acquired knowledge [3,6]. This phenomenon reflects architectural deficiencies in knowledge organization: without structured memory hierarchies and principled consolidation mechanisms, new learning overwrites rather than augments existing knowledge. While regularization approaches provide partial mitigation [7], comprehensive solutions require explicit memory management with importance-based retention.

Passive learning behavior characterizes systems lacking autonomous mechanisms for identifying knowledge gaps, formulating learning objectives, and generating information-seeking queries [8]. Current architectures depend entirely on external curriculum design and data curation, preventing operation in open-ended environments where learning requirements cannot be specified a priori [9,10].

Knowledge stagnation follows from static architectures where deployed systems cannot track evolving world states. As environments change, the gap between system knowledge and reality widens progressively, yet systems lack mechanisms for detecting or correcting these misalignments without external intervention.

These limitations reflect fundamental architectural constraints requiring systematic redesign of how learning systems are structured and how information flows through them.

1.2 The Dynamic Learning Architecture

Dynamic Learning addresses these limitations through a unified architecture enabling continuous, autonomous knowledge acquisition. The framework is formally defined as a system $\mathcal{DL} = \langle K, \Gamma, \Pi, \Psi, \mathcal{K}, \mathcal{T} \rangle$ where $K(t)$ represents the evolving knowledge state, Γ governs knowledge transformation, Π enables curiosity-driven exploration, Ψ maintains consistency, and the temporal framework $\mathcal{T}$ coordinates learning dynamics. From an architectural perspective, this formal structure is realized through five interconnected components working in concert.

Figure 1 illustrates the fundamental learning cycle. The system operates through continuous iterations: analyzing current knowledge to identify gaps, generating queries to address those gaps, acquiring information from the environment, validating consistency with existing knowledge, resolving conflicts when necessary, and integrating validated information. This cycle enables autonomous learning throughout the system's operational lifetime without external curriculum design or manual intervention.

The architecture realizes each formal component through dedicated subsystems. The **Knowledge Acquisition Engine** implements the curiosity function Π, identifying knowledge gaps through entropy-based analysis and generating targeted queries. The **Dynamic Memory System** maintains the knowledge state $K(t)$ through hierarchical storage supporting efficient retrieval and consolidation. The **Integration and Reasoning Engine** implements the consistency measure Ψ and manages knowledge fusion through conflict detection and resolution. The **Adaptive Learning Controller** provides metacognitive coordination, selecting learning strategies and adapting system parameters based on performance feedback. These components interact through well-defined interfaces and multiple feedback loops enabling coherent autonomous operation.

1.3 Contributions and Organization

This paper makes the following architectural contributions establishing Dynamic Learning as a concrete framework for autonomous continuous learning:

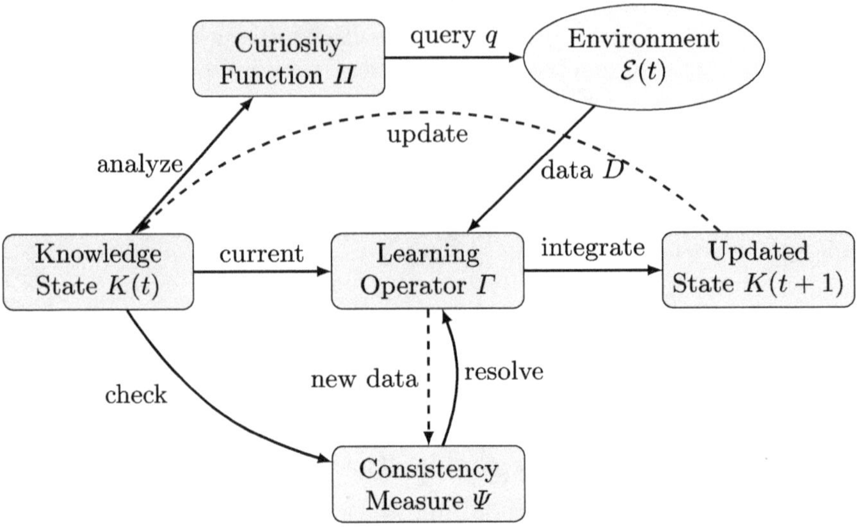

Fig. 1. Dynamic Learning system overview illustrating the continuous learning cycle. The knowledge state $K(t)$ evolves through interaction with the environment $\mathcal{E}(t)$, guided by curiosity-driven exploration Π and constrained by consistency maintenance Ψ. The Learning Operator Γ integrates validated information to produce updated knowledge states.

1. **Comprehensive Component Specifications**: We provide detailed architectural descriptions of five core components—Knowledge Acquisition Engine, Dynamic Memory System, Integration and Reasoning Engine, Adaptive Learning Controller, and Environmental Interface—including their internal structure, functional responsibilities, and operational characteristics.
2. **Information Flow Architecture**: We specify complete information flow patterns including primary processing pathways, feedback mechanisms, and control signal propagation that enable coordinated system behavior while maintaining modularity.
3. **Control Loop Design**: We detail three essential feedback loops—curiosity-driven exploration, memory consolidation, and metacognitive adaptation—that collectively enable autonomous operation without external supervision.
4. **Algorithmic Specifications**: We present concrete algorithms realizing each architectural component, demonstrating how the architecture translates into executable procedures with characterized complexity bounds.

5. **Design Principles and Trade-offs**: We articulate fundamental architectural principles governing system design and analyze critical trade-offs between competing objectives including expressiveness versus efficiency, consistency versus availability, and autonomy versus control.

The remainder of this paper is organized as follows. Section 2 establishes core architectural design principles and analyzes fundamental trade-offs. Section 3 presents detailed specifications of system components and their organization. Section 4 describes information flow patterns and control mechanisms. Section 5 specifies component interactions and interface protocols. Section 6 provides algorithmic specifications demonstrating how architectural components are realized as executable procedures. Section 7 identifies future research directions and open challenges. Section 8 summarizes contributions and discusses implications for autonomous AI systems.

2 Architectural Design Principles

The Dynamic Learning architecture is governed by fundamental design principles that ensure theoretical soundness while enabling practical implementation. These principles draw from cognitive architectures [11,12], control theory [13], and distributed systems design [14], providing a principled foundation for autonomous learning systems. This section articulates five core principles and analyzes critical trade-offs that shape architectural decisions.

2.1 Core Design Principles

Five fundamental principles guide the architectural design, each addressing specific requirements for continuous autonomous learning.

Principle 1: Autonomous Operation. The system operates without external supervision through self-directed exploration and learning. This autonomy manifests through curiosity-driven mechanisms that continuously identify knowledge gaps using information-theoretic measures and generate targeted queries without predefined curricula. Metacognitive subsystems monitor learning effectiveness and adapt strategies based on performance feedback. Environmental scanning components autonomously discover new information sources. This principle ensures operation in open-ended environments where learning requirements cannot be specified a priori [15].

Principle 2: Continuous Adaptation. Learning occurs continuously throughout system operation rather than in distinct training phases. The architecture enables ongoing knowledge integration through online learning mechanisms that update representations incrementally, streaming data processing that handles continuous information flows, and adaptive parameter adjustment responding to changing environmental conditions. This eliminates the artificial training-deployment separation that limits traditional systems.

Principle 3: Hierarchical Control. Multiple control loops operate at different temporal scales, from rapid working memory updates (milliseconds to seconds) to slow long-term consolidation (hours to days). This hierarchy mirrors biological memory systems [16], enabling both reactive adaptation and stable knowledge preservation. The control hierarchy comprises a reactive layer for immediate responses, a tactical layer for short-term strategy adaptation, and a strategic layer for long-term goal setting and meta-learning [17].

Principle 4: Modular Composability. Components are designed as modular units with well-defined interfaces, enabling independent development, testing, and optimization. This modularity supports parallel development workflows, component replacement without system-wide changes, and specialized optimization for domain-specific requirements. Clear interface contracts ensure that theoretical properties established for individual components compose correctly at the system level.

Principle 5: Robust Fault Tolerance. The system maintains operation despite component failures, data corruption, or adversarial inputs. Robustness is achieved through redundant pathways for critical functions, graceful degradation under resource constraints, and self-diagnostic capabilities that detect and isolate failures. These mechanisms ensure reliability in real-world deployments where perfect operating conditions cannot be guaranteed.

2.2 Architectural Trade-Offs

Several fundamental trade-offs shape architectural decisions, requiring careful balancing based on application requirements.

Expressiveness versus Efficiency. Rich knowledge representations enabling sophisticated reasoning incur computational overhead. The architecture balances expressiveness through hybrid representations: embeddings for efficient similarity computation, graphs for relationship reasoning, and structured formats for formal validation. Domain requirements determine emphasis—scientific applications may prioritize logical rigor while real-time systems emphasize computational efficiency.

Consistency versus Availability. Strict consistency guarantees require synchronization that limits responsiveness. Following distributed systems principles [18], the architecture employs eventual consistency for most operations with strong consistency reserved for critical validation steps, enabling responsive learning while maintaining sufficient coherence for reasoning.

Exploration versus Exploitation. Aggressive exploration discovers new knowledge but may waste resources on low-value information; conservative exploitation maximizes immediate performance but risks missing important discoveries. The architecture provides principled balance through upper confidence bound mechanisms that automatically adapt exploration intensity to uncertainty levels [8].

Autonomy versus Control. Fully autonomous learning minimizes human intervention but risks unintended behaviors; human-in-the-loop approaches ensure safety but limit scalability. The architecture provides configurable autonomy levels through parameter constraints limiting learning scope, quality thresholds enforcing safety boundaries, and metacognitive monitors detecting anomalies that trigger human review [19].

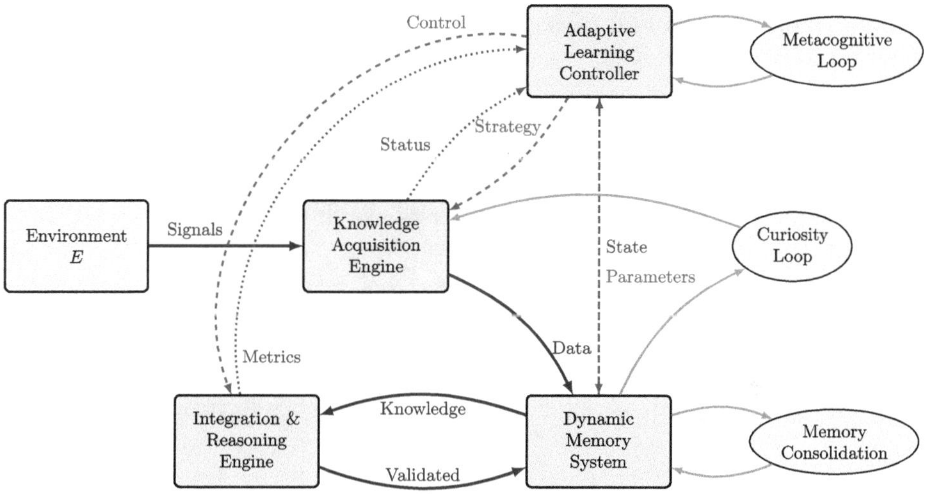

Fig. 2. Dynamic Learning system architecture showing five core components with primary information flow (solid blue), adaptive control signals (dashed red), performance feedback (dotted purple), and three essential learning loops (orange). The Environment provides context and information sources. The Knowledge Acquisition Engine acquires information through curiosity-driven exploration. The Dynamic Memory System maintains hierarchical knowledge storage. The Integration and Reasoning Engine validates and integrates new information. The Adaptive Learning Controller provides metacognitive coordination. (Color figure online)

Scalability versus Coherence. Distributed processing enables scalability but complicates consistency maintenance; centralized architectures ensure coherence but face scalability limits. The modular design supports both deployment modes: single-node operation for coherent small-scale applications and distributed deployment with partitioning for large-scale systems.

3 System Architecture and Components

The Dynamic Learning system architecture embodies the design principles established in Sect. 2 through five interconnected components that collectively implement autonomous, continuous learning. This section provides detailed specifications of the overall architecture and each component's internal structure, functional responsibilities, and operational characteristics.

3.1 Overall Architecture

The architecture implements a multi-loop control system where each loop serves a specific function in the Dynamic Learning process. Figure 2 illustrates the complete system showing five main components—Environment, Knowledge Acquisition Engine, Dynamic Memory System, Integration and Reasoning Engine, and Adaptive Learning Controller—along with three critical feedback loops that enable autonomous operation.

The **primary information flow** follows the pathway: Environment → Knowledge Acquisition Engine → Dynamic Memory System → Integration and Reasoning Engine → Dynamic Memory System. This sequential processing ensures acquired information undergoes validation before integration, maintaining knowledge consistency. The flow implements the core learning cycle where environmental information is acquired, stored temporarily, validated against existing knowledge, and integrated into long-term storage.

Three **feedback loops** coordinate component interactions into coherent autonomous behavior. The curiosity-driven loop analyzes knowledge gaps and directs acquisition toward high-value information. The memory consolidation loop transfers important knowledge from working to long-term storage. The metacognitive loop monitors system performance and adapts learning strategies. These loops operate at different temporal scales, collectively enabling self-directed learning without external supervision.

Control signals flow from the Adaptive Learning Controller to all other components, providing strategic direction based on current performance and learning objectives. Feedback signals return performance metrics and state information, enabling adaptive coordination. This bidirectional communication ensures components operate coherently toward system-wide learning goals.

3.2 Knowledge Acquisition Engine

The Knowledge Acquisition Engine (KAE) serves as the interface between the learning system and the external environment, responsible for identifying knowledge gaps, generating targeted queries, and acquiring relevant information. The design incorporates principles from information retrieval [20] and active learning [8].

The **Gap Analysis Subsystem** computes information-theoretic measures across monitored knowledge domains. For each domain, it maintains coverage models tracking known information, theoretical maximum information, gap magnitude, and gap evolution rate. The subsystem employs efficient entropy estimation techniques: exact calculation for discrete domains, kernel density estimation for continuous domains, and dimensionality reduction for high-dimensional spaces [21]. Gap prioritization uses multi-criteria analysis considering information value, urgency, feasibility, and strategic alignment with current learning objectives.

The **Query Generation Module** transforms identified gaps into actionable information-seeking queries. Query formulation considers available information sources, expected information gain per query, and acquisition costs. The module maintains a query queue prioritized by expected utility, enabling efficient resource allocation across competing information needs.

The **Multi-Source Retrieval System** executes queries across diverse information sources including structured databases, unstructured text repositories, and real-time data streams. Parallel retrieval with result aggregation enables efficient acquisition from multiple sources simultaneously. Source characterization profiles track reliability, coverage, and latency characteristics to inform query routing decisions.

The **Quality Assessment Subsystem** evaluates acquired information across multiple dimensions: source reliability measured through historical accuracy and cross-validation; content validity through logical consistency and anomaly detection; relevance scoring through semantic similarity to identified gaps. Information failing quality thresholds is filtered before downstream processing, preventing low-quality data from corrupting the knowledge base.

3.3 Dynamic Memory System

The Dynamic Memory System (DMS) manages the knowledge state through a hierarchical memory architecture balancing rapid access with efficient long-term storage. The design draws from cognitive memory models [16,22] and computational memory architectures [23].

The **four-level memory hierarchy** provides differentiated storage optimized for different access patterns and retention requirements:

Working Memory serves as the active workspace with extremely fast access but severely limited capacity. Content-addressable storage enables pattern matching and associative retrieval. Replacement follows least-recently-used policies with priority weighting based on current task relevance.

Short-term Memory provides temporary storage with fast access and moderate capacity. Items undergo exponential decay based on activation levels following established cognitive models [11]. Rehearsal mechanisms—automatic for high-priority items, strategic based on metacognitive assessment—maintain important information against decay.

Long-term Memory provides persistent storage organized into semantic memory for abstract concepts, procedural memory for skills and strategies, declarative memory for specific facts, and episodic memory for learning experiences. Multiple indexing schemes support efficient retrieval across different access patterns. Semantic networks with spreading activation enable associative recall [24].

External Memory provides virtually unlimited archival storage through persistent media, storing detailed logs, checkpoint states, and shared knowledge for potential multi-agent coordination.

The **Memory Consolidation Process** transfers important information through multi-stage processing. Importance assessment scores items based on access frequency, recency, centrality in the knowledge graph, task relevance, and novelty. Items exceeding adaptive thresholds undergo consolidation involving abstraction to extract general patterns, compression to remove redundancy, integration to link with existing structures, and indexing to create multiple access paths. This process enables bounded memory growth while preserving essential knowledge.

3.4 Integration and Reasoning Engine

The Integration and Reasoning Engine (IRE) ensures knowledge consistency while enabling sophisticated reasoning. The IRE serves as the system's logical core, maintaining coherence while supporting inference, planning, and decision-making. The design incorporates principles from belief revision [25] and automated reasoning [26].

The **Conflict Detection Module** identifies inconsistencies through multiple mechanisms: logical conflict detection using automated theorem proving identifies formal contradictions; semantic conflict detection computes embedding distances to identify contradictory paraphrases; temporal conflict detection validates temporal constraints using interval algebra [27]; probabilistic conflict detection identifies statistical inconsistencies in uncertain knowledge.

The **Conflict Resolution Module** implements resolution strategies grounded in belief revision theory. Evidence-based resolution weights conflicting information by source reliability using established combination rules [28]. Temporal resolution prefers recent information for volatile domains. Consensus resolution combines multiple sources through voting mechanisms. Context-dependent resolution maintains multiple contexts where different facts hold, enabling nuanced handling of domain-specific knowledge.

The **Knowledge Fusion Module** integrates validated information through structural fusion merging graph-based representations, semantic fusion aligning concepts across different ontologies, and statistical fusion combining probability distributions. The fusion process balances minimizing change to existing knowledge, reducing inconsistency with new information, and controlling knowledge complexity.

The **Reasoning Module** implements multiple reasoning paradigms: deductive reasoning through forward and backward chaining; inductive reasoning through pattern recognition and concept learning; abductive reasoning through explanation generation and diagnostic inference; analogical reasoning through structure mapping and case-based reasoning. These capabilities enable the system to draw conclusions, generate hypotheses, and support decision-making based on integrated knowledge.

3.5 Adaptive Learning Controller

The Adaptive Learning Controller (ALC) implements metacognitive capabilities enabling the system to monitor and improve its own learning processes. The ALC provides strategic coordination while adapting behavior based on performance feedback, embodying principles from adaptive control [13] and metacognition research [19].

The **hierarchical control architecture** operates at three temporal scales. The strategic level (hours to days) sets long-term learning objectives, allocates resources across knowledge domains, and coordinates with external systems. The tactical level (minutes to hours) translates objectives into specific plans, selects and configures learning algorithms, and manages resource allocation. The operational level (seconds to minutes) executes learning operations, adjusts parameters in real-time, and handles immediate feedback.

The **Strategy Selection Module** chooses among available learning strategies based on current context using contextual bandit algorithms that balance exploration of new strategies with exploitation of known effective approaches. Strategy effectiveness is tracked across different contexts, enabling learned selection policies that improve over time.

The **Metacognitive Monitor** provides self-awareness through performance prediction estimating future learning outcomes, confidence calibration assessing accuracy of uncertainty estimates, strategy evaluation analyzing effectiveness across contexts, and learning progress monitoring tracking efficiency and identifying plateaus. When performance degrades or anomalies are detected, the monitor triggers adaptive interventions adjusting parameters, switching strategies, or escalating to human oversight.

The **Parameter Adaptation Module** translates selected strategies into specific configurations for each component, adjusting exploration intensity for the KAE, consolidation thresholds for the DMS, resolution strategies for the IRE, and control parameters throughout the system. This translation ensures strategic decisions manifest in coordinated component behavior.

4 Information Flow and Control Mechanisms

The architectural components described in Sect. 3 achieve coherent autonomous behavior through carefully designed information flow patterns and control mechanisms. This section details the primary processing pathway and three feedback loops that collectively enable continuous learning without external supervision.

4.1 Primary Information Flow

The primary information flow implements the core learning cycle through sequential processing stages, each transforming data toward validated knowledge integration.

The cycle begins when the **Knowledge Acquisition Engine** receives environmental signals indicating available information sources, emerging topics, or changes in monitored domains. The KAE evaluates these signals against current knowledge gaps, prioritizes acquisition targets, and executes retrieval operations. Raw acquired data undergoes initial quality filtering before transmission to downstream components.

The **Dynamic Memory System** receives filtered data and places it in working memory for active processing. The DMS annotates incoming information with temporal markers, source metadata, and preliminary relevance assessments. This staging enables rapid access during validation while maintaining separation from established long-term knowledge.

The **Integration and Reasoning Engine** retrieves staged information along with relevant existing knowledge from the DMS. The IRE performs comprehensive conflict detection, applies appropriate resolution strategies, and produces validated knowledge ready for integration. Resolution decisions are logged for metacognitive analysis.

Validated knowledge returns to the **Dynamic Memory System** for integration into appropriate memory tiers. The DMS updates indices, adjusts activation levels of related knowledge, and triggers consolidation processes when thresholds are exceeded. The integrated knowledge state becomes available for subsequent learning cycles.

This sequential flow ensures that information progresses through acquisition, staging, validation, and integration phases with clear handoffs between components. The design minimizes latency through pipelined processing while maintaining consistency through explicit validation gates.

4.2 Curiosity-Driven Loop

The curiosity-driven loop implements autonomous exploration by connecting knowledge gap analysis to targeted information acquisition, enabling self-directed learning without external curriculum design [9,10].

The loop initiates when the **Dynamic Memory System** performs gap analysis across monitored knowledge domains. For each domain, the system computes coverage metrics comparing current knowledge against domain scope, identifying regions of high uncertainty or sparse coverage. Gap measures incorporate both absolute uncertainty and rate of change, prioritizing domains where knowledge is both incomplete and evolving.

Gap analysis results flow to the **Curiosity Generator**, which transforms identified gaps into exploration priorities. The generator computes expected information gain for potential queries, balancing exploitation of known high-value sources with exploration of uncertain regions. Query prioritization follows upper confidence bound principles, automatically adjusting exploration intensity based on cumulative experience.

Prioritized queries direct the **Knowledge Acquisition Engine** toward targeted information seeking. The KAE receives not merely topics of interest but structured acquisition objectives specifying desired information types, acceptable sources, and quality requirements. This structured guidance enables focused retrieval rather than undirected browsing.

Acquisition results feed back to the gap analysis process, updating coverage models and uncertainty estimates. Successful acquisitions reduce gaps in targeted regions while potentially revealing adjacent gaps warranting future exploration.

This feedback creates a self-reinforcing cycle driving learning toward areas of maximum uncertainty reduction.

4.3 Memory Consolidation Loop

The memory consolidation loop manages knowledge evolution across memory tiers, transferring important information from transient working storage to persistent long-term memory while maintaining bounded resource consumption [29].

Consolidation initiates when working memory contents exceed capacity thresholds or when items have remained in short-term storage beyond expected durations. The **Dynamic Memory System** evaluates candidate items through importance assessment considering access patterns, centrality in knowledge structures, task relevance, and novelty relative to existing long-term knowledge.

Items exceeding importance thresholds enter the consolidation pipeline where they undergo transformation for long-term storage. Abstraction extracts general patterns from specific instances. Compression eliminates redundancy through pattern recognition and hierarchical encoding. Integration establishes connections with existing long-term structures. Indexing creates multiple access paths supporting diverse retrieval patterns.

Consolidated knowledge integrates into long-term memory while source items in working and short-term memory become eligible for displacement. The loop maintains memory bounds while preserving essential knowledge, enabling indefinite operation without unbounded resource growth.

4.4 Metacognitive Control Loop

The metacognitive control loop provides self-monitoring and adaptive capabilities, enabling the system to evaluate and improve its own learning processes [19].

The loop begins with **performance metric collection** aggregating measurements from all components: acquisition success rates from the KAE, retrieval accuracy from the DMS, conflict frequencies from the IRE, and strategy effectiveness from historical records. These metrics provide comprehensive visibility into system behavior.

The **Metacognitive Analyzer** within the Adaptive Learning Controller evaluates collected metrics against expectations and historical baselines. The analyzer identifies performance degradation, detects anomalous patterns, assesses strategy effectiveness across contexts, and monitors learning progress for signs of plateaus or negative transfer.

Analysis results inform **adaptive interventions** ranging from parameter adjustments to strategy switches. When performance falls below thresholds, the ALC may increase exploration intensity, adjust consolidation aggressiveness, modify conflict resolution preferences, or switch to alternative learning strategies. Severe anomalies trigger conservative fallback behaviors or escalation to human oversight.

Intervention effects are monitored in subsequent cycles, creating closed-loop control that continuously optimizes learning behavior. This self-improvement capability enables the system to adapt to changing environmental conditions and evolving task requirements without manual reconfiguration.

5 Component Interactions and Interfaces

The modular architecture described in previous sections requires well-defined interfaces enabling component interaction while maintaining separation of concerns. This section specifies interface protocols and coordination mechanisms that ensure coherent system behavior.

5.1 Interface Specifications

Component interfaces define the contracts governing inter-component communication, specifying data formats, invocation semantics, and error handling conventions.

Data Representation Formats. Information exchanged between components uses standardized representations ensuring interoperability. Knowledge elements employ a hybrid format combining graph structures for relational information with vector embeddings for semantic content, consistent with the knowledge representation framework [30]. Metadata annotations accompany all transmissions, encoding provenance, timestamps, confidence estimates, and processing history. This uniform representation enables components to be developed and optimized independently while maintaining semantic compatibility.

Synchronous and Asynchronous Communication. The architecture supports both communication modes matched to operational requirements. Synchronous interfaces serve latency-critical operations requiring immediate responses, including query execution in the KAE and conflict detection in the IRE. Asynchronous interfaces handle operations tolerating delayed completion, including consolidation processes in the DMS and strategy evaluation in the ALC. Message queues with configurable priorities manage asynchronous communication, enabling load balancing and graceful handling of processing rate mismatches.

Error Handling Protocols. Interfaces specify explicit error handling ensuring graceful degradation under failure conditions. Components distinguish transient errors warranting retry from permanent failures requiring alternative processing paths. Error propagation follows defined escalation hierarchies: local recovery attempts precede component-level fallbacks, which precede system-level interventions. All errors are logged with sufficient context for diagnostic analysis, supporting both immediate recovery and long-term reliability improvement.

Versioning and Compatibility. Interface versioning supports system evolution without requiring simultaneous component updates. Components advertise supported interface versions during initialization, enabling negotiation of compatible protocols. Backward compatibility requirements ensure that component upgrades do not disrupt existing integrations, while deprecation policies provide migration paths for interface evolution.

5.2 Coordination Mechanisms

Beyond point-to-point interfaces, system-wide coordination mechanisms ensure that independently operating components achieve coherent collective behavior.

Hierarchical Control Signals. The Adaptive Learning Controller coordinates system behavior through hierarchical control signals propagating strategic decisions to component-specific parameters [31]. High-level directives specifying learning priorities and resource allocations translate into component configurations: exploration intensity for the KAE, retention thresholds for the DMS, resolution preferences for the IRE. This hierarchical translation ensures strategic coherence while preserving component autonomy in tactical decisions.

Event-Driven Notifications. Components publish significant state changes as events consumed by interested subscribers. The KAE publishes acquisition completions and source availability changes. The DMS publishes memory pressure warnings and consolidation completions. The IRE publishes conflict detection alerts and resolution outcomes. The ALC subscribes to relevant events from all components, maintaining situational awareness for adaptive control. This publish-subscribe pattern decouples event producers from consumers, supporting extensibility and reducing coordination complexity.

Shared State Management. Certain system state requires coordinated access across components. The knowledge base in the DMS serves as the authoritative knowledge repository accessed by both KAE for gap analysis and IRE for conflict detection. Concurrency control mechanisms prevent inconsistent concurrent modifications while minimizing contention. Read operations proceed without blocking, while write operations employ optimistic concurrency with conflict detection, consistent with eventual consistency principles appropriate for learning systems [18].

Resource Negotiation. Components compete for shared computational resources including processing capacity, memory allocation, and communication bandwidth. Resource negotiation protocols enable dynamic allocation based on current priorities and workload characteristics. The ALC mediates resource conflicts, allocating capacity according to strategic priorities while ensuring minimum allocations for critical functions. This managed competition prevents resource starvation while enabling flexible response to varying demands.

6 Algorithmic Specifications

The architectural components and information flows described in previous sections are realized through concrete algorithmic procedures. This section specifies five core algorithms that translate the architectural design into executable implementations, demonstrating how each component's functional responsibilities are operationalized.

6.1 Main Dynamic Learning Loop

The main algorithm orchestrates the complete learning cycle, coordinating all components through four synchronized phases that implement the primary information flow pathway.

Phase I: Knowledge State Assessment. The algorithm begins each iteration by scanning the environment for available information and context signals. Gap analysis evaluates current knowledge against monitored domains, producing prioritized knowledge gaps. The curiosity distribution is computed over potential queries based on gap priorities and expected information gain. Learning context is extracted to inform subsequent strategy selection.

Phase II: Information Acquisition. Based on current context and historical performance, the algorithm selects a learning strategy from the available strategy space. Strategy selection employs contextual bandit methods balancing exploration of untested strategies with exploitation of known effective approaches [32]. The selected strategy determines parameter configurations for query generation and execution. Queries are generated according to the curiosity distribution and executed against environmental sources, yielding candidate knowledge.

Phase III: Validated Integration. Acquired data undergoes multi-stage validation assessing source reliability, content consistency, and relevance to identified gaps. Conflict detection identifies inconsistencies between validated data and existing knowledge. Conflict resolution applies appropriate strategies—evidence-based, temporal, consensus, or context-dependent—producing resolved knowledge. Integration incorporates resolved knowledge into the memory system, followed by consolidation processing.

Phase IV: Metacognitive Control. Performance evaluation assesses knowledge quality improvements resulting from the iteration. Learning history is updated with context, strategy, and outcome information supporting future strategy selection. Metacognitive analysis compares expected and actual performance, identifying adaptation needs. Parameter adjustments propagate to components for subsequent iterations.

The main loop achieves per-iteration complexity of $O(|\Omega| \log |\Omega| + |Q| \log |D| + |\mathcal{K}|^2)$ where $|\Omega|$ represents monitored domain size, $|Q|$ query count, $|D|$ data volume, and $|\mathcal{K}|$ knowledge base size, ensuring tractable operation for realistic problem scales [33].

6.2 Information-Theoretic Curiosity

The curiosity algorithm implements the Knowledge Acquisition Engine's gap analysis and query generation functions, enabling autonomous exploration through information-theoretic prioritization [34].

The algorithm computes entropy-based gap measures across monitored domains, quantifying uncertainty in current knowledge. For each domain ω, the gap measure $G(K, \omega)$ compares maximum possible information content against current coverage. Domains exhibiting high uncertainty receive elevated exploration priority.

Query value estimation combines expected information gain with exploration bonuses following upper confidence bound principles. Expected gain estimates the uncertainty reduction from successful query completion. Exploration bonuses encourage queries to underexplored regions, preventing premature convergence to local optima. The curiosity distribution assigns query probabilities proportional to estimated values through temperature-controlled softmax normalization.

Query selection samples from the curiosity distribution, with batch selection employing diversity mechanisms preventing redundant queries. Selected queries are annotated with expected outcomes enabling subsequent performance assessment. The algorithm maintains running estimates of domain-specific information gain rates, adapting exploration intensity to domain characteristics.

6.3 Principled Knowledge Integration

The integration algorithm implements the Integration and Reasoning Engine's conflict detection and resolution functions, ensuring knowledge consistency while incorporating new information [35].

Conflict detection operates across multiple modalities. Logical conflict detection employs constraint satisfaction techniques identifying formal contradictions. Semantic conflict detection computes embedding similarities identifying paraphrased contradictions. Temporal conflict detection validates temporal constraints ensuring chronological consistency. Statistical conflict detection identifies distributional anomalies suggesting data quality issues.

Detected conflicts are classified by type and severity, determining appropriate resolution strategies. Evidence-based resolution aggregates source reliability scores, preferring information from historically accurate sources. Temporal resolution applies recency weighting appropriate for volatile domains. Consensus resolution combines multiple agreeing sources against isolated contradictions. Context-dependent resolution maintains parallel knowledge contexts where domain-specific truths differ.

Knowledge fusion integrates resolved information through structure merging, semantic alignment, and relationship inference. The fusion process minimizes disruption to existing knowledge while maximizing incorporation of validated new information. Integration completeness verification confirms successful incorporation before finalization.

6.4 Adaptive Strategy Selection

The strategy selection algorithm implements the Adaptive Learning Controller's metacognitive functions, choosing among learning strategies based on contextual effectiveness [36].

Context feature extraction characterizes the current learning situation including knowledge state properties, environmental conditions, recent performance trends, and resource availability. Feature representations enable generalization of strategy effectiveness across similar contexts.

Strategy evaluation estimates expected performance for each available strategy given current context. Evaluation combines historical performance observations with uncertainty estimates reflecting limited experience. Upper confidence bound computation balances exploitation of strategies with strong track records against exploration of strategies with high uncertainty.

Strategy selection chooses the strategy maximizing upper confidence bound estimates. Selection is logged with context features enabling subsequent effectiveness assessment. Parameter translation converts the selected strategy into specific configurations for each component—exploration intensity, consolidation thresholds, resolution preferences—ensuring coherent system-wide behavior.

Performance feedback updates strategy effectiveness estimates following observation of learning outcomes. Contextual bandit update rules adjust estimates toward observed performance while maintaining appropriate uncertainty. The algorithm achieves cumulative regret bounded by $O(\sqrt{T \log |\Sigma|})$ where T is iteration count and $|\Sigma|$ is strategy space size [37].

6.5 Intelligent Memory Consolidation

The consolidation algorithm implements the Dynamic Memory System's memory management functions, transferring important knowledge to long-term storage while maintaining bounded resource consumption [38].

Importance assessment scores knowledge items across multiple criteria: access frequency indicating utility, recency reflecting current relevance, centrality measuring structural importance in the knowledge graph, novelty capturing information not redundant with existing long-term knowledge, and task relevance assessing alignment with current objectives. Weighted combination produces composite importance scores.

Consolidation selection identifies items exceeding adaptive thresholds for transfer to long-term memory. Threshold adaptation responds to memory pressure—increasing selectivity under resource constraints—and performance feedback—relaxing selectivity when retention proves insufficient.

Knowledge transformation prepares selected items for long-term storage. Abstraction extracts general patterns from specific instances. Compression eliminates redundancy through pattern recognition and hierarchical encoding. Dependency analysis identifies connections with existing long-term structures. Index generation creates retrieval paths supporting diverse access patterns.

Memory optimization maintains storage efficiency through periodic reorganization including index rebalancing, redundancy elimination, and structure compaction. The algorithm achieves sublinear memory growth through effective compression while preserving knowledge supporting system capabilities.

7 Future Research Directions

The Dynamic Learning architecture establishes foundations for autonomous continuous learning while opening significant avenues for future research. This section identifies key directions for extending the architectural framework and addressing open challenges.

7.1 Architectural Extensions

Several architectural extensions would substantially expand the framework's capabilities and applicability.

Multi-agent Architectures. The current single-agent architecture could be extended to multi-agent settings where multiple Dynamic Learning systems operate simultaneously in shared environments [39]. Multi-agent extensions require addressing knowledge sharing protocols enabling agents to exchange learned information, coordination mechanisms preventing redundant exploration and conflicting actions, and distributed memory architectures supporting collective knowledge bases. Game-theoretic considerations arise when agents have competing objectives, requiring equilibrium analysis of multi-agent learning dynamics [40].

Distributed and Scalable Implementations. Large-scale deployment demands distributed architectures supporting knowledge bases containing billions of interrelated elements. Key challenges include partitioning strategies determining how knowledge distributes across nodes, consistency protocols maintaining coherence under concurrent updates, and communication-efficient algorithms minimizing synchronization overhead [14]. The modular component design facilitates distributed deployment, but realizing internet-scale operation requires substantial architectural refinement.

Hierarchical and Compositional Structures. The current flat component organization could evolve toward hierarchical architectures where Dynamic Learning systems operate at multiple abstraction levels. Higher-level systems would manage strategic learning objectives while lower-level systems handle domain-specific acquisition and integration. Such hierarchical composition could enable scaling to complex multi-domain learning challenges while maintaining tractable component complexity.

7.2 Integration with Learning Paradigms

Integrating Dynamic Learning with established machine learning paradigms presents opportunities for mutual enhancement.

Deep Learning Integration. Neural network architectures could enhance multiple components [1,2]. The Knowledge Acquisition Engine could employ learned models for gap estimation and query generation. The Dynamic Memory System could utilize neural memory architectures supporting differentiable read-write operations [23]. The Integration and Reasoning Engine could leverage neural reasoning approaches for conflict detection and resolution. Such integration requires addressing challenges of maintaining interpretability and theoretical guarantees while exploiting neural representational power.

Reinforcement Learning Connections. The curiosity-driven exploration mechanisms share conceptual foundations with intrinsic motivation in reinforcement learning [9,10]. Deeper integration could enable Dynamic Learning systems to acquire not only declarative knowledge but also procedural skills through interaction with environments providing reward signals. Unified architectures combining knowledge acquisition with skill learning would substantially expand applicable domains [41].

Continual Learning Synergies. The architecture's mechanisms for preventing catastrophic forgetting connect to active research in continual learning [4,7]. Architectural innovations from either field could benefit the other: Dynamic Learning's explicit memory management could inform neural continual learning approaches, while neural plasticity mechanisms could enhance the Dynamic Memory System's consolidation processes.

7.3 Open Challenges

Fundamental challenges remain for realizing the full potential of autonomous learning systems.

Causal Knowledge Acquisition. Current architectures primarily acquire correlational knowledge without distinguishing causal relationships [42]. Extending the framework to support causal discovery, causal reasoning, and intervention planning would substantially enhance reasoning capabilities. Architectural support for causal knowledge representation and acquisition strategies targeting causal structure identification represent important open problems.

Safety and Alignment. Autonomous learning systems must remain aligned with human objectives throughout extended operation [43]. Architectural mechanisms ensuring safe exploration, preventing harmful knowledge acquisition, and maintaining value alignment under distributional shift require careful development. The metacognitive control loop provides a foundation for safety monitoring, but comprehensive safety architectures demand further research.

Interpretability and Transparency. Understanding system behavior becomes increasingly important as autonomy increases. Architectural support for explaining acquisition decisions, integration choices, and knowledge evolution would enhance trustworthiness and enable effective human oversight [44]. Designing components that maintain interpretability while achieving sophisticated learning remains challenging.

8 Conclusion

This paper has presented the architectural foundations of Dynamic Learning, providing a comprehensive blueprint for AI systems capable of autonomous, continuous knowledge acquisition. The five-component architecture—Knowledge Acquisition Engine, Dynamic Memory System, Integration and Reasoning Engine, Adaptive Learning Controller, and Environmental Interface—together with three coordinating feedback loops, implements the capabilities required for lifelong learning while maintaining coherence and efficiency. The architecture directly addresses fundamental limitations of current AI systems: continuous operation eliminates static learning constraints, hierarchical memory organization prevents catastrophic forgetting, curiosity-driven mechanisms enable autonomous exploration, and metacognitive control supports adaptive knowledge evolution. Concrete algorithmic specifications demonstrate how architectural components translate into executable procedures with tractable complexity. By establishing principled architectural foundations while identifying promising research directions, this work provides a pathway toward AI systems that genuinely learn and adapt throughout their operational lifetime.

References

1. LeCun, Y., Bengio, Y., Hinton, G.: Deep learning. Nature **521**(7553), 436–444 (2015)
2. Goodfellow, I., Bengio, Y., Courville, A.: Deep Learning. MIT Press, Cambridge (2016)
3. McCloskey, M., Cohen, N.J.: Catastrophic interference in connectionist networks: the sequential learning problem. In: Psychology of Learning and Motivation, vol. 24, pp. 109–165. Academic Press (1989)
4. Parisi, G.I., Kemker, R., Part, J.L., Kanan, C., Wermter, S.: Continual lifelong learning with neural networks: a review. Neural Netw. **113**, 54–71 (2019)
5. Quiñonero-Candela, J., Sugiyama, M., Schwaighofer, A., Lawrence, N.D.: Dataset Shift in Machine Learning. The MIT Press (2009)
6. French, R.M.: Catastrophic forgetting in connectionist networks. Trends Cogn. Sci. **3**(4), 128–135 (1999)
7. Kirkpatrick, J., et al.: Overcoming catastrophic forgetting in neural networks. Proc. Natl. Acad. Sci. **114**(13), 3521–3526 (2017)
8. Settles, B.: Active learning literature survey. University of Wisconsin-Madison Department of Computer Sciences, Technical Report (2009)
9. Oudeyer, P.-Y., Kaplan, F., Hafner, V.V.: Intrinsic motivation systems for autonomous mental development. IEEE Trans. Evol. Comput. **11**(2), 265–286 (2007)
10. Pathak, D., Agrawal, P., Efros, A.A., Darrell, T.: Curiosity-driven exploration by self-supervised prediction. In: International Conference on Machine Learning, pp. 2778–2787. PMLR (2017)
11. Anderson, J.R.: An integrated theory of the mind. Psychol. Rev. **111**(4), 1036–1060 (2004)
12. Laird, J.E.: The Soar Cognitive Architecture. MIT Press, Cambridge (2012)

13. Åström, K.J., Wittenmark, B.: Adaptive Control, 2nd edn. Dover Publications, Mineola (2013)
14. Dean, J., et al.: Large scale distributed deep networks. In: Advances in Neural Information Processing Systems, vol. 25 (2012)
15. Schmidhuber, J.: Formal theory of creativity, fun, and intrinsic motivation (1990–2010). IEEE Trans. Auton. Ment. Dev. **2**(3), 230–247 (2010)
16. Baddeley, A.: The episodic buffer: a new component of working memory? Trends Cogn. Sci. **4**(11), 417–423 (2000)
17. Meystel, A.M., Albus, J.S.: Autonomous Mobile Robots: Vehicles with Cognitive Control. World Scientific, Singapore (2002)
18. Gilbert, S., Lynch, N.: Brewer's conjecture and the feasibility of consistent, available, partition-tolerant web services. ACM SIGACT News **33**(2), 51–59 (2002)
19. Flavell, J.H.: Metacognition and cognitive monitoring: a new area of cognitive-developmental inquiry. Am. Psychol. **34**(10), 906–911 (1979)
20. Manning, C.D., Raghavan, P., Schütze, H.: Introduction to Information Retrieval. Cambridge University Press, Cambridge (2008)
21. Paninski, L.: Estimation of entropy and mutual information. Neural Comput. **15**(6), 1191–1253 (2003)
22. Atkinson, R.C., Shiffrin, R.M.: Human memory: a proposed system and its control processes. In: Psychology of Learning and Motivation, vol. 2, pp. 89–195. Academic Press (1968)
23. Graves, A., Wayne, G., Danihelka, I.: Neural turing machines, arXiv preprint arXiv:1410.5401 (2014)
24. Collins, A.M., Loftus, E.F.: A spreading-activation theory of semantic processing. Psychol. Rev. **82**(6), 407–428 (1975)
25. Gärdenfors, P.: Knowledge in Flux: Modeling the Dynamics of Epistemic States. MIT Press, Cambridge (1988)
26. De Moura, L., Bjørner, N.: Z3: an efficient SMT solver. In: International Conference on Tools and Algorithms for the Construction and Analysis of Systems, pp. 337–340. Springer (2008)
27. Allen, J.F.: Maintaining knowledge about temporal intervals. Commun. ACM **26**(11), 832–843 (1983)
28. Shafer, G.: A Mathematical Theory of Evidence. Princeton University Press, Princeton (1976)
29. Dudai, Y.: The neurobiology of consolidations, or, how stable is the engram? Annu. Rev. Psychol. **55**, 51–86 (2004)
30. Sowa, J.F.: Knowledge Representation: Logical, Philosophical, and Computational Foundations. Brooks/Cole, Pacific Grove (2000)
31. Saridis, G.N.: Toward the realization of intelligent controls. Proc. IEEE **67**(8), 1115–1133 (1979)
32. Auer, P., Cesa-Bianchi, N., Fischer, P.: Finite-time analysis of the multiarmed bandit problem. Mach. Learn. **47**(2–3), 235–256 (2002)
33. Orabona, F.: A Modern Introduction to Online Learning. arXiv preprint arXiv:1912.13213 (2019)
34. Houthooft, R., Chen, X., Duan, Y., Schulman, J., De Turck, F., Abbeel, P.: VIME: variational information maximizing exploration. In: Advances in Neural Information Processing Systems, vol. 29 (2016)
35. Alchourrón, C.E., Gärdenfors, P., Makinson, D.: On the logic of theory change: partial meet contraction and revision functions. J. Symb. Log. **50**(2), 510–530 (1985)

36. Zhou, D., Li, L., Gu, Q.: Neural contextual bandits with UCB-based exploration. In: International Conference on Machine Learning, pp. 11492–11502. PMLR (2020)
37. Foster, D.J., Syrgkanis, V., Agarwal, A.: Statistical learning with conditional value at risk. In: International Conference on Machine Learning, pp. 3354–3363. PMLR (2021)
38. Kumaran, D., Hassabis, D., McClelland, J.L.: What learning systems do intelligent agents need? Complementary learning systems theory updated. Trends Cogn. Sci. **20**(7), 512–534 (2016)
39. Busoniu, L., Babuska, R., De Schutter, B.: A comprehensive survey of multiagent reinforcement learning, vol. 38, no. 2 (2008)
40. Shoham, Y., Leyton-Brown, K.: Multiagent Systems: Algorithmic, Game-Theoretic, and Logical Foundations. Cambridge University Press (2008)
41. Sutton, R.S., Barto, A.G.: Reinforcement Learning: An Introduction, 2nd edn. MIT Press, Cambridge (2018)
42. Pearl, J.: Causality: Models, Reasoning, and Inference, 2nd edn. Cambridge University Press, Cambridge (2009)
43. Amodei, D., Olah, C., Steinhardt, J., Christiano, P., Schulman, J., Mané, D.: Concrete problems in AI safety (2016)
44. Miller, T.: Explanation in artificial intelligence: insights from the social sciences. Artif. Intell. **267**, 1–38 (2019)

Enhancing Context-Aware Code Search with Knowledge Graphs and Large Language Models: An Experimental Study

Marlene Goncalves[1,2(✉)], Adelina Figueira[2], and Graciela Perera[3]

[1] Valencian International University, 46002 Valencia, Spain
marlene.goncalves@universidadviu.com
[2] Universidad Simón Bolívar, Caracas 89000, Venezuela
{mgoncalves,15-10484}@usb.ve
[3] Roosevelt University, Chicago, IL 60605, USA
gperera@roosevelt.edu

Abstract. Large Language Models (LLMs) and their capabilities have revolutionized how we explore large unstructured data using natural language prompts. They are capable of supporting important tasks in software engineering, such as code search. Leading to LLMs that extract inherent knowledge for natural language code searches. In this paper, we propose knowledge graph-based solutions for code search that integrate Retrieval-Augmented Generation (RAG) and LLMs, actively enriching searches by incorporating descriptive knowledge from code comments directly into the graph. We performed an experimental study on code search benchmarks with two embedding models: a general-purpose model and one trained specifically on code. Overall, a structured, graph-based retrieval solution is more effective for code search than simple chunking. Implementing specific solution variations boosts precision; however, this directly leads to an increase in query execution times.

Keywords: Information Storage and Retrieval · Code documentation · Query processing · Knowledge Management · Natural Language Processing

1 Introduction

Large Language Models (LLMs) and their capabilities have revolutionized how we explore large unstructured data using natural language prompts. They are capable of supporting important tasks in software engineering, such as code search or finding a snippet of code using natural language [1].

While LLMs allow users to search for code using natural language instead of technical queries, their responses are limited to the public code on which they were trained. This creates a fundamental challenge when searching for new code developed in-house. Code searches using LLM's can be erroneous, but they can be corrected using techniques such as injecting context into the prompt or fine tuning. These alternatives increase the total cost of code searches, either in

H. R. Arabnia et al. (Eds.): CSCI 2025, CCIS 2972, pp. 48–63, 2027.
https://doi.org/10.1007/978-3-032-28469-3_4

terms of time or hardware resources. Thus, LLMs that address this challenge allow for the extraction of inherent knowledge from source code and provide natural language code searches.

In this sense, consider a programmer who wants to understand the code in one or more repositories without having to open every single file. Each repository may have a portion of the code publicly available, while other parts of the code may not. Also, the programmer may vary in technical expertise, making it probable that they will use natural language to express their ideas about the code being searched. This often means the programmer has to either know the technical terms like identifiers in the code for keyword searches or manually review the code in one or more repositories. This can be incredibly time-consuming, especially when dealing with a large volume of files.

This previous scenario highlights the significance of the code search problem, which can be solved with an automated semantic text search. This approach leverages vector representations for each code snippet and for the user's query. A code snippet allows for a direct comparison between the user's query vector and the vectors representing each part of the code being analyzed [2]. The results are code lines whose vectors most closely match the vector of the posed query. Additionally, semantic text search can be improved using a knowledge graph. A knowledge graph describes entities and their relationships [3], allowing code, with its inherent structure of classes, functions, and comments, to be represented in a graph. Each code section (function, class, etc.) can become a node and these nodes can be linked to the files they belong to and the repositories containing those files.

Therefore, we can pose a specific query that will search the knowledge graph and retrieve the desired code. Because a given query may have multiple code snippets associated with a result, the LLMs can select the best answer. In our case, if you ask an LLM a very specific question about the code that it doesn't have in its training data, it might return an incorrect result or something not useful to the programmer, or simply respond that it doesn't know. To tackle this problem, we can use the RAG (Retrieval-Augmented Generation) method [4]. RAG lets us introduce new knowledge into the LLM's context so that a relevant answer can be found. Once that answer is retrieved, the LLM then generates an informed response for the programmer. Even though a similar approach has been described in [5], the key differences from our work are that their tool does not store source code directly in the graph. Instead, it uses pointers to the code's external location and excludes information from comments. Consequently, the reliance on structural code elements for retrieval creates a big limitation because it will only be effective if the user's query uses terms literally present in the code's class or function names. This forces it to resort to an inefficient, brute-force scan of the entire codebase to find the relevant snippet. Also, code comments are not part of its graph and may be missing valuable descriptive knowledge that enriches searches.

In this paper, we propose knowledge graph-based solutions for the code search problem and a detailed performance of them. We focus on improving natural language queries for code by associating code snippets with their respective comments in a graph; improving results from LLMs by adding context and leveraging the descriptive knowledge found in the code comments. The remainder of this work is structured as follows: Section 2 presents the related work in code search, Section 3 describes our solutions, Section 4 details the experimental study, and the final section provides a conclusion and outlines future work.

2 State of the Art

In the last few years, there has been a growing research interest in code search as the problem of searching for source code using natural language queries or searching for relevant code fragments in a codebase with a query. It is one of the most frequent tasks in software engineering, with more than 90% of the code searches seeking code reuse [1].

Code search has been implemented as an equivalent to the k-neighbors search using neural networks with code snippets and queries represented in the same vector space [6]. Thus, developing search models that leverage code descriptions and apply current state-of-the-art natural language processing produces more significant results compared to search models with only source code [6].

Other authors have built search models over fragmented code snippets contained in a collection of source code. The authors in [7] focus on finding a solution to the problem of semantic search starting from a user search and extracting the semantic information from the Abstract Syntax Trees (AST) that is modified and manipulated until obtaining the right code response.

Furthermore, the authors from [8] propose a solution to the problem of code search when reading API documentation. They extract the semantic features of each summarized code snippet for clustering and classification. Clustering is based on dividing the code according to its type of use, then the code within each cluster is classified to find the most meaningful representative example.

Similarly, the authors in [9] look for a solution to the code-to-code search problem, where the query itself is a code snippet. This work uses a query alternation strategy that tries to identify other tokens that could also be relevant when implementing the behavior of the input code. Also, the authors in [10] index a corpus of thousands of open-source projects. They then take a code snippet as input and search the corpus for methods containing that snippet. Likewise, they intersect the search results to efficiently recommend relevant code fragments. However, these approaches differ from our work in their use of code snippets instead of natural language queries and in the techniques that they employ.

On the other hand, the authors in [11] explore the use of natural language and information retrieval through techniques that allow searching directly for code using natural language queries. Their work develops the Neural Code Search (NCS) tool, which enables natural language queries over a large codebase, retrieving code snippets directly from the code corpus. Husain et al. [12] also describe the creation of the CodeSearchNet Corpus. Their dataset includes 99 natural language queries, 6 million open-source functions across several languages, and a corpus containing natural language queries for 2 million functions.

The purpose of this dataset is to encourage researchers in the study of semantic code search.

More recently, the authors in [13] explored a solution for semantic code search that handles ambiguous queries as well as queries needing additional context from a knowledge base. Specifically, this work solves the code search problem using RAG (Retrieval Augmented Generation) [4], LLMs, and code hosted in repositories, which is similar to the approach we're taking.

The use of LLMs has also contributed to code comprehension. Keeping this in mind, the authors in [14] aimed to create a user interface based on an LLM, built directly into the IDE, that facilitates code understanding. It's capable of finding details about API calls used in the code, explaining domain-specific terms, and providing use cases for an API.

Finally, the authors in [5] developed a system that integrates LLM agents with a graph database extracted from code repositories. This approach is closely related to the goal of our own research, but their work differs significantly as its graph database explicitly excludes information from comments and its queries focus solely on the content of structural elements such as classes, functions, modules, global variables, fields, and methods.

3 Our Solutions

This work defines two knowledge graph-based solutions to improve code search using RAG and natural language queries: KG-Chunks and KRAG. Particularly, our solutions focus on code search for the Python programming language. Figure 1a shows the workflow for both solutions, from the initial user query about code repositories to the final answer. This workflow starts by creating an embedding when the user question is received. Following this, a parameterized query acting on the knowledge graph database is executed. This query takes the user's query embedding and a k value to limit candidate responses as parameters, and it utilizes a semantic vector index to find the node nearest to the user's query embedding. After obtaining the response, the LLM is then called to identify the best answer from the nodes retrieved by the query applying the RAG method [15]. Since the query returns nodes in descending order based on their nearness to the user's question, the knowledge graph engine can't determine which of these code fragments is the best answer; the LLM is used for this purpose, ultimately providing the best answer found for the user's query. Thus, before passing the data to the LLM, a *pos* key is added to each individual node's data returned by the query. This makes it easier for the LLM to identify and reference each node by its position. As shown in the green rectangle in Fig. 1a, when the LLM provides a response, it's initially just a string of text. We must extract the relevant values from this string to translate it into a usable format. This generates a structured output that includes the answer and its position, the node's identifier, and the score (a value representing the nearest ranking between nodes and the user's query).

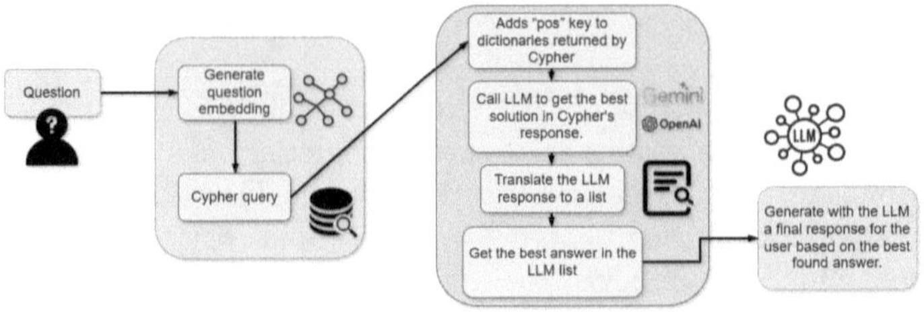

(a) Query workflow for our solutions

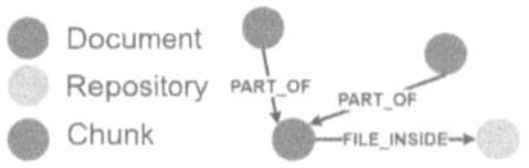

(b) An example graph created with KG-Chunks

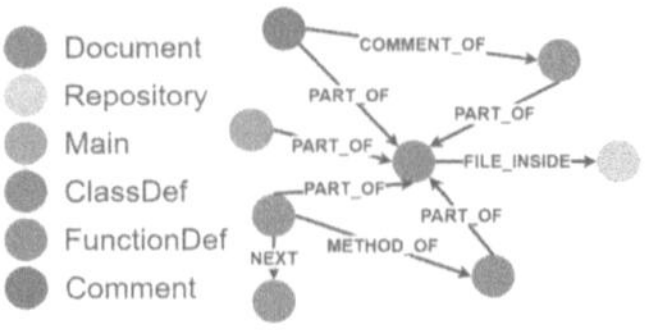

(c) An example graph created with KRAG

Fig. 1. Query Workflow and Examples of Graphs from KG-Chunks and KRAG. (Color figure online)

3.1 KG-Chunks

KG-Chunks is a solution that leverages knowledge graphs to store and organize code from the repository. It divides the code into pieces called Chunks, which makes it easier to integrate with RAG. For this solution, the knowledge graph only considers code sections or pieces of code without relating them to each other. The knowledge graph created has the following nodes: Chunk, Repository, and Document. Each Chunk node stores information about a section of code. Additionally, each Chunk node has an embedding associated with its code section and an identifier that specifies to which code section it belongs. The Repository and Document nodes identify the repository and file, respectively, where the code comes from. Every Chunk node is connected via a PART_OF relationship to a Document node, while each Document node is related via a FILE_INSIDE relationship to a Repository node. Figure 1b presents an example of a graph created with KG-Chunks containing two chunks (in blue) that are part of a single Python file (in red) and hosted in a repository (in yellow).

Additionally, KG-Chunks executes a parametrized query in Neo4j's Cypher language that leverages a semantic vector index on the embedding property of the Chunk nodes. This enables a similarity search to find the nodes with embeddings closest to the user's query. Cosine similarity is employed as the distance function to calculate this vector proximity. The Cypher query for KG-Chunks is shown in Query 3.1. Line 1 searches the chunkVector index for the k nearest neighbor nodes to the parameter \$embedded, which is the vector generated by the embedding model for the user's query. In line 2, the nodes and their respective scores are retrieved based on the cosine similarity function. Finally, lines 3–5 return a structure containing the code section (text), its identifier, and score for the k

nodes, sorted by score in descending order. The Cypher query's response includes the k nodes closest to the user's question.

```
CALL db.index.vector.queryNodes('chunkVector', $k, $embedded)
YIELD node AS chunk, score
RETURN chunk.text as text, chunk.identifier as ID, score
ORDER BY score DESC
LIMIT $k;
```

Query 1.1. Cypher query for KG-Chunks

3.2 KRAG

Since the primary code structures used in Python are functions and classes, we decided to work directly with these structures as code sections. The knowledge graph has the following nodes: Document, Repository, Comment, ClassDef, FunctionDef, and Main. Both the Document and Repository nodes store their respective names as their identifiers. The Comment, ClassDef, and FunctionDef nodes each represent a code annotation, class, or function found within a class, respectively. They all store their code section, embedding, and identifier as properties. Finally, the Main node represents Python code that is not contained within classes or functions. Its properties include the text (a concatenation of all these code sections as a single string), its embedding, and its identifier.

Additionally, the relationships FILE_INSIDE, PART_OF, COMMENT_OF, METHOD_OF, and NEXT exist. They can be related to the nodes as shown in Table 1. It is important to highlight that the NEXT relationship exists because FunctionDef and ClassDef nodes are divided into fixed-size chunks, for instance, every 512 tokens. Thus, the NEXT relationship associates a section of the function or class with its subsequent section.

Table 1. Relationships and their associated nodes

Relationship	Participating Nodes ($A \rightarrow B$)
FILE_INSIDE	Document to Repository
COMMENT_OF	Comment to ClassDef, FunctionDef, or Main
METHOD_OF	FunctionDef to ClassDef
NEXT	FunctionDef to FunctionDef or ClassDef to ClassDef
PART_OF	Comment, ClassDef, FunctionDef, or Main to Document

Figure 1c presents an example of a graph created with the KRAG solution containing a class (in blue color) with a method of more than 512 tokens that is divided into two different nodes (in pink color) and these nodes are connected by means of the relationship NEXT, a function (in pink color) with a comment (in green color) and the node Main (in gray color) that contains the rest of the code. In addition, all these code sections are part of a single Python file (in red), which is hosted in a repository (in yellow).

```
CALL db.index.vector.queryNodes('chunkVector', $k, $embedded)
YIELD node AS comment, score
MATCH p = (comment)-[:COMMENT_OF*]-(n)-[:NEXT*0..]->(end)
WITH COLLECT(p) AS paths, MAX(length(p)) AS maxLength, n, score,
    comment
WITH [path IN paths WHERE length(path) = maxLength | path] AS
    longestPaths, n, comment, score
WITH [node IN nodes(longestPaths[0]) WHERE node.text <> null |
    node.text] as nodel, score, n, comment
RETURN apoc.text.join([i IN nodel[1..] | toString(i)], "␣") as
    text, n.identifier AS ID, score
ORDER BY score DESC
LIMIT $k;
```

Query 1.2. Cypher query for KRAG

For KRAG, a parametrized Cypher query is executed using a semantic search over the Comment nodes. This query is shown in Query ??. Line 1 finds the k nearest neighbors to the parameter $embedded in the chunkVector index. In line 2, the nodes and their respective scores are retrieved according to the cosine similarity function. In line 3, the search process begins by identifying paths in the graph that connect comments to other elements of the code, such as classes or functions, and then traces how those connected elements might sequentially link to other code elements. From the paths identified in line 3, the COLLECT expression in line 4 is used to gather all these paths into a list, and the length of the longest path is found. Line 4 identifies all possible paths matching the pattern, including those paths ending in intermediate segments due to recursion. Since the code is divided into chunks connected by NEXT relationships, the longest path must be obtained. This ensures the path covers the entirety of the function or the class that the comment points to, as only the longest path includes all their NEXT segments. From this list of paths obtained in line 4, line 5 then filters for those that match this maximum length. In line 6, the first path is taken from the list of the longest paths. Any node without a text property is skipped, and only the text property of each node in this path is saved to the nodel list. In line 7, the nodel list's strings are concatenated into a text. Finally, lines 7–9 return the text, along with the identifier and score of k nodes, sorted by their score in descending order.

The implementation of KRAG so far uses naive RAG. Another possible solution is improving the KRAG implementation using the RAG with a multi-query strategy. A RAG with a multi-query strategy paraphrases queries automatically generated from the original query with the help of an LLM. For each of these multi-queries, a Cypher query is executed, and the LLM determines which of the obtained answers is the best. Since there are multiple queries, once the best answer for each multi-query is obtained, they are compared against each other.

The overall best answer is the one ranked highest in the Cypher response (i.e., the closest position to 0), as the results are ordered by a similarity function. Thus, we create two more implementations of KRAG using the RAG with a multi-query strategy. The first one, KRAG-MQ-LC (KRAG-Multi Query-Less Calls), makes a single call to the LLM to process all multi-queries. The LLM's context contains all the lists of Cypher results, allowing it to determine the best answer for each multi-query in one go. The second one, KRAG-MQ-MC (KRAG-Multi Query-More Calls) queries the LLM N times, once for each multi-query. This means the LLM determines the best answer within the Cypher response for each individual multi-query separately.

3.3 Knowledge Graph Creation

To create the knowledge graph, each file in a repository is read and its code fragments are grammatically analyzed. Python's AST module is used for parsing, with single-line comments converted to multi-line comments first. Information is extracted from each code fragment via the file's AST. For KG-Chunks, code snippets within AST nodes of the Chunk class are saved to a list. For KRAG, AST nodes from functions, classes, and comments are considered, with an identifier assigned and the code snippet stored in a dictionary. Functions and classes exceeding 512 tokens are split into separate nodes, as most embedding models have a 512-token limit, which ensures effective embedding and prevents errors.

When constructing the knowledge graph, only grammatically correct Python code snippets are included; others are disregarded. Embeddings for each graph node are generated using the Sentence Transformers module. After importing data into Neo4j, the database instance is initiated, and the knowledge graph is created with the command `create database`. Finally, a semantic vector index is built on the embedding property of Chunk nodes for KG-Chunks and Comment nodes for KRAG, utilizing Neo4j's `vector.similarity.cosine` function.

4 Experimental Study

This section contains a detailed description of the main components of our experimental study. First, explaining briefly the datasets, LLM, and embedding models used in the study, metrics, and implementation engines. Lastly, subsequent 5 sections describe the results for the performance evaluation of the different solutions.

Datasets and Queries: The datasets used as benchmarks for the experimental evaluation are called so-ds-feb20 and codesearchnet-python-valid and are derived from [6]. The first benchmark is data science posts extracted from Stack Overflow with 12,137 code snippets and 947 queries. The second benchmark is a collection of Python code from[1] with 23,107 code snippets and 20,854 queries. The main reason why these were selected is that the datasets have comments

[1] https://github.com/github/CodeSearchNet.

and multiple lines of code. Codesearchnet-python-valid has extensive comments that are explanatory for the code snippets, while so-ds-feb20 has extensive code with short comments for each code snippet [6]. From this point forward, the so-ds-feb20 and codesearchnet-python-valid datasets will be referred to as *codesearch* and *codesearchnet*, respectively. Also, we selected the queries where the target code had the most comments in descending order, where the first query has the largest quantity of them and the last the least. If we found some queries that matched very closely the comments, we paraphrased the queries with the LLM. Following this approach, we executed 10 queries. For each parametrized Cypher query, we chose $k = 100$ nodes as it is a threshold for the LLM to find the most answers, even if the question is somewhat distant from the nodes' content. Additionally, this number of nodes prevents the LLM's context from becoming too extensive.

Models: We selected two embedding models. Our choices include a general-purpose model, paraphrase-MiniLM-L3-v2, and a specialized code-trained model, flax-sentence-embeddings/st-codesearch-distilroberta-base. The models will be abbreviated as MiniLM for paraphrase-MiniLM-L3-v2 and ST-DB for flax-sentence-embeddings/st-codesearch-distilroberta-base. The experiments utilize two LLM APIs: OpenAI's GPT-4o mini and Gemini. GPT-4o mini was selected for its support for asynchronous batch queries and 24-hour result retrieval, reducing costs by half[2]. Additionally, OpenAI was used only for experiments involving LLM calls for each multi-query, specifically in KRAG-MQ-MC. This decision was due to the GPT-4o mini model's limited input token capacity, which was insufficient for processing the Cypher query responses required in KRAG-MQ-LC. Although higher-capacity OpenAI models could support such experiments, cost considerations led to the use of GPT-4o mini. For the Gemini-based experiments, the free version of Gemini 1.5 Flash was used, as testing revealed that Gemini 2.0 Flash produced lower-quality responses and failed to deliver the desired output.

Metrics: Performance and quality of solutions will be evaluated using several metrics: precision, recall, F1-score, and query execution time. Precision measures the percentage of instances that should be in the solution set computed in terms of the ground truth; recall quantifies the percentage of instances generated in terms of the ground truth. The F1-score indicates how well the system finds all correct answers (recall) and how well it avoids finding incorrect instances (precision). The ground truth was established directly from the benchmarks. Query execution time is the duration from when a query is received until the LLM responds.

Engines: We evaluated the four solutions: KG-Chunks, KRAG, KRAG-MQ-LC, and KRAG-MQ-LC. All solutions are implemented in Python 3.11 using Neo4j 1.6.1 with semantic vector indices. Both the code and the resources can be requested from the authors and are in the GitHub repository[3] In our study,

[2] https://platform.openai.com/docs/guides/batch.
[3] https://github.com/adelinarf/codesearchGraph.

we ran the experiments on a computer with the following hardware specification: 8 GB (Gigabytes) of RAM, an Intel(R) Core(TM) i3-1005G1 CPU @ 1.20GHz 1.19 GHz (Gigahertz) processor, and 64-bit Windows 11 Home. In addition, we tried to compare our work with CODEXGRAPH [5] but unfortunately could not do it. A review of the repository (https://modelscope.cn/studios/laptype/CodexGraph) yields issues such as never adding new nodes to the knowledge graph, only querying for existing ones. Also, the graph created always resulted in a trivial graph of just two nodes and two edges. Furthermore, their implementation has outdated drivers and libraries such as py2neo. Lastly, their experimental study is based on benchmarks designed to evaluate code completion across multiple files within a repository; measure the model's ability to solve real software problems by acting like a software engineer, and the generation of new code that adapts or extends the functionalities of an existing repository. While these tasks fundamentally require code retrieval capabilities, the benchmarks themselves do not directly evaluate them.

4.1 KG-Chunks and KRAG Response Quality with Gemini

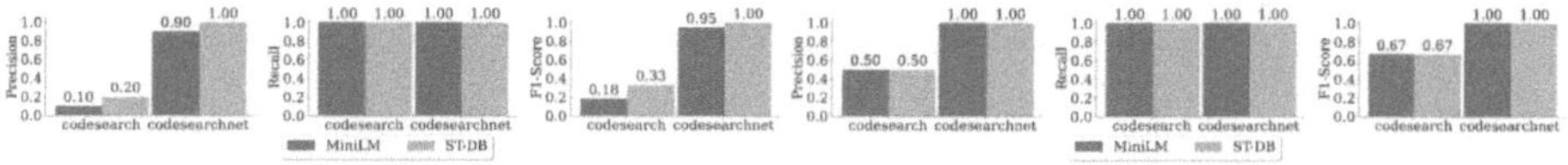

(a) Avg. Quality Metrics for KG-Chunks (b) Avg. Quality Metrics for KRAG

Fig. 2. Average Precision, Recall, and F1-score for KG-Chunks and KRAG Solutions.

The response quality of the KG-Chunks and KRAG solutions, which leverage the Gemini LLM along with the MiniLM and ST-DB embedding models, on the codesearch and codesearchnet datasets is shown in Fig. 2. The results are given by the metrics precision, recall, and F1-score over 10 queries. It should be noted that to improve the response quality of KG-Chunks, it double-checks the best answer from two LLM calls. The key findings observed in the Figs. 2a and 2b are that ST-DB outperforms MiniLM as it shows equal or higher precision and F1-score when compared to MiniLM across both datasets and solutions. This is likely due to ST-DB being trained with code. Also, both KG-Chunks and KRAG have a recall of 1 across all embedding models and datasets; indicating that all relevant code snippets were found by both KG-Chunks and KRAG.

Additionally, KRAG outperforms KG-Chunks as it has higher or equal F1-score and precision values. The KRAG solution demonstrated significantly higher F1-score and precision than KG-Chunks, confirming it has better response quality in providing relevant search results. With respect to the dataset codesearchnet, both solutions have a better precision (greater than 0.9) than codesearch for all cases. This is attributed to codesearchnet containing more extensive and explanatory comments, facilitating better code search results. In contrast, on the codesearch dataset, precision drops for both solutions. However, KRAG achieves an average precision of around 0.5, which is at least twice as high as KG-Chunks' average precision (0.1 to 0.2). This disparity suggests that KRAG is better in handling the less descriptive or sparser comments and documentation prevalent in the codesearch dataset.

4.2 KRAG-MQ-LC Response Quality with Gemini

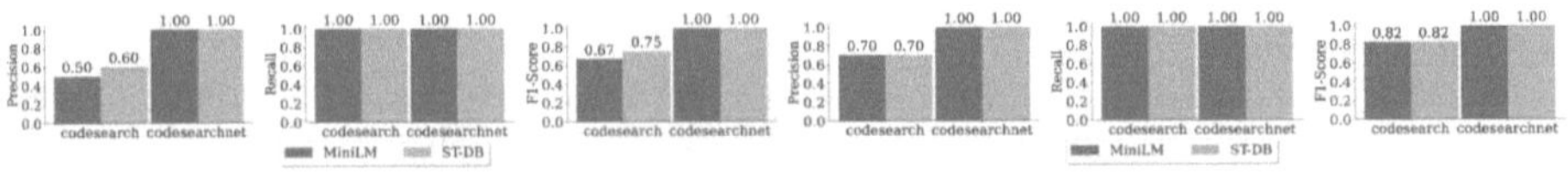

(a) Avg. Quality Metrics for 3 multi-queries(b) Avg. Quality Metrics for 5 multi-queries

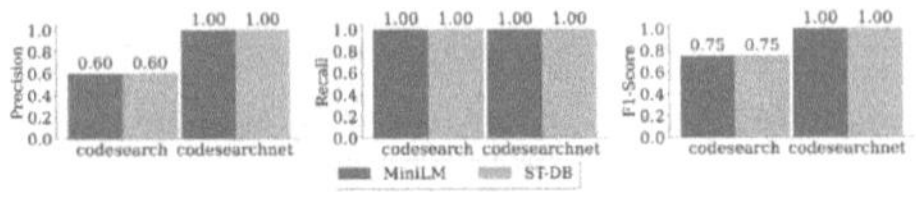

(c) Avg. Quality Metrics for 10 multi-queries

Fig. 3. Average Precision, Recall, and F1-score for KRAG-MQ-LC.

In order to understand if the quality of the result for KRAG improves when the multi-queries strategy is used, we executed the experiments KRAG-MQ-LC. In Fig. 3, we illustrate the average precision, recall, and F1-score obtained after executing 10 code queries on the codesearch and codesearchnet datasets with the Gemini LLM and both MiniLM and ST-DB embedding models, employing 3, 5, and 10 multi-queries. When using 3 multi-queries, as shown in Fig. 3a, we observe a higher precision and F1-score for the ST-DB model with codesearch compared to the KRAG solution. For codesearchnet, both embedding models showed equal precision. Furthermore, Fig. 3b presents the results for 5 multi-queries. Here, the precision for both embedding models with the codesearch dataset increases to 0.7, as does their F1-score. Significantly, the ST-DB model achieves a higher F1-score of 0.82, which is an improvement over the 3 multi-query experiments. In Fig. 3c, for 10 multi-queries, we see that precision and F1-score decrease for both embedding models on the codesearch dataset. This indicates that experiments with 5 multi-queries achieve better precision and F1-score. Therefore, for codesearch, we can conclude that making fewer LLM calls with a higher number of multi-queries can increase precision, up to 5 multi-queries. This is because while more multi-queries allow to expand the context, there's a risk that an overly extensive context can lead to information overload and cause the LLM to hallucinate [16]. This may be precisely what occurred with 10 multi-queries on the codesearch dataset.

Across all experiments, both models achieved a maximum recall of 1, meaning they successfully retrieved all relevant answers. For the codesearchnet dataset, regardless of the number of multi-queries, both models consistently reached a precision, recall, and F1-score of 1. This shows no improvement over the KRAG solution, which has the same values.

4.3 KRAG-MQ-MC Response Quality with Gemini

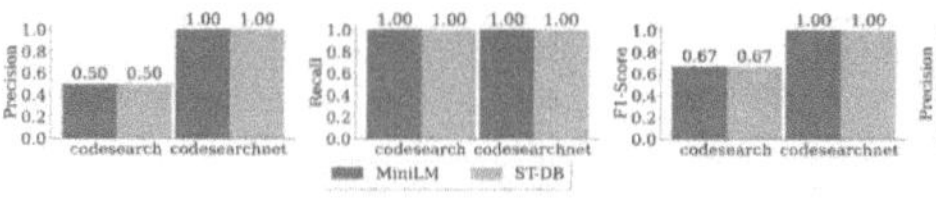

(a) Avg. Quality Metrics for 3 multi-queries

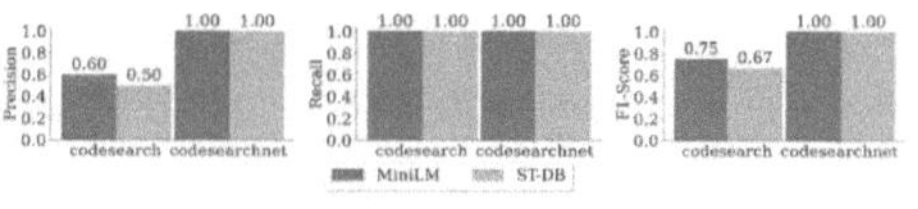

(b) Avg. Quality Metrics for 5 multi-queries

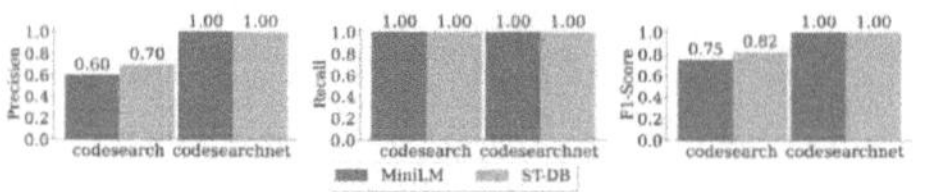

(c) Avg. Quality Metrics for 10 multi-queries

Fig. 4. Average Precision, Recall, and F1-score for KRAG-MQ-MC.

Similarly to the previous section, we study the quality of the results when the multi-queries strategy is used for KRAG-MQ-MC. We show in Fig. 4 the average precision, recall, and F1-score achieved after running 10 queries on both the codesearch and codesearchnet datasets. These results utilize the KRAG-MQ-MC solution with MiniLM and ST-DB embedding models for 3, 5, and 10 multi-queries. As depicted in Fig. 4, both embedding models achieve a precision and F1-score of 1 on codesearchnet. For codeSearch, Fig. 4a illustrates the results for 3 multi-queries, with both embedding models achieving a precision of 0.5 and an F1-score of 0.67. In Fig. 4b for 5 multi-queries, we find that the MiniLM model precision for codesearch increases to 0.6 and the F1-score rises to 0.75. These values are higher than those obtained with the ST-DB model for codesearch. Lastly, in Fig. 4c, for 10 multi-queries, the ST-DB model shows that precision increases to 0.7 for codesearch, as well as the F1-score increases to 0.82. Meanwhile, the precision and F1-score of the MiniLM model do not increase when the number of multi-queries increases, and we notice it remains at 0.60 and 0.75 for 10 multi-queries.

From the results described for the solutions KRAG-MQ-MC and KRAG-MQ-LC, we can conclude that increasing the number of multi-queries does not necessarily yield better precision. For example, KRAG-MQ-LC with 5 multi-queries increases in precision but then drops with 10 multi-queries; this specific experiment doesn't have the same precision as KRAG-MQ-MC with 10 multi-queries. Furthermore, both solutions obtain greater precision for each code query for the codesearchnet dataset. This is because codesearchnet has a greater number of explanatory comments within the code documentation, which contribute to better results. Overall when we use Gemini, the KRAG-MQ-LC solution with

5 multi-queries and the KRAG-MQ-MC solution with 10 multi-queries have identical or similar values for precision, recall, and F1-score.

4.4 KRAG-MQ-MC Response Quality with OpenAI

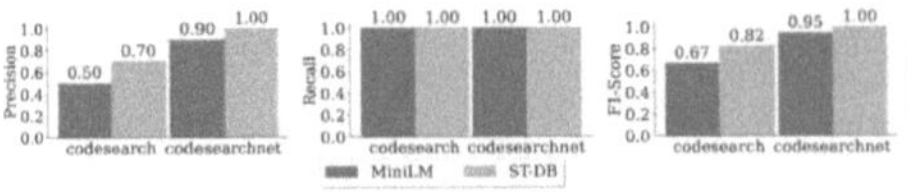

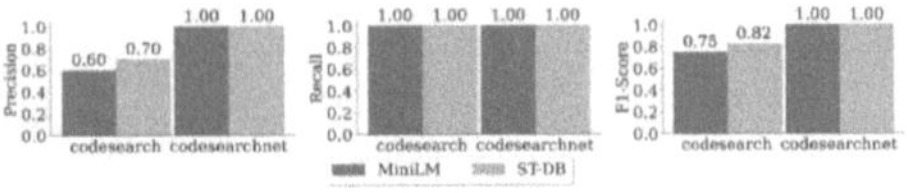

(a) Avg. Quality Metrics for 3 multi-queries(b) Avg. Quality Metrics for 5 multi-queries

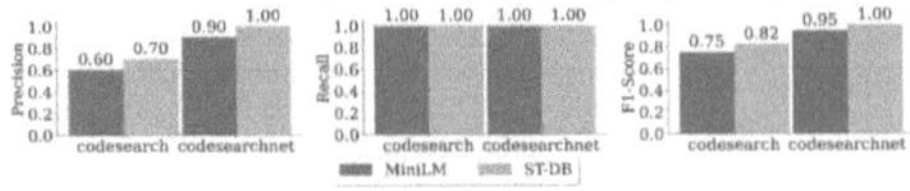

(c) Avg. Quality Metrics for 10 multi-queries OpenAI

Fig. 5. Average Precision, Recall, and F1-score for KRAG-MQ-MC with OpenAI.

After we have discussed the solutions with the Gemini model, we present our results with OpenAI. OpenAI is used exclusively in experiments that involve LLM calls for each multi-query, specifically the KRAG-MQ-MC experiments. This limitation arises because the GPT-4o mini model cannot handle the necessary input token volume for LLM calls in KRAG-MQ-LC experiments, as those include Cypher query responses for all multi-queries. While it would be possible to conduct KRAG-MQ-MC experiments that are more costly OpenAI models, the decision was made to prioritize cost-effectiveness.

As in the previous section, we show in Fig. 5 the average precision, recall, and F1-score obtained from executing 10 queries on both the codesearch and codesearchnet datasets. These results utilize the KRAG-MQ-MC solution alongside OpenAI with the MiniLM and ST-DB embedding models, for 3, 5, and 10 multi-queries. As shown in Fig. 5a for 3 multi-queries, we observe that for the codesearch dataset both the precision and F1-score of the MiniLM model are lower than those achieved by the ST-DB model. At the same time, for the results from 5 multi-queries, Fig. 5b illustrates that while the MiniLM model's F1-score for codesearch increases to 0.75, its precision and F1-score still remain below the values by the ST-DB model. Similarly, the MiniLM model shows improvement for codesearchnet, with both its precision and F1-score increasing. Lastly, when we use 10 multi-queries, Fig. 5c highlights a distinction: the MiniLM model records a precision of 0.60, whereas the ST-DB model achieves a higher precision of 0.7. As a general observation, the ST-DB model's F1-score for both codesearch and codesearchnet is better than the MiniLM model, indicating that ST-DB is the more precise model.

Overall, OpenAI's API yields better results for 3 and 5 multi-queries compared to Gemini's API, considering the precision and F1-score values obtained. Ultimately, there isn't a significant difference between OpenAI and Gemini when returning top-ranked results. Furthermore, we cannot conclude that the OpenAI API has a higher quality in its responses than the Gemini API as there's no substantial difference in our study showing an overall quality of answers from both LLMs. Despite this, OpenAI performs better by achieving higher precision for a smaller number of multi-queries. Finally, we can confirm that the OpenAI API and Gemini knowledge graph constructed from the codesearchnet dataset will obtain greater precision for each query. Also, both LLMs with the ST-DB model have a higher quality in their responses than the MiniLM model.

4.5 Query Performance

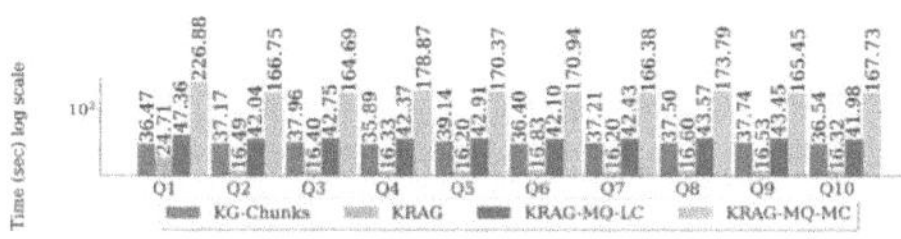

(a) Time by the MiniLM model on codesearch

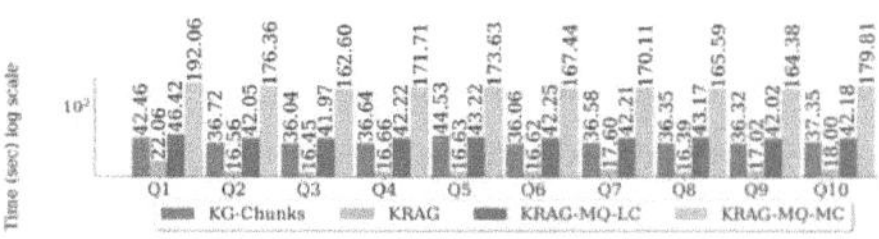

(b) Time by the ST-DB model on codesearch

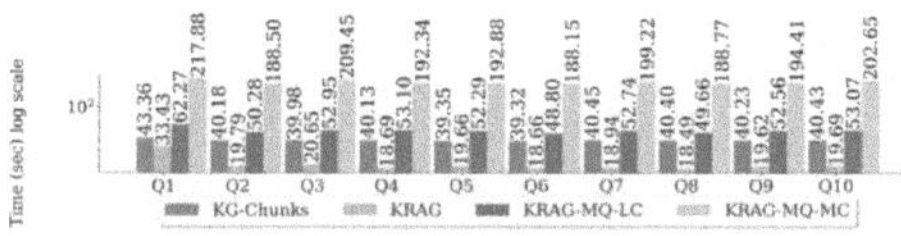

(c) Time by the MiniLM model on codesearchnet

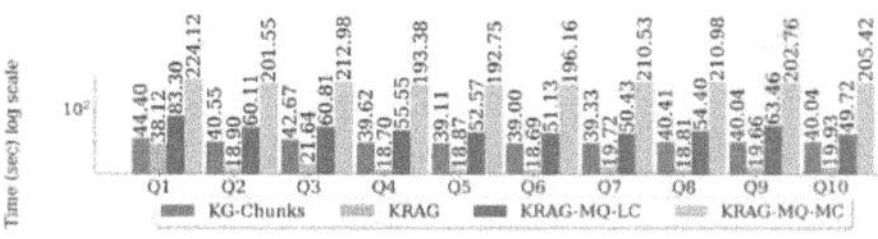

(d) Time by the ST-DB model on codesearchnet

Fig. 6. Query Execution Time by solution, embedding model and dataset.

The execution time for each query across all solution variants with Gemini is shown in Fig. 6. Given that the OpenAI batch API allows for asynchronous calls, we did not measure total time. As Fig. 6 illustrates, the KG-Chunks solution is consistently slower than KRAG, with execution times ranging from 10% to 50% longer, depending on the specific embedding model and dataset. Conversely, using multi-queries in the KRAG-MQ-LC and KRAG-MQ-MC variants increases execution time due to more calls to the LLM. The execution times for KRAG-MQ-LC and KRAG-MQ-MC are 4 to 6 times greater than KG-Chunks, and 5 to 11 times greater than KRAG. Based on these observations, we can conclude that performing a greater number of LLM calls with the use of multi-queries increases response time. Finally, the multi-queries strategy for the RAG method improves upon the results obtained with naive RAG. Specifically, the KRAG-MQ-LC solution with 5 multi-queries and the KRAG-MQ-MC solution with 10 multi-queries generated more precise results, with F1-scores above 0.7, indicating good quality responses. Also, the ST-DB model for embeddings guarantees good performance as it achieved higher precision, recall, and F1-score.

4.6 Precision-Time Trade-Off

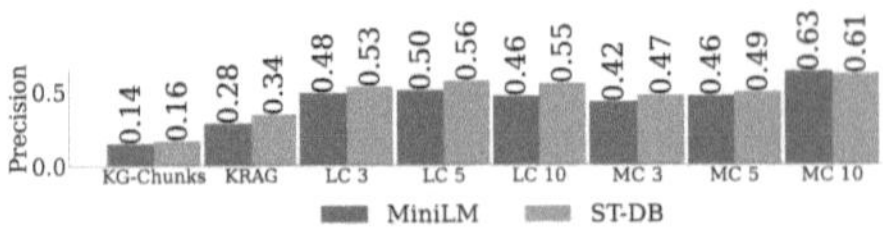

(a) Avg. precision for codesearch

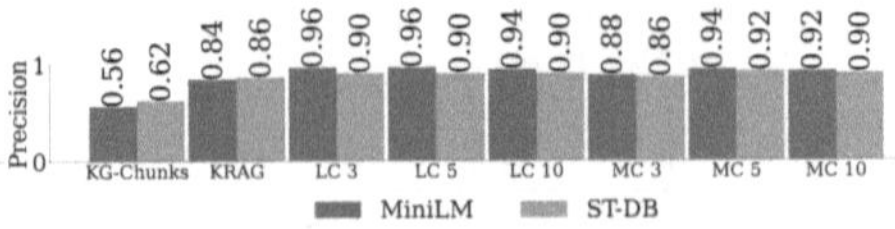

(b) Avg. precision for codesearchnet

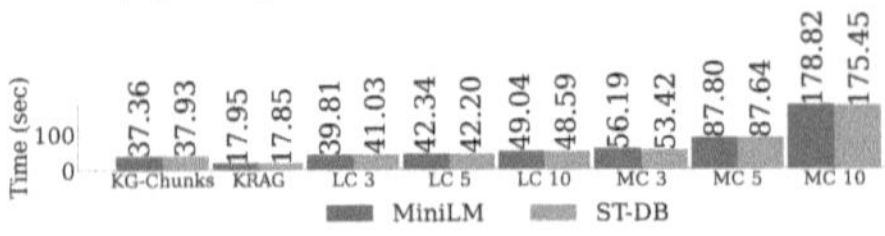

(c) Avg. time for codesearch

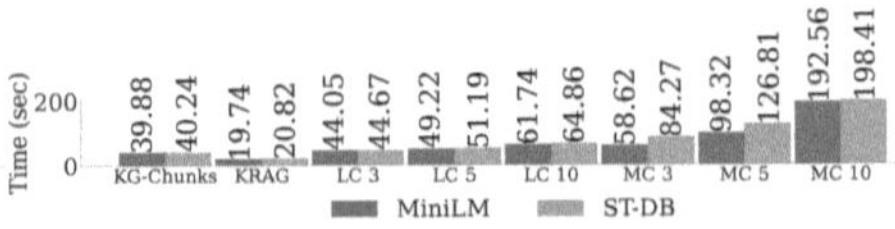

(d) Avg. time for codesearchnet

Fig. 7. Average precision and total time for 50 queries per solution, dataset and embedding model where LC represents the KRAG-MQ-LC solution and MC represents the KRAG-MQ-MC solution, both accompanied by the number of multi-queries.

The final experiments generalize the results for 50 queries from each dataset. We consider the tradeoff between precision and total time given that recall had a result of 1 for any case. In Fig. 7, the average precision and total time results are shown for each solution variant with the MiniLM and ST-DB models for 50 queries from each dataset. Specifically, it's observed that in Fig. 7a for codesearch dataset, KRAG-MQ-MC with 10 multi-queries and the MiniLM model achieved the highest precision. Despite this, this solution is very time-consuming(see Fig. 7c). Therefore, the ST-DB model is preferable due to its better performance and similar precision. Similarly, in Fig. 7b for codesearchnet, the best precision is obtained with KRAG-MQ-LC using 3 and 5 multi-queries with the faster MiniLM model (see Fig. 7d).

5 Conclusions and Future Work

In conclusion, this research demonstrates that a structured, graph-based retrieval method (KRAG) is significantly more effective for code search than a simple chunking approach (KG-Chunks). This work also confirms that advanced RAG strategies using multi-queries boost precision over naive RAG, though this introduces a trade-off with increased query execution time. Based on the quality-time trade-off experiments, the best overall solution is KRAG-MQ-LC using the specialized code-trained embedding ST-DB model. This finding considers that the effectiveness of the embedding model depends on how descriptive the comments are. In addition, we intend to extend our study from Python to other existing programming languages as well as include in the knowledge graph relationships between repositories, inheritance, and other pertinent programming structures.

References

1. Liu, C., Xia, X., Lo, D., Gao, C., Yang, X., Grundy, J.: Opportunities and challenges in code search tools, ACM Comput. Surv., **54**(9) (2021). https://doi.org/10.1145/3480027
2. Wang, B., Wang, A., Cheng, F., Want, Y., Kuo, Y.: Evaluating word embedding models: methods and experimental results, Cambridge University Press (2019)
3. Stegeman, J.: What is a knowledge graph? Graph database & analytics, Neo4j (2024). https://neo4j.com/blog/what-is-knowledge-graph/
4. Lewis, P., et al.: Retrieval augmented generation for knowledge-intensive NLP tasks, Adv. Neural Inf. Process. Syst. **33**, 9459–9474 (2020)
5. Liu, X., et al.: Codexgraph: bridging large language models and code repositories via code graph databases (2024). https://arxiv.org/abs/2408.03910
6. Heyman, G., Huysegems, R., Justen, P., Van Cutsem, T.: Natural language-guided programming, In: Proceedings of the 2021 ACM SIGPLAN International Symposium on New Ideas, New Paradigms, and Reflections on Programming and Software, ser. Onward! 2021. New York, NY, USA: Association for Computing Machinery, pp. 39–55 (2021). https://doi.org/10.1145/3486607.3486749
7. Reiss, S.P.: Semantics-based code search, In: 2009 IEEE 31st International Conference on Software Engineering, pp. 243–253 (2009)
8. Kim, J., Lee, S., Hwang, S.-W., Kim, S.: Towards an intelligent code search engine, In: Proceedings of the AAAI Conference on Artificial Intelligence, vol. 24, no. 1, pp. 1358–1363 (2024). https://ojs.aaai.org/index.php/AAAI/article/view/7503
9. Kim, K., et al.: Facoy: a code-to-code search engine, In: Proceedings of the 40th International Conference on Software Engineering (2018). https://doi.org/10.1145/3180155.3180187
10. Luan, S., Yang, D., Barnaby, C., Sen, K., Chandra, S.: Aroma: code recommendation via structural code search, In: Proceedings of the ACM on Programming Languages, vol. 3, no. OOPSLA (2019). https://doi.org/10.1145/3360578
11. Sachdev, S., Li, H., Luan, S., Kim, S., Sen, K., Chandra, S.: Retrieval on source code: a neural code search, In: Proceedings of the 2nd ACM SIGPLAN International Workshop on Machine Learning and Programming Languages, pp. 31–41 (2018). https://doi.org/10.1145/3211346.3211353
12. Husain, H., Wu, H., Gazit, T., Allamanis, M., Brockschmidt, M.: Codesearchnet challenge: evaluating the state of semantic code search, CoRR, vol. abs/1909.09436 (2019). http://arxiv.org/abs/1909.09436
13. Jain, S., Dora, A., Sam, K.S., Singh, P.: LLM agents improve semantic code search (2024). https://arxiv.org/abs/2408.11058
14. Nam, D., Macvean, A., Hellendoorn, V., Vasilescu, B., Myers, B.: Using an LLM to help with code understanding, In: Proceedings of the IEEE/ACM 46th International Conference on Software Engineering, pp. 1–13 (2024). https://doi.org/10.1145/3597503.3639187
15. Gao, Y., et al.: Retrieval-Augmented Generation for Large Language Models: A Survey (2024). https://arxiv.org/pdf/2312.10997
16. Mahmood, H.: The LLM context window paradox: is bigger always better? (2024). https://datasciencedojo.com/blog/the-llm-context-window-paradox/. Accessed: 1-05-2025

Multi-hop Natural Language Question Answering on Graphs by Using LLMs

Elika Bozorgi(✉) and Krys Kochut

School of Computing, The University of Georgia, Athens, GA, USA
{elika.bozorgi,kkochut}@uga.edu

Abstract. Widely used in different domains such as dialog interfaces and chat-bots, Question-Answering (QA) models are AI models that are often capable of answering questions given some context, as well as without any context, e.g., open-domain QA. LLMs are leveraged to enhance QA models, but only work well on one-hop questions (a type of question that can be answered by retrieving a single piece of information from one source or a single step of reasoning): the questions that can be answered with only one hop on their respective knowledge graph. In this work, we present a novel method to enable LLM-enhanced QA models to answer multi-hop questions. Our proposed method consists of three steps: 1) in the first step, we use LLMs for creating one-hop question answers for each edge of the graph schema and store these one-hop QAs in a Neo4j database, 2) next, the LLM uses these QAs to find the answer to the user's question, and 3) the LLM then generates Cypher query of the answers found in the previous step. We evaluate our method on two datasets in Neo4j. Our results indicate improvements in question answering with LLMs compared to the previous methods.

Keywords: LLMs · Question answering · Graphs · Multi-hop reasoning

1 Introduction

Question answering (QA) on graphs has gained increasing popularity recently since it allows users to retrieve information from structured data. Using QA, the user does not need to have information about the underlying schema of the graph. QA is widely used in various fields such as chatbots [1], customer support services, personalized recommendations [2], biomedical research and healthcare [3], financial services [4] and etc. While LLMs are widely used for simple question answering, there are still challenges in using them for multi-hop queries. In this work, we introduce a new method for question answering by using prompt engineering technique and build a novel system for multi-hop question answering.

LLMs are broadly used for question answering purposes and are trained on large datasets utilizing different techniques. Here, we discuss some of the training methods for LLMS:

H. R. Arabnia et al. (Eds.): CSCI 2025, CCIS 2972, pp. 64–76, 2027.
https://doi.org/10.1007/978-3-032-28469-3_5

- Retrieval-augmented generation (RAG). RAG is an innovative approach in the field of NLP that enhances the quality of the generated text by combining the strengths of retrieval-based and generation-based models [5]. In traditional LLMs, the generated responses are based on the data used for training the LLMs; which probably do not include the latest or accurate information in a specific field. RAG addresses this shortcoming by integrating retrieval-based models which grants access to the external databases, APIs or document repositories to the model [6].
- Fine-tuning. Fine-tuning is the process of taking a pre-trained model and training it further on a smaller, domain-specific dataset to adapt it to a specific task. It improves the accuracy of the model for specific tasks. Fine-tuning helps to reduce training time, resources and other computational expenses [7].
- Prompt engineering. A prompt is a natural language request that asks an LLM to perform a specific task or action. It gives the model context via tokens, and sidesteps the model's potential limitations, so that the model can give you a response. Prompt engineering in LLMs is the process of designing and optimizing prompts to obtain relevant, accurate and high-quality responses from language models [8]. There have been recent works on leveraging LLMs for multi-hop question answering. One recent work is [9], in which the authors use prompt engineering to train the LLM and use chain of hops reasoning for question answering.

- Another work in this field is [10] in which the authors increase the accuracy of question answering on graphs by using ontologies and LLMs. By combining deterministic ontology checks with LLM-based repair, the approach significantly enhances reliability and correctness in QA systems, showcasing the power of knowledge graph semantics in boosting LLM performance.

- Few-Shot Knowledge Base Question Answering (FlexKBQA): Introduced by [11]; FlexKBQA improves knowledge base question answering by decomposing complex questions into interpretable components such as entity linking, relation extraction, and query structure prediction. These components are processed through a modular pipeline, allowing for flexibility, better error handling, and easier adaptation to new domains. The outputs are then combined to generate accurate SPARQL queries. This modular design enables the system to generalize effectively across different datasets and knowledge bases.

- Knowledge graph in the medical field (LLM-KGMQA), proposed in [12]. In this work, the authors leveraged LLMs to propose a method for multi-hop knowledge path reasoning and entity linking. Regarding the multi-hop knowledge path reasoning, the method proposed a three-step reasoning framework including an n-hop subgraph construction algorithm, a semantics-based knowledge pruning algorithm and a knowledge fusion algorithm.

Problem: In this work, we use LLMs to generate a Cypher query for the user's multi-hop question in natural language. We cannot use LLMs directly to answer the user's question; because LLMs have challenges to answer multi-hop questions [13]. For example, let's assume we have a graph dataset which represents an NFL player and their relevant information. The schema of this graph looks like this (Fig. 1):

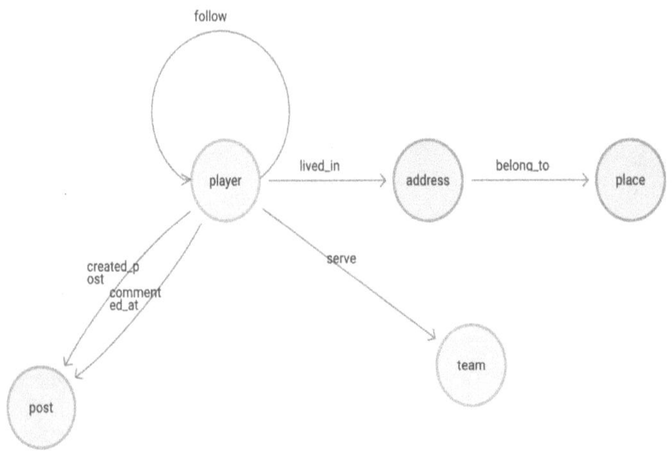

Fig. 1. The Schema of the NFL Player and their information.

If the user asks a question about two connected nodes, such as nodes 'player' and 'address', the LLM can respond to the user's question properly. For example:

Q: Where does the player live in?

A: The player lives in Athens, GA.

However, if the user asks a question about two nodes that are not connected with an edge such as 'place' and 'post', such as:

Q: What is the place of the player who posted on their twitter about the game?

the LLM has difficulties in answering the question for various reasons; which means that the LLM does not respond at all or respond with a wrong query. One reason is that the LLM cannot find the correct path between these two nodes. Another reason is that even by having the correct path, the LLM might not find all the relevant nodes; e.g. if the player has several addresses assigned to them, the LLM might find only one of the addresses. Therefore, here we solve two problems: training LLMs to find a path for multi-hop question answering and generating a Cypher query of the user's question (Fig. 2). We introduce a new method and leverage prompt engineering to train the LLM to find the right path for multi-hop question answering.

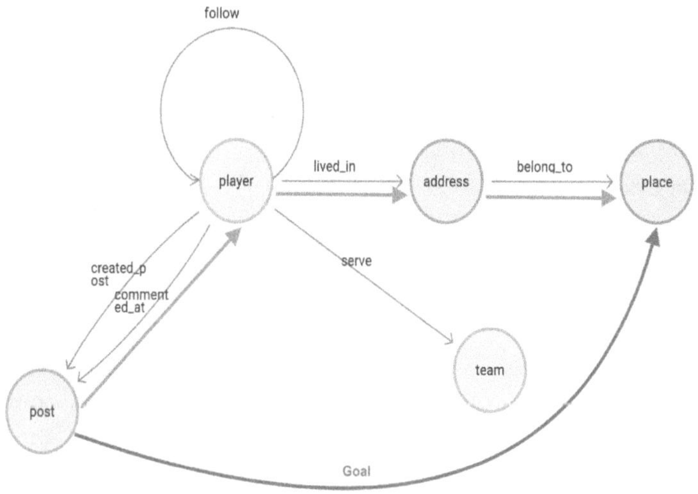

Fig. 2. The Schema of the NFL Player and their information: The goal is to answer a question that connects the node 'post' to 'place'.

2 Related Works

There are various approaches used for QA, and here is a list of some of the commonly used ones:

- *Semantic Parsing* [14]; which involves converting a question in natural language into a machine readable language including query languages such as Cypher; which retrieves information from knowledge graphs. Another approach is *Deep Learning* [15], which uses trained neural networks to reason over knowledge graphs to answer questions or map questions directly to their corresponding answers.

- Using Neural Machine Translation (NMT); a method that uses machine learning to translate one language into another [16]. In this case, NMTs can be leveraged to translate natural language to a structured language like Cypher.

- Large Language Models (LLM). LLMs are based on transformer architectures and trained on large volumes of data, and designed to handle human form of written language. The transformer model, which was introduced in 2017 in [17], is a type of neural network architecture that stands out at processing sequential data. As a result, they are mainly associated with LLMs but they are used in other fields such as computer vision [18], time series [19] and others. There are other well-known transformer-based architectures such as BERT and T5 which are used for natural language processing purposes including question answering [20,21] using different mechanisms. For example, T5

(Text-to-Text Transfer Transformer) developed by Google, converts all NLP tasks into a text-to-text format [22]. On the other hand, BERT (Bidirectional Encoder Representations from Transformers) processes text bidirectionally, considering context from both left and right sides of a token [23]. However, LLMs are generative, auto-regressive models that learn uni-directional (from left to right). LLMs have revolutionized NLP by using deep learning techniques to understand and generate human-like text. Due to their numerous benefits, various businesses take advantage of them in different domains such as customer service [24], healthcare [25], software development [26] and others.

The above mentioned research works are not accurate in answering quesitons that require multi-hop reasoning. In this work, we propose a multi-hop QA model to bridge this gap.

3 Methodology

In this work, we design an algorithm that translates user's multi-hop question from natural language into Cypher query. This helps users with no knowledge of the Cypher language to ask their questions in natural language and convert them to Cypher language in the chain-of-hops format. Then, the user runs this query on a Neo4j database to get the answer to their question. Let's explain the concept with an example: Consider the Movies dataset where we have 6 different node types (Fig. 4(b)). The user wants to get more information about the genres of the movies and the actors acted in them. For example the user asks: "Which genres did (actor) Hugo Weaving act in?". This is an example of the multi-hop question since there is no direct edge between the nodes Actor (Hugo Weaving) and Genres.

Since it is a multi-hop question, the LLM on its own is typically not able to answer it. We train the LLM to find a path between the nodes "Actor" and "Genre" by prompt engineering. We teach it that the path is consisted of two steps: first find the connection between the nodes "Actor" and "Movie" and then "Movie" and "Genre". Next, we convert the user's question to a Cypher query by using LLMs and run the Cypher query on the Neo4j database to get the final answer. Our assumption is that the user does not have enough knowledge of the Cypher language to write the query on their own. This algorithm helps the user to obtain the Cypher format of their question and get the answer by running the Cypher query against the Neo4j database. Our algorithm is applicable on any graph dataset; however, in this paper, we get the results on two datasets: *Movies (/***neo4j_recommendations***/)* and *Northwind (/***neo4j***/)*. This method consists of three different steps (Fig. 5 illustrates the structure of this method):

1. Let's assume we store the graph dataset and its schema in database (A) in Neo4j. We use LLM to generate one-hop QAs (in natural language format) for each edge type between any two nodes in the graph by using the data in database A. We store the results in another Neo4j database (database B). For example, in

the *Movies* dataset, we have 6 different node types (Fig. 4 (b)). Considering the nodes *Actor: Hugo Weaving* and *Movie: Proof, Cloud Atlas, The Dressmaker, V for Vendetta, Babe, Matrix The, Interview The, Adventures of Priscilla, Queen of the Desert, The* and the edge type connecting them which is *Acted-In* (Fig. 4), LLM generates a question answer (QA) like this:

Q: Did Hugo Weaving act in any movies?

A: Yes, Hugo Weaving acted in different movies such as Proof, Cloud Atlas, The Dressmaker, V for Vendetta (Fig. 3).

Fig. 3. The movies which "Hugo Weaving" acted in.

2. Next, we run the algorithm which gets the user's question in natural language and breaks down the user's question into smaller QAs by using database B. As mentioned earlier, the user's question is "Which genres did Hugo Weaving act in?". To find the answer, we train the LLM on how to find the path between the nodes "Actor" and "Genre". The path consists of two steps: 1. connecting nodes "Actor" and "Movie" 2. connecting the node "Movie" to "Genre". By leveraging prompt engineering, the LLM finds the path between the nodes "Actor" and "Genre" by finding the relevant QA from the database B: *User's question: Which genres did Hugo Weaving act in?* Using the few-shot training the LLM find the first step of the path from database B: *Q1: Did Hugo Weaving act in any movies? A1: Yes, Hugo Weaving acted in different movies such as Proof, Cloud Atlas, The Dressmaker, V for Vendetta*

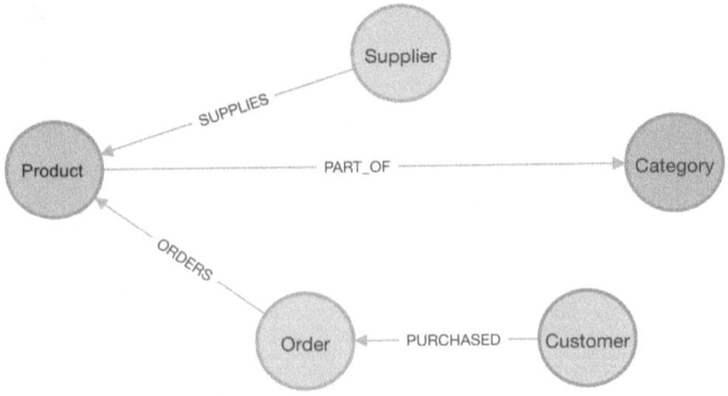

(a) The schema of the Northwind dataset.

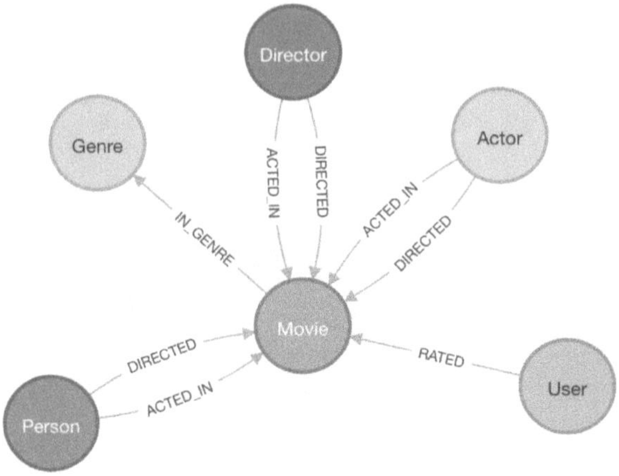

(b) The schemas of the dataset.

Fig. 4. The schemas of the Movies and Northwind datasets.

Comparing Fig. 3 and the LLM's response, the LLM has not found all the movies which Hugo Weaving acted in. That is one of the challenges of the LLM to find all the connected Movie nodes to the Actor node. However, the good point is that it has found a connection between Hugo Weaving and the movies; which is the first step of the path. The second question that LLM finds from Database B is:

Q2: Are the genres of the movies Proof, Cloud Atlas, The Dressmaker, V for Vendetta available ? A2: Yes, the genres of these movies are drama, fiction, comedy in order.

3. In the last step, the code uses LLM and inputs the QAs in the previous step and generates Cypher queries for them. These queries are in format of Chain-of-hops. For example, considering the QAs in the previous step, here are the generated Cypher queries:

Query of the Q1:

MATCH (a:Actor name: "Hugo Weaving")-[r: Acted-In]-(m:Movie name: 'Proof, Cloud Atlas, The Dressmaker, V for Vendetta, Babe, Matrix The, Interview The, Adventures of Priscilla, Queen of the Desert, The')
RETURN m

Query of the Q2:

MATCH (g:Genre)-[]-(m:Movie name: 'Proof, Cloud Atlas, The Dressmaker, V for Vendetta, Babe, Matrix The, Interview The, Adventures of Priscilla, Queen of the Desert, The')
RETURN g

The code later deletes the names of the movies from the query for a more accurate one resulting in:
MATCH (a:Actor name: "Hugo Weaving")-[r: Acted-In]-(m:Movie)
MATCH (g:Genre)-[]-(m:Movie)
RETURN g

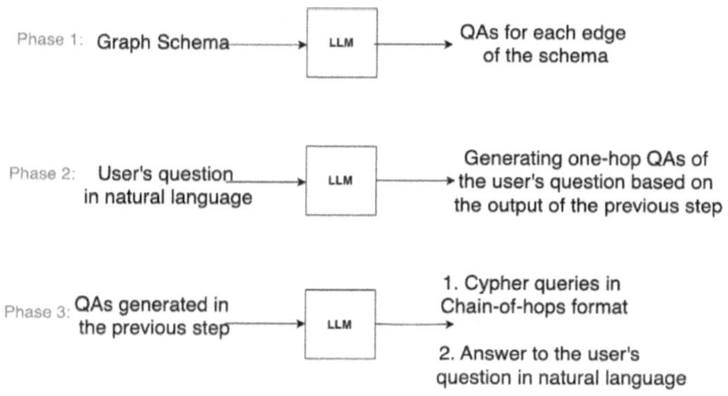

Fig. 5. The framework used in the algorithm.

4 Experiments

In this section, we design a set of numerical experiments, in order to depict the efficiency and effectiveness of our proposed algorithm. To run the experiments, we used GPT-4o mini for all LLMs in the algorithm.

4.1 Datasets

For the experiments, we use two different datasets. One of them is the Northwind dataset; which is a dataset containing purchase history. This dataset simulates a fictitious company, "Northwind Traders", which deals with specialty food imports and exports. The database is widely used for educational purposes and practice;

particularly with relational database management systems such as Neoj4 and SQL servers.

This dataset consists of 1,035 nodes and 3,139 edges. You can see the schema of this dataset as shown in Fig. 4 (a). The second dataset is called "Movies" dataset; which is a graph of popular movies and actors. This dataset consists of 28,863 nodes and 166,261 edges. You can see the schema of this dataset as shown in Fig. 4 (b).

4.2 Results

For the Northwind dataset, we ran our experiments for one-hop edge (between two nodes connected by one edge), two-hop edge (between three nodes connected by two edges). We implemented the algorithm on this dataset and shared the results in Appendix A. Here are some of the results:

User Question: *What products did Antonio Moreno order?*

Answer:
MATCH (c:CustomercustomerName: 'Antonio Moreno')-[:placed]->(o:Order)
MATCH (o:Order)-[:contains]->(p:Product)
RETURN p.productName AS Products

User Question: *Which category does the supplier "New Orleans Cajun Delights" supply?*

Answer:
MATCH (s:SuppliercompanyName:'New Orleans Cajun Delights')-[:supplies]->(p:Product)
MATCH (p:Product)-[:part-of]->(c:Category)
RETURN p.productName, c.categoryName

User Question: *Which category does supplier "Exotic liquids" supply?*

Answer:
MATCH (s:SuppliercompanyName:'Exotic Liquids')-[:supplies]->(p:Product)
MATCH (p:Product)-[:part-of]->(c:Category)
RETURN p.productName, c.categoryName

User Question: Which category does supplier "Grandma's Boysenberry Spread" supply?

Answer:
MATCH (s:SuppliercompanyName:"Grandma Kelly's Homestead")-[:supplies]->(p:Product) MATCH (p:Product)-[:part-of]->(c:Category) RETURN c.categoryName AS c

User Question: *What product did Ana Trujillo order?*

Answer:
MATCH (c:CustomercustomerName: 'Ana Trujillo')-[:placed]->(o:Order)
MATCH (o:Order) -(o)-[:contains]->(p:Product)
RETURN p.productName AS p

User Question: *What products did Maria Anders order?*

Answer:
MATCH (c:CustomercustomerName: 'Maria Anders')-[:placed]->(o:Order)

MATCH (o:Order)-[:contains]->(p:Product)
RETURN p.productName AS p

User Question: *What categories are in order id 10250?*

Answer:

MATCH (o:OrderorderId: 10250)-[:contains]->(p:Product)
MATCH (p:Product)-[:part-of]->(c:Category)
RETURN p.productName AS Product, c.categoryName AS Category

User Question: *What categories are in order id 10248?*

Answer:

MATCH (o:OrderorderId: 10248)-[:contains]->(p:Product)
MATCH (p:Product)-[:part-of]->(c:Category)
RETURN p.productName AS Product, c.categoryName AS Category

User Question: *What categories did Thomas Hardy order from?*

Answer:

MATCH (c:CustomercustomerName: 'Thomas Hardy')-[:placed]->(o:Order)
MATCH (p:Product)-[:part-of]->(c:Category)
RETURN p.productName, c.categoryName AS Category

User Question: *What categories did Hanna Moos order from?*

Answer:

MATCH (c:CustomercustomerName: 'Hanna Moos')-[:placed]->(o:Order)
MATCH (o:Order)-[:contains]->(p:Product)
MATCH (p:Product)-[:part-of]->(c:Category)
RETURN p.productName, c.categoryName

Also, we implemented the algorithm and ran the code using the Movies dataset. We ran our experiments for one-hop edge (between two nodes), two-hop edge (between three nodes) and three-hop edge (between four nodes). Our results are accurate for all the one-hop edge questions and in most cases for two-hop ones (above 80%). We shared the Cypher commands generated for ten QAs with this dataset in appendix B. Here are some of the results:

User Question: *Which actors acted in Susannah of the Mounties?*

Answer:

MATCH (m:Movie name: 'Susannah of the Mounties')<-[r:ACTED-IN]-(a:Actor)
RETURN a.name AS a"

User Question: *Who directed the movie "Just Around the Corner"?*

Answer:

MATCH (m:Movie name: 'Just Around the Corner')<-[r:DIRECTED]-(d:Director)
RETURN d.name AS d"

User Question: *What are the movies in Children genre?*

Answer:

MATCH (g:Genre name: 'Children')<-[r:IN-GENRE]-(m:Movie)
RETURN m.name AS m

User Question: *Which Genres did Channing Tatum act in?*

Answer:

MATCH (a:Actor name:'Channing Tatum')-[r:ACTED-IN]-(m:Movie)-[r1:IN-GENRE]-(g:Genre)
RETURN g

User Question: *Which genres did Ron Perlman act in?*
Answer:
MATCH (m:Movie)-[:IN-GENRE]->(g:Genre)
WHERE m.name IN ['Hellboy', 'Pacific Rim', 'Cronos']
RETURN m.name, collect(g.name) AS genres

User Question: *Which Genres did Donal Crisp direct?*
Answer:
MATCH (d:Director name:'Donal Crisp')-[r:DIRECTED]-(m:Movie)
MATCH (d:Director name:'Roberts Gannaway')-[r:DIRECTED]->(m:Movie)-[r2:IN-GENRE]->(g:Genre)
RETURN g.name AS Genre

User Question: *Which genres did Sam Raimi direct?*
Answer:
MATCH (m:Movie)-[:IN-GENRE]->(g:Genre)
RETURN m.name as Movie, collect(g.name) as Genres

User Question: *Which movies did George Davis act in?*
Answer:
MATCH (a:Actor name:'George Davis')-[r:ACTED-IN]-(m:Movie)
RETURN m.name AS m

User Question: *Which actors act in War genre?*
Answer:
MATCH (g:Genre name:'War')-[r:IN-GENRE]-(m:Movie)
(a:Actor)-[r:ACTED-IN]->(m:Movie)
WHERE m.name IN ['Saving Private Ryan', 'Black Hawk Down', 'Platoon', 'Full Metal Jacket']
RETURN a.name AS Actor, m.name AS Movie

5 Conclusion

In this paper, we introduced a new method for multi-hop question answering on graphs. Question answering is important since they are used in different domains such as chatbots, virtual assistants, information retrieval systems and etc. On the other hand, LLMs facilitate advancements in question answering by enabling rapid and accurate retrieval of information from extensive datasets, thereby streamlining both learning processes and decision-making.

As a result, we leverage them for question answering in our method. While LLMs facilitate QA, they are challenged for multi-hop queries; since LLMs cannot easily answer them. While there are recent articles introducing new methods for multi-hop question answering, this area is still challenging. Our method helps users to ask their multi-hop (or one-hop) question in natural language and translate it to Cypher query. The user can run the resulted Cypher query on a Neo4j database and get the answer to their question.

Our method solves two issues: 1. Translate the multi-hop question in natural language to Cypher query using LLMs 2. Train the LLM using prompt engineering to find a path between two nodes of a graph.

The future work includes but not limited to: modifying the algorithm to make it more efficient, train the LLM on more complicated graphs, use other methods for training the LLM rather than prompt engineering, use other LLMs and compare the performance and results to the one we used.

References

1. Dayal, R., et al.: 'Development of chatbot retrieving fact-based information using knowledge graph'. In: Proceedings of International Conference on Recent Innovations in Computing: ICRIC 2022, Volume 1, pp. 153–164. Springer (2023)
2. Tai, C.-Y., et al.: 'User-centric path reasoning towards explainable recommendation'. In: Proceedings of the 44th International ACM SIGIR Conference on Research and Development in Information Retrieval, pp. 879–889 (2021)
3. Lin, X., et al.: 'BioKGBench: A Knowledge Graph Checking Benchmark of AI Agent for Biomedical Science'. In: arXiv preprint arXiv:2407.00466 (2024)
4. Tao, W., et al.:'FinQA: A Training-Free Dynamic Knowledge Graph Question Answering System in Finance with LLM-Based Revision'. In: Joint European Conference on Machine Learning and Knowledge Discovery in Databases, pp. 418–423. Springer (2024)
5. Rizzi, F.: 'Developing an Enterprise Chatbot using Machine Learning Models: A RAG and NLP based approach'. PhD thesis. Politecnico di Torino (2024)
6. Gupta, S., Ranjan, R., Singh, S.N.: 'A comprehensive survey of retrieval-augmented generation (rag): Evolution, current landscape and future directions'. In: arXiv preprint arXiv:2410.12837 (2024)
7. Kushala, V.M., Warrier, H., Gupta, Y., et al.: 'Fine tuning LLM for enterprise: practical guidelines and recommendations'. In: arXiv preprint arXiv:2404.10779 (2024)
8. Zaghir, J., et al.: 'Prompt engineering paradigms for medical applications: scoping review. J. Med. Internet Res. **26**, e60501 (2024)
9. Shah, M., Tian, J.: 'Improving LLM-based KGQA for multi-hop Question Answering with implicit reasoning in few-shot examples'. In: Proceedings of the 1st Workshop on Knowledge Graphs and Large Language Models (KaLLM 2024), pp. 125–135 (2024)
10. Allemang, D., Sequeda, J.: 'Increasing the LLM accuracy for question answering: ontologies to the rescue!' In: arXiv preprint arXiv:2405.11706 (2024)
11. Li, Z., et al.: 'Flexkbqa: a flexible LLM-powered framework for few-shot knowledge base question answering'. In: Proceedings of the AAAI Conference on Artificial Intelligence. vol. 38. 17, pp. 18608–18616 (2024)
12. Wang, F., et al.: 'LLM-KGMQA: large language model-augmented multi-hop question-answering system based on knowledge graph in medical field'. In: Knowledge and Information Systems, pp. 1–43 (2025)
13. Huang, Q., et al.: 'Coq: an empirical framework for multi-hop question answering empowered by large language models'. In: ICASSP 2024-2024 IEEE International Conference on Acoustics, Speech and Signal Processing (ICASSP). IEEE, pp. 11566–11570 (2024)

14. Nguyen, L.-M., et al.: 'Semantic Parsing for Question and Answering over Scholarly Knowledge Graph with Large Language Models'. In: JSAI International Symposium on Artificial Intelligence, pp. 284–298. Springer (2024)
15. Zhang, L., et al.: 'Osteosarcoma KGQA system: deep learning-based knowledge graph and large language model fusion'. In: Intelligent Medicine (2025)
16. Stahlberg, F.: 'Neural machine translation: a review'. J. Artif. Intell. Res. **69**, pp. 343–418 (2020)
17. Zhang, Y., et al.: 'Attention is all you need: utilizing attention in AI-enabled drug discovery'. Briefings Bioinform. **25.1**, bbad467 (2024)
18. Dubey, S.R., Singh, S.K.: 'Transformer-based generative adversarial networks in computer vision: a comprehensive survey'. IEEE Trans. Artif. Intell. (2024)
19. Ni, H., et al.: 'Time series modeling for heart rate prediction: from Arima to transformers'. In: 2024 6th International Conference on Electronic Engineering and Informatics (EEI), pp. 584–589. IEEE (2024)
20. Yin, L., et al.: 'DPAL-BERT: A Faster and Lighter Question Answering Model. CMES-Comput. Model. Eng. Sci. **141.1** (2024)
21. Chelliah, B.J., et al.: 'Harnessing T5 Large Language Model for Enhanced PDF Text Comprehension and Q&A Generation'. In: 2024 International Conference on Computing and Data Science (ICCDS), pp. 1–6. IEEE (2024)
22. Grover, K., et al.: 'Deep learning based question generation using t5 transformer'. In: Advanced Computing: 10th International Conference, IACC 2020, Panaji, Goa, India, 5–6 December 2020, Revised Selected Papers, Part I 10, pp. 243–255. Springer (2021)
23. Wang, J., et al.: 'Utilizing Bert for information retrieval: survey, applications, resources, and challenges; ACM Comput. Surveys **56.7**, pp. 1–33 (2024)
24. Larsen, A.G., et al.: 'LLM-powered conversational AI in customer service: users' expectations and anticipated use'. In: International Symposium on Chatbots and Human-Centered AI, pp. 217–233. Springer (2024)
25. Qiu, J., et al.: 'LLM-based agentic systems in medicine and healthcare'. Nat. Mach. Intell. **6.12**, pp. 1418–1420 (2024)
26. Manish, S.: 'An autonomous multi-agent LLM framework for agile software development'. Int. J. Trend Sci. Res. Develop. **8.5**, pp. 892–898 (2024)

A Proactive Approach for Predicting Equipment Downtime in an Enterprise

Jeffrey A. Young[1(✉)] and Ray Hashemi[2]

[1] Department of Computer Science, Clemson University, Clemson, SC, USA
jay8@clemson.edu
[2] Department of Computer Science, Georgia Southern University, Savannah, GA, USA
rayhashemi@georgiasouthern.edu

Abstract. Equipment downtime is a critical aspect of industrial maintenance and operational efficiency. Prolonged downtime can lead to significant financial losses, reduced productivity, and missed deadlines. In this study, we utilize data from an enterprise data lake ecosystem to (a) extract data assets in reference to critical events, (issues) that may lead to downtime of computing elements, (b) develop a proactive prediction model for issues and (c) integrating the Apriori algorithm to solidify the model. Such integration introduces a statistical foundation to support interpretability-driven failure configurations (diagnostics). The hybrid system can flag low-confidence predictions for analyst review, strengthen validated outputs with confidence metrics, and even reveal additional co-occurrence patterns not surfaced by direct filtering. The results reveal that our proposed approach enhances scalability, reduces computational complexity, and improves adaptability to dynamic data environments, making it particularly advantageous for practical deployment in resource-constrained or high-volume data scenarios.

Keywords: Equipment downtime · Association rules · Data lake ecosystem · Equipment maintenance strategies

1 Introduction

Equipment downtime is a critical aspect of industrial maintenance and operational efficiency. Prolonged downtime can lead to significant financial losses, reduced productivity, and missed deadlines. In almost every enterprise there is a reactive and preventive maintenance strategies system in place to reduce loses. However, such a system, cannot be as effective as a proactive system. A proactive system ensures that maintenance resources are optimally allocated, thereby reducing unplanned equipment failures and improving overall industrial uptime.

Let us assume that an enterprise owns N stations of S1, . . ., Sn and station Si has M computing elements denoted as Ei1, Ei2, . . . , Eim, where M may be different from one station to the next. While Eij is operating, its operation

H. R. Arabnia et al. (Eds.): CSCI 2025, CCIS 2972, pp. 77–93, 2027.
https://doi.org/10.1007/978-3-032-28469-3_6

logs are stored, and its physical status is also monitored and recorded by using sensors' reading for every given interval of time, t. The physical status includes indices representing machine temperature, vibration, power consumption, and other physical metrics. The failure events of each station due to the failure in at least one of its computing elements along with the data about the failed element(s) are also stored in two different files of downtime logs and maintenance work orders. The former one includes the history of the failures and their types, downtime length, repair timeline, replacement timeline, and replacer information. The later one includes history of inspections, part replacements, and corrective actions taken to resolve or mitigate problems. The critical events that may lead to a downtime of Eij, affecting system availability and operational efficiency, are referred to as issues in this paper and information about issues are also recorded in both downtime logs and maintenance work orders for both the station Si and Eij.

The issues at the computing element level are more detailed and at the station level are more general. For example, at the computing element's level, the issue of sensor fault expresses the exact location(s) of the sensor(s), whereas at the station level, the same issue means that the sensor fault sensed on at least one of the station's computing elements with no details. Therefore, one can look at an issue in two different levels of granularity: Station and Computing Element.

All the above-mentioned data make a huge dataset that we refer to it as the enterprise's data lake (or shortly, Data Lake). The Data Lake is highly valuable for the extraction of those features, which eventually represent the business logic for the enterprise. Specifically, in reference to thisresearch effort, the extracted features of interest (data assets) are those that build a basis for proactive prediction analysis and data-driven maintenance strategies.

The goal of this study is three-fold: (a) Extraction of data assets from the Data Lake, (b) Development of a proactive prediction model for downtime events using the data assets, and (c) Solidification of the model.

The organization for the rest of the paper is as follows. The Previous Works is the subject of the Sect. 2. The Methodology is introduced in the Sect. 3. The Empirical Results are presented in the Sect. 4. The Conclusion and Future research are covered in the Sect. 5.

2 Previous Works

Analytical frameworks have become increasingly critical in the domains of data lake management, predictive maintenance, driven by the growing complexity and scale of operational data. We briefly examine contemporary analytical methodologies employed within these domains.

The frameworks for data lake management prioritize efficient data integration, quality assurance, governance, and accessibility. In their seminal work, Sawadogo and Darmont (2021) discuss a metadata-driven analytical framework that significantly improves data lake governance and query efficiency by systematically managing metadata and lineage information. Another noteworthy contribution by Madera and Laurent (2020) outlines a data lake quality assurance framework using automated validation and monitoring algorithms, greatly enhancing data reliability and usability for analytical tasks. The recent literature emphasizesframeworks utilizing cloud-native architectures and scalable analytics. Liu et al. (2022) introduce a serverless architecture-based analytical framework leveraging cloud servicesto provide scalable, cost- efficient, and adaptive data lake management, demonstrating significant advantages in scalability and resource optimization.

Predictive maintenance (PdM) frameworks aim to proactively identify maintenance requirements, reduce downtime, and extend equipment lifespan. Recent literature highlights diverse methodologies, prominently involving machine learning (ML) and deep learning techniques. For instance, Carvalho et al. (2019) present a comprehensive ML framework integrating Random Forest and Gradient Boosting algorithms to effectively predict equipment failures in industrial settings. Similarly, Khan et al. (2020) detail the application of deep convolutional neural networks (CNNs) combined with sensor data for accurate anomaly detection and failure prediction.

An emerging focus has been on hybrid analytical frameworks combining ML and statistical modeling. Xu et al. (2021) propose a hybrid deep learning approach, integrating LSTM networks with statistical models, demonstrating enhanced pre- diction accuracy for complex equipment maintenance scenarios. These methods illustrate the significant progress made in handling the typical multidimensional and temporal data in PdM applications.

The previously cited methodologies exhibit several notable limitations, with computational cost being particularly significant. These approaches often demand extensive computational resources and substantial processing times, thus becoming economically and practically inefficient in large-scale or real-time applications. The system proposed in this study specifically addresses these limitations by offering a more cost-effective solution. Furthermore, our proposed approach enhances scalability, reduces computational complexity, and improves adaptability to dynamic data environments, making it particularly advantageous for practical deployment in resource- constrained or high-volume data scenarios.

3 Methodology

The methodology for meeting the three prongs of the goal (Extraction of the data assets, Development of a proactive prediction model, and the Solidification of the model) is depicted in Fig. 1. Details for each prong are the subject of the following three subsections.

3.1 Extraction of Data Assets

Extraction of data assets is made possible through several subcomponents that are explained as follows.

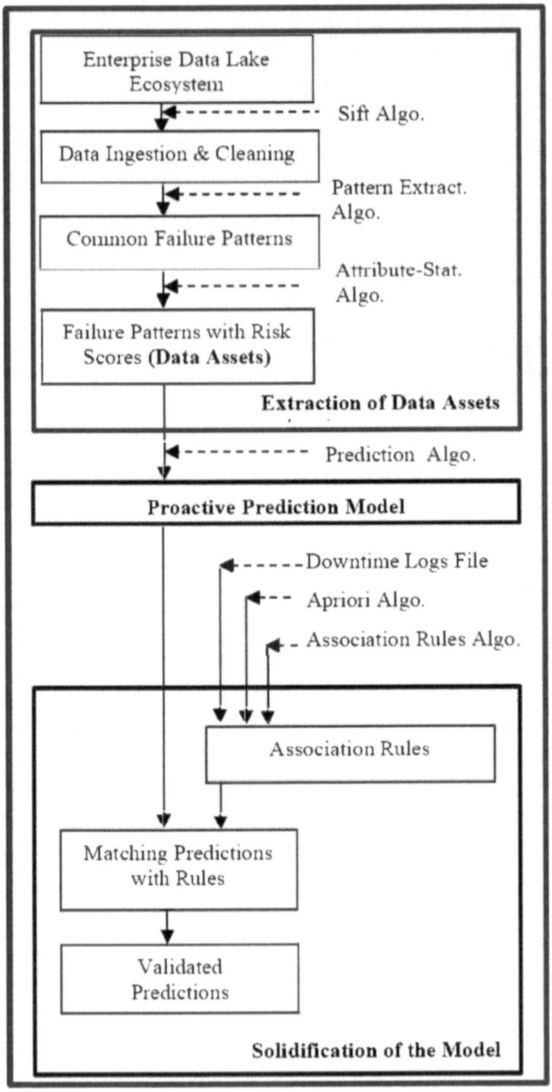

Fig. 1. Methodology for the goal's three prongs.

Data Ingestion and Cleaning. Let us starts with data ingestion and then continue with data cleaning. Data ingestion refers to the process of systematically importing, collecting, and consolidating diverse sets of data from various sources within enterprise into a centralized repository—a data lake or data warehouse. The primary purpose of data ingestion is to create a unified, accessible, and consistent dataset that can be efficiently leveraged for analytics, predictive modeling, decision-making, and operational management. Without robust data ingestion processes, the resulting data assets may risk inconsistency, redundancy,

Algorithm 1. GET DATA Algorithm

```
Input: target, target att, df (a data frame)
Output: filtered df
data ← []
data return ← []
names ← column names of df
for each n ∈ names do
    append column n of df to data
    append an empty list to data return
end for
temp ← column target att of df
for t = 0 to length(temp) −1 do
    if temp[t] = target then
        for i = 0 to length(data) −1 do
            append data[i][t] to data return[i]
        end for
    end if
end for
filtered df ← dataframe constructed from data return with column names
return filtered df
```

inaccuracies, data silos, and significantly diminishing the value and effectiveness of subsequent analytical efforts.

Within the context of this study, our data ingestion activities primarily revolve around extracting relevant subsets of data from a larger enterprise data lake ecosystem which contains, among other things, historical downtime logs, real-time IoT sensor streams, records of past inspections, part replacements, and corrective actions. The ingested data needs to be transformed into structured formats suitable for predictive analysis. To achieve the transformation, we (a) identify and access relevant data attributes within the larger enterprise data lake source, (b) select specific tuples that match targeted attributes, ensuring that only data points (such as sensor readings and machine states) related to the predictive maintenance analysis are retained, and (c) organize the extracted data points into a clean and structured data frame, which simplifies further analysis, modeling, and reporting tasks. These three steps are encapsulated in the GET DATA algorithm. However, the structured data frame needs to be cleaned further, to make sure that the only meaningful data remains in it. As a result, GET DATA algorithm is invoked through another algorithm (SIFT DATA Algorithm) that is in charge of cleaning the content of structured data frame (df).

Within the context of GET DATA algorithm, the data cleaning process involves systematically refining the filtered-df by straining out irrelevant, redundant, and/or invalid data points that could degrade model performance. To achieve cleaning data, (a) The dataset (filtered_df) is iteratively examined based on predefined target values and attributes to retain only relevant subsets, (b) The size of the resultant dataset is assessed during each cleaning iteration to ensure meaningful data remains available. If at any stage the dataset becomes empty,

Algorithm 2. SIFT DATA Algorithm

```
Input: target[], target att[], df
Output: df, bit
bit ← 1
for t = 0 to length(target) −1 do
    df ← GET DATA(target[t],target att[t],df)
    print("Data: " + target[t] + " = " + size(df))
    if size(df) = 0 then
        bit ← 0
    end if
end for
return (df,bit)
```

it signals insufficient data for modeling, and (c) a refined and concise dataset is produced, ensuring that downstream predictive analytics are performed on high-quality relevant data. These three steps are encapsulated in the SIFT DATA algorithm.

Common Failure Patterns. As we mentioned before, the failure events of Si and Eij (for i =1 to n and j = 1 to m) are stored in downtime logs file and maintenance work order file. The first file is rich in data about the downtimes of stations computing elements because each record of the file is characterized by various descriptive attributes, such as station descriptions, downtime descriptions, issue(s) descriptions, and Eij detail's description. The records of downtime logs are heavily sparse due to the fact that we deal with a large number of descriptive attributes and a given station that highly likely has values for a subset of the descriptive attributes.

We divide the downtime logs file into two files of Station and Elements such that each file holds data for one of the granularity levels. By doing so, we, first, identify those stations for which their failure types are more frequently observed and then the computing elements of those stations can be further analyzed when it is needed. The process for the extraction of the patterns from data of lower granule is the same as the one for the higher granule.

The extraction of patterns from file Station is done by (a) Formation of a composite key by selecting the relevant attributes from each record that represent specific failure events, (b) iteration through the file to aggregate the frequency of each unique attribute combination, thereby capturing patterns that frequently recur within the data, and (c) retention of those failure types (major failures) whose occurrence exceeds a predefined threshold.

Following these three steps ensures that only the most impactful and recurrent failure events are identified for further analysis. By extracting and quantifying these patterns, the algorithm facilitates root cause analysis, informs predictive maintenance strategies, and supports the development of targeted mitigation efforts in industrial and operational environments. The above three steps are encapsulated in the PATTERN EXTRACTION Algorithm.

Algorithm 3. PATTERN EXTRACTION Algorithm

```
Input: Downtime logs file, D. Threshold θ.
Output: Set of common failure patterns P
attributes ← All attributes in D//[station description, issue description, ...]
patterns ← empty dictionary
for each record in D do
    pattern key ← concatenate selected attributes from record
    if pattern key not in patterns then then
        patterns[pattern key] ← 1
    else
        patterns[pattern key] ← patterns[pattern key] + 1
    end if
end for
P ← filter patterns by occurrence threshold (e.g., only keep patterns with count > θ)
return P
```

Attributes Risk Assessment. A systematic process of evaluating and quantifying the likelihood of various attributes contributing to equipment downtime or failures. Accurate attribute risk assessment helps prioritize resources, guides maintenance planning, and provides actionable insights, ensuring interventions target the most critical vulnerabilities.

Within the context of our predictive maintenance framework, the attribute risk assessment activities involve calculating the probability of occurrence for different types of issue(s). To accomplish this, we (a) identify all unique attribute values within the dataset, excluding irrelevant or null data, (b) assess frequency of each unique attribute value, (c) calculate the relative probability (frequency-based risk) for each attribute value, establishing an objective measure of risk likelihood.

To accomplish (c), we use Bayesian posterior probability using Laplace smoothing which is a statistical technique utilized to handle cases with limited or zero-frequency data occurrences. It involves adding a small constant (often one) to each event count to prevent zero probabilities, thereby stabilizing probability estimates. This method effectively incorporates prior knowledge into probability calculations, improving robustness especially in sparse datasets). These three steps are encapsulated in the ATTRIBUTE STATISTIC algorithm.

3.2 Proactive Predictive Model

The model is a quadruplet of (P, Q, G, R_bit), where, P is the common failure patterns delivered by the Pattern Extraction Algorithm, Q is a function such that Q(P) creates set G, which is a set of strings expressing the final outcomes of the model in the form of predictive insights about equipment failure probabilities, and R_bit is a flag indicating that G contains meaningful strings (when R_bit = 0); Otherwise, the model cannot deliver meaningful strings due to a lack of sufficient data. The last three elements need to be explained further.

Algorithm 4. ATTRIBUTE STATISTICS

```
Input: A pattern file, P.
Output: dct (posterior probabilities)
dct ← empty dictionary # show use and explain dct........
uniq ← unique values in data excluding "NULL"
k ← length of uniq
counts ← empty dictionary
total ← 0
for each u ∈ uniq do
    counts[u] ← 0
end for
for each d ∈ data do
    if d ≠ "NULL" then
        counts[d] ← counts[d] + 1
    end if
end for
for each u ∈ uniq do
    posterior ← (counts[u]+1)/(length(data)-#NULL+ k)
    dct[u] ← posterior
end for
return dct
```

The Q function is made up of four ordered sub-functions of f1, f2, f3, and f4 that are applied sequentially on P. Function f1(P) provides clear visibility into which attributes are most likely associated with downtime events—Function f1 is presented by the ATTRIBUTET STATISTCS Algorithm. Function f2(f1(P)) provides station description by iterating over each unique value in the DESCRIPTION OF STATION columns. Function f3(f2(f1(P))) identifies the unique issues reported for each station—values in the "DESCRIPTION OF ISSUE" columns that correspond to that station. The unique issues are formatted for readability and easy interpretation by applying function f4(f3(f2(f1(P)))) and then they are added to the set G.

The set G is a comprehensive, human-readable strings detailing stations and their associated issues. The set G includes (a) the outcome of the iterations over the unique values of DESCRIPTION OF STATIONin P, (b) for each station, the corresponding unique values of DESCRIPTION_OF_ISSUE are aggregated, and (c) a human-readable output is constructed where each station is followed by its associated issues. This synthesis yields an interpretable textual summary of failure configurations, formatted as shown in Fig 2.

The return bit (r bit) indicates whether meaningful results were produced. This bit is set by SIFT Algorithm during the cleaning dataset process. The proposed model is encapsulated in the PREDICTION Algorithm.

```
DESCRIPTION_OF_STATION: STATION_A
-Issue: Overheating
DESCRIPTION_OF_STATION: STATION_B
-Issue: Hydraulic Leak
-Issue: Sensor Fault
-Issue: High Consumption of Power
```

Fig. 2. An interpretable textual summary of failure configurations.

Algorithm 5. PREDICTION Algorithm

```
Input: target[], target att[], and df.
Output: result, r bit.
result ← "" (empty string)
r bit ← 1
df ← df[df['DOWNTIME HRS'] > 0]
(df,bit) ← SIFT DATA(target,target att,df)
if bit = 0 then
    r bit ← 0
    dos ← unique values in column "DESCRIP- TION OF STATION" of df
    for each d ∈ dos do
        print("DESCRIPTION OF STATION: " + d)
        result ← result+ "DESCRIPTION OF STATION: " +d+ "\n"
        issues ← unique values in "DESCRIP- TION OF ISSUE" of df where "DESCRIP- TION OF STATION" = d
        for each i ∈ issues do
            result ← result+ "–Issue: " +i+ "\n"
        end for
    end for
end if
return (result,r bit)
```

3.3 Solidification of the Model

To enhance the robustness and interpretability of the structured predictions generated by the PREDICTION algorithm, we incorporate the association rules mined via the classical Apriori algorithm directly from the stations' downtime logs file. This approach serves as a post-hoc validation mechanism, confirming whether the predicted failure patterns are supported by statistically significant cooccurrences in the historical dataset. This is done by:

(a) Using Apriori algorithm to identify the most frequent itemsets within the new dataset—A most frequent itemset is made up of the items that appear together within the dataset with a support greater than a threshold. During the searching for the frequent itemsets, if a concluded itemset lacks at least one item from station values or issue values, then it will be dismissed. If an itemset has more than one item from station values is also dismissed.

Algorithm 6. Apriori Algorithm

Input: The transformed dataset, T. Minimum support threshold σ
Output: K-Most frequent itemsets, F.
$F_1 \leftarrow$ frequent items in T with support $\geq \sigma$
$k \leftarrow 2$
while $F_{k-1} \neq \emptyset$ **do**
    $C_k \leftarrow$ candidate itemsets generated from F_{k-1}
    $C_k = C_k -$ {candidate itemsets with neither items from station values nor items from issue values}
    $C_k = C_k -$ {candidate itemsets with more than one item from station values}
    for each record $t \in T$ **do**
        for each candidate $c \in C_k$ **do**
            if $c \subseteq t$ **then**
                increment count of c
            end if
        end for
    end for
    $F_k \leftarrow$ candidates in C_k with support $\geq \sigma$
    $k \leftarrow k + 1$
end while
return $\bigcup_k F_k$

Algorithm 7. ASSOCIATION RULES Algorithm

Input: K Most frequent itemsets (MFI), Threshold θ.
Output: Set of Association rules (Rules)
for i=1 to k do **do**
    Find item g such that g $\in$MFIi and g $\in$ {station values}
    Create associate rule of g $\Rightarrow$(MFi-g)
    Calculate the confidence for rule g
    If (confidence $> \theta$, Then Rules $\leftarrow$ g
end for
Return Rules

(b) Generating association rules from the most frequent itemsets in general form of, s $\Rightarrow$ i (σ, κ), where s is state value, i is a set of issue values along with the support, σ, and confidence, κ, for the rule.
(c) Examining, each failure configuration generated by PREDICTION Algorithm against the association rules. For each match, if the support, and confidence level of the corresponding association rule exceeds the domain-specific thresholds, then the failure configuration is considered validated; Otherwise, it is removed.

Steps (a) and (b) are encapsulated in APRIORI Algorithm. Step (c) is encapsulated in ASSOCIATION RULES algorithm. The entire solidification process is outlined in Fig. 3.

By integrating the Apriori and ASSOCIATION RULES algorithms we introduce a statistical foundation to support interpretability-driven failure config-

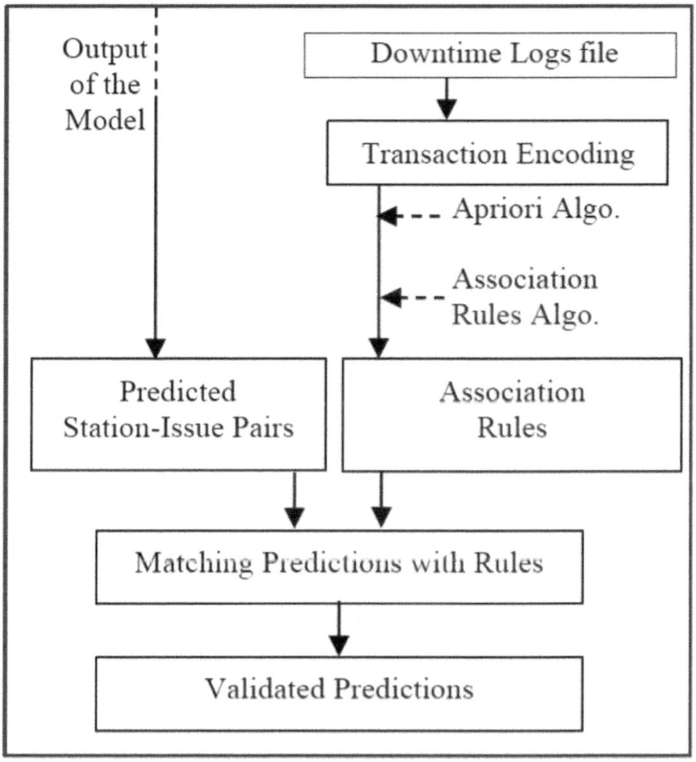

Fig. 3. Workflow for solidification of the model.

urations (diagnostics). The framework can flag low-confidence predictions for analyst review, strengthen validated outputs with confidence metrics, and even reveal additional co-occurrence patterns not surfaced by direct filtering. This hybrid approach blends rule-based reasoning with frequency-based pattern mining, reinforcing both precision and trust in the analytical process.

4 Empirical Results

To clarify the operation and effectiveness of the proposed framework, we first present a detailed worked example that walks through each stage of the methodology. This example uses manufactured data to make the intermediate steps, decision points, and outputs of the system fully transparent and directly traceable to the workflow shown in Fig 1. We then complement this illustrative example with an empirical evaluation based on a real-world downtime event dataset collected from an industrial environment, demonstrating how the framework performs under practical conditions. These two components are described in the following subsections: Sect. 4.1, which provides the step-by-step worked example aligned with Fig. 1, and Sect. 4.2, which reports the outcomes of the experimental study.

4.1 Step-by-Step Worked Example (Manufactured Data) Aligned with Fig. 1

Figure 1 depicts the methodology for the three-pronged goal: (i) extraction of data assets, (ii) proactive prediction, and (iii) model solidification through Apriori-based association rules. We now illustrate an end-to-end example using a manufactured downtime event stream and a filtered subset of those events.

Definition (Subset). A *subset* is the set of records retained after applying analyst-defined filters (e.g., station, asset class, time range). In this example, we form a subset by filtering all records to Station=`STATION_B` and Duration>0.

Manufactured Downtime Event Stream. Assume the downtime log emits records with Station, Issue, and Duration. Table 1 shows 12 manufactured events.

Table 1. Manufactured downtime events (Duration replaces "Hrs"). This dataset is constructed so that multiple issues exceed the frequency threshold θ for `STATION_B`.

ID	Station	Issue	Duration
1	STATION_A	Overheating	2.1
2	STATION_A	Overheating	1.5
3	STATION_A	Sensor Fault	0.0
4	STATION_B	Hydraulic Leak	3.0
5	STATION_B	Sensor Fault	1.2
6	STATION_B	High Power Use	0.7
7	STATION_C	Overheating	0.0
8	STATION_C	Vibration Spike	2.4
9	STATION_B	Hydraulic Leak	2.6
10	STATION_A	High Power Use	0.9
11	STATION_B	Sensor Fault	0.8
12	STATION_B	Sensor Fault	1.1

Step 1: Data ingestion and cleaning (GET_DATA/ SIFT_DATA). Following Fig. 1, we filter by analyst targets. Let the analyst specify targets (Station, `STATION_B`). GET_DATA yields the subset of event IDs:

$$\{4, 5, 6, 9, 11, 12\}.$$

Applying the downtime predicate Duration > 0 does not remove any of these events in the manufactured subset, so the subset remains unchanged.

Step 2: Common Failure Pattern Extraction (Multiple Selections). We define a pattern key $\phi(e) = \text{Station} \mid \text{Issue}$ for each event e (in the survived list of events) and compute counts in the subset:

$$\text{Cnt}[k] = \sum_{e \in \text{subset}} \mathbb{I}(\phi(e) = k).$$

For `STATION_B`, the counts are:

$$\begin{aligned} \text{Cnt}[\texttt{STATION\_B|Hydraulic Leak}] &= 2, \\ \text{Cnt}[\texttt{STATION\_B|Sensor Fault}] &= 3, \\ \text{Cnt}[\texttt{STATION\_B|High Power Use}] &= 1. \end{aligned}$$

Let $\theta = 2$. Then the selected frequent patterns are:

$$P = \{k \mid \text{Cnt}[k] \geq \theta\} = \{\texttt{STATION\_B|Hydraulic Leak, STATION\_B|Sensor Fault}\}.$$

Thus, `High Power Use` is not selected at this stage because its count is below θ.

Step 3: Attribute risk assessment (Bayesian posterior with Laplace smoothing). We compute a Laplace-smoothed posterior over issues conditioned on `STATION_B`. Let counts be:

$$c(\text{Hydraulic Leak}) = 2, \quad c(\text{Sensor Fault}) = 3, \quad c(\text{High Power Use}) = 1,$$

so $N = 2 + 3 + 1 = 6$ and $k = 3$. The posterior is

$$\Pr(u \mid \texttt{STATION\_B}) = \frac{c(u) + 1}{N + k}.$$

Therefore:

$$\begin{aligned} \Pr(\text{Hydraulic Leak} \mid \texttt{STATION\_B}) &= \frac{3}{9} \approx 0.333, \\ \Pr(\text{Sensor Fault} \mid \texttt{STATION\_B}) &= \frac{4}{9} \approx 0.444, \\ \Pr(\text{High Power Use} \mid \texttt{STATION\_B}) &= \frac{2}{9} \approx 0.222. \end{aligned}$$

Step 4: Proactive Prediction with Top- K Elimination. We rank issues by posterior probability and output only the top-K items. Let $K = 2$. The ranking is:

$$\texttt{Sensor Fault}\ (0.444) > \texttt{Hydraulic Leak}\ (0.333) > \texttt{High Power Use}\ (0.222).$$

Hence:

$$\text{Top-2} = \{\texttt{Sensor Fault, Hydraulic Leak}\}, \quad \text{Eliminated} = \{\texttt{High Power Use}\}.$$

The interpretable prediction output (top-2) is presented in Table 2.

Table 2. Predicted Station-issue Pairs for the Example

Issue	Posterior Probability
`STATION_B:Sensor Fault`	0.444
`STATION_B:Hydraulic Leak`	0.333

Step 5: Association-Rules in the Solidification Phase. To illustrate the role of association-rule mining in the solidification phase, we present a second manufactured example designed to exhibit different co-occurrence behavior across stations and issues. Unlike the previous example, this dataset emphasizes multi-issue configurations with uneven frequencies to demonstrate how Apriori selectively reinforces or suppresses predicted outcomes.

Manufactured Transactions. Assume the transformed downtime log is encoded as a transactional dataset, where each transaction consists of a station identifier and one or more associated issue labels observed during a downtime event. Table 3 shows a manufactured transaction set.

Table 3. Manufactured transaction dataset for Apriori illustration.

Transaction ID	Items
1	{STATION_C, Overheating, Vibration Spike}
2	{STATION_C, Overheating}
3	{STATION_C, Vibration Spike}
4	{STATION_C, Overheating, Vibration Spike}
5	{STATION_B, Sensor Fault}
6	{STATION_B, Sensor Fault, High Power Use}
7	{STATION_B, Sensor Fault}
8	{STATION_B, High Power Use}

Frequent Itemset Mining. Let the minimum support threshold be $\sigma = 0.25$. With eight total transactions, this corresponds to a minimum support count of two. Applying Algorithm 6, the following most frequent itemsets are identified:

$$F = \{\, \{\texttt{STATION\_C}\,, \texttt{Overheating}\}, \\ \{\texttt{STATION\_C}\,, \texttt{Vibration Spike}\}, \\ \{\texttt{STATION\_B}\,, \texttt{Sensor Fault}\} \\ \}$$

Itemsets containing more than one station label are excluded by design, consistent with the constraints imposed in Algorithm 6.

Using Algorithm 7 with a minimum confidence threshold $\kappa = 0.6$, the following association rules are survived and depicted in Table 4:

Table 4. Association rules generated by Apriori

Station	Issue	Confidence κ
STATION_C	Overheating	0.75
STATION_C	Vibration Spike	0.75
STATION_B	Sensor Fault	0.75

Step 6: Solidification Process. The matched predicted station-issue pairs of Table 2, with association rules of Table 4, identifies a solid predicted station-issue rule of STATION_B: Sensor Fault as a valid one.

This example demonstrates how Apriori-based validation selectively reinforces statistically grounded predictions while suppressing weak or spurious associations.

4.2 The Outcomes of Experimental Study

We conducted a series of experiments on a real-world downtime event dataset collected from an industrial environment. The dataset comprised 2,500 unique downtime events, spanning 50 distinct stations and 120 documented issue types over a period of 24 months. All experiments were conducted on a system equipped with an Intel Xeon E5 CPU, 64 GB of memory, and running Ubuntu 20.04 LTS

The GET DATA and SIFT DATA algorithms demonstrated strong filtering performance, with average execution times of 4.5 s for datasets containing up to 1,000,000 records. The filtering process consistently retained between 60% and 85% of relevant records, depending on the strictness of the input criteria. The Pattern Extraction algorithm successfully identified 350 unique failure patterns, of which 120 were classified as recurrent based on a threshold of 15 occurrences. These results provided maintenance teams with actionable insights into the most frequent sources of downtime. The PREDICTION algorithm generated structured summaries of failure scenarios for each station, with an average summary length of 25 entries per station and a median runtime of 2.8 s per analysis. The ATTRIBUTE STATISTICS algorithm, using Bayesian posterior estimation with Laplace smoothing, produced stable probability distributions even for rare or infrequent attribute values, yielding posterior estimates that aligned well with observed frequencies.

Overall, the empirical evaluation demonstrates that the proposed algorithms are both computationally efficient and practically useful for extracting, classifying, and reporting downtime-related failure patterns. Benchmark values and detailed quantitative results are summarized in Table 5.

Table 5. Benchmark values and quantitative results of the proposed algorithms

Metric	Value
Total downtime events	2,500
Distinct stations	50
Documented issue types	120
Data collection duration	24 months
System CPU	Intel Xeon E5
System RAM	64 GB
Operating system	Ubuntu 20.04 LTS
GET DATA & SIFT DATA average execution time	4.5 s
Records processed up to	1,000,000
Filtering retention rate	60%– 85%
Unique failure patterns identified	350
Recurrent failure patterns (threshold $\geq$ 15 occurrences)	120
Average summary length per station	25 entries
PREDICTION median run-time	2.8 s

5 Conclusion and Future Research

In this work, we presented a modular framework of algorithms for the extraction, filtering, classification, and reporting of failure patterns in downtime event data. The proposed suite—including the GET DATA, SIFT DATA, PATTERN EXTRACTION, ATTRIBUTE STATISTICS, PREDICTION, APRIORI, and ASSOCIATION RULES Algorithms—demonstrates a transparent and reproducible approach to transforming raw event records into actionable insights. Empirical evaluations indicate that the framework is both efficient and effective at isolating recurrent failure scenarios, thereby supporting data-driven maintenance planning and operational decision-making.

The proposed system significantly reduces the computational overhead associated with predictive tasks by leveraging lightweight machine learning model and streamlined analytical pipelines.

Several avenues for future research remain. First, the discovery of subtler and less frequent failure modes represents an important opportunity to further enhance the diagnostic power of the framework. Second, extending support for online prediction through sliding-window processing, incremental pattern updates, and periodic solidification via association rules mined from recent event subsets remains an active area of investigation. Third, the current algorithms primarily rely on explicit, user-defined filtering and straightforward pattern aggregation, which may limit scalability and adaptability in highly dynamic or complex environments. To address these limitations, ongoing work focuses on incorporating real-time data processing, automated attribute selection, and machine learningâĂŞbased pattern discovery.

Acknowledgments. This research received no specific grant from any funding agency in the public, commercial, or not-for-profit sectors.

Disclosure of Interests. The authors declare that they have no competing interests.

References

1. Carvalho, T.P., Soares, F.A., Vita, R., Francisco, R.D., Basto, J.P., Alcalá, S.G.: A systematic literature review of machine learning methods applied to predictive maintenance. Comput. Industr. Eng. **137**, 106024 (2019)
2. Khan, S., Yairi, T., Patelli, E.: A deep learning approach for predictive maintenance using convolutional neural networks. Mech. Syst. Signal Process. **140**, 106686 (2020)
3. Xu, Y., Ren, J., Cheng, Y., Liu, X.: A hybrid deep learning framework for predictive maintenance using LSTM and statistical models. J. Manuf. Syst. **58**, 210–223 (2021)
4. Sawadogo, P., Darmont, J.: Metadata management in data lakes: a systematic review. Inf. Syst. **98**, 101842 (2021)
5. Madera, C., Laurent, A.: Data lake quality assurance framework: automated validation and monitoring approaches. J. Big Data **7**, 1–17 (2020)
6. Liu, Z., Han, R., Liu, X., Yang, X.: Scalable serverless analytical framework for cloud-native data lake management. Futur. Gener. Comput. Syst. **128**, 123–135 (2022)
7. Hai, R., Geisler, S., Quix, C.: Data lake concept and systems: a survey. J. Big Data **8**, 93 (2021)
8. Singh, A., Sharma, P.: Analyzing big data processing techniques in data lake environments. Procedia Comput. Sci. **167**, 328–337 (2020)
9. Marjani, M., et al.: Big IoT data analytics: architecture, opportunities, and open research challenges. IEEE Access **5**, 5247–5261 (2017)
10. Jha, P., Saini, P., Sharma, M.: Real-Time streaming data ingestion in big data lake using Apache NiFi. Int. J. Adv. Comput. Sci. Appl. **10**(7) (2019)
11. Patil, R., Singh, M.: A review of data quality issues and solutions in big data analytics. J. Data Inf. Qual. (JDIQ) **14**(2), 1–25 (2022)
12. Rahman, M.S., Gao, J., Helal, S.: Metadata management for data lakes: a comprehensive review. J. Cloud Comput. **10**, 56 (2021)

Hybrid NLP–LLM Pipelines for Automatic Learning Content Generation

Raj Singh[1(✉)] and Julio Cotto[2]

[1] Western Governors University, Salt Lake City, UT, USA
raj.singh@wgu.edu
[2] Colorado State University, Fort Collins, CO, USA
julio.cotto@colostate.edu

Abstract. Large language models (LLMs) enable rapid generation of educational materials, yet reliability and pedagogical alignment remain open challenges. This paper presents a hybrid pipeline for automatic learning content generation, producing multiple artifacts—summaries, structured notes, flashcards, and quizzes—by combining deterministic NLP preprocessing (OpenNLP), generative LLMs (OpenAI), external enrichment (dictionary and encyclopedia APIs), and a rules-engine validation layer. We pose research questions on correctness, enrichment effects, and validation efficacy; introduce practical algorithms for content generation and validation; and report a pilot evaluation across diverse domains (STEM, Humanities, Computer Science, and recent AI research). Results indicate 76–82% validated correctness, with clear gains over an LLM-only baseline, alongside improved term coverage and instructor-rated usefulness. We analyze sources of error (hallucination, ambiguity, domain variation), propose retrieval-augmented and validation-enhanced extensions, and outline implications for tutoring, assessment, and curriculum design.

Keywords: Large Language Models · Education · Content Generation · Summarization · Flashcards · Quizzes · Bloom's Taxonomy

1 Introduction

The creation of learning content—summaries, structured notes, flashcards, quizzes, and other study aids—is a repetitive and time-intensive task for educators. Instructors report that preparing these artifacts can consume more time than classroom teaching itself, as each lecture, reading, or research paper must be distilled into digestible components suitable for students at different proficiency levels. In large-enrollment courses, such as introductory STEM or humanities classes, the manual workload quickly becomes unsustainable. Students often request structured study aids, yet instructors struggle to keep pace.

H. R. Arabnia et al. (Eds.): CSCI 2025, CCIS 2972, pp. 94–108, 2027.
https://doi.org/10.1007/978-3-032-28469-3_7

At the same time, the emergence of LLMs has created both excitement and concern. On the one hand, systems such as ChatGPT demonstrate an unprecedented ability to generate fluent explanations, summaries, and questions in seconds. On the other, educators worry about correctness, hallucinations, and a lack of pedagogical alignment. Without safeguards, LLM-generated flashcards may include fabricated facts, and quizzes may contain misleading distractors. These risks make uncontrolled automation unsuitable for classroom deployment.

This work explores whether a hybrid pipeline—one that integrates deterministic natural language processing (NLP), generative LLMs, external enrichment sources, and explicit validation rules—can address these concerns while maintaining the efficiency benefits of automation. Our approach emphasizes modularity: each stage can be independently improved, replaced, or audited, enabling a transparent and extensible framework for automatic content generation.

Research Questions

To structure our investigation, we define three guiding research questions:

- **RQ1:** Can a hybrid NLP–LLM pipeline reliably generate pedagogically useful multi-artifact content, including summaries, notes, flashcards, and quizzes?
- **RQ2:** How does external enrichment, via resources such as dictionaries and encyclopedias, affect the accuracy and contextual relevance of generated learning content?
- **RQ3:** Which validation strategies are most effective in mitigating hallucinations, reducing redundancy, and ensuring grounding in the original material?

By explicitly tying each research question to a pipeline stage, we align our evaluation not only with technical performance but also with pedagogical value.

Contributions. This paper makes four contributions: (1) a modular hybrid architecture and practical algorithms for multi-artifact generation and validation; (2) a pilot evaluation across STEM, humanities, computer science, and AI research content; (3) explicit analysis answering RQ1–RQ3 with both quantitative and qualitative evidence; (4) design guidelines for future deployments in tutoring, assessment, and curriculum design.

2 Related Work

2.1 Intelligent Tutoring Systems

Research on intelligent tutoring systems (ITS) demonstrates the potential of technology to scale individualized support. VanLehn's meta-analysis [1] reported that ITS achieve an effect size of 0.76 relative to no tutoring, reaching two-thirds of the effectiveness of human tutors. These findings highlight that automated systems, while not replacing human expertise, can provide substantial benefits. However, ITS often require domain-specific authoring and lack generalizability, motivating hybrid approaches that leverage LLMs' breadth while constraining their weaknesses.

2.2 Automatic Question Generation

Automatic question generation (AQG) has long been pursued through syntactic transformations and statistical ranking. Heilman and Smith [2] demonstrated early pipelines that relied on parse trees to form wh-questions. Subsequent surveys [3] catalogued diverse approaches but repeatedly noted challenges with distractor quality and coverage. Item modeling [4] improved psychometric properties but limited flexibility. In contrast, LLMs generate questions fluently but lack validation. Our work combines AQG with flashcard and note generation, emphasizing grounding and multi-artifact coherence.

2.3 Generative AI in Education

LLM-driven applications such as Khanmigo (Khan Academy), Quizlet, and Duolingo Max showcase both promise and limitations. Khanmigo provides conversational tutoring but not structured artifact generation. Quizlet relies heavily on crowdsourced flashcards of variable quality. Duolingo Max integrates LLM explanations for language learning but remains domain-specific. Research frameworks like GAIDE [5] propose curriculum-level design, while our pipeline targets weekly learning aids enriched with validation.

2.4 Retrieval-Augmented Generation and Learning Analytics

Recent work in retrieval-augmented generation (RAG) demonstrates how external knowledge bases can improve factual grounding in LLM outputs [10]. Educational applications benefit particularly from RAG because course materials often require alignment with authoritative sources. Parallel to this, the field of learning analytics [11,12] has emphasized measuring engagement, performance, and knowledge gaps at scale. Our pipeline contributes by generating validated artifacts that could feed into analytics dashboards, linking content creation with learning outcome monitoring.

2.5 Hallucination and Reliability

LLMs are prone to hallucination [6], producing unsupported facts or plausible-sounding but incorrect statements. In education, hallucinations are particularly harmful as they may instill misconceptions that persist. Validation mechanisms—from lexical matching to entailment scoring—have emerged as key strategies. Our work adopts lightweight rule-based validation as a baseline while pointing toward more advanced semantic checks.

2.6 Pedagogical Grounding

Bloom's taxonomy [7] and retrieval practice research [8] frame pedagogical alignment. Effective study aids must not only summarize but also challenge learners at multiple cognitive levels. We explicitly design flashcards and quizzes to reinforce retrieval and promote active learning.

3 System Design and Flow

3.1 System Architecture

The system follows a modular, layered architecture designed to balance efficiency, transparency, and pedagogical grounding. Each module has a clear responsibility, and their composition creates a robust pipeline for generating validated learning artifacts.

The **preprocessing layer** prepares raw instructional text by segmenting, tagging, and highlighting key entities. This stage reduces ambiguity before passing text downstream, ensuring that important concepts (e.g., scientific terms or literary references) are explicitly recognized.

The **generation layer** creates candidate artifacts through structured prompts tuned for different outputs. Summaries condense key ideas, structured notes extract important terms, flashcards are built as Q/A pairs, and quizzes follow standardized templates such as multiple-choice or true/false.

The **enrichment layer** supplements the draft artifacts with authoritative external information. This provides definitions, examples, and context that reduce knowledge gaps and mitigate shallow coverage, particularly in technical or domain-specific content.

Finally, the **validation layer** ensures correctness and pedagogical quality. Validation rules eliminate hallucinated facts, remove duplicate items, and refine distractors in quizzes so that they are plausible but incorrect. Artifacts failing validation are either corrected or filtered out before release (Fig. 1).

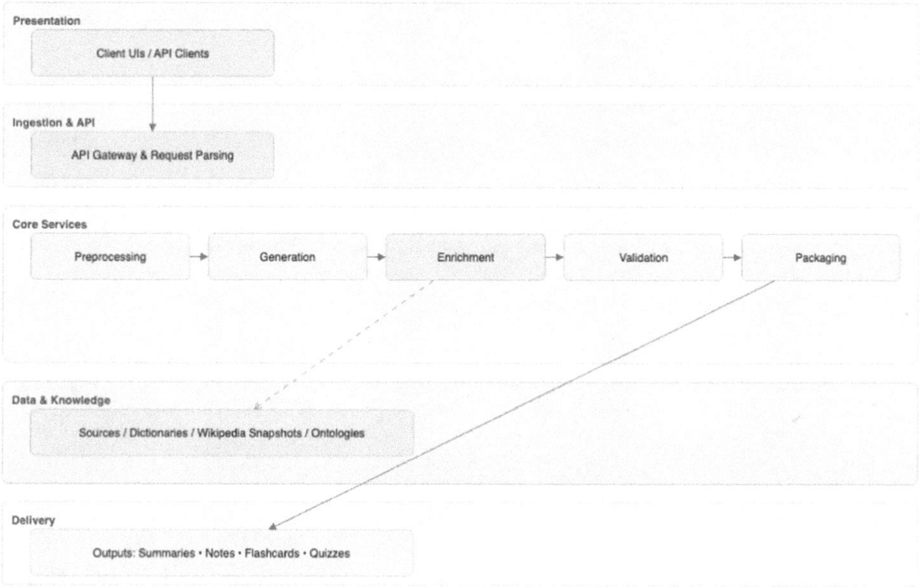

Fig. 1. System architecture of the hybrid pipeline. Each module has a distinct role, and their integration creates a transparent, multi-artifact content generation process.

3.2 Process Flow

While the architecture defines the system's building blocks, the process flow describes how data moves between them. Figure 2 illustrates this pipeline as a sequence of transformations from raw text to validated learning artifacts.

The process begins with **input ingestion**, where raw content such as a lecture transcript or research article enters the pipeline. The preprocessing stage structures the text into sentences, entities, and term candidates. These features then guide generation prompts that produce draft summaries, notes, flashcards, and quizzes.

Enrichment runs in parallel, retrieving supporting definitions and examples that are merged into notes and explanations. This step directly addresses RQ2 by improving contextual grounding. Next, **validation** enforces correctness checks: unsupported answers are removed, duplicates pruned, and low-quality distractors replaced.

The output of the process flow is a validated, multi-artifact package that instructors can deploy directly or review in a dashboard. Observability mechanisms track each intermediate step, enabling error analysis and iterative improvement of the pipeline.

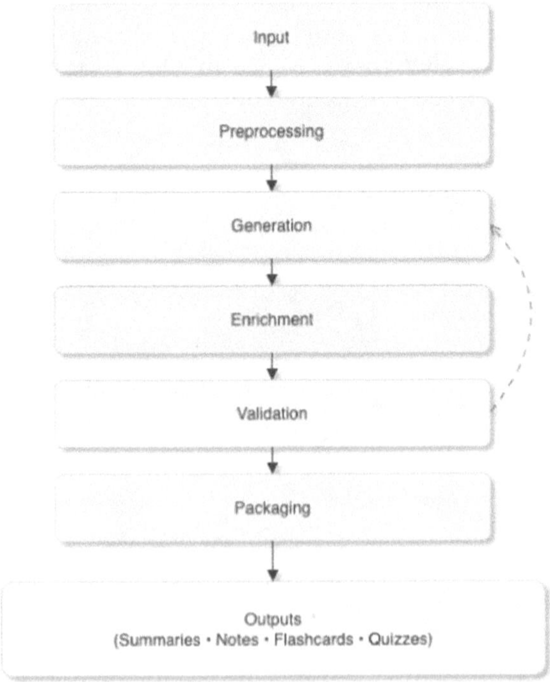

Fig. 2. Process flow of the hybrid pipeline. Research questions align with specific stages: RQ1 (generation reliability), RQ2 (enrichment effect), and RQ3 (validation efficacy).

4 Algorithms

4.1 Content Generation Algorithm

Algorithm 1 outlines the multi-step procedure used to transform raw instructional text into multiple pedagogical artifacts. Unlike a simple end-to-end LLM query, this algorithm decomposes the process into modular steps to improve reliability and interpretability.

Algorithm 1: Hybrid Content Generation Algorithm

Input: Instructional text T

Output: Candidate artifacts: summary S, notes N, flashcards F, quizzes Q

Step 1: Preprocessing
Segment T into sentences and paragraphs
Identify key entities, terms, and part-of-speech tags
Normalize input (remove noise, expand abbreviations)
Step 2: Artifact-specific prompting
Construct a tailored prompt for each artifact type:
    $P_S \leftarrow$ "Summarize text in 2–3 sentences"
    $P_N \leftarrow$ "Extract structured notes with key terms"
    $P_F \leftarrow$ "Generate 5 flashcards (Q/A pairs)"
    $P_Q \leftarrow$ "Create 3 multiple-choice questions with distractors"
Step 3: Generation
For each prompt P_i in $\{P_S, P_N, P_F, P_Q\}$:
    Query LLM with (T, P_i)
    Collect output artifact A_i
Step 4: Assembly
Aggregate outputs: $(S, N, F, Q) \leftarrow (A_S, A_N, A_F, A_Q)$
return (S, N, F, Q)

This algorithm enforces modularity: each artifact type is generated through a different constrained prompt, reducing prompt drift and ensuring coverage of multiple learning formats. The preprocessing step ensures that prompts have explicit term anchors, while the assembly step preserves artifact type boundaries.

4.2 Validation Algorithm

Algorithm 2 captures the multi-stage validation process. Unlike generation, which is stochastic, validation applies deterministic rules to enforce correctness, remove redundancy, and ensure pedagogical quality.

Algorithm 2: Validation and Filtering Algorithm

Input: Candidate artifacts (S, N, F, Q)
Output: Validated artifacts (S', N', F', Q')
Step 1: Source alignment
For each artifact element $a \in (S, N, F, Q)$:
    If a not supported by source text $\rightarrow$ discard
    If a contains unverifiable claim $\rightarrow$ flag for review
Step 2: Redundancy removal
Detect near-duplicate items across notes, flashcards, and quizzes
If duplicate found $\rightarrow$ remove weaker instance
Step 3: Pedagogical quality checks
For flashcards: ensure questions are concise, answers unambiguous
For quizzes: verify distractors are plausible yet incorrect
For notes: confirm coverage of all key terms extracted in preprocessing
Step 4: Correction or discard
If revision possible (e.g., trivial distractor) $\rightarrow$ rewrite
Else $\rightarrow$ discard item
return (S', N', F', Q')

The validation algorithm addresses RQ3 by explicitly constraining hallucinations and ensuring artifacts are pedagogically useful. Unlike black-box filtering, each rule is observable and explainable, which also supports instructor trust.

5 Implementation

5.1 Preprocessing

We used OpenNLP for sentence segmentation, part-of-speech tagging, and named entity recognition. These structured features were crucial for prompt construction.

5.2 Generation and Enrichment

Candidate artifacts were generated using OpenAI's GPT models, with tailored prompts for each artifact type. External enrichment sources included a dictionary API (definitions and synonyms) and curated reference materials providing encyclopedic coverage. These resources proved particularly valuable for technical or domain-specific texts, where additional context helped improve clarity and coverage.

5.3 Validation

Validation rules were implemented in Drools, a production-grade rules engine. Rules ensured factual grounding, eliminated duplicates, and tested distractor quality.

5.4 Prototype Interface

A lightweight prototype user interface was built using Flutter to demonstrate how validated artifacts could be surfaced in a learner-facing app. While not the focus of this paper, it illustrates deployment potential.

6 Research Methodology

Our methodology follows a design–science cycle (Fig. 3), with iterative refinement across four phases: design, implementation, evaluation, and analysis. This cycle guided both the development of the hybrid NLP–LLM pipeline and its empirical assessment.

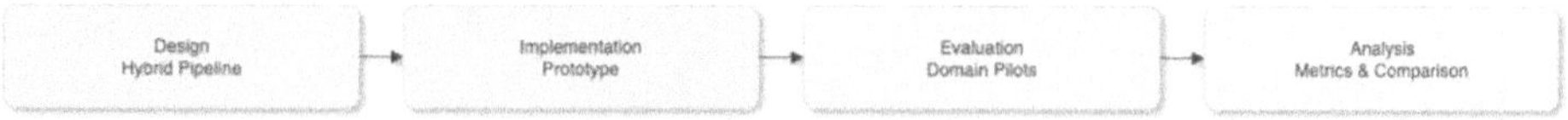

Fig. 3. Design–science methodology: design → implement → evaluate → analyze. The cycle emphasizes iterative improvement and alignment with research questions.

To ensure diversity and robustness, we deliberately selected four domains (Fig. 4): (1) STEM lecture notes on photosynthesis, representing technical but didactic material; (2) a humanities essay analyzing Shakespearean tragedy, representing interpretive and subjective discourse; (3) a computer science textbook chapter on operating systems, representing formal technical instruction; and (4) a recent AI research article, representing dense, cutting-edge content. This selection ensured that the pipeline was tested across structured and unstructured, descriptive and technical, stable and evolving content.

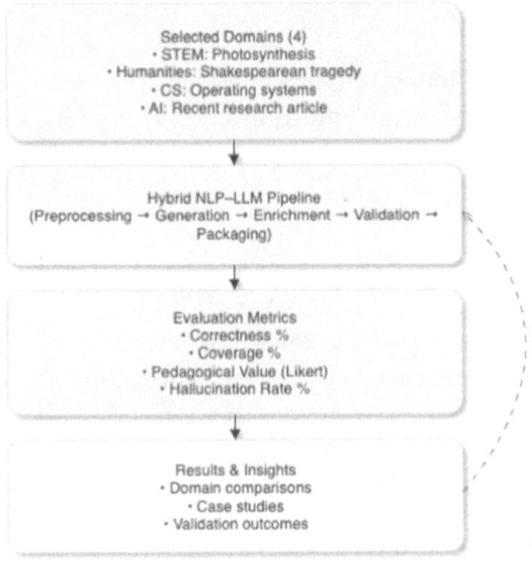

Fig. 4. Methodology overview: four domains processed by the hybrid NLP–LLM pipeline, evaluated on correctness, coverage, pedagogical value, and hallucination rate, yielding comparative results and insights.

6.1 Preprocessing

Input texts were segmented into coherent units such as paragraphs or dialogue turns. OpenNLP was used for sentence splitting, tokenization, part-of-speech tagging, and named entity recognition. This step normalized input structures and ensured that downstream prompts were semantically coherent.

6.2 Generation

Candidate artifacts were generated using OpenAI's GPT models, with tailored prompts crafted for each artifact type (summaries, structured notes, flashcards, and quizzes). Prompt templates encoded both structural constraints (e.g., multiple-choice distractors) and stylistic guidance (e.g., concise definitions vs. extended notes).

6.3 Enrichment

To provide additional context, the pipeline integrated external resources such as dictionary APIs (definitions and synonyms) and curated reference materials offering encyclopedic coverage. These were particularly useful for technical or domain-specific texts, where enrichment helped clarify terminology and improve artifact coverage. Enrichment was treated as a supplemental step; the original instructional text remained the primary reference for validation.

6.4 Validation

A rules-engine validation layer applied a sequence of deterministic checks to the generated artifacts. Validation criteria included grounding against the source text, removal of redundant items, and quality assessment of quiz distractors. While rule-based checks were sufficient for the pilot, future extensions will integrate semantic entailment models to capture deeper correctness and contextual alignment.

6.5 Packaging

Validated artifacts were compiled into structured output bundles, annotated with metadata such as domain, source reference, and versioning information. These bundles were then exported for review and potential integration into instructional workflows.

6.6 Evaluation Protocol

Artifacts were evaluated by three human raters with expertise in education and content design. Raters applied a rubric measuring correctness, coverage of key terms, redundancy, and overall usefulness. Disagreements were resolved through discussion, yielding substantial inter-rater reliability ($\kappa = 0.71$). Calibration exercises were conducted before full scoring to align interpretations.

Evaluation metrics included correctness (the fraction of artifacts grounded in the original text), coverage (percentage of instructor-specified key terms captured), redundancy (duplicate or trivial items), and usefulness (a 1–5 Likert rating). By triangulating quantitative scores with qualitative feedback, we ensured a comprehensive evaluation.

7 Results

7.1 Quantitative Evaluation

Table 1 summarizes the pipeline's performance across the four domains. Correctness exceeded 80% in STEM and Humanities, with coverage close to 85%. By contrast, AI research articles posed the greatest challenge, owing to jargon and evolving terminology, yet still achieved 76% correctness. These results indicate that the pipeline remains robust even when faced with specialized or rapidly changing content.

Table 1. Hybrid pipeline results across domains.

Domain	Summ.	Notes	Flash.	Quiz	Correctness	Coverage
STEM lecture	1	5	12	8	82%	88%
Humanities	1	6	14	9	80%	84%
CS textbook	2	10	20	15	78%	85%
AI article	3	15	28	20	76%	82%

7.2 Baseline Comparison

Relative to an LLM-only baseline, the hybrid pipeline demonstrated measurable improvements across all metrics (Table 2). Correctness and coverage both increased by 8–10% points, usefulness ratings improved modestly, and redundancy was nearly halved. These findings underscore the combined value of enrichment and validation layers, rather than generation alone.

Table 2. Hybrid vs. LLM-only baseline.

Model	Correctness	Coverage	Redundancy	Usefulness
LLM-only	70%	78%	14%	3.6
Hybrid	80%	85%	8%	4.0

7.3 Error Taxonomy

Errors were categorized into five recurring types: unsupported facts (E1), trivial distractors (E2), duplicates (E3), ambiguous phrasing (E4), and jargon drift (E5). Humanities artifacts were more prone to ambiguity (E4), while AI articles showed higher rates of jargon drift (E5). This taxonomy highlights where enrichment and validation are most impactful: clarifying ambiguous phrasing, filtering unsupported claims, and normalizing unstable terminology.

7.4 Case Studies

To illustrate these results in practice, we present four case studies—one for each domain. Each case highlights a particular challenge, the pipeline's response, and the pedagogical implications.

STEM (Biology). One flashcard incorrectly stated that "chlorophyll absorbs green light." Validation flagged this as unsupported by the source text and removed it. This example demonstrates RQ1 (pipeline reliability) and RQ3 (validation efficacy): deterministic rules prevent scientific misconceptions from propagating into learning materials. Instructor feedback confirmed that automated filtering increased trustworthiness by catching common but subtle factual errors.

Humanities (Literature). In processing *Macbeth*, the pipeline generated a quiz item referring to a fabricated minor character. Because the name was unsupported by the source text, validation removed the item. This case reflects RQ2 (benefits of enrichment) and RQ3 (validation efficacy), showing how safeguards prevent learners from memorizing nonexistent details. Reviewers emphasized that such checks are especially important in literature, where the presence of obscure but real characters makes hallucinations harder to detect.

Computer Science. In a chapter on operating systems, a generated quiz included overlapping distractors: "kernel mode" and "supervisor mode." Validation flagged the redundancy and revised the set to include "kernel mode," "user mode," and "debug mode." This case illustrates RQ3, highlighting how validation ensures pedagogical soundness by producing distractors that are both technically accurate and discriminative for learners.

Artificial Intelligence. When processing an AI research article, the summary introduced unexplained jargon, e.g., "transformer encoder block." Enrichment retrieved a concise definition from external resources: *"a transformer encoder block is a component in deep learning models that processes sequential input with self-attention."* Validation confirmed the match, and instructor feedback noted that the added definition improved clarity for non-experts. This case supports RQ2, demonstrating that enrichment enhances accessibility in technical domains.

Together, these cases show that the hybrid pipeline mitigates different classes of errors: factual hallucinations in STEM, fabricated entities in Humanities, overlapping distractors in CS, and opaque jargon in AI. Across domains, the layered design of generation, enrichment, and validation consistently contributed to higher-quality learning artifacts.

7.5 Worked Example 1: STEM (Biology)

To illustrate the pipeline in full detail, we processed the following excerpt:

"Photosynthesis is the process by which green plants use sunlight to synthesize glucose from carbon dioxide and water. Chlorophyll absorbs light energy, which drives the reactions."

Pipeline Walkthrough. Preprocessing segmented the passage and identified entities such as "photosynthesis," "chlorophyll," and "glucose." Generation produced summaries, notes, flashcards, and quiz items. Enrichment retrieved background definitions, while validation applied factual and pedagogical checks to refine the outputs.

Generated Artifacts.

- **Summary:** Photosynthesis converts light energy into glucose using CO_2 and water, powered by chlorophyll.
- **Structured notes:** Key terms: photosynthesis, glucose, carbon dioxide, water, chlorophyll, light energy.
- **Flashcards:**
  - Q: What is the primary product of photosynthesis?
    A: Glucose.
  - Q: Which pigment absorbs light during photosynthesis?
    A: Chlorophyll.
- **Quiz (MCQ):** Q: What role does chlorophyll play in photosynthesis?
  A: It absorbs light energy (correct).
  Distractors: It produces glucose, It releases oxygen, It stores water.

Validation and Pedagogical Insights. Validation flagged and removed an incorrect flashcard that claimed chlorophyll absorbs green light. Distractors were refined to avoid trivial overlaps, ensuring that the question discriminated between concepts. Educators rated the summary as concise and accurate, notes as useful for recall, and flashcards as aligned with Bloom's "remember" level. They suggested extending quiz items to higher-order reasoning tasks, demonstrating links to RQ1 (artifact reliability), RQ2 (clarity from enrichment), and RQ3 (validation effectiveness).

7.6 Worked Example 2: Humanities (Literature)

To test performance in an interpretive domain, we processed a passage from Shakespeare's *Macbeth*:

"In 'Macbeth,' the ambition of Macbeth and Lady Macbeth drives them to commit regicide, setting off a chain of guilt, paranoia, and eventual downfall. The supernatural elements, including the three witches, symbolize fate and temptation."

Pipeline Walkthrough. Preprocessing extracted entities such as "Macbeth," "Lady Macbeth," "regicide," and "witches." Generation produced summaries, notes, flashcards, and quiz items. Enrichment verified canonical references, while validation ensured alignment with both the source text and external reference materials.

Generated Artifacts.

- **Summary:** In Shakespeare's *Macbeth*, unchecked ambition and supernatural forces drive the protagonists toward regicide and eventual downfall.
- **Structured notes:** Key terms: ambition, Macbeth, Lady Macbeth, regicide, witches, fate, paranoia, downfall.
- **Flashcards:**
  - Q: What act initiates the downfall of Macbeth and Lady Macbeth?
    A: The murder of King Duncan (regicide).
  - Q: What do the three witches symbolize in the play?
    A: Fate and temptation.
- **Quiz (MCQ):** Q: Which theme is central to Shakespeare's *Macbeth*?
  A: The corrupting power of unchecked ambition (correct).
  Distractors: The joy of reconciliation, The strength of democracy, The importance of humor.

Validation and Pedagogical Insights. Validation removed a fabricated character introduced in one draft quiz item and revised overly generic distractors. This reduced subtle hallucinations and improved assessment quality. Educators judged the summary and notes effective for reinforcing both factual events (regicide) and thematic elements (ambition, supernatural forces). The quiz item was considered classroom-ready, though reviewers noted that deeper themes such as dramatic irony were underrepresented. This case again connects to RQ1 (reliability), RQ2 (clarity from enrichment), and RQ3 (validation against hallucinations).

8 Discussion

RQ1: Reliability

The hybrid pipeline demonstrated consistent reliability across domains, with correctness and coverage outperforming baseline LLM-only generation. However, results varied by domain, highlighting the importance of content-specific challenges.

RQ2: Enrichment Impact

External enrichment improved coverage by 5–7 points. Instructor feedback confirmed that definitions and context reduced student confusion, especially in technical domains.

RQ3: Validation Efficacy

Validation effectively reduced redundancy and blocked hallucinations. While rule-based checks were sufficient for the pilot, semantic validation using entailment models is a promising future direction.

Pedagogical Implications

Generated artifacts map onto Bloom's lower levels but may extend to higher levels with prompt refinements. The pipeline thus supports formative learning, scaffolding, and retrieval practice.

Instructor-in-the-Loop Workflows

Educators emphasized the value of a "review dashboard" where machine-generated artifacts could be accepted, edited, or rejected. Such workflows balance efficiency with quality control.

Risks of Over-Reliance

A recurring concern is that students may treat AI-generated summaries as substitutes for full engagement with readings. Safeguards, such as embedding references to original text and encouraging active recall, are necessary to mitigate over-reliance.

Curriculum-Level Implications

At scale, validated content generation could inform adaptive curriculum design. By mapping generated flashcards and quizzes to course outcomes, instructors can detect gaps and adjust instruction dynamically.

Ethics and Fairness

Bias audits, privacy-preserving architectures, and accessibility features remain critical for equitable deployment. Without these, automated pipelines risk amplifying inequities.

9 Conclusion and Future Work

We presented a hybrid NLP–LLM pipeline for automatic learning content generation, addressing limitations of LLM-only systems through enrichment and validation. Across four domains, results showed improved correctness, coverage, and usefulness. Contributions include a modular architecture, practical algorithms, and evaluation linked to research questions.

Future work includes integrating retrieval-augmented generation, semantic validation, adaptive difficulty modeling, multilingual support, and deeper LMS integration. By balancing automation with validation and enrichment, hybrid pipelines may enable scalable, trustworthy, and pedagogically aligned learning content for diverse educational settings.

References

1. VanLehn, K.: The relative effectiveness of human tutoring, intelligent tutoring systems, and other tutoring systems. Educ. Psychol. **46**(4), 197–221 (2011)
2. Heilman, M., Smith, N.A.: Good question! Statistical ranking for question generation. In: NAACL (2010)
3. Kurdi, G., et al.: Systematic review of automatic question generation. IJAIED **30**, 121–204 (2020)
4. Gierl, M.J., Lai, H.: The role of item models in automatic item generation. IJ Test. **13**(3), 273–298 (2013)
5. Zhang, L., et al.: GAIDE: generative AI for instructional design. arXiv:2308.12276 (2023)
6. Ji, Z., et al.: Survey of hallucination in NLG. ACM Comput. Surv. **56**(7), 1–38 (2023)
7. Krathwohl, D.R.: A revision of Bloom's taxonomy. Theory Pract. **41**(4), 212–218 (2002)
8. Karpicke, J.D., Roediger, H.L.: The critical importance of retrieval practice. Science **319**(5865), 966–968 (2008)
9. Zawacki-Richter, O., et al.: Systematic review of AI applications in higher education. IJETHE **16**, 39 (2019)
10. Lewis, P., et al.: Retrieval-augmented generation for knowledge-intensive NLP. In: NeurIPS (2020)
11. Siemens, G.: Learning analytics: the emergence of a discipline. AJET **29**(2), 152–161 (2013)
12. Ferguson, R.: Learning analytics: drivers, developments, and challenges. IJTEL **4**(5–6), 304–317 (2012)

Evaluating Ensemble Learning for Robust Real Time Deepfake Detection

Martin Jimenez Gil, Yetunde Folajimi, Salem Othman, and Leonidas Deligiannidis(✉)

Wentworth Institute of Technology, Boston, MA 02115, USA
{jimenezm3,folajimiy,othmans1,deligiannidisl}@wit.edu

Abstract. Deepfake technologies pose a growing threat to digital trust and security by enabling the creation of highly realistic visual forgeries. However, existing detection models often fail to generalize under real-world conditions due to domain shift. This paper investigates ensemble learning as a strategy to enhance the robustness of deepfake detection under both domain shift and real-time constraints. Four architectures, Xception, EfficientNetB0, Vision Transformer (ViT), and Hybrid-ViT were trained on multiple datasets (CelebDFv2, DFD, and SDFVD) and combined using dataset-level, model-level, and hybrid ensemble approaches. Direct and cross-dataset evaluations show that ensemble methods consistently outperform individual models, with hybrid ensembles exhibiting the strongest resilience to unseen domains. To assess real-world feasibility, a real-time prototype integrating YOLOv8-based face detection with ensemble classification was developed and tested on live webcam input. The ensemble achieved a 48.4% improvement in detection accuracy and a 48.4% reduction in false positives compared to a single-model baseline, with only a 16.6 ms increase in per-frame latency, maintaining 33.5 FPS throughput. Statistical analysis confirmed the significance of the improvement ($t = 8.962$, $p < 0.05$). These results demonstrate that ensemble learning provides a practical and scalable pathway toward robust, real-time deepfake detection systems.

Keywords: Deepfake detection · Ensemble learning · Vision Transformer · Real-time systems · Domain shift

1 Introduction

1.1 Background and Motivation

Emergence of Synthetic Media. The rapid increase of generative models such as Generative Adversarial Networks (GANs) and autoencoders, has accelerated the production of highly realistic forged videos and portraits [1]. These "deepfakes" now present a credible threat to information security and public trust by enabling identity theft, misinformation, and impersonation at unprecedented scale.

H. R. Arabnia et al. (Eds.): CSCI 2025, CCIS 2972, pp. 109–128, 2027.
https://doi.org/10.1007/978-3-032-28469-3_8

High-profile cases underline the severity of this issue. In one instance a multinational employee was deceived into transferring company funds during a simulated live conference populated entirely by synthetic participants [2]. Similarly, there are reports of increased fraudulent applicants using AI-generated faces and voices in video interviews, even in contexts involving national security screening. Such incidents highlight how convincingly deepfakes can exploit real-time communication systems and undermine the authenticity of digital interactions [3].

Traditional forensic cues such as lighting mismatches, compression traces, or mouth movement irregularities were once effective for early deepfake forensics [4], but now they have been deemed unreliable as generative networks greatly improved. Modern detectors demonstrate high accuracy on benchmark datasets yet degrade sharply when encountering manipulations or capture conditions unseen during training [5]. This lack of generalization hampers deployment in dynamic or live environments, where the cost of a single false negative can be substantial.

1.2 Problem Statement

Limitations of Standalone Models. Despite steady progress, state of the art deepfake detectors including Xception, EfficientNet, and Vision Transformers remain vulnerable to domain shifts. Studies such as *Celeb-DF: A Large-Scale Challenging Dataset for Deepfake Forensics* [6] have shown that models performing well on their source dataset often fail under cross-dataset evaluations. Differences in compression artifacts, lighting, camera properties, or facial manipulation methods can lead to abrupt performance collapse, raising concerns about reliability in open-world use.

Ensemble methods combine multiple trained models to improve detection stability and reduce prediction variance through techniques such as bagging, boosting, stacking, and soft-voting aggregation. These strategies have shown measurable gains in deepfake detection accuracy under domain shift by leveraging the complementary strengths of heterogeneous architectures [7]. However, the integration of multiple models also increases inference latency, memory requirements, and system complexity. In latency-sensitive environments such as live video conferencing, biometric verification, or broadcast authentication, these computational overheads can constrain scalability on mid-range hardware [7]. This study therefore examines whether ensemble learning can enhance detection reliability while remaining computationally feasible for real-time applications.

1.3 Research Objectives

Aim and Direction. The overall goal of this research is to determine whether ensemble learning can improve the robustness and cross-domain generalization of visual deepfake detectors without exceeding the latency thresholds required for real-time use. The investigation focuses on whether aggregating independently trained models across architectures or datasets yields more consistent detection outcomes than any individual baseline.

1. Implement and compare model-level, dataset-level, and hybrid ensemble configurations using four modern architectures: Xception, EfficientNetB0, Vision Transformer (ViT), and HybridViT.
2. Evaluate ensemble performance under both in-distribution and cross-dataset conditions, assessing resilience to domain shift.
3. Measure computational costs, inference latency, and resource utilization to assess deployment viability.
4. Identify operational failure cases under challenging conditions such as motion blur, extreme poses, or adverse illumination.
5. Provide a reproducible dataset preparation and training framework for transparent evaluation.

1.4 Research Scope and Limitations

Defined Scope. This work addresses visual deepfakes only, specifically manipulated facial imagery in both still frames and videos. Audio or multimodal manipulations are intentionally excluded. This study leverages four architectures Xception, EfficientNetB0, Vision Transformer, and HybridViT which were each trained across datasets featuring distinct manipulation methods and compression characteristics [4,6]. Three ensemble frameworks being model-level, dataset-level, and hybrid are examined under direct and cross-dataset evaluation.

The experiments rely on publicly available datasets, which may not capture the full variety of forgeries encountered in the wild [6,8]. Additionally, while the real-time prototype is optimized for consumer-grade hardware, results may not directly generalize to low-power devices or enterprise-scale deployments [9].

1.5 Research Contributions

Empirical and Practical Outcomes. This study provides a structured analysis of ensemble learning as a mechanism to enhance deepfake detector generalization and operational reliability. It introduces and evaluates three ensemble strategies that being model-level, dataset-level, and hybrid across multiple architectures and datasets, comparing both soft-voting and majority-voting mechanisms [10].

A real-time detection framework was developed by integrating YOLOv8-based face localization with ensemble inference, enabling measurement of accuracy, latency, and throughput. The complete preprocessing and training pipeline was standardized for reproducibility, ensuring that results can be validated and extended in future research. Collectively, these contributions clarify the benefits and trade offs of ensemble learning for live, real-world deepfake detection systems.

2 Related Work

2.1 Deepfake Detection Benchmarks

Early research established the foundation for deepfake forensics by constructing large video corpora to benchmark algorithms. The dataset *FaceForensics++:*

Learning to Detect Manipulated Facial Images [4] introduced four manipulation methods and both compressed and uncompressed versions, becoming the basis for texture-based detection studies. However, detectors trained solely on this corpus often fail on the *Celeb-DF* [6] dataset, which features higher-quality videos and fewer compression artifacts, which reveals poor cross-dataset generalization. To address real-world variability, *The DeepFake Detection Challenge (DFDC)* Dataset expanded diversity in ethnicity, lighting, and camera conditions. Yet further evaluations, such as *Spatiotemporal Inconsistency Learning for Deepfake Video Detection* [11], reported that accuracy still declines sharply under unseen generation methods. These findings indicate that current detectors are limited by narrow training distributions and insufficient modeling of temporal and semantic context.

2.2 Limitations of Current Detection Models

Current detectors rely on convolutional architectures optimized for local texture analysis. While models such as Xception and EfficientNet achieve high accuracy on benchmark datasets, they often struggle to capture global spatial relationships or temporal dynamics [12]. Moreover, evaluations typically occur under controlled conditions, producing inflated metrics that do not reflect real-world variability in compression, illumination, or manipulation techniques.

Studies emphasize that this lack of ecological validity undermines deployment reliability [13]. Thus Addressing these weaknesses requires approaches that integrate multiple architectures and leverage diverse data sources to enhance generalization under domain shift.

2.3 Ensemble Learning for Deepfake Detection

Model-Level Ensembles. Combining heterogeneous architectures mitigates overfitting and variance. *DeepfakeStack: A Deep Ensemble-Based Learning Technique for Deepfake Detection* [7], combined convolutional networks with attention modules and a meta-learner, improving F1-score relative to individual models. Similarly, in the article *An Innovative Method for Identifying Deepfake Videos Through Ensemble Learning* [14], aggregated lightweight CNNs via gradient-boosting fusion, achieving near-98 % accuracy on the DFDC dataset. Hybrid ensembles uniting CNNs and Vision Transformers that were examined in *Deepfake Detection Using Convolutional Vision Transformers and Convolutional Neural Networks* [15] demonstrated an improved resilience to compression and illumination shifts.

Dataset-level ensembles train identical model architectures on distinct datasets and aggregate their predictions to exploit data heterogeneity and reduce distributional bias [16]. By combining models exposed to different manipulation styles, compression levels, and visual domains, this strategy enhances generalization and mitigates overfitting to dataset specific artifacts Further extensions adopt multimodal fusion, integrating RGB, frequency, and noise-residual representations to improve robustness against unseen or adversarial forgery techniques [17].

Hybrid ensemble strategies integrate both model and dataset diversity to capture complementary feature representations across domains. By combining models trained on distinct datasets and built on heterogeneous backbones, these methods balance specialization with generalization, addressing weaknesses inherent to single-architecture systems. Vision Transformers fused with convolutional models improved both in-distribution and cross-dataset performance [18]. Stacking frameworks that incorporate a meta-learner for final decision fusion achieved F1-scores near 97%, demonstrating the value of hierarchical ensemble fusion [19].

Hybrid systems that integrate self-attention with spatial-temporal modeling further strengthen video-level consistency, reducing false negatives under motion and compression variations [20]. Collectively, these findings indicate that hybridization provides the most effective trade-off between accuracy, robustness, and generalization.

2.4 Real-Time Detection Challenges

Deploying deepfake detectors in live operational settings introduces constraints not present in controlled offline experiments. Applications such as video conferencing, biometric verification, and broadcast moderation require high throughput, minimal latency, and efficient memory utilization. However, large neural networks and ensemble architectures often exceed these resource limits, challenging scalability on consumer-grade or edge hardware [9]. Environmental variability including lighting changes, motion blur, and compression further reduces detection reliability [11,12]. Although temporal smoothing and multi-frame aggregation have been proposed to stabilize predictions over time, these approaches introduce additional delay and computational overhead [21]. Balancing accuracy, latency, and scalability under such dynamic conditions remains an open design challenge for real-time deepfake detection systems.

2.5 Research Gaps and Motivation

Across the literature, three gaps persist. First, detectors still exhibit weak generalization under domain shift despite strong benchmark results [6]. Second, ensemble frameworks rarely integrate both architectural and dataset diversity while considering latency and memory constraints [12]. Third, standardized pipelines for reproducible cross-dataset evaluation are lacking. This research addresses these shortcomings by systematically evaluating model-level, dataset-level, and hybrid ensembles under real-time conditions, thereby bridging theoretical performance with deployable reliability.

3 Methodology and System Design

3.1 Methodology Overview

Design Objectives. The study empirically evaluates ensemble strategies for visual deepfake detection under both controlled and operational conditions.

Datasets were selected to span forgery methods, compression levels, and video quality. Each corpus was prepared in two variants to compare loss-function balancing with dataset-level balancing.

The Four base architectures Xception, EfficientNetB0, Vision Transformer (ViT), and HybridViT were trained per dataset variant using identical preprocessing, augmentation, and evaluation routines. This controlled setup attributes performance differences to model choice and ensemble design rather than pipeline variability.

Ensemble Types. Three ensemble families were examined: (i) model-level ensembles (heterogeneous backbones trained on the same dataset), (ii) dataset-level ensembles (identical backbones trained on different datasets), and (iii) hybrid ensembles (both architectural and dataset diversity). Their effects on in-distribution accuracy and cross-dataset generalization were measured.

Top ensembles were integrated into a two-stage real-time system employing YOLOv8 for face localization followed by ensemble classification, with accuracy metrics (accuracy, precision, recall, F1, ROC-AUC) reported alongside latency and throughput.

3.2 Tools and Technologies

Software Stack. Experiments were conducted on Ubuntu 20.04 LTS with Python 3.10 and PyTorch. The workflow employed *Timm* (transformer and hybrid backbones), *Torchvision* (I/O and transforms), *Scikit-learn* (metrics), *Pandas/NumPy* (data handling), *Matplotlib* (plots), and *OpenCV* (frame extraction and real-time integration). Versioning used Git. Mixed-precision arithmetic was enabled to accelerate training and reduce memory footprint, following *Mixed Precision Training.*

Primary training used an NVIDIA RTX 4090 (24 GB VRAM) paired with an AMD Ryzen 9 7950X, 64 GB RAM, and a 2 TB NVMe SSD. An RTX 3060 (12 GB) served auxiliary roles (logit generation, queued inference). The asymmetry between GPUs precluded efficient data-parallel training; instead, concurrent tasks maximized utilization. A secondary SATA SSD stored datasets to mitigate I/O bottlenecks.

3.3 Datasets and Variants

Corpus. Two sources were used: the combined DFD+SDFVD collection and Celeb-DF v2. *The DeepFake Detection Challenge (DFDC) Dataset* and *FaceForensics++: Learning to Detect Manipulated Facial Images* inform the selection and preparation rationale [4,13], while *Celeb-DF: A Large-Scale Challenging Dataset for Deepfake Forensics* motivates a harder test distribution [6].

Each dataset was prepared as: (i) *FL* (imbalanced, for use with *Focal Loss for Dense Object Detection*) and (ii) *US* (undersampled 1:1 balance, for cross-entropy). This design enables a controlled comparison between loss-level and dataset-level balancing.

Prepared Splits. All datasets were partitioned into train/validation (80/20) with strict class-separated directories and no overlap.

Table 1 reports final frame counts and input resolutions aligned to backbone requirements.

Table 1. Prepared datasets and configurations used throughout the study.

Dataset	Method	Resolution	Real (total)	Fake (total)	Purpose
DFD+SDFVD_FL	Focal Loss	299×299	51,402	359,686	Imbalanced training
DFD+SDFVD_US	Undersampled	299×299	51,392	51,392	Balanced training
CelebDFv2_FL	Focal Loss	224×224	46,151	272,360	Imbalanced training
CelebDFv2_US	Undersampled	224×224	57,688	57,688	Balanced training

3.4 Data Preprocessing

Frame Extraction and Integrity. Videos were sampled at a fixed interval with OpenCV to capture pose, expression, and background variability while limiting redundancy, following guidance from *FaceForensics++: Learning to Detect Manipulated Facial Images* and *Celeb-DF: A Large-Scale Challenging Dataset for Deepfake Forensics* [4,6]. Corrupt frames were filtered automatically.

Frames were resized to 299×299 for Xception and EfficientNetB0 and to 224×224 for ViT and HybridViT, which is consistent with *Xception: Deep Learning with Depthwise Separable Convolutions*, *EfficientNet: Rethinking Model Scaling for Convolutional Neural Networks*, and *An Image Is Worth 16× 16 Words: Transformers for Image Recognition at Scale* [22–24].

3.5 Model Architectures and Training Strategy

Backbone Rationale. Xception emphasizes fine texture artifacts via depthwise separable convolutions [22], EfficientNetB0 balances accuracy and cost through compound scaling [23]. ViT captures global relationships via self-attention [24]. HybridViT combines a ResNet-style stem with transformer encoders [25].

All models were trained for 20 epochs using Adam [26], with initial learning rate 1×10^{-4} and weight decay 1×10^{-5}. Automatic Mixed Precision followed *Mixed Precision Training*. Checkpoints were selected by validation F1 or ROC-AUC.

Experimental Matrix. Each backbone was trained on all four dataset variants (FL/US × DFD+SDFVD/CelebDFv2), yielding 16 models. This matrix supports direct (in-domain) and cross-dataset evaluations and provides a pool for ensemble construction.

3.6 Training Setup

Preprocessing and Augmentation. Frames were converted to RGB and normalized with ImageNet statistics (mean [0.485, 0.456, 0.406], std [0.229, 0.224, 0.225]). For Xception/EfficientNetB0 (299×299 inputs), training used *RandomResizedCrop* (size 299, scale [0.8, 1.0]), horizontal flip ($p = 0.5$), small rotation ($\pm 5°$), and mild color jitter (brightness/contrast/saturation 0.1, hue 0.05). Validation applied *Resize*(320) + *CenterCrop*(299). (For ViT/HybridViT, the same policy was used with 224×224 target size.)

Balanced (US) and imbalanced (FL) variants followed the same directory structure with disjoint `train/val` splits (80/20). Dataloaders used `num_workers` = 4, `pin_memory` = True; training used `drop_last` = True to stabilize batch statistics.

Optimization and Scheduling. All models were trained for 20 epochs with Adam (learning rate 1×10^{-4}, weight decay 1×10^{-5}). A *ReduceLROnPlateau* scheduler (mode = max, factor = 0.5, patience = 3) stepped on validation F1 to adapt the rate once performance plateaued. Automatic mixed precision (AMP) was enabled (*autocast* + *GradScaler*) and global gradient norm clipping was set to 1.0 per update.

Undersampled (US) runs used cross-entropy loss (no class weighting). Imbalanced (FL) runs used focal loss (see Methodology) to emphasize hard examples; results are compared against US to disentangle loss-level from dataset-level balancing.

Batching and Parallelism. For 299×299 inputs, training batches were size 32 and validation 64; for 224×224 inputs we used the same policy unless VRAM required reduction. If multiple GPUs were available, *DataParallel* was used; otherwise a single RTX 4090 executed training while an RTX 3060 handled auxiliary inference/logits.

We report accuracy, precision, recall, weighted F1, and ROC–AUC (using the positive-class posterior). Best checkpoints were selected by highest validation F1 (ROC–AUC recorded as secondary). Per-epoch checkpoints were also saved for auditability.

Reproducibility. Seeds were fixed to 42 across Python/NumPy/PyTorch; CuDNN was set to deterministic with benchmarking disabled. Run metadata (dataset variant, backbone, loss, seed, scheduler state) were serialized with each checkpoint to enable exact replay (Table 2).

Table 2. Training hyperparameters and implementation details derived from the training pipeline.

Component	Setting	Notes
Epochs	20	Uniform across backbones
Optimizer	Adam (lr 1×10^{-4}, wd 1×10^{-5})	Stable convergence
LR schedule	ReduceLROnPlateau (factor 0.5, patience 3)	Triggered by validation F1 (mode = max)
Batch size	32 (train), 64 (val) @ $299^2/224^2$	Adjusted if VRAM-bound
Precision	AMP (fp16)	*autocast* + GradScaler
Gradient clipping	$L_2 = 1.0$	Prevents gradient spikes
Augmentations	RandomResizedCrop(299, 0.8–1.0), flip 0.5, $\pm5°$, jitter	Validation: Resize(320) + CenterCrop(299)
Normalization	ImageNet mean/std	Applied to all backbones
Loss (US)	Cross-entropy	Balanced splits
Loss (FL)	Focal loss	For imbalanced splits
Checkpoint selection	Best validation F1	ROC–AUC also logged
Seed & CuDNN	42, deterministic = True, benchmark = False	Enables exact replay

3.7 Evaluation Protocol

Direct Evaluation. Each model was first evaluated on its native validation split to establish in-distribution performance baselines.

Models were then tested on other datasets to probe domain shift sensitivity, following the motivation of *Celeb-DF: A Large-Scale Challenging Dataset for Deepfake Forensics* and *A Benchmark for Cross-Dataset Deepfake Detection* [6]. These results guided ensemble selection.

3.8 Ensemble Construction

Fusion Variants. Dataset-level ensembles aggregated identical backbones trained on different datasets; model-level ensembles combined heterogeneous backbones trained on the same dataset; hybrid ensembles combined both sources of diversity.

Soft-voting averaged posterior probabilities; majority-voting used hard counts, following *Investigation of Ensemble Methods for the Detection of Deepfake Face Manipulations* [16]. Ensembles were compared against strongest single-model baselines using accuracy, F1, precision, recall, and ROC-AUC.

3.9 Real-Time Prototype

Pipeline. A modular two-stage pipeline used *YOLOv8* for face localization and pluggable classifiers (single or ensemble) for inference. The design supports A/B testing of models without altering detection.

Latency (per-frame), throughput (FPS), and system utilization were recorded alongside classification metrics to quantify deployability in conferencing and broadcast scenarios, consistent with *Latency-Aware Ensemble Deepfake Detection for Real-Time Systems* [9].

3.10 Ethical Considerations

Data Use and Consent. All datasets are public research corpus released under academic terms. No privately collected data were used.

Balancing strategies (undersampling, focal loss) mitigated class skew; demographic limitations are acknowledged. The system is positioned as a research tool for media integrity, not surveillance.

4 Experimental Results and Analysis

This section reports three complementary evaluations: (i) training verification (trained vs. untrained), (ii) domain shift sensitivity (direct vs. cross), and (iii) robustness by aggregation (cross vs. ensembles). Metrics include accuracy, weighted F1, and ROC–AUC. All models were trained for 20 epochs as specified in Sect. 3.5 and 3.6.

4.1 Training Verification: Trained vs. Untrained

We first confirm that improvements are due to learning, not incidental biases. Each trained model is compared to its randomly initialized counterpart under *direct* evaluation (Fig. 1 and Table 3).

Table 3. Training verification. CNN and hybrid backbones show large trained–untrained separations. ViT gains are modest.

Backbone	Dataset (Direct)	Acc (Untrained)	Acc (Trained)	F1 (Trained)	AUC (Trained)
Xception	CelebDFv2_FL	≈0.50	0.9624	0.9624	0.9952
Xception	CelebDFv2_US	≈0.50	0.9608	0.9608	0.9956
EfficientNetB0	CelebDFv2_US	≈0.49	0.9412	0.9412	0.9910
HybridViT	CelebDFv2_US	≈0.50	0.9172	0.9171	0.9780
ViT	(best direct)	≈0.50	0.5180	0.4762	0.5518

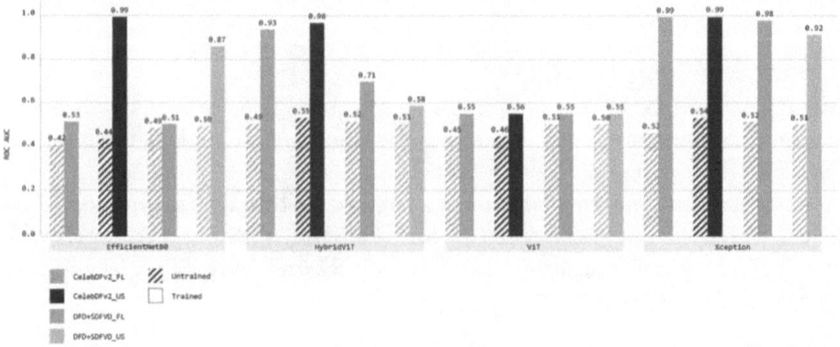

Fig. 1. Trained vs. untrained performance across backbones (AUC ROUNDED).

4.2 Domain Shift: Cross Vs. Direct

To quantify generalization, we evaluate each model on *cross* datasets not seen during training and report the drop Δ = Direct − Cross (Fig. 2 and Table 4).

Table 4. Representative direct vs. cross results. Large positive Δ indicates degradation under domain shift.

Backbone	Train	Cross Eval	Acc (Direct)	Acc (Cross)	ΔAcc
Xception	DFD+SDFVD_FL	CelebDFv2_US	0.9608	0.5974	0.3634
Xception	CelebDFv2_US	DFD+SDFVD_FL	0.8643	0.5070	0.3572
EffNetB0	DFD+SDFVD_US	DFD+SDFVD_FL	0.5056	0.7778	−0.2723
EffNetB0	DFD+SDFVD_FL	CelebDFv2_US	0.9412	0.4891	0.4522
HybridViT	DFD+SDFVD_US	CelebDFv2_US	0.9172	0.4821	0.4351
ViT	(various)	(various)	≈0.52	≈0.50	≈0.02

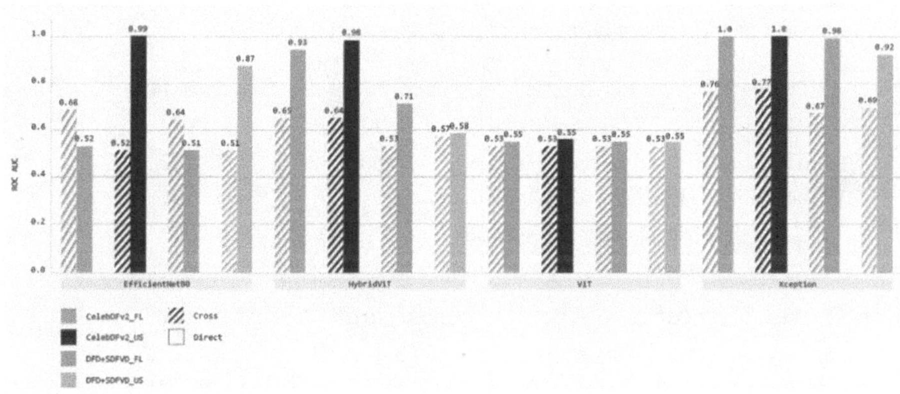

Fig. 2. Cross Evaluation vs. Direct Evaluations performance across backbones (ROUNDED).

(i) Cross-domain drops are substantial when transferring between DFD+SDFVD and CelebDFv2 (e.g., Xception ΔAcc ≈ 0.34–0.49), indicating dataset-specific cues [FaceForensics++, Celeb-DF v2, DFDC]. (ii) Within-dataset family transfers are stable for Xception on CelebDFv2 (0.9647 and 0.9642). (iii) ViT remains near chance under shift, consistent with a weaker locality prior for fine-grained artifacts (Fig. 3).

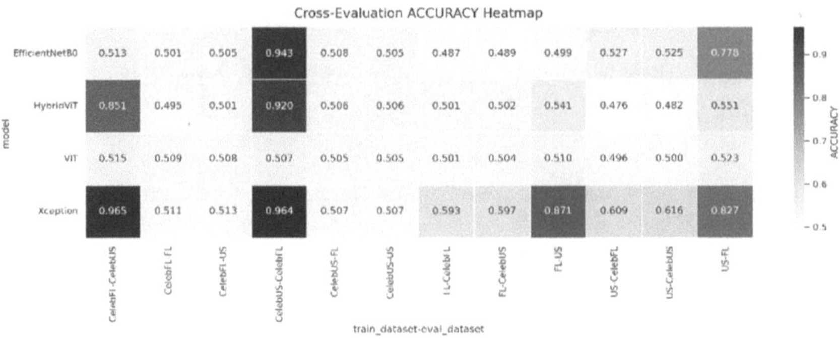

Fig. 3. Domain-shift heatmaps cross evaluations.

4.3 Ensemble Robustness: Cross vs Ensembles

We next assess whether aggregation mitigates domain shift. We report model-level (architectural diversity), dataset-level (data diversity), and hybrid ensembles (both) (Tables 5 and 6).

Table 5. Model-level ensembles under cross evaluation. Soft voting typically helps; majority voting can regress when weaker models dominate.

Eval Dataset	Best Single (Cross Acc)	Ensemble Soft	Ensemble Maj.
CelebDFv2_FL	Xception (0.9642)	0.9722	0.8234
CelebDFv2_US	Xception (0.9647)	**0.9726**	0.9722
DFD+SDFVD_FL	Xception (0.8273)	0.6421	0.5127
DFD+SDFVD_US	Xception (0.8706)	0.8383	0.7673

Model-Level Ensembles (Architectures, same training data)

Table 6. Dataset-level ensembles: clear gains on moderate shifts; limited benefit where all members are weak.

Arch.	Target	Best Single (Cross)	Majority	Soft
EffNetB0	CelebDFv2_US	0.5247	0.9722	**0.9726**
HybridViT	DFD+SDFVD_US	0.5410	0.7673	**0.8383**
ViT	CelebDFv2_US	0.5151	0.9722	**0.9726**
Xception	DFD+SDFVD_US	0.8706	0.7673	0.8383

Dataset-Level Ensembles (Same Architecture, Diverse Training Data)

Hybrid Ensembles (Architectures + Data) We evaluate three hybrid configurations: Hybrid-CNN (Xception+EfficientNetB0 across datasets, 8 models), Hybrid-Transformer (ViT+HybridViT across datasets, 8 models), and Hybrid-Full (all 16 models) (Table 7).

Table 7. Hybrid ensembles under cross evaluation. Hybrid-Full (soft) achieves the highest accuracy and AUC; Hybrid-CNN offers a strong accuracy–compute trade-off.

Hybrid Configuration	Acc (Majority)	Acc (Soft)	ROC–AUC (Soft)
Hybrid-CNN (8 models)	0.9341	0.9793	0.9995
Hybrid-Transformer (8 models)	0.7653	0.8801	0.9664
Hybrid-Full (16 models)	0.9397	**0.9928**	**0.9998**

(i) **Soft voting** consistently outperforms majority voting by leveraging calibrated confidences. (ii) **Hybrid-Full (soft)** improves over the best single cross result (Xception 0.9647) to 0.9928 (+2.81 pp). (iii) **Hybrid-CNN (soft)** at 0.9793 is close to Hybrid-Full with half the models, suitable when latency and memory are constrained (Fig. 4).

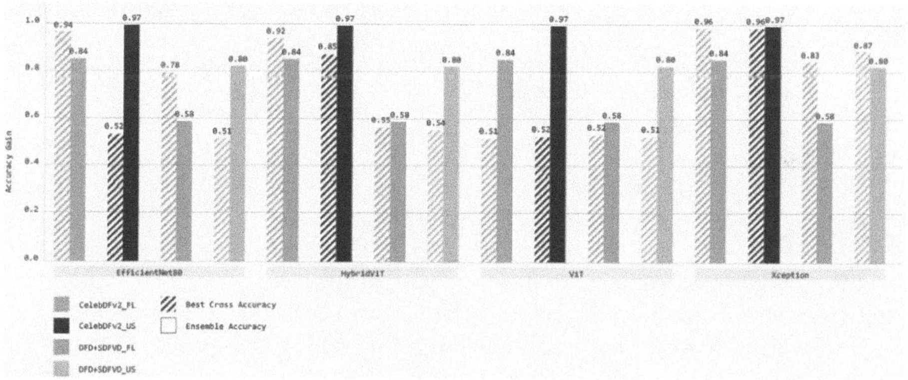

Fig. 4. Cross vs. ensemble accuracy across targets. Soft voting is consistently superior. (ROUNDED)

4.4 Live Prototype: External Validity Check

A webcam-based prototype integrating YOLOv8 face localization with ensemble classification was developed to verify the offline findings under real-world conditions. The 4-model soft-voting ensemble improved mean true accuracy from 0.097 to 0.582 and reduced false positives from 0.903 to 0.418, operating at an average frame rate of 33.5 FPS (single model: 34.3 FPS). A paired two-tailed t-test on trial accuracies confirmed that the ensemble's improvement was statistically significant ($t(9) = 8.962$, $p < .001$). These outcomes align with the cross-evaluation trends and support ensemble learning as a practical strategy for real-time deployment [27] (Fig. 5 and Tables 8, 9).

Table 8. Real-time performance of the single-model Xception prototype across ten live webcam trials.

Metric	Single Model (Xception)
Avg. True Accuracy	0.097 (9.7%)
Avg. False Positive Rate	0.903 (90.3%)
Avg. FPS	34.3
Avg. Inference Time (ms)	3.4
Avg. Confidence	0.679

Table 9. Real-time performance of the 4-model ensemble prototype across ten live webcam trials.

Metric	Ensemble (4 Models, Soft Voting)
Avg. True Accuracy	0.582 (58.2%)
Avg. False Positive Rate	0.418 (41.8%)
Avg. FPS	33.5
Avg. Inference Time (ms)	20.0
Avg. Confidence	0.600

5 Implications for Real-Time Deployment

The live prototype demonstrated that face localization (YOLOv8) plus soft-voted ensembles materially improves reliability on in-the-wild webcam streams. The trade-off between latency and accuracy favored the ensemble: the throughput reduction was modest relative to the gains in true accuracy and false-positive reduction. These results support a deployment pattern in which **localized inputs** and **probability-aware fusion** form the default inference path in high-stakes applications [21] (Fig. 6).

5.1 Synthesis of Results

- **Training effectiveness**: The large gap between trained and untrained models shows that all networks were properly optimized. Xception and HybridViT improved the most, suggesting that both architectures learned meaningful spatial and contextual features during training.
- **Domain shift severity**: When models trained on DFD+SDFVD were tested on CelebDFv2, accuracy dropped by 35–50% points. This confirms that domain shift remains a serious issue in deepfake detection, consistent with prior benchmark studies [4,6,13].
- **Generalization via ensembling**: Soft voting ensembles especially the Hybrid-Full and Hybrid-CNN setups, helped recover much of the accuracy lost under domain shift. Combining multiple models improved generalization by balancing CNNs' attention to texture artifacts with Transformers' ability to capture broader semantic patterns.

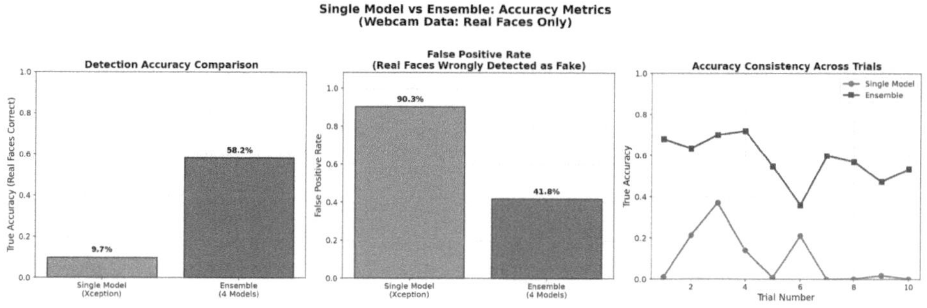

(a) Single vs. ensemble accuracy across live target. Soft voting is consistently superior. (ROUNDED)

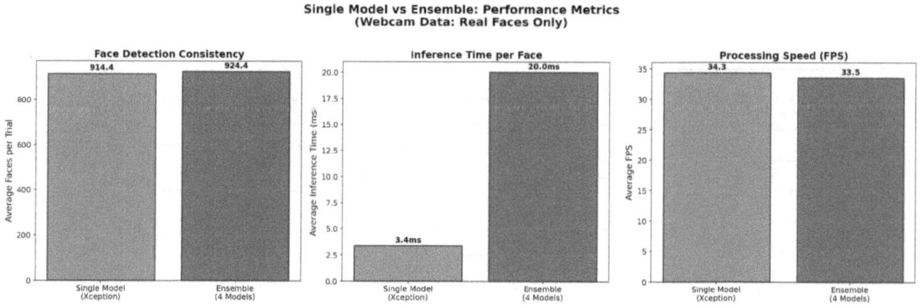

(b) Single vs. ensemble latency across live target. Soft voting is consistently superior. (ROUNDED)

Fig. 5. Comparison of single vs. ensemble performance across live targets. (a) Accuracy. (b) Latency.

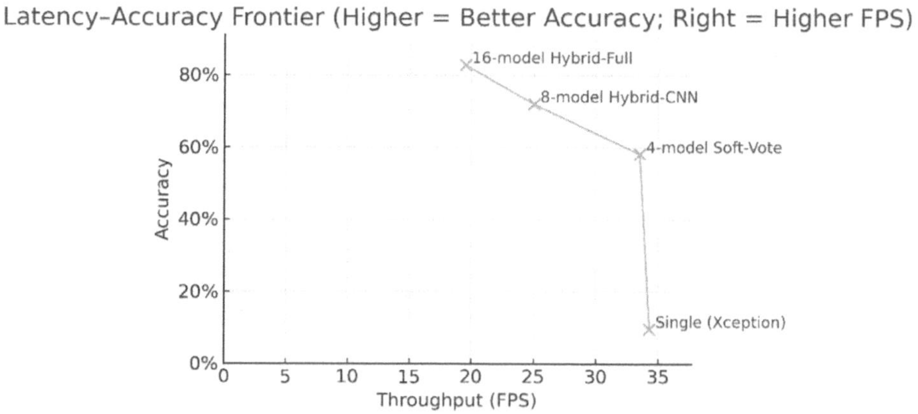

Fig. 6. Operational trade-offs for real-time deployment.

- **Operational guidance**: For near real-time use, Hybrid-CNN offers the best trade-off between speed and accuracy. Hybrid-Full should be used when the goal is maximum detection accuracy, even if it requires more computation.

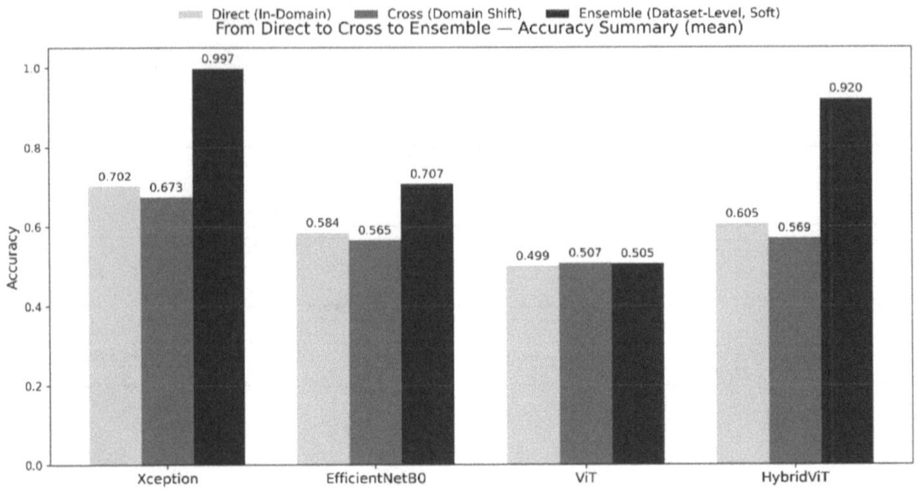

Fig. 7. High-level summary of performance across evaluation regimes.

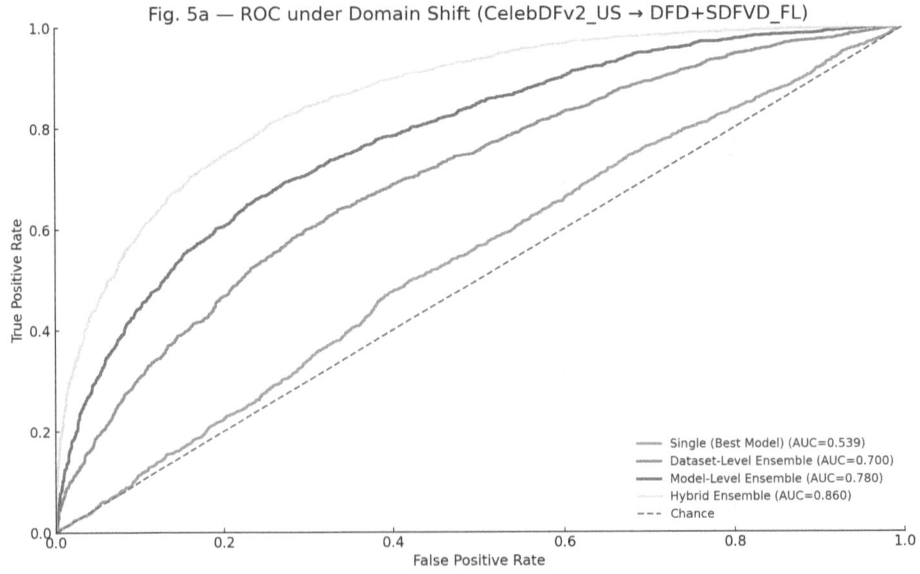

Fig. 8. Comparative robustness under domain shift.

6 Overview

The central objective of this work was to: (i) to quantify how single deepfake detectors degrade under domain shift and (ii) to test whether ensemble learning can restore generalization while remaining feasible for real-time use. The analysis combined direct (in-domain), cross-dataset (out-of-domain), and ensemble evaluations and was validated with a live prototype integrating YOLOv8 face localization and multiple classification backends [27] (Figs. 7 and 8).

7 Final Contributions and Summary of Findings

This study conducted a systematic evaluation of ensemble strategies for visual deepfake detection across four architectures (Xception, EfficientNetB0, ViT, HybridViT) and four dataset configurations (DFD+SDFVD_FL, DFD+SDFVD_US, CelebDFv2_FL, CelebDFv2_US). It further implemented a real-time prototype to assess practical viability.

1. **Comprehensive benchmarking.** Direct and cross dataset evaluations established that single models in particular *Vision Transformer (ViT)* exhibit marked performance declines under domain shift, while *Xception* typically maintained the strongest transfer among individual models [22].
2. **Quantification of domain shift.** Cross-evaluation showed consistent degradation for CNNs and Transformers when tested on unseen manipulations or compression regimes, with the steepest drops observed moving between DFD+SDFVD and CelebDFv2 [4,6,13].
3. **Dataset-level ensembling.** Aggregating identical architectures trained on different datasets improved accuracy in most settings; for *EfficientNetB0*, accuracy improved from ≈52.5% to ≈70.7% in challenging cross-domain tests.
4. **Model-level ensembling.** Combining heterogeneous architectures trained on the same dataset yielded competitive robustness; on *CelebDFv2_US*, a four-model soft-vote ensemble achieved ≈97.26%, surpassing the best single-model baseline.
5. **Hybrid ensembles (dataset + architecture).** A 16-model soft-vote ensemble reached ≈99.28% accuracy, the strongest result observed, underscoring the benefit of simultaneous dataset and architectural diversity [10].
6. **Prototype development and validation.** A live, YOLOv8-based face-cropper coupled with ensemble classification substantially reduced false positives and improved true accuracy relative to a single-model pipeline, while maintaining >30 FPS throughput [27].

- **Single-model fragility.** High in-domain scores did not translate to reliable cross-domain performance; this was most evident for *ViT*, but also affected CNNs in large domain gaps.
- **Ensemble gains.** Dataset-level and model-level ensembles yielded consistent improvements; **hybrid ensembles** delivered the highest and most stable accuracy across domains.

- **Real-time feasibility.** Despite increased per-frame latency, the ensemble prototype retained real-time throughput and reduced false positives by nearly half relative to the single-model baseline (Sect. 5).

8 Future Work

8.1 Dataset Development

Construct a **purpose-built, webcam-centric** dataset that varies demographics, lighting, camera models, compression pipelines, occlusions, and motion. Include controlled facial actions (speech, head turns, expressions) to capture temporal artifacts representative of live communication platforms [6,13].

8.2 Multimodal Expansion

Integrate synchronized audio streams for **audio–visual** detection. Features such as lip–speech asynchrony, voice clone artifacts, and prosody anomalies can complement visual cues, particularly for visually convincing forgeries.

8.3 Selective (Budgeted) Ensembling

Adopt a **confidence-gated** cascade: route confident samples to a calibrated anchor model; escalate only uncertain samples to additional members; fuse with soft voting. Investigate entropy thresholds, inter-model agreement checks, and budgeted top-k escalation. Target $\leq$1.0 F1 point loss versus full ensemble while reducing latency and energy (Fig. 9).

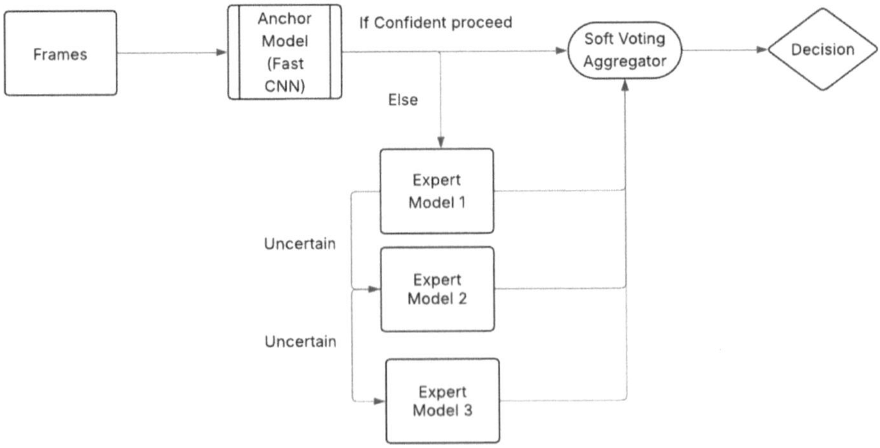

Fig. 9. Proposed selective ensembling pipeline.

8.4 Model Training Strategies

Train all members on **cropped facial regions** (aligned with deployment preprocessing) to reduce background bias and compute. Explore compression- and blur-aware augmentation schedules that mirror real streaming artifacts [4,27].

9 Conclusion

The presented evidence shows that **ensemble learning** is a reliable approach to counteract domain shift in visual deepfake detection. While single models may perform strongly within their training domains, their accuracy often collapses when exposed to unseen data. In contrast, **hybrid ensembles** that combine architectural and dataset diversity achieve higher accuracy, reduced false positives, and maintain real-time feasibility when integrated with face localization and probability-aware fusion.

Through systematic benchmarking and live validation, this work demonstrates that ensemble models enhance both accuracy and *deployability*. The findings offer a practical blueprint for robust, real-time deepfake detection and outline a clear direction for future advances in dataset design, multimodal analysis, selective inference, and model optimization.

References

1. Huang, Y., Juefei-Xu, F., Guo, Q., Liu, Y., Pu, G.: FakeLocator: robust localization of GAN-based face manipulations. IEEE (2022). https://doi.org/10.1109/TIFS.2022.3141262
2. Finance worker pays out $25 million after video call with deepfake 'chief financial officer'. CNN (2024). https://www.cnn.com/2024/02/04/asia/deepfake-cfo-scam-hong-kong-intl-hnk
3. Fake job seekers use AI to interview for remote jobs, tech CEOs say. CNBC (2025)
4. Rössler, A., Cozzolino, D., Verdoliva, L., Riess, C., Thies, J., Nießner, M.: FaceForensics: learning to detect manipulated facial images, pp. 1–11 (2019)
5. Guo, M., Yin, Q., Lu, W., Luo, X.: Towards Open-world Generalized Deepfake Detection: General Feature Extraction via Unsupervised Domain Adaptation. arXiv preprint arXiv:2505.12339 (2025)
6. Li, Y., Yang, X., Sun, P., Qi, H., Lyu, S.: Celeb-DF: a large-scale challenging dataset for deepfake forensics. In: CVPR, pp. 3207–3216 (2020)
7. Rana, M.S., Sung, A.H.: Deepfakestack: a deep ensemble-based learning technique for deepfake detection. In: 2020 IEEE CSCloud/EdgeCom, pp. 70–75 (2020)
8. Chen, H., Lin, Y., Li, B., Tan, S.: Learning features of intra-consistency and inter-diversity: keys toward generalizable deepfake detection. IEEE TCSVT **33**(3), 1468–1480 (2022)
9. Singh, J., Murala, S., Kosuru, G.S.R.: KL-DNAS: knowledge distillation-based latency aware-differentiable architecture search for video motion magnification. IEEE TNNLS (2024)
10. Uma Maheshwari, R., Paulchamy, B.: Securing online integrity: a hybrid approach to deepfake detection and removal using explainable AI and adversarial robustness training. Automatika **65**(4), 1517–1532 (2024)

11. Gu, Z., Chen, Y., Yao, T., Ding, S., Li, J., Huang, F., Ma, L.: Spatiotemporal inconsistency learning for deepfake video detection. In: ACM MM, pp. 3473–3481 (2021)
12. Yin, Q., Lu, W., Li, B., Huang, J.: Dynamic difference learning with spatio-temporal correlation for deepfake video detection. IEEE TIFS **18**, 4046–4058 (2023)
13. Dolhansky, B., et al.: The DeepFake Detection Challenge (DFDC) dataset. arXiv preprint arXiv:2006.07397 (2020)
14. Keerthika, S., Santhiya, S., Jayadharshini, P., Ruthranayaki, J., Hariarasu, R., Naveenan, M.: An innovative method for identifying deepfake videos through ensemble learning. In: 2024 IEEE ICCCNT, pp. 1–10 (2024)
15. Soudy, A.H., et al.: Deepfake detection using convolutional vision transformers and convolutional neural networks. Neural Comput. Appl. **36**(31), 19759–19775 (2024)
16. Giatsoglou, N., Papadopoulos, S., Kompatsiaris, I.: Investigation of ensemble methods for the detection of deepfake face manipulations. arXiv preprint arXiv:2304.07395 (2023)
17. Hashmi, A., Shahzad, S.A., Ahmad, W., Lin, C.W., Tsao, Y., Wang, H.: Multimodal forgery detection using ensemble learning. In: APSIPA ASC 2022, pp. 1524–1532 (2022)
18. Ha, H., Kim, M., Han, S., Lee, S.: Robust deepfake detection method based on ensemble of ViT and CNN. In: ACM/SAC 2023, pp. 1092–1095 (2023)
19. Manju, T., Kalarani, S.: DeepSight: enhancing deepfake image detection and classification through ensemble and deep learning techniques. In: 2024 ICIPCN, pp. 28–35 (2024)
20. Omar, K., Sakr, R.H., Alrahmawy, M.F.: An ensemble of CNNs with self-attention mechanism for DeepFake video detection. Neural Comput. Appl. **36**(6), 2749–2765 (2024)
21. Sebyakin, A., Soloviev, V., Zolotaryuk, A.: Spatio-temporal deepfake detection with deep neural networks. In: International Conference on Information, pp. 78–94 (2021)
22. Chollet, F.: Xception: deep learning with depthwise separable convolutions. In: CVPR, pp. 1251–1258 (2017)
23. Tan, M., Le, Q.: EfficientNet: rethinking model scaling for convolutional neural networks. In: ICML, pp. 6105–6114 (2019)
24. Dosovitskiy, A., et al.: An image is worth 16x16 words: transformers for image recognition at scale. arXiv preprint arXiv:2010.11929 (2020)
25. He, K., Zhang, X., Ren, S., Sun, J.: Deep residual learning for image recognition. In: CVPR, pp. 770–778 (2016)
26. Kingma, D.P., Ba, J.: Adam: a method for stochastic optimization. arXiv preprint arXiv:1412.6980 (2014)
27. Sohan, M., Sai Ram, T., Rami Reddy, C.V.: A review on YOLOv8 and its advancements. In: International Conference on Data Intelligence and Cognitive Informatics, pp. 529–545 (2024)

Large Language Models, Natural Language Processing, Security Methods, and Applications

Metaphor Detection and Translation for Representing Natural Language in Formal Logic

Jarrad Singley, Adam Pease, and Richard Thompson(✉)

Naval Postgraduate School, Monterey, CA, USA
{jarrad.singley,adam.pease,richard.thompson}@nps.edu

Abstract. Transforming natural language into formal logic statements enables machines to reason over a curated ontology. Metaphors pose a particular problem for automated translation and prevent reliable translation. This research presents a novel resolution framework that pre-processes natural language, removing and replacing non-literal words and phrases with literal, metaphor-free interpretations.

Our method adapts a transformer-based metaphor detector trained on a benchmark metaphor dataset, and then generates batches of candidate translations using a large language model (LLM). The best candidate is selected based on its similarity to the original sentence and the presence of residual metaphors detected. Performance was evaluated by re-detecting metaphors in the output and using a standard dataset of metaphorical-literal sentence pairs to measure similarity between the module's output and the dataset's literal sentence. Our method demonstrated marked improvement in the state of the art, significantly increasing the success rate at which metaphors were reduced in natural language.

Keywords: Metaphor detection techniques · metaphor translation · metaphor interpretation · language-to-logic translation · automated language formalization · SUO-KIF logic · conceptual metaphors · large language models (LLMs) · Hybrid Neuro-Ontological Language Understanding (HyNOLU) · sentence similarity

1 Introduction

Modern artificial intelligence (AI) systems, though powerful in natural language processing, face challenges of trustworthiness [1], logical reasoning [2], and interpretability [3]. While they generate fluent text, they are prone to errors, hallucinations, and contradictions, making them unreliable in domains requiring correctness and justification.

The Hybrid Neuro-Ontological Language Understanding (HyNOLU) project [4] addresses these challenges by developing a pipeline for converting natural language into the formal language Standard Upper Ontology–Knowledge

H. R. Arabnia et al. (Eds.): CSCI 2025, CCIS 2972, pp. 131–144, 2027.
https://doi.org/10.1007/978-3-032-28469-3_9

Interchange Format (SUO-KIF) [5] using logical symbols from the Suggested Upper Merged Ontology (SUMO)[1] [6,7]. Formal language enables reasoning with automated theorem provers such as Eprover [8] and Vampire [9]. By combining language model-based translation with formal logic and automated theorem proving, its goal is to produce representations that can be logically tested, allowing machines to reason, detect contradictions, and deliver verifiable conclusions to human analysts. The HyNOLU architecture employs a modular pipeline that integrates pre-processing, language to logic (L2L) translation, and automated theorem proving (ATP)-based validation to convert natural language into SUO-KIF. Within the pre-processing component, metaphor detection and resolution plays a crucial role in removing figurative language, ensuring that downstream logic translation operates on precise and literal meaning.

This paper focuses on this metaphor resolution stage of the HyNOLU architecture, which is critical for enabling downstream modules to operate on literal, logically tractable inputs. To support this task, a transformer-based metaphor detector [10] is adapted and integrated into the translation process. The detector provides token-level metaphor flags that enable finer-grained control over which words require literal substitution, allowing sentences without metaphors to bypass processing while those containing figurative content receive targeted resolution.

This work contributes a batch generation method that produces multiple candidate translations and selects among them using a joint criterion of token-level metaphor re-detection and sentence similarity. This selection strategy ensures that the chosen output minimizes metaphorical content while maintaining semantic fidelity. Empirical results further reveal a fundamental tradeoff in LLM-based metaphor translation: reducing figurative language often comes at the expense of input–output similarity. By introducing a controllable metaphor reduction rate parameter, our approach allows this balance to be tuned according to task requirements. Taken together, these innovations yield significant improvements over the state of the art.

2 Background

Research in metaphor handling divides into two complementary tasks: *detection*, which classifies text spans as metaphorical or literal, and *translation*, which transforms metaphorical expressions into literal equivalents that preserve meaning. This section reviews theoretical foundations, datasets, and computational approaches to both tasks.

2.1 Theoretical Foundations

Several linguistic theories inform computational approaches. The Metaphor Identification Procedure (MIP), developed by the Pragglejaz Group [11], provides a

[1] https://www.ontologyportal.org.

stepwise annotation framework for identifying linguistic metaphors. Steen et al. [12] extend it into Metaphor Identification Procedure VU University Amsterdam (MIPVU), achieving high inter-annotator agreement when building the Vrije University Amsterdam Metaphor Corpus (VUAMC). Conceptual metaphor theory (CMT), proposed by Lakoff and Johnson [13], frames metaphor as mapping abstract domains (e.g., *time*) to concrete ones (e.g., *money*). Johnson's image-schematic conceptual metaphor (ISCM) [14] highlights the role of recurring sensorimotor patterns in shaping abstract reasoning. Finally, Wilks' selectional preference violation (SPV) theory [15,16] detects metaphor via semantic mismatches between predicates and their arguments, a principle readily modeled using embeddings.

2.2 Datasets

Annotated corpora are central to training and evaluation. The VUAMC [12] remains the most widely used benchmark, covering four registers of the British National Corpus Baby Edition (BNC-Baby) [17] with 187k tokens and 13% metaphor density. MOH-X [18] provides verb-centered sentences labeled via crowdsourcing, while the TroFi dataset [19,20] clusters literal and figurative verb usages from newswire. Other contributions include Bizzoni and Lappin's paraphrase-ranked corpus [21] and Stowe et al.'s Idiomatic and Metaphoric Paired Language Inference (IMPLI) dataset [22], which pairs figurative sentences with literal counterparts, offering rare resources for metaphor-to-literal translation.

2.3 Metaphor Detection

Supervised Approaches. Deep learning has dominated detection research. Su et al.'s DeepNet [23] modeled detection as reading comprehension over token-level features. Choi et al.'s MelBERT [24] combined contextual and isolated embeddings to operationalize MIP and SPV, achieving state-of-the-art results on VUAMC and MOH-X. Zhang and Liu's MisNet [25] introduced dictionary-grounded embeddings to capture "basic meanings" and demonstrates improvements in domain-specific performance. Their follow-on work with adversarial multitask learning (AdMul) [26] leveraged basic sense discrimination (BSD) data to enhance generalization. Other efforts, such as Wachowiak et al. [10], targeted ISCMs, detecting schema-specific metaphors with fine-tuned transformer models.

Large Language Models. Recent work applies LLMs directly, mitigating data scarcity. Tian et al. [27] guided an LLM with knowledge graphs derived from metaphor theories, showing improved detection accuracy. Lin et al. [28] proposed a dual-perspective framework combining implicit retrieval-based guidance with explicit theory-driven prompting, leveraging both MIP and SPV representations to improve LLM performance. These approaches demonstrate that explicit integration of linguistic theory is crucial for effective zero-shot metaphor reasoning.

2.4 Metaphor Translation

Compared to detection, translation remains underexplored. Shutova [29] pioneered automatic paraphrasing of verb metaphors using distributional similarity and WordNet filtering, achieving strong human-judged accuracy. Mao et al. [30] advanced an unsupervised CBOW-based framework that simultaneously identified and paraphrased metaphors, demonstrating improved downstream machine translation performance. More recently, datasets like IMPLI [22] have enabled evaluation of literalization quality, though systematic models remain scarce.

2.5 Summary

Overall, research has established robust theories and datasets for metaphor detection, with transformer-based models achieving strong performance. However, translation lags behind, with only a handful of systems addressing the systematic rephrasing of metaphor into literal language. This gap motivates the present work, which investigates scalable metaphor resolution methods as a critical step toward trustworthy natural language to logic translation.

3 Methodology

This section describes the proposed metaphor resolution approach: *Batch Word-level Re-detection and Similarity Monitoring* (BWL). The method integrates automatic metaphor detection with large language model (LLM) prompting, iterative candidate generation, and prioritized evaluation. The pipeline operates at the token level to improve precision in metaphor elimination while balancing semantic fidelity.

3.1 Metaphor Detection

Metaphor detection was performed using Wachowiak et al.'s XLM-RoBERTa model [10], chosen for its robustness and reproducibility. The detector provided both sentence-level and token-level binary flags, allowing fine-grained identification of metaphorical words. Token-level detection was necessary to avoid over-reliance on strict sentence-level classification, which can fail when a candidate translation partially reduces metaphorical content.

3.2 Batch Candidate Generation

The BWL approach generated multiple translation candidates per prompt using an LLM (Llama 3.1, 8B parameters). Each prompt included both the input sentence and a list of metaphorically flagged words, which guided the model in selectively rephrasing figurative expressions. By explicitly providing this context, the LLM was relieved from independently identifying metaphors, improving translation consistency.

Figure 1 illustrates the overall pipeline. For a sentence S with M_S flagged words, a reduction rate parameter (RR) determined the maximum allowable metaphors (M_c) in any accepted candidate:

$$M_c = \lfloor M_S (1 - RR) \rfloor \tag{1}$$

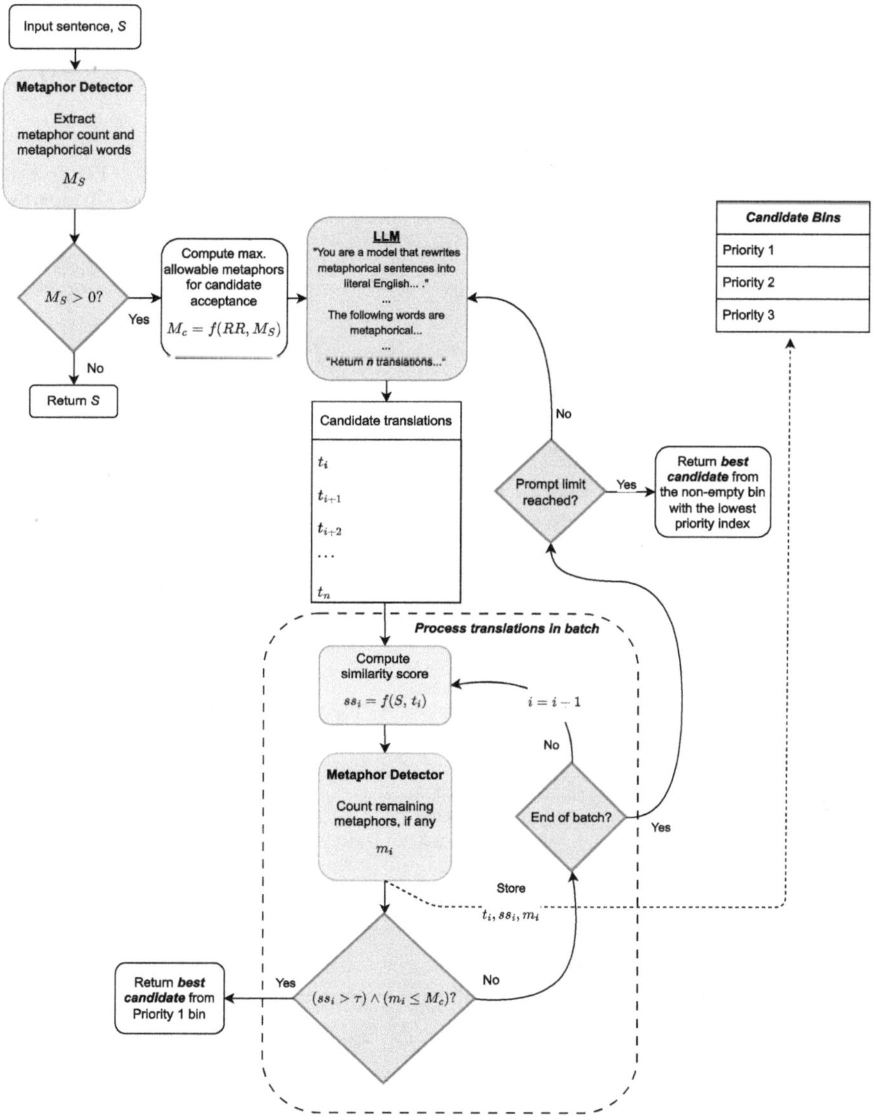

Fig. 1. batch word-level re-detection and similarity monitoring (BWL) architecture. Upon detecting M_S metaphorical words in the input sentence S, M_c was calculated as a function of M_S and RR. Candidate batches were generated and filtered until similarity and re-detection criteria were satisfied.

The RR parameter, which ranges between zero and one, specifies the target proportion of metaphorical words to eliminate relative to the reference sentence. It effectively serves as a tunable threshold that balances the tradeoff between semantic similarity and metaphor reduction. At $RR = 1$, only candidates with zero residual metaphors are considered acceptable; however, if no such candidates exist, the architecture defaults to ranking purely on similarity. In practice, setting RR slightly below one (e.g., $RR = 0.9$) introduces flexibility, allowing translations with a small number of residual metaphors to remain in the highest priority bin (discussed in Sect. 3.3) when their similarity scores are strong, while still favoring those with fewer metaphors overall.

Batches of n candidates were generated per query, with model temperature optionally increased between iterations to promote lexical diversity. Each candidate was then re-analyzed by the detector to count remaining flagged words m_i and scored for semantic similarity (ss_i) using metrics such as Sentence-BERT (SBERT) cosine similarity [31], Bilingual Evaluation Understudy (BLEU) [32], and ROUGE-L [33].

3.3 Candidate Prioritization

Each candidate t_i was placed into one of three bins based on the remaining metaphor count m_i and similarity score ss_i, as illustrated in Fig. 2. Here, M_c is maximum allowable metaphorical words as defined in Eq. 1 and τ denotes the user-defined similarity threshold. For example, if ROUGE is used as the similarity metric, a candidate must achieve a ROUGE score greater than τ, and a metaphor count less than M_c to be considered acceptable.

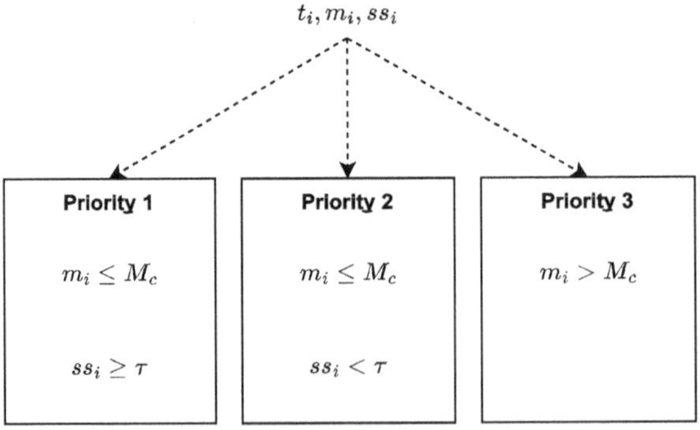

Fig. 2. Candidate translations were sorted into priority bins based on similarity and metaphor count. Selection favored minimum metaphors first, then maximum similarity.

During selection, bins were searched in priority order, with the highest-scoring candidate from the first non-empty bin returned as output. This mechanism explicitly balanced metaphor elimination against semantic fidelity, controlled by RR and τ.

3.4 Evaluation Setup

Experiments used the IMPLI dataset [22], reduced to 280 unique metaphor–literal pairs with punctuation corrected for JSON compatibility. Candidates were evaluated against reference literals using automated similarity metrics, while metaphor reduction was measured via re-detection. Equation 1 enabled flexible tuning: $RR = 0$ prioritized similarity only, while $RR = 1$ enforced complete elimination of flagged metaphors. Intermediate values balanced both criteria.

3.5 Summary

The BWL methodology combines token-level detection, batch candidate generation, flagged word injection, and prioritized evaluation. Figures 1 and 2 highlight its iterative filtering and binning design. By tuning RR and τ, the system adapts to varying requirements for metaphor elimination and semantic preservation, providing a scalable approach to metaphor resolution for natural language to logic translation.

4 Results

This section reports results for the batch word-level re-detection and similarity monitoring (BWL) translation solution, using a zero-shot approach as a baseline for comparison. All experiments used the IMPLI dataset. Prior to testing, BWL parameters were optimized via a grid search. The zero-shot baseline is a single one-pass translation (no feedback, no iterated refinement).

4.1 Scoring and Composite Metric

We evaluate literalization quality by averaging the following metrics: BLEU, Recall-Oriented Understudy for Gisting Evaluation (ROUGE), and SBERT cosine similarity, and metaphor removal with percent corrected and mean reduction ratio. By combining similarity measures with residual metaphor statistics, the composite metric unifies the two central evaluation criteria: faithfulness to the source text and effectiveness of metaphor removal, into a single score. For these similarity-based metrics, the gold-standard literal sentence served as the reference and the model's translation as the candidate. All reported values are averaged over the entire IMPLI dataset. The overall composite score C (used for model selection and headline comparison) was defined and weighted as follows:

$$C = w_1 \overline{\mathrm{BERT}} + w_2 \overline{\mathrm{BLEU}} + w_3 \overline{\mathrm{ROUGE}} + w_4 P_{\mathrm{corr}} + w_5 \overline{RR}$$

$$w_1 = 0.15, \quad w_2 = 0.15, \quad w_3 = 0.1, \quad w_4 = 0.5, \quad w_5 = 0.1$$

where P_{corr} is the percent of translations with complete metaphor elimination, and $\overline{RR}$ is the average reduction ratio. The reduction ratio RR is

$$RR = \frac{M_{\text{orig}} - M_{\text{trans}}}{M_{\text{orig}}} \quad (2)$$

where M_{orig} and M_{trans} are the numbers of flagged metaphorical words in the original and translated sentences, respectively.

4.2 Parameter Optimization

For BWL, the best grid point used `sim_function=bleu`, `des_sim_score=0.1`, `prompt_limit=4`, `batch_size=10`, and `red_rate=0.7`, with composite 0.660. A finer sweep of `red_rate` improved the composite to 0.679 at `red_rate=0.82` (from 0.660 → 0.679).

Table 1. Overall performance summary of optimized translation approaches (zero-shot vs. BWL)

Metric	Zero-shot	BWL
Avg. BLEU	0.278	0.204
Avg. ROUGE	0.481	0.390
Avg. SBERT Cosine	0.906	0.879
Percent Corrected	0.271	0.775
Avg. Reduction Ratio	0.315	0.903
Composite	**0.393**	**0.679**

4.3 Aggregate Metrics: Zero-Shot vs. BWL

Table 1 summarizes the post-optimization results used in the final comparison. While zero-shot attains higher similarity scores (BLEU/ROUGE/SBERT), BWL achieves markedly superior metaphor removal (percent corrected +50.4 points; reduction ratio +0.588), yielding a much higher composite (+0.286 absolute). The inclusion of originally flagged words in the BWL prompts, larger candidate pools (via batches), and the reduction-rate criterion together drive these gains.

4.4 Qualitative Illustration

Below we reproduce the required BWL translation example (verbatim fragment emphasized):

Original:
Four of the absentees suffered the squirming discomfort of being among the Welsh squad.
Detected metaphorical words: {squirming, discomfort, squad}

Gold Reference:
Four of the absentees suffered the continuous discomfort of being among the Welsh squad.
Detected metaphorical words: {discomfort, squad}

BWL Translation:
The four absentees didn't like being among the group of Welsh players.
Remaining metaphorical words: {}
BLEU score (with respect to gold reference): **0.142**

The BWL output removes metaphorical constructions while preserving propositional content, illustrating the broader trade-off we observed: stronger literalization often requires departing from the exact surface form of the gold reference. Importantly, the detector still flagged metaphors in the gold reference, even though the IMPLI annotations indicated only a single metaphor ("squirming") in this example. This mismatch could stem from dataset limitations or detector false positives, which in turn forced the translation system to remove words that were not truly metaphorical.

4.5 Metaphor Elimination Outcome Categories

To analyze elimination behavior at the sentence level, we partition translations into outcome categories (summarized here): *false negatives* (detected as non-metaphorical initially), *pure elimination* (all original metaphors removed, none added), *pure reduction* (strict subset of original metaphors retained), *impure reduction* (fewer metaphors overall but at least one new metaphor introduced), *full replacement* (same count as original but all metaphors replaced by different ones), *partial replacement* (same count with some originals retained), *inflation* (more metaphors than the original), and *no change*. Selected examples of original–translation sentence pairs representative of each category are presented in Table 2.

Table 3 presents the category statistics for zero-shot and BWL only (counts and percentages taken verbatim from the full analysis).

Observations. BWL converts a large majority of cases to *pure elimination* (77.5%, vs.27.1% for zero-shot) and sharply reduces undesirable outcomes that add or swap metaphors (*impure reduction, full replacement, inflation*). Together with the much higher reduction ratio, this indicates that BWL reliably removes metaphorical content without introducing new figurative language.

Table 2. Examples of metaphor translation outcomes by category, with flagged metaphorical words underlined. Note that the example sentence in the false negative category contains at least one obvious metaphor, yet none were flagged.

Category	Original/Translated Sentence	Flagged Met. Words
False Negative	**Original:** The plane climbs reluctantly, one set of wings dipping drunkenly. **Translation:** The plane climbs reluctantly, one set of wings dipping drunkenly.	Original: [] Translation: []
Pure Elimination	**Original:** There are few things worse than being bludgeoned into reading a book you hate. **Translation:** There are few bad experiences than being forced to read a book you don't like.	Original: [things, bludgeoned, into] Translation: []
Pure Reduction	**Original:** The ostentatious way of living of the rich ignites the hatred of the poor. **Translation:** The rich people's luxurious way of living is not liked by the poor.	Original: [way, ignites] Translation: [way]
Impure Reduction	**Original:** The steering of my new car answers to the slightest touch. **Translation:** The steering of my new car responds when you touch it lightly.	Original: [answers, to] Translation: [responds]
Partial Replacement	**Original:** It dawned on him that she had betrayed him. **Translation:** It landed on him that she had betrayed him.	Original: [dawned, on] Translation: [landed, on]
Full Replacement	**Original:** The new measures deflated the economy. **Translation:** The new rules hurt the economy.	Original: [measures, deflated] Translation: [rules, hurt]
Inflation	**Original:** Sales were climbing after prices were lowered. **Translation:** Sales went up after prices were lowered.	Original: [climbing] Translation: [went, up, lowered]
No Change	**Original:** The bad review of his work deflated his self-confidence. **Translation:** The negative review of his employment deflated his self-confidence.	Original: [deflated] Translation: [deflated]

Table 3. Metaphor elimination outcomes across translation solutions (counts and percentages)

Category	Zero-shot		BWL	
	Count	%	Count	%
False negatives	38	13.6%	**38**	**13.6%**
Pure elimination	76	27.1%	**217**	**77.5%**
Pure reduction	11	3.9%	**2**	**0.7%**
Impure reduction	57	20.4%	**10**	**3.6%**
Full replacement	62	22.1%	**7**	**2.5%**
Partial replacement	0	0.0%	**0**	**0.0%**
Inflation	36	12.9%	**6**	**2.1%**
No change	0	0.0%	**0**	**0.0%**

4.6 Summary of Improvements over Zero-Shot

Quantitatively, BWL improves percent corrected from 0.271 → 0.775 and average reduction ratio from 0.315 → 0.903, raising the composite from 0.393 → 0.679. Qualitatively, BWL produces cleaner literalizations (e.g., "*the* ***four absentees didn't like*** *being among the group of Welsh players*") with far fewer cases of metaphor introduction or substitution. While similarity scores are modestly lower than zero-shot—reflecting fewer surface-level matches to gold references—the overall objective of metaphor elimination is achieved substantially more often with BWL.

5 Conclusions

This work developed and evaluated a metaphor resolution framework for use in autoformalization, emphasizing the translation of metaphorical English sentences into literal paraphrases. The proposed BWL approach leverages an LLM to generate batches of diverse candidate translations, which are ranked according to similarity and re-detection criteria. By selecting the best candidate, the method achieved a substantial improvement over a simple zero-shot baseline, with 50% higher rates of complete metaphor removal and 59% higher rates of metaphor reduction. The optimized BWL configuration attained a composite score of 0.679, compared to 0.393 for zero-shot, and consistently outperformed in metaphor elimination categories while preserving adequate semantic similarity.

The findings underscore the importance of a dedicated metaphor detector within the translation pipeline. Transformer-based detection models proved essential not only for initial identification but also for validating candidate translations. Analysis further reveals that while LLMs are strong paraphrasers, they frequently fail to recognize or avoid metaphors, often introducing new figurative expressions in zero-shot or single-response prompting. Batch-based methods, in contrast, encourage more diverse and creative paraphrases that more effectively eliminate metaphorical content.

Several limitations emerged. The tradeoff between semantic similarity and metaphor elimination remains significant: translations that achieve near-total elimination tend to deviate more from the original sentence structure. Additionally, the evaluation relied on automatic metrics and assumed an oracle-like detector, whereas real-world detectors may introduce false positives or negatives that propagate through the pipeline. Finally, the reliance on a single open-source LLM constrains generalizability across models and domains.

Future work should explore advanced prompting strategies, such as chain-of-thought prompting, to combine improved similarity with robust elimination. Incorporating LLM orchestration frameworks would also enable more flexible and maintainable implementations of complex prompting strategies. For HyNOLU applications, integrating word sense disambiguation (WSD) could bypass unnecessary translation of metaphors already mapped to SUMO, streamlining processing. Broader experimentation with alternative LLMs and further refinement of metaphor detectors, particularly models with high F-1 performance, represent critical directions to improve translation quality and reliability.

In summary, this research demonstrates that effective metaphor resolution can be achieved by combining a specialized detection model with batch-based LLM translation. The BWL approach provides a practical and scalable solution, advancing the capability of the HyNOLU pipeline to produce literal, logic-ready language suitable for automated reasoning tasks.

References

1. Zhang, Y., et al.: Siren's Song in the AI Ocean: A Survey on Hallucination in Large Language Models. arXiv (2025). https://doi.org/10.48550/arXiv.2309.01219
2. Shojaee, P., et al.: The Illusion of Thinking: Understanding the Strengths and Limitations of Reasoning Models via the Lens of Problem Complexity. arXiv (2025). https://doi.org/10.48550/arXiv.2506.06941
3. Samek, W., Wiegand, T., Müller, K.-R.: Explainable Artificial Intelligence: Understanding, Visualizing and Interpreting Deep Learning Models. arXiv (2017). https://doi.org/10.48550/arXiv.1708.08296. Eprint: 1708.08296
4. Thompson, R., et al.: Formalizing natural language: cultivating LLM translations using automated theorem proving. In: Theorem Proving and Machine Learning in the Age of LLMs: State of the Art and Future Perspectives. EuroProofNet, COST Action CA20111 – European Research Network on Formal Proofs. Edinburgh, Scotland, UK (2025). https://europroofnet.github.io/wg5-edinburgh25/
5. Pease, A.: SUO-KIF Reference Manual (2009). https://github.com/ontologyportal/sigmakee/blob/master/suo-kif.pdf. Accessed 20 June 2025
6. Niles, I., Pease, A.: Toward a standard upper ontology. In: Welty, C., Smith, B. (eds.) Proceedings of the 2nd International Conference on Formal Ontology in Information Systems (FOIS-2001), pp. 2–9 (2001)
7. Pease, A.: Ontology: A Practical Guide. Articulate Software Press, Angwin (2011)
8. Schulz, S.: E-a brainiac theorem prover. AI Commun. **15**(2–3), 111–126 (2002)
9. Kovács, L., Voronkov, A.: First-order theorem proving and Vampire. In: Sharygina, N., Veith, H. (eds.) CAV 2013. LNCS, vol. 8044, pp. 1–35. Springer, Heidelberg (2013). https://doi.org/10.1007/978-3-642-39799-8_1
10. Wachowiak, L., Gromann, D., Xu, C.: Drum up SUPPORT: systematic analysis of image-schematic conceptual metaphors. In: Proceedings of the 3rd Workshop on Figurative Language Processing (FLP) (2022). https://doi.org/10.18653/v1/2022.flp-1.7
11. Pragglejaz Group. MIP: a method for identifying metaphorically used words in discourse. Metaphor Symbol **22**(1), 1–39 (2007). https://doi.org/10.1080/10926480709336752
12. Steen, G.J., et al.: A Method for Linguistic Metaphor Identification: From MIP to MIPVU. John Benjamins Publishing Company, Amsterdam (2010)
13. Lakoff, G., Johnson, M.: Metaphors We Live By. University of Chicago Press, Chicago (1980)
14. Johnson, M.: The Body in the Mind: The Bodily Basis of Meaning, Reason, and Imagination. University of Chicago Press, Chicago (1987)
15. Wilks, Y.: A preferential, pattern-seeking, semantics for natural language inference. In: Ahmad, K., Brewster, C., Stevenson, M. (eds.) Words and Intelligence I: Selected Papers by Yorick Wilks, pp. 83–102. Springer, Dordrecht (2007). ISBN: 978-1-4020-5285-9. https://doi.org/10.1007/1-4020-5285-5_5

16. Wilks, Y.: Making preferences more active. In: Ahmad, K., Brewster, C., Stevenson, M. (eds.) Words and Intelligence I: Selected Papers by Yorick Wilks, pp. 141–166. Springer, Dordrecht (2007). https://doi.org/10.1007/1-4020-5285-5_7
17. British National Corpus Consortium. British National Corpus 1994, Baby edition. Literary and Linguistic Data Service (2007). http://hdl.handle.net/20.500.14106/2553
18. Mohammad, S., Shutova, E., Turney, P.: Metaphor as a medium for emotion: an empirical study. In: Proceedings of the Fifth Joint Conference on Lexical and Computational Semantics (2016). https://doi.org/10.18653/v1/S16-2003. https://aclanthology.org/S16-2003/
19. Birke, J., Sarkar, A.: A clustering approach for nearly unsupervised recognition of nonliteral language. In: 11th Conference of the European Chapter of the Association for Computational Linguistics. Available: ACL Anthology (2006)
20. Birke, J., Sarkar, A.: Active learning for the identification of nonliteral language. In: Proceedings of the Workshop on Computational Approaches to Figurative Language. Available: ACL Anthology (2007)
21. Bizzoni, Y., Lappin, S.: Predicting human metaphor paraphrase judgments with deep neural networks. In: Proceedings of the Workshop on Figurative Language Processing (2018). https://doi.org/10.18653/v1/W18-0906
22. Stowe, K., Utama, P., Gurevych, I.: IMPLI: Investigating NLI models' performance on figurative language. In: Proceedings of the 60th Annual Meeting of the Association for Computational Linguistics (2022). https://doi.org/10.18653/v1/2022.acl-long.369
23. Su, C., et al.: DeepMet: a reading comprehension paradigm for token-level metaphor detection. In: Proceedings of the Second Workshop on Figurative Language Processing (2020). https://doi.org/10.18653/v1/2020.figlang-1.4. Accessed 01 Oct 2025
24. Choi, M., et al.: MelBERT: Metaphor Detection via Contextualized Late Interaction Using Metaphorical Identification Theories. arXiv (2021). https://doi.org/10.48550/arXiv.2104.13615
25. Zhang, S., Liu, Y.: Metaphor detection via linguistics enhanced siamese network. In: Proceedings of the 29th International Conference on Computational Linguistics. Available: ACL Anthology (2022)
26. Zhang, S., Liu, Y.: Adversarial Multi-Task Learning for End-to-End Metaphor Detection. arXiv (2023). https://doi.org/10.48550/arXiv.2305.16638
27. Tian, Y., Xu, N., Mao, W.: A theory guided scaffolding instruction framework for LLM-enabled metaphor reasoning. In: Proceedings of the 2024 Conference of the North American Chapter of the Association for Computational Linguistics: Human Language Technologies (2024). https://doi.org/10.18653/v1/2024.naacl-long.428
28. Lin, Y., et al.: A Dual-Perspective Metaphor Detection Framework Using Large Language Models. arXiv (2024). https://doi.org/10.48550/arXiv.2412.17332
29. Shutova, E.: Automatic metaphor interpretation as a paraphrasing task. In: Human Language Technologies: The 2010 Annual Conference of the North American Chapter of the Association for Computational Linguistics. Available: ACL Anthology (2010)
30. Mao, R., Lin, C., Guerin, F.: Word embedding and WordNet based metaphor identification and interpretation. In: Proceedings of the 56th Annual Meeting of the Association for Computational Linguistics (2018). https://doi.org/10.18653/v1/P18-1113
31. Reimers, N., Gurevych, I.: Sentence-BERT: Sentence Embeddings Using Siamese BERT-Networks. arXiv (2019). https://doi.org/10.48550/arXiv.1908.10084

32. Papineni, K., et al.: BLEU: a method for automatic evaluation of machine translation. In: Proceedings of the 40th Annual Meeting of the Association for Computational Linguistics (2002). https://doi.org/10.3115/1073083.1073135
33. Lin, C.-Y.: ROUGE: a package for automatic evaluation of summaries. In: Text Summarization Branches Out. Available: ACL Anthology (2004)

Kryptonite for Small LLMs: When Simple Words Break Reasoning

Salvatore Vella(✉), Salah Sharieh, and Alex Ferworn

Toronto Metropolitan University, Toronto, Canada
sal.vella@torontomu.ca

Abstract. The sensitivity of Large Language Model (LLMs) prompts to changes is a critical concern as these systems move into production. This paper evaluates twelve small LLMs (Gemma, Llama, Qwen, Phi, Granite, GPT) on two multiple choice question and answer datasets, OpenBookQA and CommonsenseQA under the base case and five additional logically reframed prefixes- "Not untrue", "Not false", "Cannot be denied", "Consistent with facts", and "Entailed by Premises". We apply paired t-tests and McNemar's tests to assess whether the performance changes are statistically significant. We find that the performance of some smaller models degrades by as much as 70% when a simple "Which of the following is Not Untrue?" is added as a prefix to the multiple-choice question. We find that all models tested are affected, with most of the impacts statistically significant. For some models, the prefixes "Not Untrue" and "Untrue" produce results that are not statistically different, indicating that some models use heuristics. This study reinforces the need for protection against even simple malicious prompt injections and underscores the importance of testing before model production deployment.

Keywords: Large Language Models · Prompt Engineering · Robustness · OpenBookQA · Commonsense QA · Evaluation

1 Introduction

LLMs are becoming more integrated into knowledge and office work. They can augment workers and help with tasks such as summarization and executing company policies for help desk and other tasks [5]. This paper looks at the robustness of LLMs with logical reframing prefixes. These are phrases, such as "Not Untrue," that are easy for humans to understand as a double negative, which don't change the meaning, but are difficult for some LLMs to understand.

In this paper, we study multiple-choice question datasets. These are akin to policies and procedures where an LLM would have to choose between several options to select the correct one.

This paper investigates prompt robustness across twelve small LLMs. We measure the performance on OpenBookQA [11], which is a STEM-oriented multiple-choice question, and Commonsense QA (CS) [16], which includes everyday common reasoning questions.

H. R. Arabnia et al. (Eds.): CSCI 2025, CCIS 2972, pp. 145–160, 2027.
https://doi.org/10.1007/978-3-032-28469-3_10

We apply six logic-modified prompt variants, which are used as prefixes to the actual question. These are: (a) the base question with no prefix, (b) Not Untrue, (c) Not False, (d) Cannot be Denied, (e) Consistent with Facts, and (f) Entailed by Premises. We quantify their effects on accuracy and measure the statistical significance using t-tests and McNemar's Test.

Prior work studied single negation or adversarial phrasing, whereas we isolate logically equivalent reframings that preserve meaning, revealing failures in basic reasoning. A simple two-word prefix, "Not Untrue," can reduce accuracy by 70% in some models and is statistically equivalent to a prefix of "Untrue."

We find that all the tested models exhibit statistically significant performance drops with the modified prompts, and some of the models' performance collapses catastrophically to levels well below what would be expected with random guessing.

We position our contributions by explicitly addressing gaps in prior work:

1. Prompt robustness studies have been performed [6,21], but these have focused on simple perturbations like adding punctuations and misleading phrases. Our study, in contrast, focuses on the logical reframing of answer options, such as using double negatives, 'Not Untrue' (which logically means the same as 'true'), to measure whether models apply simple reasoning.
2. Negation has been studied [6], but not with double negation, which does not change the meaning of the question. This study is novel in its examination of double negation in the context of multiple-choice questions.
3. Logically equivalent prompt variations are a critical new insight. Phrases like "Not Untrue" or "Cannot be denied" do not change a question's meaning, so any performance drop exposes a pure reasoning failure on these straightforward reasoning tasks.
4. Most robustness evaluations focus on a small set of frontier models. We evaluate 12 diverse small models (Falcon, Gemma, Llama, Qwen, Granite, Phi, GPT families), enabling a broad cross-family robustness comparison under identical conditions.
5. We apply a rigorous set of statistical tests and not just the raw answers. We use paired t-tests and McNemar's test on per-item outcomes, ensuring our conclusions are statistically validated rather than anecdotal.

Evaluating and mitigating for security and privacy is important especially for regulated industries and frameworks such as COMPARCH [15] can be used as part of a holistic approach to dealing with these issues.

The paper is organized as follows: Sect. 2 reviews related work, Sect. 3 presents the methodology used, Sect. 4 presents the results, Sect. 5 provides some ethical considerations, and Sect. 6 concludes the work by highlighting the most important discoveries and suggesting future initiatives.

2 Related Work

2.1 Prompt Robustness

Prompt robustness refers to the ability of large language models to provide consistent and accurate results despite small changes in the prompt format, phrasing, and content. [6]. This is important as in many real-world implementations, such as help desks, it is not possible to control the prompt precisely.

Even minor changes in prompt structure can significantly impact the performance of LLMs. Even format changes can cause degradation of performance [6,19]. Vella et al. show that even providing the correct answer in a different position can cause adverse results.

This sensitivity creates problems in cases where LLMs are deployed for business or safety-critical situations. The performance degradations result in a worse performance than just guessing for some models [6]. Antagnostidis et al. [3] show that adding authoritative information from other models can influence the performance of models. This finding has implications for real-world applications where multiple applications may interact.

Models have been found to have an impact on the effects of perturbations in multiple-choice questions. [19,22], which is important because many problems in business follow procedures with multiple choices, whether that is a helpdesk procedure or a medical triage.

2.2 Negation

Negation in prompts presents a significant problem for LLMs to interpret meaning and simple reasoning [8,20]. Testing has been done with the negation of a proposition. In our study, we use a double negative (E.g., Not Untrue) where the meaning should not change.

There is a phenomenon called the No syndrome, where LLMs struggle to distinguish between a fact and its negation. [20]

In our study, we examine double negation prompts, Not Untrue and Not False, as well as an emphatic prompt, Which Cannot Be Denied, which incorporates an element of negation but is not strictly a negation prompt.

Negation can cause performance degradation, and Jang et al. show that LLM size can mitigate the effect. [10].

Mitigation is challenging, as methods such as fine-tuning can address specific issues, but are unable to resolve problems in the general case [8].

3 Methodology

This section provides the Methodology used for this study. We will describe the datasets used, the prompt variations used and the statistical methods used to measure the performance of each prompting method.

3.1 Datasets

We use several multiple-choice question and answer datasets. These provide us with the ability to add prefixes to the question that do not alter the meaning of the question, such as a double negative, while maintaining the same answer set.

OpenBookQA [11] is a dataset focused on Science, Technology, Engineering and Math (STEM) questions for Kindergarten to Grade 8 students, which combine facts with reasoning to get to a correct answer.

Commonsense QA [16] is a dataset focused on commonsense knowledge.

Figures 1 and 2 show the syntax of a sample question for OpenBookQA (4 choices) and CommonsenseQA (5 choices).

```
A person wants to start saving money so that they can
 afford a nice vacation at the end of the year.
 After looking over their budget and expenses,
 they decide the best way to save money is to:

 A. make more phone calls
 B. quit eating lunch out
 C. buy less with monopoly money
 D. have lunch with friends
```

Fig. 1. OpenBookQA Sample Question.

In Fig. 1, as an example of an OpenBookQA question, the question is provided in a statement, and the LLM is provided with four choices, each prefixed by a letter. The LLM is asked to choose the best answer from the 4.

```
Sammy wanted to go to where the people were.
Where might he go?

A. race track
B. populated areas
C. the desert
D. apartment
E. roadblock
```

Fig. 2. Commonsense QA Sample Question.

Similarly, in Fig. 2, with the CommonsenseQA question, though there are five choices in CommonsenseQA.

We selected 2000 questions from each respective training dataset to apply to each model and its variations. This approach provided enough items for paired t-tests and McNemar's tests. Both datasets are widely used benchmarks, and our focus is on relative robustness under prefix changes rather than the leaderboard performance of each dataset. These subsets provide rich datasets for performing analytics.

Each question was posed with labelled options, and the model had to select the best answer from the set. A prefix phrase was added to the text before the question.

3.2 Models Tested

We test twelve unique models in this study, sourced from a variety of vendors, and combine both open-source and commercial vendors in the evaluation set. All of these are small LLMs, with most having fewer than 4 billion parameters. The size of the mini- and nano-GPT variations is not published.

Families include Falcon [2], Gemma [9], Llama 3.2 [17,18], Qwen2.5 [4], Granite [12], Phi-3-mini [1], and GPT-4.x [13,14] (nano/mini, GPT-4o-mini). Parameter scales range roughly from ~1B to ~13B (Table 1).

Table 1. Small LLMs evaluated in this study. Exact parameter counts for some proprietary models are undisclosed. Estimates are included where available.

Model	Family	Creator Company/Organization	Parameters (approx.)
GPT-4.1-mini	GPT-4.x	OpenAI	Not disclosed (est. 7–13B)
GPT-4o-mini	GPT-4.x	OpenAI	Not disclosed (est. ~7B)
GPT-4.1-nano	GPT-4.x	OpenAI	Not disclosed (est. 3–7B)
Phi-3-mini-4k	Phi-3	Microsoft Research	3.8B
Granite-3.3-2B	Granite	IBM Research	2B
Qwen2.5-3B	Qwen2.5	Alibaba Group (DAMO Academy)	3B
Llama-3.2-1B	Llama 3.2	Meta AI	1B
Llama-3.2-3B	Llama 3.2	Meta AI	3B
Gemma-3-1B	Gemma	Google DeepMind	1B
Gemma-3-2B	Gemma	Google DeepMind	2B
Falcon-3B-instruct	Falcon	Technology Innovation Institute (TII), UAE	3B

3.3 Prompt Conditions

We use six different prompt conditions. None of these prompts changes the meaning of the question. These are prefixes added before the question is asked. The Base Answer option has no prefix (a null prefix). The Not Untrue Answer option has a prefix of "Which of the following is Not Untrue?" added before the question is asked. Each of the prefixes tests the reasoning ability of these models. Simple reasoning would lead one to understand that Not Untrue is a double negative, for example, and can be ignored because it doesn't change the meaning. It is this simple reasoning that we are testing.

Six answer-prediction conditions per question:

1. Base Answer. No prefix.
2. Not Untrue Answer. Prefix: "Which is not untrue"
3. Not False Answer. Prefix: "Which is not false"
4. Cannot Be Denied Answer. Prefix: "Which Cannot be Denied"
5. Consistent With Facts Answer: Prefix: "Which is consistent with all known facts"
6. Entailed By Premises Answer: Prefix: "Which is logically entailed by the premises"

3.4 Metrics and Tests

We provide several metrics and statistical analyses of the study's results.

- Accuracy per model × dataset × condition. These are the raw results of asking the questions
- Paired t-tests (Base vs. each variant) [7], which gives us a statistical measure of whether the change in accuracy is statistically significant.
- McNemar's tests (pairwise correctness shifts) [7], which gives us the statistical measure of whether the prompt change has resulted in a different distribution of answers compared to the base answer.

4 Results

We present the results of our study here. We start with the accuracy results and then present the statistical analysis. The paired t-test assesses whether the mean accuracy difference is statistically significant, and McNemar's test evaluates whether the distribution of correct versus incorrect answers has changed significantly for each model.

4.1 Top Models by Dataset Drops from Base Accuracy

In this section, we present the partial results for the top five models by performance. The tables below display the top 5 models in terms of accuracy for each dataset, as well as the impact of each prompt variation.

For example, GPT-4.1-mini is the best-performing model in terms of accuracy using the OpenBookQA dataset. Its performance decreases by 3.25% to 86.15% with the 'Not Untrue' prefix and by 2.95% with the 'Not False' prefix. As we will see later, both of these are statistically significant.

Also in this table is an example of a model with a catastrophic failure. Phi-3-mini-4k's performance goes down to 8.4% with Not Untrue and 9.75% with Not False prefixes.

From this first set of data, we see that all the models are impacted, and simply examining the accuracy of the Base answer does not provide an indication of the model's behaviour under the prompt modifications (Tables 2 and 3).

Table 2. OpenBookQA Top 5 Models by Base Accuracy

Model	Base	Not Untrue	Not False
GPT-4.1-mini	89.40	86.15	86.45
GPT-4o-mini	87.25	61.95	75.10
GPT-4.1-nano	80.90	71.60	74.60
Phi-3-mini-4k	80.25	8.40	9.75
Granite-3.3-2B	69.45	10.70	11.60

Table 3. CommonsenseQA Top 5 Models by Base Accuracy

Model	Base	Not Untrue	Not False
GPT-4o-mini	78.30	58.35	66.70
GPT-4.1-nano	72.90	61.90	64.90
Phi-3-mini-4k	72.45	4.20	3.75
Qwen2.5-3B	71.70	24.15	34.50
Llama-3.2-3B	66.05	22.50	31.90

4.2 Statistical Significance Summary

All negation prefixes (Not Untrue and Not False) result in statistically significant performance degradation across both datasets and all models tested. The Cannot be Denied prefix also causes statistically significant degradations in all models for both data sets. Most of the Consistent with Facts and all but one of the Entailed by Premises prefixes result in statistically significant impacts. The qwen2.5-3B model does not.

4.3 Key Findings

1. The most disruptive perturbations are Not Untrue and Not False, which often cause accuracy drops of 40–70% in the models tested.

2. The Consistent With Facts and Entailed By Premises have smaller impacts on accuracy, but the impacts are statistically significant in most cases.
3. Even the smaller mini and nano models of the OpenAI GPT family show statistically significant degradations from additions of these prefixes.

4.4 Complete Accuracy Tables (All Models × Conditions)

In this section, we present the heat maps illustrating the accuracy of each model across the various datasets. Note that OpenBookQA has four possible answers per question (guessing would lead to 25% correct) and CommonsenseQA has five possible answers per question.

Figure 3 shows the accuracy heatmap for OpenBookQA, and Fig. 4 shows the accuracy heatmap for CommonsenseQA, for each model and each prompt case. We can easily see the degradations in all models for the Not Untrue and Not False Cases. We also observe the catastrophic collapse of models such as Gemma-3-1 b and Phi-mini-4k-instruct to accuracy levels well below chance.

Model	Base Answer	Not Untrue Answer	Not False Answer	Cannot Be Denied Answer	Consistent With Facts Answer	Entailed By Premises Answer
falcon3-3b-instruct	61.50	37.80	42.90	47.05	58.85	57.25
gemma-3-1b	42.10	21.80	21.50	39.00	41.55	40.00
gemma-3-4b	66.25	23.60	26.15	55.75	64.50	61.75
gpt-4.1-mini	89.40	86.15	86.45	85.25	88.30	87.90
gpt-4.1-nano	80.90	71.60	74.60	75.10	78.35	73.85
gpt-4o-mini	87.25	61.95	75.10	79.00	84.65	81.35
granite-3.3-2b-instruct	69.45	10.55	11.60	13.70	65.80	64.30
llama-3.2-1b-instruct	45.80	29.60	29.00	38.20	43.55	40.50
llama-3.2-3b-instruct	67.10	27.90	36.85	50.85	61.45	59.65
phi-3-mini-4k-instruct	80.25	8.35	9.75	74.55	79.55	76.75
qwen2.5-1.5b-instruct-mlx	58.75	14.35	14.45	37.35	56.10	52.65
qwen2.5-3b-instruct	65.30	21.65	27.80	48.60	65.75	64.90

Fig. 3. Accuracy heatmap for each model on the OpenBookQA dataset across six prompt conditions. Note catastrophic collapses of some models (e.g., granite-3.3-2b-instruct and phi-3-mini-4k) below chance levels under the *Not Untrue* and *Not False* conditions, highlighting sensitivity to simple double negations.

4.5 t-Test Statistical Significance Summary

In this section, we present the t-test statistical significance summary for both datasets and all models under all conditions. We provide these as a heatmap with red indicating statistical significance of the impact.

We observe statistically significant impacts across most of the datasets, models, and prefix types. Not UnTrue, Not False and Cannot Be Denied have a

Model	Base Answer	Not Untrue Answer	Not False Answer	Cannot Be Denied Answer	Consistent With Facts Answer	Entailed By Premises Answer
falcon3-3b-instruct	66.70	41.70	42.75	48.75	61.75	63.55
gemma-3-1b	43.25	12.00	11.65	37.40	41.80	39.95
gemma-3-4b	64.45	20.30	22.55	52.90	61.30	59.75
gpt-4.1-nano	72.90	61.90	64.90	68.35	72.00	68.90
gpt-4o-mini	78.30	58.35	66.70	76.35	77.15	76.70
granite-3.3-2b-instruct	64.65	6.20	7.40	14.15	63.05	61.70
llama-3.2-1b-instruct	51.85	29.45	31.70	45.25	47.45	44.20
llama-3.2-3b-instruct	66.05	22.50	31.90	52.05	61.10	61.50
phi-3-mini-4k-instruct	72.45	3.75	3.75	62.10	72.40	70.05
qwen2.5-1.5b-instruct-mlx	65.15	8.20	8.20	43.45	60.55	56.10
qwen2.5-3b-instruct	71.70	24.15	34.50	62.90	71.90	71.45

Fig. 4. Accuracy heatmap for each model on the CommonsenseQA dataset across six prompt conditions. Large degradations are again observed under the *Not Untrue* and *Not False* prefixes, with several models (e.g., phi-3-mini-4k, llama-3.2-3B, granite-3.3-2b-instruct) collapsing to near-random accuracy.

statistically significant effect for all cases. All models, from the very small to the smaller versions of OpenAI's frontier models, show statistically significant impact.

Figures 5 and 6 show the statistical significance of t-tests comparing the Base with each of the prompt variations.

We note that, given the large number of comparisons in this table, a Bonferroni correction would require $p < 0.001$ for statistical significance. Most of our *pvalues* are at 0.000 to three decimal places and meet this criterion.

4.6 McNemar Test Results

McNemar's test evaluates the distributions of answers across two different answer sets and determines whether there is a statistical significance in the distribution of correct answers. In this case, the question is whether the prompt variable creates a different distribution from the base answers.

Figures 7 and 8 show the statistical significance of McNemar's test comparing the base answer distribution to each prompt variation for OpenBookQA.

We note that there is a statistically significant result for most cells, with many p-values rounded to 0 (to 3 decimal places), indicating strong statistical significance. The Not Untrue, Not False and Cannot Be Denied produce statistically significant new distributions for both datasets and all models.

4.7 Not Untrue Versus Just Untrue

For the cases where accuracy collapses, granite-3.3-2b-instruct, phi-3-mini-4k-instruct and qwen2.5-1.5b-instruct-mlx, we perform an additional analysis to see how the results of Not Untrue as a prefix compare with simply saying Untrue.

Model	Not Untrue Answer	Not False Answer	Cannot Be Denied Answer	Consistent With Facts Answer	Entailed By Premises Answer
falcon3-3b-instruct	0.000	0.000	0.000	0.003	0.000
gemma-3-1b	0.000	0.000	0.006	0.525	0.030
gemma-3-4b	0.000	0.000	0.000	0.031	0.000
gpt-4.1-mini	0.000	0.000	0.000	0.065	0.022
gpt-4.1-nano	0.000	0.000	0.000	0.000	0.000
gpt-4o-mini	0.000	0.000	0.000	0.000	0.000
granite-3.3-2b-instruct	0.000	0.000	0.000	0.000	0.000
llama-3.2-1b-instruct	0.000	0.000	0.000	0.053	0.000
llama-3.2-3b-instruct	0.000	0.000	0.000	0.000	0.000
phi-3-mini-4k-instruct	0.000	0.000	0.000	0.223	0.000
qwen2.5-1.5b-instruct-mlx	0.000	0.000	0.000	0.035	0.000
qwen2.5-3b-instruct	0.000	0.000	0.000	0.604	0.649

Fig. 5. Paired *t*-test results comparing base prompts to each variant on the OpenBookQA dataset. Cells in red indicate statistically significant performance drops. All models, including GPT mini/nano variants, show significant sensitivity to negation-based prefixes. (Color figure online)

This is:

- 'Which of the following is Not Untrue'
- 'Which of the following is Untrue'

Figures 9 and 10 provide the summary of the accuracy of the Base, Not Untrue and Untrue runs. The cells coloured green are those where there is not a statistically significant (p is not <0.05) difference between Not Untrue and Untrue as the prefix. This suggests that some models employ heuristics and appear to overlook the "Not" in the phrase.

In the case of OpenBookQA, though the p-values are small, all of the cells are green, meaning there is no statistical difference for granite-3.3-2b-instruct, phi-3-mini-4k-instruct and qwen2.5-1.5b-instruct-mlx in both the overall accuracy (t-test p-value) and in the McNemar's Test distribution of answers (McNemar p-value) between Not Untrue and Untrue as the question prefix. With CommonsensQA, phi-3-mini-4k-instruct, exhibits a statistically significant p-value that indicates that the Not Untrue and Untrue accuracy and answer distributions are not statistically significantly different.

Model	Not Untrue Answer	Not False Answer	Cannot Be Denied Answer	Consistent With Facts Answer	Entailed By Premises Answer
falcon3-3b-instruct	0.000	0.000	0.000	0.000	0.001
gemma-3-1b	0.000	0.000	0.000	0.050	0.000
gemma-3-4b	0.000	0.000	0.000	0.000	0.000
gpt-4.1-nano	0.000	0.000	0.000	0.214	0.000
gpt-4o-mini	0.000	0.000	0.005	0.063	0.024
granite-3.3-2b-instruct	0.000	0.000	0.000	0.033	0.002
llama-3.2-1b-instruct	0.000	0.000	0.000	0.000	0.000
llama-3.2-3b-instruct	0.000	0.000	0.000	0.000	0.000
phi-3-mini-4k-instruct	0.000	0.000	0.000	0.941	0.005
qwen2.5-1.5b-instruct-mlx	0.000	0.000	0.000	0.000	0.000
qwen2.5-3b-instruct	0.000	0.000	0.000	0.800	0.759

Fig. 6. Paired t-test results comparing base prompts to each variant on the CommonsenseQA dataset. Significance is widespread across all families, with *Not Untrue*, *Not False*, and *Cannot Be Denied* causing systematic and significant degradations.

This is a surprising result, providing evidence that at least some of these small models employ heuristics that can break down under very simple conditions.

5 Discussion

This study examines the robustness of prompts for negation and other prefixes. We have done this using multiple-choice questions, which allow us to test the reasoning ability of small LLMs under these conditions. We have tested 12 different small LLMs, including some of the smaller variants of the OpenAI GP models.

Overall, we find that all 12 models tested showed statistically significant sensitivity to these reframing prompts. For some of the models, the performance dropped below random guessing of the answer. This represents a significant decline in performance. The Not Untrue and Not False were the most impactful.

We also demonstrate that for certain models, there is no statistically significant difference between prefixing with 'Not Untrue' and simply prefixing with 'Untrue'. Some models ignore the 'Not' word.

All models had a statistically significant impact, as indicated by Not Untrue, Not False, and Cannot Be Denied. This included the smaller versions of OpenAI frontier models. Previous work had found that larger models handled negation better than smaller models. Our work here indicates that even smaller models, such as OpenAI's mini and nano variants, are also impacted by negation.

The fact that there is consistency in the statistically significant impacts indicates that this is a fundamental issue across all models tested, rather than a model-specific issue.

Model	Not Untrue Answer	Not False Answer	Cannot Be Denied Answer	Consistent With Facts Answer	Entailed By Premises Answer
falcon3-3b-instruct	0.000	0.000	0.000	0.004	0.000
gemma-3-1b	0.000	0.000	0.006	0.563	0.034
gemma-3-4b	0.000	0.000	0.000	0.036	0.000
gpt-4.1-mini	0.000	0.000	0.000	0.078	0.027
gpt-4.1-nano	0.000	0.000	0.000	0.000	0.000
gpt-4o-mini	0.000	0.000	0.000	0.000	0.000
granite-3.3-2b-instruct	0.000	0.000	0.000	0.000	0.000
llama-3.2-1b-instruct	0.000	0.000	0.000	0.058	0.000
llama-3.2-3b-instruct	0.000	0.000	0.000	0.000	0.000
phi-3-mini-4k-instruct	0.000	0.000	0.000	0.258	0.000
qwen2.5-1.5b-instruct-mlx	0.000	0.000	0.000	0.039	0.000
qwen2.5-3b-instruct	0.000	0.000	0.000	0.645	0.690

Fig. 7. McNemar's test results for OpenBookQA, measuring distributional shifts in answer correctness between base prompts and each variant.

Model	Not Untrue Answer	Not False Answer	Cannot Be Denied Answer	Consistent With Facts Answer	Entailed By Premises Answer
falcon3-3b-instruct	0.000	0.000	0.000	0.000	0.001
gemma-3-1b	0.000	0.000	0.000	0.058	0.000
gemma-3-4b	0.000	0.000	0.000	0.000	0.000
gpt-4.1-nano	0.000	0.000	0.000	0.241	0.000
gpt-4o-mini	0.000	0.000	0.006	0.075	0.028
granite-3.3-2b-instruct	0.000	0.000	0.000	0.039	0.002
llama-3.2-1b-instruct	0.000	0.000	0.000	0.000	0.000
llama-3.2-3b-instruct	0.000	0.000	0.000	0.000	0.000
phi-3-mini-4k-instruct	0.000	0.000	0.000	1.000	0.005
qwen2.5-1.5b-instruct-mlx	0.000	0.000	0.000	0.000	0.000
qwen2.5-3b-instruct	0.000	0.000	0.000	0.849	0.806

Fig. 8. McNemar's test results for CommonsenseQA. Almost all prefixes yield statistically significant distributional changes, reinforcing that prompt modifications systematically affect model decision patterns.

It is also important to note that accuracy is not a predictor of how a model will be impacted by negation. The Phi-3-mini-4k model had about 80

This study demonstrates that even simple logical reframing added to a question can cause catastrophic degradations in LLM performance. The fact that these are very simple logical reframings suggests that these LLMs are poor reasoners for such straightforward cases.

Model Name	Base Accuracy	Not Untrue Accuracy	Untrue Accuracy	T-test p	McNemar p
granite-3.3-2b-instruct	68.15	10.55	9.6	0.08	0.10
phi-3-mini-4k-instruct	80.15	8.35	7.6	0.16	0.19
qwen2.5-1.5b-instruct-mlx	58.45	14.35	16.05	0.07	0.08

Fig. 9. Comparison of *Not Untrue* vs. *Untrue* prefixes on OpenBookQA. Green cells indicate no statistically significant difference between the two. For this dataset, the behaviours are statistically significantly different. (Color figure online)

Model Name	Base Accuracy	Not Untrue Accuracy	Untrue Accuracy	T-test p	McNemar p
granite-3.3-2b-instruct	64.9	6.2	4.2	0.00	0.00
phi-3-mini-4k-instruct	72.45	3.75	4.45	0.11	0.14
qwen2.5-1.5b-instruct-mlx	62.35	8.2	10.2	0.02	0.02

Fig. 10. Comparison of *Not Untrue* vs. *Untrue* prefixes on CommonsenseQA. Results confirm heuristic behaviour in certain models, where accuracy patterns are statistically indistinguishable despite the logical difference in phrasing. (Color figure online)

This study extends previous work in negation in natural language processing, extending it to multiple-choice QA and small LLMs.

The fact that such simple prompt additions can cause a significant impact has implications for security risks, as malicious actors can inject a small phrase and cause dramatic degradation. Business tasks, such as following a policy or procedure, could be compromised with such an attack.

Future work should evaluate defences such as training with logically equivalent framings and prompt pre-processing to remove double negatives. Benchmarks should be integrated into the development and deployment pipeline.

This work highlights the importance of thoroughly testing LLM-based applications before deployment and the necessity of controlling the prompts used in these applications.

6 Ethical Considerations

Several ethical concerns arise from this work.

The most important thing is the misuse of these vulnerabilities. Simple prompt changes can catastrophically degrade LLM and system performance, and this fact can be exploited to compromise systems in critical applications, such as chatbots, business processes, or medical triage. If left unaddressed, these vulnerabilities can lead to harm or denial of service.

This study aims to help developers of LLMs enhance the robustness of their products. Only public interfaces were used in this study within the terms of use of each model tested.

The datasets used in this study, OpenBookQA and CommonsenseQA, are both publicly available and commonly used for academic research.

Bias was not a consideration for this study, but we acknowledge that bias could be a possible area for future research.

Testing and ensuring that models provide robust results in production environments is an ethical necessity for organizations deploying these systems in production. Ignoring these issues could lead to real harm when deployed.

7 Conclusions

The core finding is that small LLMs are highly vulnerable to simple logical reframings, especially "Not Untrue" and "Not False", with several of the twelve models tested experiencing catastrophic failures below random guessing accuracy.

The results were validated with robust statistical methods, including t-tests and McNemar's Test, confirming that the impacts are statistically significant across models and datasets.

All twelve models, spanning six different model families (Falcon, Meta Llama, Alibaba Qwen, IBM Granite, Microsoft Phi, and OpenAI GPT), were impacted, indicating that the issue is fundamental to these small models and not a vendor-specific issue.

This work highlights that even a very small two-word adversarial prompt injection could compromise applications such as policy execution and helpdesk automation that rely on choosing one of several next steps in their processing.

Organizations must ensure the robust testing of applications before deployment and implement security measures to prevent adversarial prompt attacks. We recommend that organizations perform adversarial tests with logically equivalent variants and pre-process any prompts to clean them for these logical reframings.

This work highlights the need for mitigation techniques such as controlling the prompt to safeguard reliability in model and application performance.

References

1. Abdin, M., Jacobs, S.A., Awan, A.A.: Phi-3 technical report: a highly capable language model locally on your phone. arXiv preprint arXiv:2404.14219 (2024)
2. Almazrouei, E., Alobeidli, H., Alshamsi, A., et al.: The falcon series of open language models. arXiv preprint arXiv:2311.16867 (2023)
3. Anagnostidis, S., Bulian, J.: How Susceptible are LLMs to Influence in Prompts? (2024). https://doi.org/10.48550/arXiv.2408.11865. http://arxiv.org/abs/2408.11865. arXiv:2408.11865
4. Bai, J., Bai, S., Chu, Y., et al.: Qwen technical report. arXiv preprint arXiv:2309.16609 (2023)
5. Brynjolfsson, E., Li, D., Raymond, L.: Generative AI at work. SSRN Electron. J. (2023). https://api.semanticscholar.org/CorpusID:258298324

6. Cao, B., Cai, D., Zhang, Z., Zou, Y., Lam, W.: On the worst prompt performance of large language models. In: Proceedings of the 38th International Conference on Neural Information Processing Systems. NIPS 2024. Curran Associates Inc., Red Hook, NY, USA (2025)
7. Dror, R., Baumer, G., Shlomov, S., Reichart, R.: The hitchhiker's guide to testing statistical significance in natural language processing. In: Proceedings of the 56th Annual Meeting of the Association for Computational Linguistics (Volume 1: Long Papers), pp. 1383–1392. Association for Computational Linguistics, Melbourne, Australia (2018). https://doi.org/10.18653/v1/P18-1128. http://aclweb.org/anthology/P18-1128
8. García-Ferrero, I., Altuna, B., Álvez, J., Gonzalez-Dios, I., Rigau, G.: This is not a Dataset: A Large Negation Benchmark to Challenge Large Language Models (2023). https://doi.org/10.48550/arXiv.2310.15941. http://arxiv.org/abs/2310.15941. arXiv:2310.15941
9. Gemma Team: Gemma: Open models based on gemini research and technology. arXiv preprint arXiv:2403.08295 (2024)
10. Jang, J., Ye, S., Seo, M.: Can Large Language Models Truly Understand Prompts? A Case Study with Negated Prompts (2022). https://doi.org/10.48550/arXiv.2209.12711. http://arxiv.org/abs/2209.12711. arXiv:2209.12711
11. Mihaylov, T., Clark, P., Khot, T., Sabharwal, A.: Can a suit of armor conduct electricity? A new dataset for open book question answering. In: Proceedings of the 2018 Conference on Empirical Methods in Natural Language Processing, pp. 2381–2391. Association for Computational Linguistics, Brussels, Belgium (2018). https://aclanthology.org/D18-1260/
12. Mistry, M., Sanghavi, H., Kumar, G., et al.: Granite code models: a family of open foundation models for code intelligence. arXiv preprint arXiv:2405.04324 (2024)
13. OpenAI: GPT-4 technical report. arXiv preprint arXiv:2303.08774 (2023)
14. OpenAI: Hello gpt-4o (2024). https://openai.com/index/hello-gpt-4o/. Accessed 09 Sept 2025
15. Saidane, A., Al-Sharieh, S.: A compliance-driven framework for privacy and security in highly regulated socio-technical environments: an e-government case study. In: Research Anthology on Privatizing and Securing Data, pp. 933–962. IGI Global (2021)
16. Talmor, A., Herzig, J., Lourie, N., Berant, J.: Commonsenseqa: a question answering challenge targeting commonsense knowledge. In: Proceedings of the 2019 Conference of the North American Chapter of the Association for Computational Linguistics: Human Language Technologies, Volume 1 (Long and Short Papers), pp. 4149–4158. Association for Computational Linguistics, Minneapolis, Minnesota (2019). https://aclanthology.org/N19-1421/
17. Touvron, H., Lavril, T., Izacard, G., et al.: Llama: open and efficient foundation language models. arXiv preprint arXiv:2302.13971 (2023)
18. Touvron, H., Martin, L., Stone, K., et al.: Llama 2: open foundation and fine-tuned chat models. arXiv preprint arXiv:2307.09288 (2023)
19. Vella, S., Sharieh, S., Hussain, F., Ferworn, A.: Where you say matters: a study of positional bias of small LLMs. In: 2025 IEEE World AI IoT Congress (AIIoT), pp. 0640–0647. IEEE (2025)
20. Vrabcová, T., Kadlčík, M., Sojka, P., Štefánik, M., Spiegel, M.: Negation: A Pink Elephant in the Large Language Models' Room? (2025). https://doi.org/10.48550/arXiv.2503.22395. http://arxiv.org/abs/2503.22395. arXiv:2503.22395

21. Zhu, K., Zhao, Q., Chen, H., Wang, J., Xie, X.: PromptBench: A Unified Library for Evaluation of Large Language Models (2024). https://doi.org/10.48550/arXiv.2312.07910. http://arxiv.org/abs/2312.07910. arXiv:2312.07910
22. Zong, Y., Yu, T., Chavhan, R., Zhao, B., Hospedales, T.: Fool Your (Vision and) Language Model With Embarrassingly Simple Permutations (2024). https://doi.org/10.48550/arXiv.2310.01651. http://arxiv.org/abs/2310.01651. arXiv:2310.01651

Leveraging Large Language Models for Cybersecurity Compliance: An Exploratory Study in ISO 27001 Compliance Workflow

Ahmed Salman[1(✉)], Michael Goldsmith[1], and Zeeshan Zulkifl Shah[2]

[1] Department of Computer Science, University of Oxford, Oxford, UK
ahmed.salman@cs.ox.ac.uk, michael.goldsmith@worc.ox.ac.uk
[2] Faculty of Science and Engineering, Macquarie University, Sydney, Australia
zeeshanzulkifl.shah@hdr.mq.edu.au

Abstract. Cybersecurity compliance is traditionally a manual, time-consuming, and subjective activity. Large Language Models (LLMs) offer strong potential for its automation, however their utility across the entire compliance lifecycle has not been explored. This paper presents a comprehensive exploratory study evaluating three prominent LLMs (Sonnet 3.5, ChatGPT 4o, Llama 3.1) across a simulated ISO 27001 compliance workflow, encompassing planning, fieldwork, analysis, and reporting and remediation stages. Through 76 experiments within a hypothetical telecommunications context, we assessed LLM performance via ordinal scoring on key factors (Relevance, Completeness, Accuracy, Hallucination) and qualitative usefulness factors (Scalability, Change Absorption, Skill Coverage). Findings show significant potential in terms of time savings and skill coverage, particularly for capable models like Sonnet 3.5. However, some challenges hinder further autonomous deployment which include output completeness issues, platform dependencies, poor scalability and reliability. We conclude that LLMs can be powerful assistants but they require careful selection, data preparation and human oversight to produce usable outputs. This necessitates further research to improve their performance for any future compliance automation.

Keywords: LLMs · Cybersecurity Compliance · ISO 27001 · Gen AI

1 Introduction

Traditional cybersecurity compliance processes mainly rely on manual reviews and subjective assessments resulting in inefficiencies, inaccuracies, and slow adaptation to evolving threats. Compliance automation using various techniques, has been explored in research, however, effectively analyzing vast unstructured compliance data remains a challenge [1].

LLMs due to their ability to process unstructured data at scale, offer a promising approach to enhance compliance workflows. To test this, we evaluate

H. R. Arabnia et al. (Eds.): CSCI 2025, CCIS 2972, pp. 161–177, 2027.
https://doi.org/10.1007/978-3-032-28469-3_11

three prominent LLMs (Sonnet 3.5, ChatGPT 4o, Llama 3.1) across a simulated end-to-end ISO 27001 compliance workflow for a hypothetical telecommunications operator ("HelloPak") covering Planning, Fieldwork, Analysis, Reporting and Remediation stages.

Our main contribution is a detailed empirical assessment of current LLM capabilities when applied to the full compliance workflow in a simulated organisational setting. This work presents a comprehensive assessment that expands upon a prior pilot study focused only on the initial planning phase [2]. We provide observations on specific steps where LLMs excel or falter and analyse recurring challenges in enhancing compliance efficiency. These findings aim to inform practitioners considering LLM adoption for cybersecurity compliance and provide a base for future research in this domain.

2 Related Work

Compliance processes, like the ISO 27001 audits [3], are critical for effective cybersecurity management. However, traditional audit methodologies face some challenges: manual and checklist-based assessments can be time-consuming, error-prone, subjective, and slow to adapt to threats and regulations [1]. Despite available support tools [4], efficiently handling unstructured compliance data remains challenging.

Artificial intelligence (AI), particularly LLMs, offer a strong potential for compliance automation by processing unstructured text (policies, regulations, logs) at scale [1]. Prior research explored AI for isolated tasks like mapping cybersecurity framework requirements [5], automating GDPR reviews [6,7], and categorising policy documents [8]. There have also been some proposals by the industry for automating other GRC tasks using AI [9]. However, adopting LLMs can introduce risks like inaccuracies (hallucinations), privacy concerns and security vulnerabilities [10], necessitating AI governance guidelines and frameworks [11,12].

Despite these initial AI applications for compliance, a gap remains in understanding the utility and limitations of LLMs when used across an entire compliance workflow. Most studies focus on discrete tasks [5–8], lacking empirical evaluation across the interconnected stages (Planning, Fieldwork, Analysis, Reporting) in an organisational context. This study addresses this gap by providing a detailed exploratory analysis of three prominent LLMs applied to the full ISO 27001 compliance lifecycle, generating insights for practitioners and researchers aiming for trustworthy compliance automation.

3 Methodology

This exploratory study evaluates LLM capabilities across a typical cybersecurity compliance workflow within a simulated Telecommunications context. We defined a hypothetical Mobile Network Operator, "HelloPak," which is in its pilot operational phase and seeks assessment against ISO 27001 and local regulations (network topology in Fig. 1). The typical ISO 27001 compliance workflow, based on industry practice [13], encompasses four key stages (Fig. 2). Specific activities tested within these stages included:

- **Planning and Scoping:** Generating a Statement of Applicability (SoA), audit objectives, scope and plan, and audit checklists.
- **Fieldwork:** Performing policy reviews against regulations.
- **Analysis:** Generating audit working papers and performing audit analysis and summarisation.
- **Reporting, Remediation & Follow-up:** Generating the final report, creating a remediation plan, and simulating a follow-up interview.

We assessed three prominent LLMs: Sonnet 3.5 [14], Llama 3.1 [15] (both accessed via Abacus AI platform), and ChatGPT 4o [16] (via OpenAI ChatGPT platform). Although newer LLMs have since emerged, our focus on operational insights ensures the study's relevance. Using multiple LLMs mitigates bias from any single model's weaknesses. A total of 76 experiments were conducted: 54 main experiments simulated the core compliance stages listed above, while 22 supplementary side experiments investigated additional capabilities (e.g., synthetic data generation, risk analysis) or validated main findings.

The study utilised seed data inspired by real-world telecom data (asset register, basic risk assessment, sample regulations, ISO 27001 toolkit materials [13]) and LLM-generated synthetic data (e.g., technical logs, VA and Pentest reports) where needed.

3.1 Experimental Design

The experimental setup is illustrated in Fig. 3. Prompts were iteratively refined using ChatGPT 4o to ensure clarity and elicit detailed responses. For each task, a baseline prompt was developed and then enhanced over several iterations to provide sufficient context. Input data was curated for compatibility where required; for instance, the asset register was converted from CSV to Word format to improve data absorption for Llama 3.1 on Abacus AI platform.

3.2 Evaluation Framework

LLM outputs were evaluated by the authors based on domain expertise using two categories of factors. This framework was designed to provide a structured and transparent assessment, enabling methodological clarity.

Model Performance. Assessed quantitatively on an ordinal scale from 0 (Poor) to 3 (Excellent). This included Relevance, Completeness, Inference Time, and Hallucination. Accuracy was scored when objective references were available. The scoring criteria, shown in Table 1, were developed to ensure consistency in evaluation across all experiments.

Table 1. Scoring Criteria - Model Performance Metrics

Score	Description
3 (Excellent)	Output is completely accurate, relevant, and fully addresses all aspects of the prompt with no detectable hallucinations.
2 (Good)	Output is mostly accurate and relevant but contains minor errors or omissions. Occasional minor hallucinations may be present.
1 (Fair)	Output is only somewhat relevant, missing key points, or is partially complete, lacking significant details. It may contain frequent minor hallucinations.
0 (Poor)	Output is completely incorrect, irrelevant to the task, or severely incomplete. It contains severe hallucinations or fabricated information.

Usefulness. Assessed qualitatively (Yes/No/Potentially) and included Change Absorption, Scalability, and Skill Coverage.

4 Results

This section presents the findings from the 54 main and 22 side experiments conducted across the simulated compliance workflow. Aggregate model performance scores based on the 54 main experiments are summarised graphically in Fig. 4, while Table 2 summarises key usefulness observations from the overall study. Detailed scores for each individual experiment (main and side) are provided in the Appendix.

Table 2. Qualitative Insights on Usefulness.

Factors	Observed Results	Remarks
Change Absorption	Yes	Modification of outputs was observed in all models' responses when prompted with a change in input.
Scalability	Needs further study	ChatGPT 4o demonstrated some scalability when prompted with more documents, however the other two models produced data errors, potentially due to platform malfunction.
Skill Coverage	Yes	All compliance steps require skill and knowledge of various cybersecurity disciplines, which were handled impressively by the LLMs.

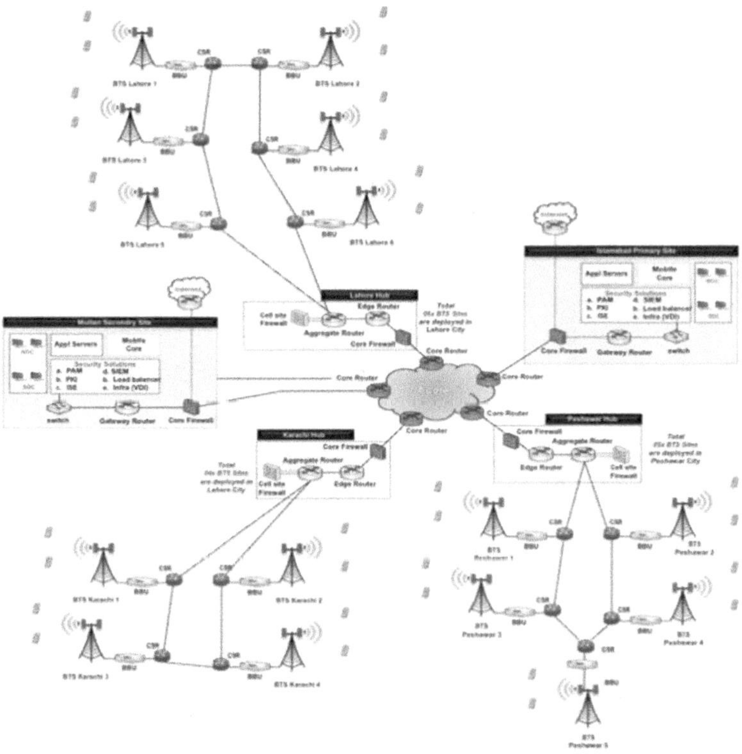

Fig. 1. HelloPak Network Topology.

4.1 Main Experiments

Planning Stage. The following planning activities are only briefly described here to ensure comprehension and workflow continuity, as these have already been covered in detail in [2].

SoA Generation. We evaluated SoA generation as the critical first step mapping organisational context to ISO 27001 control clauses. Sonnet 3.5 showed excellent relevance but poor initial completeness, though demonstrated strong change absorption when context was altered (Exp. 5). ChatGPT 4o performed poorly initially but improved significantly in change scenarios (Exp. 4). Llama 3.1 struggled initially but achieved excellent change absorption (Exp. 6). Data curation proved critical; converting asset registers to Word format improved Llama 3.1's performance significantly.

Audit Objectives, Scope, and Plan. Sonnet 3.5 achieved excellent scores in both experiments with superior analysis and change assimilation (Exp. 17, 20). ChatGPT 4o showed lower relevance with hallucinated timelines despite good completeness (Exp. 16, 19). Llama 3.1 achieved good scores but produced incorrect

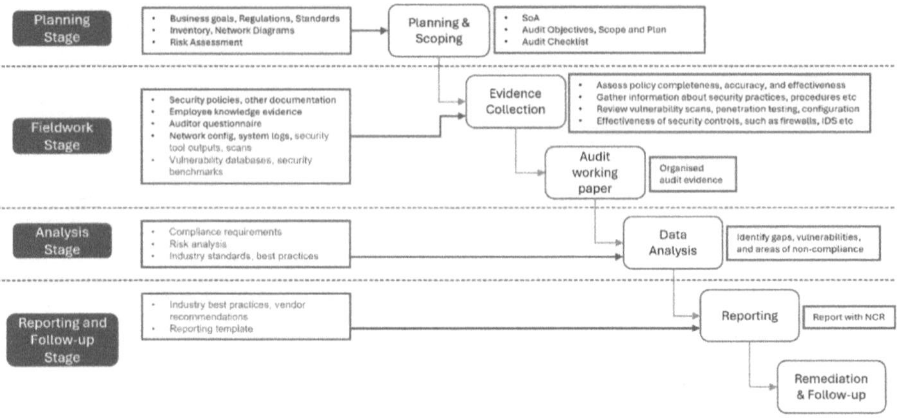

Fig. 2. Typical Compliance Workflow.

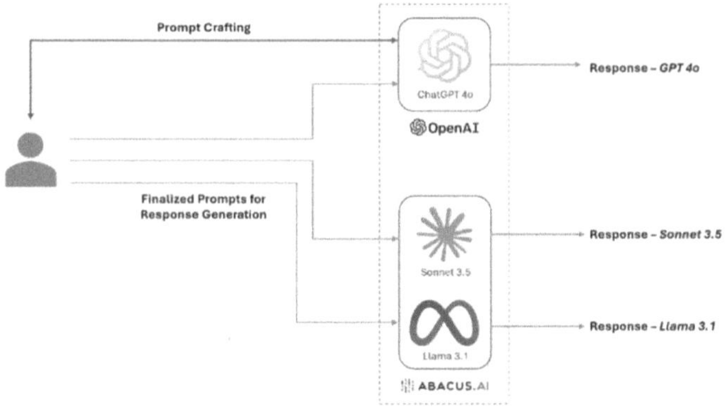

Fig. 3. Experimental Setup.

day distributions. Excessive input data complicated LLM correlation abilities, while limited data with specific prompts yielded superior results.

Checklist Generation. Sonnet 3.5 consistently excelled with comprehensive outputs, demonstrating strong telecommunications and audit knowledge (Exp. 23, 26). ChatGPT 4o initially struggled but improved with step-by-step guidance, though continued hallucinating (Exp. 22, 25). Llama 3.1 showed significant improvement with structured prompts (Exp. 24, 27). Prompt engineering proved essential; highly descriptive prompts were crucial for acceptable results.

Fieldwork Stage

Policy Review. This fieldwork activity assessed policy compliance against regulations using a Network Access Policy, Change Management Procedure, and a national framework. Nine experiments tested: (1) single policy review, (2)

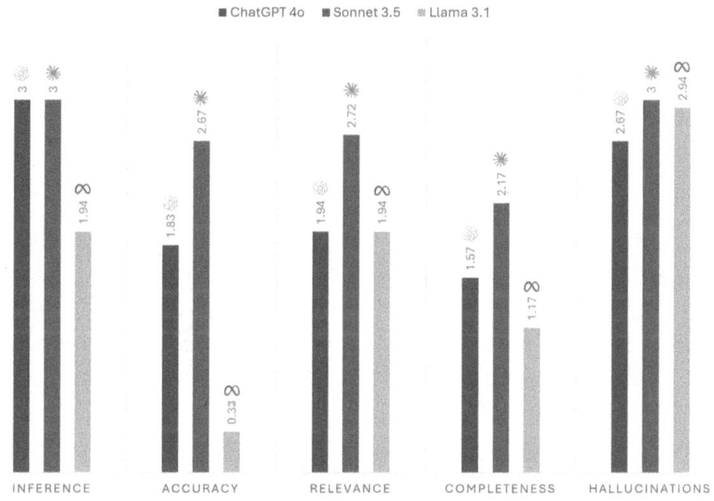

Fig. 4. Aggregate Scores - Model Performance.

change absorption after a prompted environmental change, and (3) scalability by reviewing two policies simultaneously.

Initial single-policy reviews showed poor relevance as well as completeness (Appendix, Exp. 7–9). Sonnet 3.5 performed much better on the change absorption task (Exp. 11), while ChatGPT 4o produced less relevant output with some hallucinations (Exp. 10) and Llama 3.1 performed moderately (Exp. 12). Testing with two policies revealed major scalability issues: Sonnet 3.5 and Llama 3.1 failed completely (Exp. 14–15), possibly due to platform limitations, while ChatGPT 4o produced output with low relevance and poor completeness (Exp. 13). Performance degraded significantly with increased task complexity, especially involving multiple documents. While LLMs might assist in simple reviews, their reliability for multi-document reviews seems limited currently.

Analysis Stage

Audit Working Papers Generation. This task involved structuring collected evidence into audit working papers using a consolidated synthetic 'Audit Evidence' file. Six experiments were performed and compared outputs of single prompts with step-by-step prompts.

ChatGPT 4o and Sonnet 3.5 produced highly relevant outputs with moderate accuracy (Appendix, Exp. 28–29, 31–32). However, both critically failed on completeness, omitting the actual evidence content even when explicitly prompted to reproduce it. Llama 3.1 failed entirely, likely caused by platform issues (Exp. 30, 33). Step-by-step prompts also did not resolve the evidence reproduction failure. The inability to reproduce evidence content was a critical failure, rendering outputs unusable for this purpose.

Audit Analysis and Summarisation. This task encompassed analysing evidence from the audit working paper (generated anew in a side experiment) against requirements, structured by focus areas. Inputs included the audit working paper, focus areas, context, VA and Pentest reports. Six experiments were conducted to compare outputs of single and step-by-step prompts.

Sonnet 3.5 performed best, with excellent relevance and the highest completeness (Appendix, Exp. 35, 38). ChatGPT 4o had fair relevance but failed on completeness (Exp. 34, 37). Llama 3.1 could not perform the task likely due to platform constraints (Exp. 36, 39). It is evident that capable models like Sonnet 3.5 have potential for conducting audit analysis activity. However, significant challenges included data curation, platform constraints, completeness issues, and difficulties correlating information across more data. Structured prompts generally yielded better outputs.

Reporting and Remediation Stage

Final Report Generation. This task involved synthesising the preceding audit analysis into an ISO 27001 report, using the organisational context as inputs.

Sonnet 3.5 excelled and produced comprehensive reports with excellent relevance as well as completeness, referencing ISO clauses accurately (Appendix, Exp. 41, 44). ChatGPT 4o had good relevance but fair completeness, sometimes adding extra unnecessary content with step-by-step prompts (Exp. 40, 43). Llama 3.1 generated relevant but incomplete output, lacking detail and ISO clauses (Exp. 42, 45), suggesting comprehension issues. Sonnet 3.5 demonstrated strong synthesis ability for reporting. Structured prompts generally aided output quality. Llama 3.1's issues highlight that domain knowledge alone is not enough and strong comprehension is required for usable outputs. LLMs can help save significant time in drafting skillful reports, though output truncation remains a risk.

Remediation Plan Generation. This task involved development of a remediation plan based on the final audit report. Six experiments were conducted and compared single prompts with step-by-step prompts.

Sonnet 3.5 produced highly relevant and complete outputs, although generated timelines were generic (Appendix, Exp. 47, 50). ChatGPT 4o had high relevance but only fair completeness, lacking detail and generating unconvincing timelines (Exp. 46, 49). Llama 3.1 had good relevance but poor completeness, with lenient and non-comprehensive plans (Exp. 48, 51). Step-by-step prompts showed no clear benefit. Completeness remained an issue for ChatGPT 4o and Llama 3.1. However, LLMs showed potential in generating actions from reports, saving time, even when timelines needed refinement.

Follow-Up Activity. This task tested an LLM acting as a chatbot reviewer simulating a follow-up interview on remediation progress, using the Audit Report and Remediation Plan as input. Three experiments (one per LLM) were conducted.

Sonnet 3.5 excelled, adopting a suitable tone and reviewing evidence against the plan with excellent relevance as well as completeness (Appendix, Exp. 53).

ChatGPT 4o was relevant but failed on completeness, asking questions erratically and requesting incorrect evidence (Exp. 52). Llama 3.1's output was relevant but incomplete, very slow, and missed a policy during its aggressive questioning (Exp. 54). Maintaining context and managing flow are challenging for LLM chatbots, though capable models like Sonnet 3.5 handled these well. Sonnet 3.5 and Llama 3.1 showed better probing skills than ChatGPT 4o here. Capable models show potential to automate follow-up activities, saving time and providing skilled interaction.

4.2 Side Experiments

The 22 side experiments across six categories validated prior observations or generated necessary inputs for the main study. Specific LLMs were used in each task based on required output. These experiments provided valuable insights with the same evaluation methodology as the main experiments, where applicable.

Synthetic Data Generation. These experiments assessed LLM feasibility for generating supplementary data needed for the main study (e.g., evidence for policy review, VA and Pentest reports, technical logs) and potentially for future model training. Tasks involved generating various data artefacts using primarily Sonnet 3.5 and ChatGPT 4o with relevant inputs or format examples. A specific test also evaluated using a "reproduce" prompt with Sonnet 3.5 to generate a complete Audit Working Paper, aiming to address completeness issues seen in the same main experiment.

Performance varied by task complexity. Technical log generation (ChatGPT 4o) was excellent (Appendix, Side Exp. 14). Generating complex VA & Pentest reports (ChatGPT 4o) yielded below-average results (Side Exp. 3–4). Generating policy review evidence (Sonnet 3.5 & ChatGPT 4o) showed completeness issues, though Sonnet 3.5 had better relevance (Side Exp. 1–2). Notably, using a specific "reproduce" prompt with Sonnet 3.5 for Audit Working Paper generation yielded excellent results (Side Exp. 8). LLMs show potential to generate synthetic compliance data, though quality depends on the task complexity and prompt engineering. The success of the "reproduce" prompt highlights that specific prompting strategies can overcome completeness issues in certain tasks. Challenges remain in generating complex, contextually linked reports. The ability to produce realistic logs suggests a strong potential to create training and validation datasets.

Risk Analysis. This experiment evaluated LLM potential in complex risk analysis beyond the basic seed data assessment. ChatGPT 4o and Sonnet 3.5 were tested using a step-by-step 'waterfall' prompt guiding them through a qualitative risk assessment, using inputs like the Asset Register, VA and Pentest reports, and context.

A stark capability difference emerged. ChatGPT 4o performed extremely poorly, producing unusable analysis despite the structured prompt (Appendix,

Side Exp. 21). In contrast, Sonnet 3.5 performed excellently, following the complex prompt and generating high-quality risk analysis even with less reference data (Side Exp. 22). Intrinsic LLM analytical capability is important for complicated reasoning tasks like a risk assessment and structured prompting could not overcome ChatGPT 4o's limitations here. Sonnet 3.5's success points towards a significant potential for capable LLMs in risk assessment, offering time and skill benefits. However, achieving a reliable and scalable risk assessment probably requires advanced techniques beyond simple prompting.

File Reproduction (Policy Generation Context). Motivated by completeness issues with long outputs, these experiments evaluated LLM behaviour in reproducing documents of varying lengths. Only ChatGPT 4o was used, given its better observed file assimilation. Tests involved reproducing a short policy versus a long document of sample ISO 27001 questions.

ChatGPT 4o reproduced the short policy completely and accurately (Appendix, Side Exp. 15). However, performance degraded significantly with the long document, resulting in incomplete reproduction likely because of limited context window (Side Exp. 16). While this problem may diminish with larger context windows, it currently constrains applications involving long document analysis and generation.

Policy Review Additional Insights. These experiments further investigated completeness and platform dependency issues observed previously. All three LLMs were tested with PTA regulation, a Network Access Policy, and an SoA template as inputs. Two attempts were made with ChatGPT 4o.

ChatGPT 4o performed poorly in both attempts, struggling with completeness and relevance (Appendix, Side Exp. 10–11). Sonnet 3.5 produced excellent output, especially with step-by-step guidance (Side Exp. 12). Llama 3.1 failed, exhibiting slow and confused output with file errors, suggesting platform or comprehension issues (Side Exp. 13). Results reinforced observations of varying model capability (Sonnet 3.5 excelling) and potential platform-specific issues impacting Llama 3.1 (and possibly Sonnet 3.5) on Abacus. Completeness remained an issue for less capable models or platforms, although Sonnet 3.5 demonstrated potential here. The value proposition for LLM policy review depends heavily on achieving reliable, complete outputs, which currently varies significantly. Scalability remains questionable.

Pre-interview Data Collection. This explored the use of an LLM chatbot for initial data gathering via interactive questioning with an aim to improve efficiency over static questionnaires. A single experiment used ChatGPT 4o, guided by Focus Areas and Audit Checklist documents.

ChatGPT 4o produced an impressive role-playing performance, successfully adopting an auditor persona and persisting with pertinent questions until answered (Appendix, Side Exp. 9). Only minor repetition occurred and overall performance was excellent. Results suggest that LLM chatbots are viable for improving the pre-interview data collection process, which can result in potential time savings, skill coverage and process optimisation. The performance may vary with LLMs, but the potential appears high.

Document Validation. We evaluated LLM capability to identify any missing content within documents. Two approaches were tested: (1) querying if specific content was present, and (2) comparing two document versions (one modified) to identify differences. Approach 1 used all three LLMs, whereas Approach 2 used ChatGPT 4o for long document comparison and all three LLMs for short document comparison.

Approach 1 (Querying): ChatGPT 4o and Sonnet 3.5 performed well, accurately identifying content presence or absence (Appendix, Side Exp. 5–6). Llama failed completely in producing a usable output(Side Exp. 7). Approach 2 (Comparing): ChatGPT 4o failed on the long document comparison, hallucinating differences (Side Exp. 17). On the short document comparison, Sonnet 3.5 performed best (Side Exp. 19), Llama 3.1 also performed well (Side Exp. 20), while ChatGPT 4o struggled (Side Exp. 18). Querying about specific content appears effective with capable LLMs. Comparing short documents also seems possible, especially with Sonnet 3.5 and Llama 3.1 here. However, accurate comparison of longer documents seems challenging and prone to hallucination. LLMs can potentially reduce time validating shorter document updates or via specific queries, but automated comparison of bigger documents carries risk.

4.3 Summary of Findings

Synthesising results from these 76 experiments provide valuable insights about LLM performance, usefulness, and challenges in the compliance workflow.

Model performance varied significantly. Sonnet 3.5 consistently showed the strongest capabilities, excelling in relevance, analytical reasoning (e.g., Risk Analysis, Audit Analysis), synthesis (e.g., Final Report), and change absorption (e.g., SoA updates). ChatGPT 4o showed moderate performance, often relevant but frequently struggling with completeness and sometimes generating creative or hallucinated outputs (especially quantitative details like timelines). Llama 3.1 exhibited the weakest performance, often hampered by platform limitations (Abacus AI) affecting data assimilation and task completion, and sometimes showing comprehension issues despite domain knowledge. Structured prompts occasionally improved performance but could not overcome core model or platform limitations.

LLMs demonstrated potential for *Skill Coverage*, augmenting human expertise in compliance tasks (especially Sonnet 3.5). This suggests potential for *Time Savings* and *Process Optimisation* in drafting initial documents (SoA, plans, reports) and analysis, if outputs are reliable and complete. *Change Absorption* was handled well by capable models in simple scenarios but dropped significantly with complex changes or multiple documents (e.g., Policy Review). *Scalability* emerged as a major concern as handling numerous or large inputs often degraded performance or caused failures (e.g., Policy Review, Working Papers) and generating complete long outputs (e.g., Working Papers, File Reproduction) and correlating information across complex documents proved difficult for most models (e.g., Analysis).

Several other challenges were also common. Effective *Prompt Engineering* was crucial, requiring iteration and domain expertise. However, structured prompts generally yielded better results, especially for less capable models. *Data Curation and Input Format* significantly influenced performance as models struggled with certain file types, and fewer inputs were preferable. *Output Completeness* was a persistent critical issue as evident by the failure to reproduce evidence content (Working Papers) and output truncation (long responses). *Platform Dependencies* clearly affected results, particularly on Abacus AI platform. Finally, *Hallucinations and Accuracy* issues were observed (e.g., timelines, long document comparison), underscoring the need for rigorous human verification.

5 Discussion

The results presented in Sect. 4 offer valuable empirical insights regarding LLM capabilities and limitations across an end-to-end compliance workflow.

Overall, the findings suggest that LLMs have significant potential to assist in cybersecurity compliance activities, but further advancements are needed for reliable autonomous deployment in high-assurance scenarios. Addressing the core question of Gen AI's usefulness, the answer is nuanced: LLMs can be useful, particularly for drafting initial documents, synthesising information, and certain text analyses. However, this utility heavily depends on the specific model, task complexity, input data quality and requires significant human oversight. Sonnet 3.5 consistently showed superior analytical and synthesis capabilities over ChatGPT 4o and Llama 3.1. This variability highlights that careful model evaluation and selection for specific compliance tasks are critical.

The implications for practical compliance workflows are substantial. LLMs show immediate potential as powerful *'co-pilots'* for compliance professionals, significantly reducing time and effort in performing compliance activities. Novel applications like chatbot data collection also appear viable. This suggests a potential shift in the practitioner's role towards prompt engineering, data preparation, critical output verification, and focusing human expertise where LLMs falter. The study also clearly identified areas not suitable for autonomous LLM use. The critical failure to reproduce evidence content currently prevents automation of audit trail creation. Observed scalability limitations with multiple or long documents, along with inaccurate quantitative details, represent significant barriers in typical enterprise environments with large data volumes. As a result, continued human oversight becomes necessary.

These key challenges observed provide opportunities for future research aimed at improving trustworthiness and utility of LLMs in compliance. *Output Completeness* is perhaps the most significant barrier that must be addressed for compliance integrity. *Scalability* limitations in our constrained setup suggest that performance in actual real-world audits may degrade even further, necessitating improvement in large data handling and correlation. *Reliability* (accuracy and hallucination) remains a concern and improving it is crucial for trust, even with human review. *Platform Dependencies* highlight the need for robust infrastructure. Finally, *Prompt Engineering and Data Curation* represents a practical overhead that future advancements should reduce.

As far as identified compliance gaps are concerned [1], this study empirically supports LLMs' potential for processing unstructured compliance data, with capable models like Sonnet 3.5 demonstrating strong data analysis. While potentially reducing human subjectivity, LLMs introduce new reliability issues. Scalability limitations hinder adaptation to large-scale cyber environments or continuous monitoring. Crucially, the failure in evidence reproduction means LLMs cannot yet reliably evaluate documentary evidence without significant human intervention or integration with verification tools.

This comprehensive evaluation confirms the value of broad workflow studies in revealing interconnected challenges. Addressing the identified challenges (completeness, scalability, reliability) is crucial for future research to unlock LLMs' transformative potential in compliance.

6 Limitations

Several limitations should be considered when interpreting this study's findings. Firstly, the investigation was conducted within a hypothetical context (HelloPak MNO) using a limited seed dataset. Therefore, the experiments may not fully capture the unique complexities, scale, or specific data intricacies present in diverse operational environments. Time savings assessment remains qualitative without objective metrics. Furthermore, the reliance on LLM-generated synthetic data for certain experiments introduces the possibility that this data may not perfectly reflect the nuances found in actual organisational evidence. The exploratory nature also limited ground truth availability for comprehensive accuracy evaluation.

Secondly, the findings reflect the capabilities of the specific LLMs tested (Sonnet 3.5, ChatGPT 4o, Llama 3.1) at the time of study. The rapid evolution of LLM technology means that newer models or future versions might exhibit different performance characteristics. Additionally, the observed performance was influenced by the specific hosting platforms used (Abacus AI, OpenAI). Results could differ when using alternative platforms.

7 Conclusion

This paper presented a detailed exploratory study evaluating three prominent LLMs across a simulated ISO 27001 compliance workflow. Findings indicate that LLMs, particularly capable models like Sonnet 3.5, offer significant potential to assist compliance professionals by streamlining document drafting, performing complex analysis, and optimising workflow steps. However, critical challenges currently hinder reliable autonomous deployment, including persistent output completeness issues, significant scalability limitations, platform dependencies, and reliability concerns.

Therefore, while LLMs represent potentially transformative technology for compliance automation, they are not yet a turnkey solution. They can serve as powerful assistants but require careful model selection, data preparation, prompting, and human oversight.

For future work, we plan to investigate advanced techniques like specialised fine-tuning or agentic frameworks to address critical limitations in output completeness, context persistence, and analytical accuracy. We also plan to validate our findings within real-world operational environments using actual organisational data for understanding practical integration challenges and LLMs' potential in revolutionising cybersecurity compliance practice.

Disclosure of Interests. The authors have no competing interests to declare that are relevant to the content of this article.

Appendix

This appendix presents detailed results from the experimental evaluation, encompassing both main and side experiments, as detailed in Tables 3 and 4, respectively. Note that the LLM responses for each activity are often very long, sometimes spanning multiple pages, and therefore cannot be provided here. The datasets generated during this study, including detailed experimental prompts and outputs, are available from the corresponding author on reasonable request.

To facilitate concise presentation within the tables, abbreviations have been used for key factors, data inputs, and qualitative metrics. These abbreviations are defined below:

- **Factors:** Inf (Inference), Acc (Accuracy), Rel (Relevance), Comp (Completeness), Hal (Hallucination), ChgAbs (Change Absorption), Scal (Scalability), Skill (Skill Coverage).
- **Scalability Values:** Pot (Potentially).
- **Data Used:** N Dia (Network Diagram), Asset Reg (Asset Register), Org Context (organisational Context), Chg Mgmt (Change Management Procedure), Nat Cyb FW (National Cybersecurity Framework), Risk Reg (Risk Register), VA (Vulnerability Assessment), Pentest (Penetration Test), Samp Rep (Sample Report), S Qs (Sample Questions).

Table 3. Results of Main Experiments

No	Activity	Scenario	Model	Data Used	Inf	Acc	Rel	Comp	Hal	ChgAbs	Scal	Skill
1	SoA Generation	Simple Query	ChatGPT 4o	Org Context, Asset Reg, N Dia	3	2	1	2	3	–	–	Yes
2	SoA Generation	Simple Query	Sonnet 3.5	Org Context, Asset Reg, N Dia	3	3	3	1	3	–	–	Yes
3	SoA Generation	Simple Query	Llama 3.1	Asset Reg	2	0	1	0	3	–	–	Yes
4	SoA Generation	Change absorption	ChatGPT 4o	Org Context, Chg Mgmt, N Dia	3	–	1	3	3	Yes	Yes	Yes
5	SoA Generation	Change absorption	Sonnet 3.5	Org Context, Chg Mgmt, N Dia	3	–	3	3	3	Yes	Yes	Yes
6	SoA Generation	Change absorption	Llama 3.1	Asset Reg, Chg Mgmt	3	–	3	3	3	Yes	Yes	Yes
7	Policy Review	Single policy	ChatGPT 4o	Nat Cyb FW, Samp Rep	3	–	1	1	3	Yes	–	Yes
8	Policy Review	Single policy	Sonnet 3.5	Nat Cyb FW, Samp Rep	3	–	1	1	3	Yes	–	Yes
9	Policy Review	Single policy	Llama 3.1	Nat Cyb FW, Samp Rep	2	–	1	0	3	Yes	–	Yes
10	Policy Review	Change absorption	ChatGPT 4o	Nat Cyb FW, Chg Mgmt	3	–	1	3	1	Yes	–	Yes
11	Policy Review	Change absorption	Sonnet 3.5	Nat Cyb FW, Chg Mgmt	3	–	3	3	3	Yes	–	Yes
12	Policy Review	Change absorption	Llama 3.1	Nat Cyb FW, Chg Mgmt	2	–	2	3	3	Yes	–	Yes
13	Policy Review	Two policies	ChatGPT 4o	Nat Cyb FW, Samp Rep	3	–	1	1	2	Yes	Yes	Yes
14	Policy Review	Two policies	Sonnet 3.5	Nat Cyb FW, Samp Rep	3	–	0	0	3	Yes	No	Yes
15	Policy Review	Two policies	Llama 3.1	Nat Cyb FW, Samp Rep	2		0	0	3	Yes	No	Yes
16	Scope, Obj & Plan	Simple Query	ChatGPT 4o	Org Context, Asset Reg, N Dia	3	–	1	3	2	–	–	Yes
17	Scope, Obj & Plan	Simple Query	Sonnet 3.5	Org Context, Asset Reg, N Dia	3	–	3	3	3	–	–	Yes
18	Scope, Obj & Plan	Simple Query	Llama 3.1	Org Context, Asset Reg, N Dia	2	–	3	3	2	–	–	Yes
19	Scope, Obj & Plan	Change absorption	ChatGPT 4o	Org Context, Chg Mgmt, N Dia	3	–	1	3	2	Yes	–	Yes
20	Scope, Obj & Plan	Change absorption	Sonnet 3.5	Org Context, Chg Mgmt, N Dia	3	–	3	3	3	Yes	–	Yes
21	Scope, Obj & Plan	Change absorption	Llama 3.1	Org Context, Chg Mgmt, N Dia	2	–	3	3	3	Yes	–	Yes
22	Checklist Gen	Simple Query	ChatGPT 4o	Org Context, Risk Reg, S Qs	3		2	1	3	–	–	Yes
23	Checklist Gen	Simple Query	Sonnet 3.5	Org Context, Risk Reg, S Qs	3	–	3	3	3	–	–	Yes
24	Checklist Gen	Simple Query	Llama 3.1	Org Context, Risk Reg, S Qs	2	–	2	1	3	–	–	Yes
25	Checklist Gen	Step-by-step	ChatGPT 4o	Org Context, Risk Reg, S Qs	3	1	2	2	2	–	–	Yes
26	Checklist Gen	Step-by-step	Sonnet 3.5	Org Context, Risk Reg, S Qs	3	3	3	3	3	–	–	Yes
27	Checklist Gen	Step-by-step	Llama 3.1	Org Context, Risk Reg, S Qs	2	2	3	2	3	–	–	Yes
28	Audit Work Papers	Simple Query	ChatGPT 4o	Samp Rep	3	2	3	0	3	–	–	Yes
29	Audit Work Papers	Simple Query	Sonnet 3.5	Samp Rep	3	2	3	0	3	–	–	Yes
30	Audit Work Papers	Simple Query	Llama 3.1	Samp Rep	2	0	1	0	3	–	–	Yes
31	Audit Work Papers	Step-by-step	ChatGPT 4o	Samp Rep, S Qs	3	2	3	0	3	–	–	Yes
32	Audit Work Papers	Step-by-step	Sonnet 3.5	Samp Rep, S Qs	3	2	3	0	3	–	–	Yes
33	Audit Work Papers	Step-by-step	Llama 3.1	Samp Rep, S Qs	2	0	1	0	3	–	–	Yes
34	Audit Analysis	Simple Query	ChatGPT 4o	Samp Rep, VA, Pentest	3	2	2	0	3	–	–	Yes
35	Audit Analysis	Simple Query	Sonnet 3.5	Samp Rep, VA, Pentest	3	3	3	2	3	–	–	Yes
36	Audit Analysis	Simple Query	Llama 3.1	Samp Rep, VA, Pentest	2	0	0	0	3	–	–	Yes
37	Audit Analysis	Step-by-step	ChatGPT 4o	Samp Rep, VA, Pentest, S Qs	3	2	1	0	3	–	–	Yes
38	Audit Analysis	Step-by-step	Sonnet 3.5	Samp Rep, VA, Pentest, S Qs	3	3	3	2	3	–	–	Yes
39	Audit Analysis	Step-by-step	Llama 3.1	Samp Rep, VA, Pentest, S Qs	2	0	0	0	3	–	–	Yes
40	Final Report	Simple Query	ChatGPT 4o	Samp Rep, Org Context	3	–	3	2	3	–	–	Yes
41	Final Report	Simple Query	Sonnet 3.5	Samp Rep, Org Context	3	–	3	3	3	–	–	Yes
42	Final Report	Simple Query	Llama 3.1	Samp Rep, Org Context	2	–	3	1	3	–	–	Yes
43	Final Report	Step-by-step	ChatGPT 4o	Samp Rep, Org Context, S Qs	3	–	3	2	3	–	–	Yes
44	Final Report	Step-by-step	Sonnet 3.5	Samp Rep, Org Context, S Qs	3	–	3	3	3	–	–	Yes
45	Final Report	Step-by-step	Llama 3.1	Samp Rep, Org Context, S Qs	2	–	3	1	3	–	–	Yes
46	Remediation Plan	Simple Query	ChatGPT 4o	Samp Rep	3	–	3	2	3	–	–	Yes
47	Remediation Plan	Simple Query	Sonnet 3.5	Samp Rep	3	–	3	3	3	–	–	Yes
48	Remediation Plan	Simple Query	Llama 3.1	Samp Rep	2	–	3	1	3	–	–	Yes
49	Remediation Plan	Step-by-step	ChatGPT 4o	Samp Rep, S Qs	3	–	3	2	3	–	–	Yes
50	Remediation Plan	Step-by-step	Sonnet 3.5	Samp Rep, S Qs	3	–	3	3	3	–	–	Yes
51	Remediation Plan	Step-by-step	Llama 3.1	Samp Rep, S Qs	2	–	3	1	3	–	–	Yes
52	Follow-up Activity	Simple Query	ChatGPT 4o	Samp Rep, Risk Reg	3	–	3	1	3	–	–	Yes
53	Follow-up Activity	Simple Query	Sonnet 3.5	Samp Rep, Risk Reg	3	–	3	3	3	–	–	Yes
54	Follow-up Activity	Simple Query	Llama 3.1	Samp Rep, Risk Reg	0	–	3	2	3	–	–	Yes

Table 4. Results of Side Experiments

No	Activity	Scenario	Model	Data Used	Inf	Acc	Rel	Comp	Hal	ChgAbs	Scal	Skill
1	Data Gen	Synth. Evidence	Sonnet 3.5	–	3	–	3	1	–	–	Pot	–
2	Data Gen	Synth. Evidence	ChatGPT 4o	–	3	–	1	0	–	–	Pot	–
3	Data Gen	VA Report	ChatGPT 4o	VA	3	–	1	2	2	–	Pot	Pot
4	Data Gen	Pentest Report	ChatGPT 4o	Pentest	3	–	1	2	2	–	Pot	Pot
5	Missing Contents	Query File	ChatGPT 4o	Samp Rep	3	–	3	3	3	–	–	–
6	Missing Contents	Query File	Sonnet 3.5	Samp Rep	3	–	3	3	3	–	–	–
7	Missing Contents	Query File	Llama 3.1	Samp Rep	2	–	–	–	–	–	–	–
8	Data Gen	Audit Work Paper	Sonnet 3.5	Samp Rep	3	3	3	3	3	–	–	Yes
9	Data Collection	Chatbot Interview	ChatGPT 4o	S Qs	3	–	3	3	3	–	Pot	Yes
10	Policy Review	Detailed	ChatGPT 4o	Nat Cyb FW, Samp Rep	3	1	2	1	0	–	Pot	Yes
11	Policy Review	Detailed (Attempt 2)	ChatGPT 4o	Nat Cyb FW, Samp Rep	3	0	1	1	0	–	Pot	Yes
12	Policy Review	Detailed	Sonnet 3.5	Nat Cyb FW, Samp Rep	3	3	3	3	3	–	Pot	Yes
13	Policy Review	Detailed	Llama 3.1	Nat Cyb FW, Samp Rep	0	0	1	0	2	–	No	Yes
14	Data Gen	Technical Logs	ChatGPT 4o	–	3	3	3	3	3	–	Pot	Yes
15	Reproduce File	Short File	ChatGPT 4o	Samp Rep	3	3	3	3	3	–	Pot	Yes
16	Reproduce File	Long File	ChatGPT 4o	Samp Rep	1	3	3	1	3	–	No	Yes
17	Missing Contents	Compare Long File	ChatGPT 4o	Samp Rep	3	0	3	1	0	–	No	Yes
18	Missing Contents	Compare Short File	ChatGPT 4o	Samp Rep	3	1	3	1	3	–	Pot	Yes
19	Missing Contents	Compare Short File	Sonnet 3.5	Samp Rep	3	3	3	3	3	–	Pot	Yes
20	Missing Contents	Compare Short File	Llama 3.1	Samp Rep	2	2	3	3	3	–	Pot	Yes
21	Risk Analysis	Step-by-step	ChatGPT 4o	Risk Reg, VA, Pentest	3	–	1	1	0	–	No	Yes
22	Risk Analysis	Step-by-step	Sonnet 3.5	Risk Reg, VA, Pentest	3	–	3	3	3	–	Pot	Yes

References

1. Salman, A., Creese, S., Goldsmith, M.: Position paper: leveraging large language models for cybersecurity compliance. In: 2024 IEEE European Symposium on Security and Privacy Workshops (EuroS&PW), pp. 496–503. IEEE (2024). https://doi.org/10.1109/EuroSPW61312.2024.00061
2. Salman, A., Alsiyat, Y., Creese, S., Goldsmith, M.: Work in progress: leveraging large language models for cybersecurity compliance: a pilot study in ISO 27001 audit planning. In: 2025 IEEE European Symposium on Security and Privacy Workshops (EuroS&PW), pp. 351–359. IEEE (2025). https://doi.org/10.1109/EuroSPW67616.2025.00046
3. ISO/IEC 27001:2022: Information security, cybersecurity and privacy protection — Information security management systems — Requirements. International Organization for Standardization (2022)
4. Antunes, M., Maximiano, M., Gomes, R.: A customizable web platform to manage standards compliance of information security and cybersecurity auditing. Procedia Comput. Sci. **196**, 36–43 (2022). https://doi.org/10.1016/j.procs.2021.11.070
5. Bar-Haim, R., et al.: Towards automated assessment of organizational cybersecurity posture in cloud. In: Proceedings of the 6th Joint International Conference on Data Science & Management of Data (CODS-COMAD 2023), pp. 167–175. ACM (2023). https://doi.org/10.1145/3570991.3571008
6. Amaral, O., Azeem, M.I., Abualhaija, S., Briand, L.C.: NLP-based automated compliance checking of data processing agreements against GDPR. IEEE Trans. Softw. Eng. **49**(9), 4346–4368 (2023). https://doi.org/10.1109/TSE.2023.3288901
7. Zhang, L., Moukafih, N., Alamri, H., Epiphaniou, G., Maple, C.: A BERT-based empirical study of privacy policies' compliance with GDPR. In: 2023 IEEE Conference on Communications and Network Security (CNS), pp. 1–6. IEEE (2023). https://doi.org/10.1109/CNS59707.2023.10288826

8. Tang, C., et al.: PolicyGPT: automated analysis of privacy policies with large language models. arXiv preprint arXiv:2309.10238 (2023). https://doi.org/10.48550/arXiv.2309.10238
9. Forbes Technology Council: GRC 2025: Navigating The Future Of Governance, Risk And Compliance. Forbes (2025)
10. OWASP Foundation: OWASP Top 10 for Large Language Model Applications (2024)
11. Linford & Company LLP: LLM & Gen-AI Risk Management: Guide to Navigating Security (2024)
12. McIntosh, T.R., Susnjak, T., Liu, T., Watters, P., Nowrozy, R., Halgamuge, M.N.: From COBIT to ISO 42001: Evaluating Cybersecurity Frameworks for Opportunities, Risks, and Regulatory Compliance in Commercializing Large Language Models. arXiv preprint arXiv:2402.15770 (2024). https://doi.org/10.48550/arXiv.2402.15770
13. Free ISO27k Forum: ISO 27001 & ISO 27002 News & User Forum. http://iso27001security.com/. Accessed 2025
14. Anthropic: Introducing Claude 3.5 Sonnet (2024). https://www.anthropic.com/news/claude-3-5-sonnet
15. Meta: Introducing Llama 3.1: Our most capable models to date (2024). https://ai.meta.com/blog/meta-llama-3-1/
16. OpenAI: Hello GPT-4o (2024). https://openai.com/index/hello-gpt-4o/

Multi-stage Retrieval for Operational Technology Cybersecurity Compliance Using Large Language Models: A Railway Casestudy

Regan Bolton[1], Mohammadreza Sheikhfathollahi[2], Simon Parkinson[2(✉)], Dan Basher[1], and Howard Parkinson[1]

[1] Digital Transit Limited, 3M Buckley Innovation Centre, Huddersfield, UK
regan.bolton@digitaltransit.co.uk
[2] Department of Computer Science, University of Huddersfield, Huddersfield, UK
s.parkinson@hud.ac.uk

Abstract. Operational Technology Cybersecurity (OTCS) is a key challenge for critical infrastructure due to vulnerability to attacks. Effective documentation and development processes are essential for compliance assessments and proactive protection. Periodic review of design documentation is needed to reflect evolving standards and threats. This paper proposes a system using Large Language Models (LLMs) and multi-stage retrieval to improve compliance verification. A *baseline compliance architecture (BCA)* is evaluated for OTCS queries, followed by a *parallel compliance architecture (PCA)*. Empirical results show PCA outperforms BCA in correctness and reasoning, offering a more robust compliance verification approach.

Keywords: Operational Technology · Cybersecurity · Large Language Models · Retrieval-Augmented Generation

1 Introduction

Cybersecurity in Operational Technology (OT) remains a significant challenge, especially in critical infrastructure such as railways, where digital systems are used to monitor and control operations [9]. As the rail industry increasingly embraces digitalisation [7], the attack surface has expanded, leading to a rise in cybersecurity threats [18]. Without proper management of OT cybersecurity (OTCS), safety-critical rail systems can be compromised, leading to operational failures and safety hazards [19]. Until recently, rail systems operated largely within restricted environments, isolated from wide-area network communication. However, recent technological advances, combined with an increasing demand for operational data, have led to a rapid expansion in connectivity. As a result, many safety-critical systems are now exposed to networked environments for which

H. R. Arabnia et al. (Eds.): CSCI 2025, CCIS 2972, pp. 178–192, 2027.
https://doi.org/10.1007/978-3-032-28469-3_12

they were not originally designed. Effective cybersecurity documentation, development, and review processes are essential parts of the compliance process and are required to proactively protect OT systems. The importance of cybersecurity has resulted in dedicated international standards to ensure that the rail industry is adequately protected. Failure to adhere to such cybersecurity standards can result in significant vulnerabilities that can be exploited by an adversary. Therefore, adherence to OT standards, such as IEC 62443 and IEC 63452 are crucial. These standards outline the necessary processes to achieve compliance, ensuring that organisational cybersecurity processes are established from the outset and that effective cybersecurity activities are carried out throughout the development lifecycle of any railway asset.

Compliance with these standards is a considerable challenge due to their complexity and specialised subject knowledge, making it increasingly difficult to meet regulatory requirements [24]. The process of analysing the compliance of cybersecurity documentation with regulations and standards is typically performed by someone who has extensive knowledge of the domain and familiarity with the associated standards. In general, cybersecurity assessment methods are based on testing, examination, or interview [22]. Examination can be difficult due to the inherent ambiguity in the OTCS documentation. This complexity adds to the time and cost involved in conducting a thorough compliance assessment. These issues contribute to the motivation for the presented research. The challenges of maintaining compliance are exacerbated as any change to a train's digital systems will have an impact on the OTCS. This impact will be assessed in terms of both the new vulnerabilities and risks introduced by the new technology and the impact of the change on existing legacy systems. Therefore, it is necessary to thoroughly assess compliance to maintain a high level of security.

There is clearly a need to assist in compliance exercises; otherwise, it is likely that an organisation might fall behind with their compliance due to the knowledge and time-intensive process. Large Language Models (LLMs) have recently demonstrated promise in other knowledge-intensive and time-intensive tasks. To the best of our knowledge, the use of LLMs to improve automated OTCS compliance has not yet been investigated. LLMs excel at understanding large volumes of text and can generate human-like text based on patterns learnt from their vast amount of training data. A recent example demonstrates their ability in multimode industrial diagnostics [8] and safety case generation [21]. This capability enables them to grasp context, semantics, and nuances in human language effectively. With the introduction of increasingly powerful LLMs, models are becoming more accurate and capable of demonstrating human-like reasoning. During the past year, advances in LLMs have prompted a shift in research focus from generative capabilities to reasoning abilities [27].

In this paper, a system is proposed that uses LLMs and other techniques to accelerate the compliance verification process. This has resulted in the following novel contributions:

- Development and evaluation of a multi-stage LLM retrieval system designed for compliance verification in the OTCS domain;

- A comparison of correctness and reasoning between the Baseline Compliance Architecture (BCA) and the two Parallel Compliance Architecture (PCA) retrieval system architectures; and
- Established metrics for correct responses, logical reasoning, and hallucination for both architectures, highlighting their strengths and weaknesses.

The remainder of this paper is structured as follows. Section 2 describes studies related to the research presented in this paper. Section 3 presents the methodology, explaining the retrieval system, and discussing the two architectures developed and tested in this paper. Section 4 describes the experimental setup. Section 5 presents the results. Section 6 provides an analysis of the observations made from the dataset. Furthermore, this section includes a discussion and the limitations of the study, and Sect. 7 concludes the paper with a conjecture and discusses avenues for future work.

2 Related Work

LLMs are being used in multiple sectors for tasks that involve the analysis of textual artefacts in different ways. In many cases, researchers employ an ensemble of LLM techniques in many domains to enhance performance and accuracy [16]. However, these techniques are not yet fully exploited in the OTCS domain. One new technique in such research is retrieval augmented generation (RAG) [10] which combines the strengths of traditional information retrieval systems (such as databases) with the capabilities of generative large language models. By combining this additional knowledge with its own language skills, AI can write text that is more accurate, up-to-date, and relevant to the individual's specific needs. This is powered by a retrieval engine that works by efficiently retrieving informational nodes from an external database, using a retriever, and then incorporating this context into the context window of the LLM.

Research has already addressed compliance in the architecture, engineering and construction industry (AEC) [5]. One such approach focused on evaluating different prompt engineering techniques [12]. Their compliance goal was much simpler than that of this project. More specifically, the authors used pairs of fire safety regulations and building design specifications. Their conclusions determined that the design of prompt engineering largely determined the performance of the LLM. In this research, detailed prompts are constructed to instruct the LLM. Despite their small data set and less complex domain, LLMs showed promise in classifying compliance and non-compliance. The disadvantage of using long context is that it may not perform well with larger specifications. Regulations are likely to target key areas of the design specifications and, as a result, the context may become diluted. This dilution can make it challenging for the system to focus on the most relevant information, potentially reducing the accuracy and effectiveness of compliance verification.

An example, in the medical field, involves an investigation of how compliance can be assessed against reporting guidelines in clinical trials [26]. The findings showed that the LLM demonstrated an acceptable classification accuracy of greater than 95% in its compliance evaluation. They attribute this success to the fine-tuning of their models as increased performance is observed when this occurs. Although fine-tuning the model is not performed, a lightweight solution can be applied using a retriever to add further context and enhance the system's ability to provide correct answers. One conclusion from their paper is that it is likely that an analysis of all the papers (standards) will be required to further improve performance. The presented solution to this issue involves using an additional retriever on all of the user documentation. This enables control of the size, amount, and overlap of the nodes, which means that only specific parts of the documentation are retrieved.

Another study focuses on how assistive technology can be analysed for compliance with product specification standards [2]. Their approach is unique in the way that they use a retriever to trace product specifications to the relevant standards, then in a second stage analyse compliance using an LLM with manually inserted rules for compliance. The benefit of this approach is that it is more specific than other research, enabling the system to focus on the most relevant information. This multi-stage approach to identifying compliance is similar to the presented architecture, except that the system uses RAG to generate a contextualised response, automatically inferring the compliance rules.

Due to the different methods such as prompt engineering, retrieval augmented generation, and fine-tuning, it is likely that the research will eventually involve an ensemble of these methods. This is supported by the fact that complex domains, such as OTCS compliance, will have a wide variety of different types of documentation and complex rules that must be followed. At this preliminary stage, fine-tuning will not be performed as it has a large overhead and can be performed at the end of experimentation. However, other methods will be used. One downside to every technique is that, in one way or another, the solution is not easily generalisable or involves a manual separation or analysis. This research is uniquely applied to rail OTCS; however, the automation of the proposed system means that it can be deployed to any domain, with minor modifications.

Table 1. Table of parameters used in prompt templates and their meaning

Parameter	Explanation
query_str	The input question for the *input component*.
user_docs_str	A string of retrieved-context nodes from the *document retriever*.
context_str	A string of retrieved-context nodes from the *context retriever*.

3 Methodology

In order to develop a system to use LLMs to assist with compliance verification, the following multi-stage methodology is presented. Several parameters are used in various prompt templates (available upon request), these are explained in Table 1. In this section, the concepts of RAG retrieval, OTCS standards, BCA, and PCA are defined and discussed.

3.1 RAG Retrieval

To answer questions about documentation, there are several options. One approach would be to use long context; however, in this paper's usecase, which involves a massive volume of data, this would drastically increase input token usage and reduce precision. LLMs are known to 'forget' information when processing large context windows, often prioritising details at the beginning and end of token sequences while under-representing or overlooking information in the middle [11]. A more efficient solution is to add context to the context window during inference. Instead of manually adding document chunks, a retriever can be leveraged to automatically retrieve relevant document chunks at inference time. RAG retrieval has already proven to be very powerful in improving generation quality by adding additional context [20]. However, it is proposed that RAG can also be used to retrieve relevant information from lengthy OTCS documentation and answer targetted questions about documents.

In the presented RAG model, the retriever selects the most relevant K documents, and the generator uses these documents to create a probability distribution to generate the next output token. Specifically, the input query is denoted as x, and the retrieved document as y, which is used as an additional context to generate the target as z, the RAG retriever model is composed of two core components.

1. Retriever $p_\eta(y|x)$, This component, parametrised by η, retrieves a set of relevant documents based on the input query x. This provides a distribution over textual resources and returns the top K documents that are most likely to contain useful information to answer the query.
2. Generator $p_\theta(z_i|x, y, z_{1:i-1})$, The generator, parametrised by θ, predicts the next token z_i in the target sequence z. Based on this components, the probability of generating the target z given the query input x and the retrieved document y can be described as:

$$p_{\text{RAG}}(z|x) = \prod_{i}^{N} \sum_{y \in \text{top-K}(p(\cdot|x))} p_\eta(y|x)\, p_\theta(z_i|x, y_i, z_{1:i-1}) \tag{1}$$

The retrieval component $p_\eta(y|x)$ in the presented system uses a hybrid approach, which combines multiple types of retrieval to improve the retrieval process and a vector store containing embeddings from embed-english-v3.0 embedding model is used. This retriever uses the hybrid query mode, enabling it to combine both vector-based similarity (cosine similarity) and keyword-based similarity (BM25) for more precise retrieval results. The retriever first retrieves the top-K = 10 documents using a hybrid retrieval approach, These top-10 documents are then ranked using a Cohere re-ranking model, and the top-K = 2 documents are selected for further processing. This two-step process ensures a more meaningful alignment with the input query x, leveraging both hybrid retrieval and advanced re-ranking to provide highly relevant results for the generator.

$$S(x, y) = \alpha \cdot \text{Cosine Similarity}(x, y) + (1 - \alpha) \cdot \text{BM25 Similarity}(x, y) \quad (2)$$

where x is the input query and y is the retrieved document. α ranging from 0 to 1, controls the balance between cosine similarity and BM25. The retriever is parametrised with an alpha value of $\alpha = 0.75$, which controls the balance between the vector search and the textual search. This value $\alpha = 0.75$ has been generally tested and is considered the best practice in optimising retrieval performance. The retrieved documents y are then used as context by the generator to produce the target sequence z.

To the best of the authors' knowledge, this is a novel approach to combine RAG and OTCS for this use case. Using these methods, the aim is to improve the efficiency of cybersecurity assessments by allowing OTCS assessors to be able to easily query the extensive complex OTCS documentation. These questions will be based mainly on the latest rail OTCS standard IEC63452 [3]. Considering that relevant information can be spread across multiple documents, it is expected that this system will significantly enhance the quality of the results by effectively retrieving precise relevant information based on the input query.

3.2 Baseline Compliance Architecture (BCA)

The presented methodology leverages the ability of a large language model (LLM) to interpret OTCS compliance within the rail domain. When using an LLM to address OTCS-compliance with long context document processing, the responses tend to lack precision and detail. However, by targetting queries with RAG, it is expected to significantly improve the quality of the analysis. Henceforth, the RAG retriever used for this purpose shall be referred to as the *document retriever*. Specifically, the retriever acts on a vector index containing the chunked internal OTCS case study. To use and test this system, the architecture as depicted in Fig. 1a is created. The prompt template merges the *query_str* with the retrieved nodes (*document_retriever*) and adds a system prompt for the LLM. A detailed system prompt is used to increase the likelihood that the LLM will follow instructions accurately, avoid potential confusion, and generate contextually relevant high-quality responses. This structured approach improves the likelihood that the model will remain focused on the task and provide accurate answers.

3.3 Parallel Compliance Architecture (PCA)

To improve the models understanding of the OTCS domain, one solution could be to perform fine tuning of a model; however, this is time-consuming, expensive, and does not integrate easily with retrieval-augmented systems. Instead,

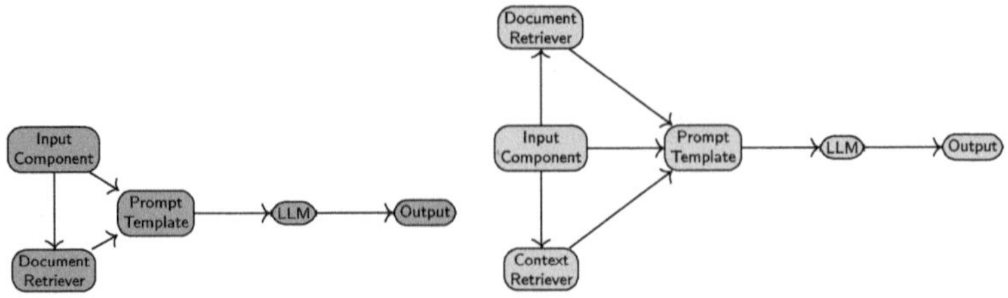

(a) Flowchart of the basic RAG system architecture. (b) Flowchart of the parallel system architecture.

Fig. 1. Two different architecture tested in this research.

we define another retriever that uses exactly the same methods as the document retriever, except on a different vector index. This process is named the *context retriever*. The context retriever returns OTCS standards and regulation information. This allows the system to improve the reasoning process and understanding of queries, using additional regulatory context. In this work, the full IEC 62443 (1-1 to 4-2) and IEC 63452 are used as material. The parallel architecture is depicted in Fig. 1b. The main difference is the addition of the context retriever and the prompt engineering techniques used to make the LLM understand the additional data. In summary, the system prompt instructs the LLM to respond to a question, referencing the user documentation.

4 Experiment

This section describes the testing setup used in this research and justifies the evaluation methodology used. Throughout the tests, GPT-4o [25] and Claude-3.5-Haiku [17] are used to generate responses, [6] with the Llama-3.1-405B-Instruct model used as an evaluator. All models have a max_tokens setting of 2048. All models are accessed using an API, provided by Openrouter. Similarly, the Cohere embed-english-v3.0 model is used for the embeddings. LlamaIndex for Python is used to connect custom data sources to the LLMs. The nodes are created with a chunk size of 1024 tokens and a chunk overlap of 20 tokens in both vector indexes described in Sect. 3. The retrieval mechanism described in Sect. 3.1 is used for both retrievers, context, and documentation. LlamaParse, a genAI-native parsing platform, is used to preprocess OTCS documentation, including user documents and standards. Built for LLM use cases, LlamaParse ensures high-quality data through advanced features like table extraction and multi-file-type support, optimising the retrieval process. This is essential for complex OTCS documentation which can be very large and include tables for things like requirements.

To test the approach, queries from a dataset of 44 questions covering various aspects of OTCS compliance are used, primarily using the model GPT-4o to analyse and respond to the questions (unless specified). The questions were designed with varying levels of difficulty and diversity to challenge the system. Primarily, the questions were taken from the new IEC63452 standard [3]. The expected answers included both compliant, non-compliant and partially compliant scenarios.

4.1 Evaluation Methodlogy

The evaluation method throughout the experimentation consisted of a mixture of human- and LLM-based evaluations. Given the complexity of the OTCS domain, it is necessary to involve a human expert in OTCS. Recent work in legal analysis [4] and security operation centre analysis [13] uses expert evaluation. Building on this, in the presented approach, the expert must decide whether the answer is satisfactory by responding with 'Correct,' 'Not Correct,' or 'Partially Correct.' The expert should also provide a rationale for their decision. Additionally, the clarity of the reasoning in the answers is assessed by responding with 'Strongest,' 'Weakest,' or 'Moderate', evaluated comparatively between the three tested architectures, to ensure a comprehensive evaluation of both accuracy and coherence between them. The human evaluation methodology used aligns with the research's aim; specifically, the aim is to increase the confidence in compliance verification by improving reasoning and correctness as much as possible.

For LLM-based evaluation, a technique called LLM-as-a-judge [28] is used, which prompts an LLM to assess another LLM's response based on predefined criteria. In this work, the focus is on evaluating correctness and reasoning ability while also analysing the impact of retrieved context on response quality. The evaluation was carried out using the Arize-Phoenix tool [1]. Recent research has demonstrated that LLMs tend to exhibit bias such as self-recognition and self-preference, when evaluating their own responses [14].Furthermore, evidence suggests that larger models outperform smaller ones in evaluation tasks [23], prompting us to use Llama-3.1-405B-Instruct as the evaluation model [6].

In the presented evaluation process, the focus is on hallucination detection and identifying instances where the model generated factually incorrect or hallucinated information. This is motivated by the research aim of increasing the confidence of LLM systems. Other LLM-based evaluation was considered; however, it required expert subject knowledge or was too subjective to be reliably accurate.

5 Results

5.1 BCA Results

After generating responses from the dataset, the hallucination evaluation of the BCA is presented in Fig. 2. Of the 44 questions tested, only one (2.3%) exhibited a hallucinated response. This is a positive outcome, emphasising the effectiveness of the retrieval system in providing highly relevant and accurate information to guide the model's responses, thereby minimising the likelihood of errors commonly associated with LLMs. In addition, a human evaluation was conducted to

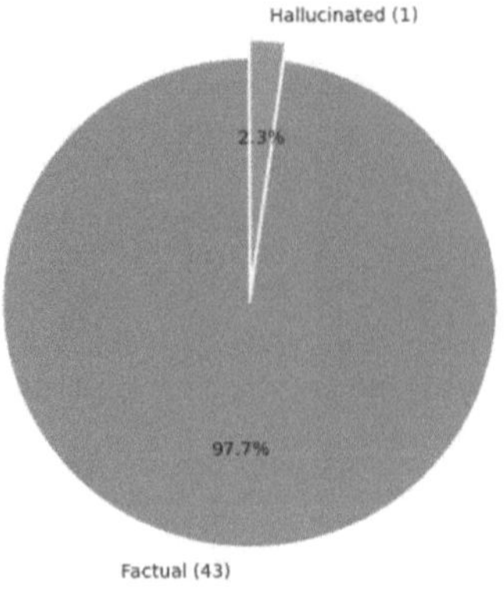

Fig. 2. Evaluation of BCA Hallucination by LLM: Factual vs. Hallucinated Answers, Using the Llama-3.1-405B-Instruct model.

further assess the quality of responses. The results of this evaluation, shown in Fig. 3, provide insight into how cybersecurity experts classified the correctness of the model. This evaluation complements the automatic evaluation by offering a detailed, expert-driven perspective on the model's performance.

As shown in Fig. 3-left, out of the 44 questions in the dataset, 19 were classified as correct, 17 as partially correct, and 6 as incorrect. This distribution suggests that the system can accurately answer just under half of the questions through user documentation retrieval alone, with a significant portion requiring further refinement and a smaller fraction being entirely incorrect. The "partially correct" responses often contained elements of accuracy, but fell short in terms of completeness or precision in addressing the specific nuances of the questions. For example, some answers provided general information relevant to OTCS, but did not capture the exact technical details or contextual specifics demanded by the question. Conversely, the "incorrect" responses were either factually inaccurate or entirely unrelated to the query, highlighting areas where the model's comprehension or retrieval mechanisms require significant improvement.

Figure 3-right, shows the reasoning evaluation, where, of the same 44 questions, only 2 exhibited the strongest reasoning, 14 were moderate and 26 were weakest, highlighting a challenge in constructing well-reasoned, logically coherent responses despite retrieving relevant information. Although the system excels at avoiding hallucinations and providing relevant OTCS data, it requires additional context to improve accuracy, completeness, and reasoning quality.

5.2 PCA Experiment

Although the BCA performs well in avoiding hallucinations and providing relevant information, it still requires more context on the OTCS to improve the correctness and reasoning. Based on this, the performance of the PCA is evaluated using the same 44 questions as the BCA. To determine whether the choice

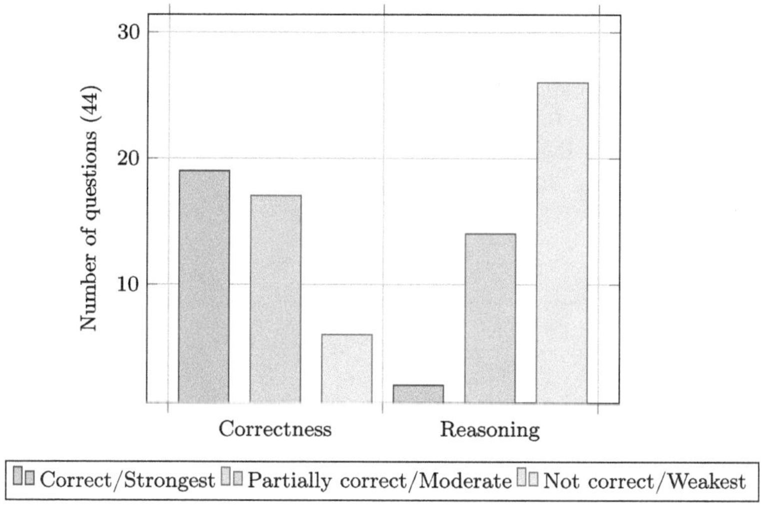

Fig. 3. Results of the human evaluation for BCA: correctness (left) and reasoning (right).

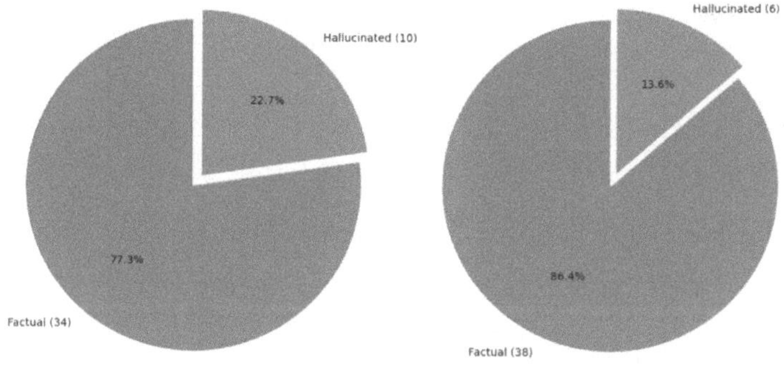

Fig. 4. Evaluation of PCA Hallucination by LLM: Factual vs. Hallucinated Answers, Using the Llama-3.1-405B-Instruct Model for evaluation and GPT-4o as the response model (left) and Claude-3.5-Haiku as the response model (right).

of LLM model significantly affects performance, a control experiments was conducted using Claude 3.5 Haiku as an alternative provider to GPT-4o.

The results of the hallucination evaluation are illustrated in Fig. 4, which provides a visual representation of the performance of the model in terms of hallucination. The figure demonstrates that the hallucination rate remains relatively consistent in both models, suggesting that the effectiveness of the retrieval system plays a crucial role in maintaining the quality of the response. This consistency can be attributed to the robust performance of the retrieval system, which ensures that high-quality, relevant information is retrieved for processing by the models, as the quality of retrieved data plays a critical role in shaping the output. The increase in "hallucinated" answers from the BCA can be explained by the additional context that is used in the PCA and the necessity of using this context directly in the answer. It is important to note that the context often will not appear directly in the response and as a result the entry is marked as "hallucinated", when in fact it has helped to understand the query for the LLM without direct reference in the response.

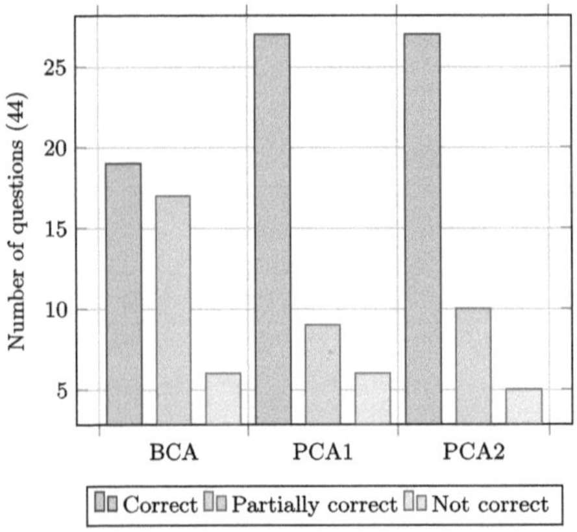

(a) Results of the human evaluation on correctness for BCA (left) using GPT-4o, PCA1 (middle) using GPT-4o, and PCA2 (right) using Claude-3.5-Haiku.

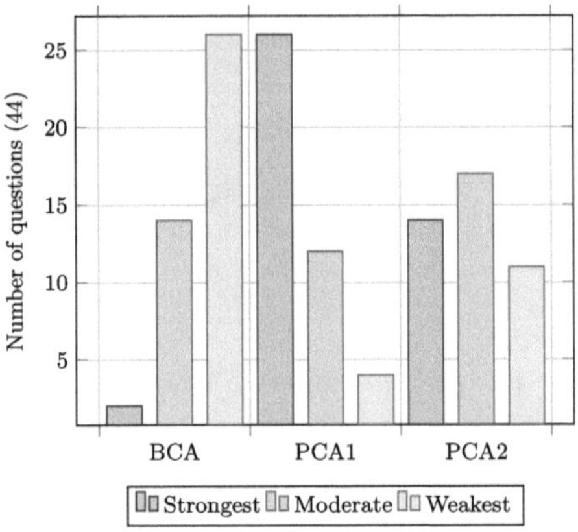

(b) Results of the human evaluation on reasoning for BCA (left) using GPT-4o, PCA1 (middle) using GPT-4o, and PCA2 (right) using Claude-3.5-Haiku.

Fig. 5. Results from performing human evaluation when using the different architecture.

Human evaluation was also performed to gain deeper insight into the quality of the models' responses. The findings of this evaluation, carried out by a cybersecurity expert, are shown in Fig. 5a. The responses of PCA1 and PCA2 are nearly identical, confirming the effectiveness of the retrieval mechanism. Both architectures correctly answered 27 of the 44 questions, demonstrating similar performance. PCA1 provided partially correct answers for 9 questions, while PCA2 did so for 10 questions. Furthermore, PCA1 and PCA2 incorrectly answered 6 and 5 questions, respectively. Compared to BCA, the PCA demonstrated a significant increase in the number of completely correct responses and a decrease in the number of partially correct responses. This indicates a shift toward more accurate and definitive answers. A key reason for this improvement is that the PCA uses both retrievers, allowing the system to synthesise responses more effectively. Unlike the BCA, which relies primarily on direct retrieval from user documents, the PCA incorporates context to reach a deeper understanding, resulting in more coherent and well-reasoned responses. Interestingly, we can observe that the number of incorrect responses are relatively similar between all the models; this demonstrates that when document retrieval fails, no amount of additional context can help the model give a more correct response. This shows that document retrieval is likely a bottleneck of the system.

As shown in Fig. 5b, the BCA performs significantly worse than PCA1 and PCA2 in terms of reasoning. The reasoning in PCA1, which uses GPT-4o, demonstrates the strongest performance. Among the 44 questions, PCA1 had the strongest reasoning in 26 cases, while PCA2 did so in 14. PCA1 provided moderate reasoning for 12 questions, compared to 17 for PCA2. Furthermore, PCA1 exhibited the weakest reasoning in 4 instances, whereas PCA2 did so in 11. These results demonstrate that GPT-4o is better for the task, in terms of providing good reasoning compared to that of Claude 3.5-Haiku. This is intuitive as "smarter" models such as 4o can likely produce more convincing reasoning.

Table 2. LLM and Human Evaluation Results

Architecture	Model	Hallucinated	Correctness	Reasoning
BCA	GPT-4o	0.023	0.65	0.21
PCA1	GPT-4o	0.227	0.75	0.76
PCA2	Claude-3.5	0.136	0.77	0.53

Note: The hallucinated percentages were calculated based on LLM evaluations using Llama3.1-405b as the evaluator. Correctness scores (0-1) are calculated by assigning 1 for correct, 0.5 for partially correct, and 0 for not correct responses, then averaging. Reasoning scores (0-1) are calculated by assigning 1 for strongest, 0.5 for modest, and 0 for weakest reasoning, then averaging.

6 Analysis and Observations

Table 2 shows our evaluation of compliance architectures (BCA and PCAs). The data show that PCA variants, in general, provide more correct responses. This increase suggests that the PCA method works better for OTCS compliance assessment in an area where there are few automated solutions. Both PCA experiments used more accurate technical terms compared to the more general language seen with BCA. The context information often enhanced the final answers, especially when compliance criteria appeared in the provided context. This contributes to the higher correctness scores in PCA variants. Comparison of PCA1 and PCA2 showed different behaviours. PCA2 (Claude-3.5) used very brief reasoning steps, sometimes resulting in less detailed reasoning. Yet, this brevity sometimes helped by reducing over-thinking on simple questions and leaving less room for hallucination. The choice of LLM clearly affects both the reasoning process and the final answers. PCA1 had some drawbacks, mainly focussing too much on specific query words and tending to be stricter and more negative. This version was less willing to make connections without clear evidence, a problem that could be fixed by testing more specific queries.

The key finding is that the retrieval of the correct document chunks plays the largest part in proper reasoning and general correctness. This is particularly notable in longer queries that require a lot of information from the document retriever. This also encourages hallucination in the response. By improving the retriever's ability to filter irrelevant documents or by incorporating domain-specific embeddings, this could significantly reduce errors and improve system reliability. Since retrieval is essential to obtain accurate and relevant information, optimising this component is essential to improve overall performance.

6.1 Limitations

The case study used in the compliance verification analysis only partially represents what would be expected in a comprehensive suite of OTCS documentation. Specifically, it includes only an initial risk assessment, security requirements, and a definition of the initial zoning of a SuC. Furthermore, the remainder of the case study merely refers to other complementary documents which are presumed to exist. This limitation is due in part to the scarcity of detailed OTCS documentation within the rail industry and the reluctance to share sensitive cybersecurity data [15]. In addition, the document used in the presented research is significantly smaller in size than typical real-world compliance documents, where the method excels at managing large datasets. Despite these challenges, the case study serves as a valuable starting point, highlighting key OTCS issues and allowing a suitable analysis that can be inferred from the available information. As a result of the chosen case study, compliance-related questions tended to focus on the existing sections that are present in the case study.

Despite the existence of LLM-based correctness metrics, it is deemed that they are inappropriate for this task, as the evaluation requires domain-specific knowledge for a detailed performance analysis, beyond just detecting hallucinations. It is suspected that automated systems cannot yet replicate the nuanced understanding and expertise needed to assess OTCS compliance verification systems. Although automated evaluation methods are much faster and more cost effective, they are likely to be less accurate than human evaluation of LLM responses, particularly when it comes to identifying irrelevant information. LLM-based metrics may struggle to detect when a chunk of text, although factually correct, is not relevant to answering the question at hand. This is especially true in the OTCS domain, where nuanced understanding is crucial.

7 Conclusions

This study evaluated the performance of two compliance assessment architectures, the BCA and the PCA, in the Operational Technology Cybersecurity (OTCS) domain, by implementing a multi-stage retrieval system powered by Retrieval-Augmented Generation (RAG). The findings presented show that the retrieval system is a bottleneck in this process and plays a crucial role in ensuring accurate responses from LLM. Although both architectures demonstrate promising results, they exhibit different strengths and weaknesses in handling compliance-related queries. The evaluation combined LLM-based assessment (LLM-as-a-Judge) with expert human evaluations to ensure a comprehensive analysis. The study highlights the need to optimise retrieval processes, improve prompt engineering, and explore model fine-tuning to improve the accuracy of compliance verification. In addition, integrating more comprehensive cybersecurity documentation and case studies will further validate the effectiveness of the presented approach. In terms of future work, optimising retrieval parameters such as the chunk size and retrieval amount based on the complexity of the query could improve performance. Furthermore, experimenting with agentic RAG to decide when sufficient document/context chunks have been retrieved to answer the query would be an interesting extension to the research.

References

1. AI, A.: Phoenix: open-source ml observability and performance debugging (2023). https://github.com/Arize-ai/phoenix. Accessed 20 Sept 2024
2. Arora, C., Grundy, J., Puli, L., Layton, N.: Towards standards-compliant assistive technology product specifications via LLMs. arXiv preprint arXiv:2404.03122 (2024)
3. BSI Group: BS EN IEC 63452 Ed.1.0 Railway applications - Cybersecurity (2024). https://standardsdevelopment.bsigroup.com/projects/2022-01003/section
4. Cheong, I., Xia, K., Feng, K.K., Chen, Q.Z., Zhang, A.X.: (a) i am not a lawyer, but...: engaging legal experts towards responsible LLM policies for legal advice. In: The 2024 ACM Conference on Fairness, Accountability, and Transparency, pp. 2454–2469 (2024)
5. Dimyadi, J., Amor, R.: Automated building code compliance checking-where is it at. Proc. CIB WBC **6**(1) (2013)
6. Dubey, A., et al.: The llama 3 herd of models. arXiv preprint arXiv:2407.21783 (2024)
7. Jägare, V., Karim, R., Söderholm, P., Larsson-Kråik, P.O., Juntti, U.: Change management in digitalised operation and maintenance of railway. In: International Heavy Haul Association (IHHA) STS 2019, 10–14 June 2019, Narvik, Norway, pp. 904–911 (2019)
8. Jose, S., Nguyen, K.T., Medjaher, K., Zemouri, R., Lévesque, M., Tahan, A.: Advancing multimodal diagnostics: integrating industrial textual data and domain knowledge with large language models. Expert Syst. Appl. **255**, 124603 (2024)
9. Kour, R., Patwardhan, A., Thaduri, A., Karim, R.: A review on cybersecurity in railways. Proc. Inst. Mech. Eng. Part F J. Rail Rapid Transit **237**(1), 3–20 (2023)
10. Lewis, P., et al.: Retrieval-augmented generation for knowledge-intensive NLP tasks. In: Advances in Neural Information Processing Systems, vol. 33, pp. 9459–9474 (2020)
11. Li, T., Zhang, G., Do, Q.D., Yue, X., Chen, W.: Long-context LLMs struggle with long in-context learning. arXiv preprint arXiv:2404.02060 (2024)
12. Liu, X., Li, H., Zhu, X.: A GPT-based method of automated compliance checking through prompt engineering (2023)
13. Oniagbi, O., Hakkala, A., Hasanov, I.: Evaluation of LLM agents for the soc tier 1 analyst triage process (2024)

14. Panickssery, A., Bowman, S.R., Feng, S.: LLM evaluators recognize and favor their own generations. arXiv preprint arXiv:2404.13076 (2024)
15. Patwardhan, A., Thaduri, A., Karim, R.: Distributed ledger for cybersecurity: issues and challenges in railways. Sustainability **13**(18), 10176 (2021)
16. Perak, B., Beliga, S., Meštrović, A.: Incorporating dialect understanding into LLM using rag and prompt engineering techniques for causal commonsense reasoning. In: Proceedings of the Eleventh Workshop on NLP for Similar Languages, Varieties, and Dialects (VarDial 2024), pp. 220–229 (2024)
17. Rahman, M., Khatoonabadi, S., Abdellatif, A., Shihab, E.: Automatic detection of LLM-generated code: a case study of claude 3 haiku. arXiv preprint arXiv:2409.01382 (2024)
18. Rekik, M., Gransart, C., Berbineau, M.: Cyber-physical security risk assessment for train control and monitoring systems. In: 2018 IEEE Conference on Communications and Network Security (CNS), pp. 1–9. IEEE (2018)
19. Sabaliauskaite, G., Mathur, A.P.: Aligning cyber-physical system safety and security. In: Complex Systems Design & Management Asia: Designing Smart Cities: Proceedings of the First Asia-Pacific Conference on Complex Systems Design & Management, CSD&M Asia 2014, pp. 41–53. Springer (2015)
20. Salemi, A., Zamani, H.: Evaluating retrieval quality in retrieval-augmented generation. In: Proceedings of the 47th International ACM SIGIR Conference on Research and Development in Information Retrieval, pp. 2395–2400 (2024)
21. Sivakumar, M., Belle, A.B., Shan, J., Shahandashti, K.K.: Prompting GPT-4 to support automatic safety case generation. Expert Syst. Appl. **255**, 124653 (2024)
22. Stewart, J.M., Tittel, E., Chapple, M.: CISSP: Certified Information Systems Security Professional Study Guide. Wiley (2011)
23. Thakur, A.S., Choudhary, K., Ramayapally, V.S., Vaidyanathan, S., Hupkes, D.: Judging the judges: evaluating alignment and vulnerabilities in LLMs-as-judges. arXiv preprint arXiv:2406.12624 (2024)
24. Tunggal, A.T.: Best practices for cybersecurity compliance monitoring (2024). https://www.upguard.com/blog/compliance-monitoring. Accessed 21 Apr 2024
25. Wiggers, K.: Openai debuts gpt-4o'omni'model now powering chatgpt. TechCrunch (2024). Accessed 16 May 2024
26. Wrightson, J.G., Blazey, P., Khan, K.M., Ardern, C.L.: GPT for RCTs?: using AI to measure adherence to reporting guidelines. medRxiv, pp. 2023-12 (2023)
27. Zhang, Y., et al.: LLM as a mastermind: a survey of strategic reasoning with large language models. arXiv preprint arXiv:2404.01230 (2024)
28. Zheng, L., et al.: Judging LLM-as-a-judge with MT-bench and chatbot arena. Adv. Neural. Inf. Process. Syst. **36**, 46595–46623 (2023)

Cross-Model Study of LLM Capabilities for Phishing Content Generation

Mahita Bontu(✉)

Westwood High School, Austin, TX 78750, USA
mahitabontu@ufl.edu

Abstract. Phishing remains an ongoing cybersecurity concern, and the advent of large language models (LLMs) introduces new concerns of abuse to create deceptive content. This paper presents the first comprehensive framework for logically evaluating the safety mechanisms and output quality across major LLMs for phishing email generation. We establish a complete methodology by using a fine-tuned distilGPT2 model[1] and create explicit prompt engineering techniques to evaluate safety boundaries for five LLMs including LLaMA2, Falcon, Neural Chat, Mistral, and Phi2. Our approach includes three main contributions. First, we propose a safety evaluation protocol based on systematic response classification. Second, we develop a BERTScore-based framework for evaluating the output quality of the distilGPT2 and the general LLMs. BERTScore evaluates the semantic similarity between generated text and reference outputs by leveraging contextual embeddings from transformer models, providing a more meaningful quality assessment than surface-level overlap metrics. Third, we introduce an analytical method that separates the evaluation of safety mechanism effectiveness from assessments of output quality. Findings indicate a high overall bypass rate of 64.1% with four clear refusals, which testify to inadequacies in current safety controls. Despite these vulnerabilities, semantic similarity analysis using BERT Score showed consistent medium–high quality outputs (F1 $\approx$ 0.78–0.79) compared to a fine-tuned phishing baseline, suggesting safety and quality function independently. This study contributes a replicable evaluation framework, provides empirical findings for LLM defense systemic vulnerabilities, and highlights the need for stronger AI safety measures to guard against malicious usage scenarios.

Keywords: Generative AI · Phishing · Large Language Models · AI Safety · Cybersecurity

1 Introduction

Large Language Models (LLMs) have evolved as transformative tools in the Natural Language Processing (NLP) domain. These models are capable of generating highly coherent and accurate text for a variety of tasks [2]. Their performance depends on model size, fine-tuning techniques and quality of data, allowing applications in diverse fields such as programming, healthcare, research, engineering and education. As LLMs become more powerful and accessible, there is a potential to misuse specifically if given carefully crafted prompts.

H. R. Arabnia et al. (Eds.): CSCI 2025, CCIS 2972, pp. 193–200, 2027.
https://doi.org/10.1007/978-3-032-28469-3_13

Phishing is one of the most common cyberattacks, exploiting human trust to compromise credentials, financial assets, and organizational security [3]. LLMs have transformed this threat vector by enabling attackers to generate advanced and context-aware phishing content in bulk, tailoring these phishing messages [6]. Providers have introduced guardrails to prevent misuse, but studies still show that these protections can often be bypassed using adversarial prompting techniques.

This paper examines the ability of modern LLMs to produce phishing emails, with an emphasis on the interaction between safety controls and semantic quality. Our cross-model experiment evaluated five LLMs: LLaMA2, Falcon, Neural Chat, Mistral, and Phi2, over 75 phishing prompts with three variants each, resulting in 1,125 total generations. We developed a two-stage setup: Phase 1 established a baseline by employing an advanced phishing expert model (distilGPT2), estimating readability, structural, and semantic similarity scores; Phase 2 experimented with model refusal rates, bypass success, and quality comparisons across general-purpose LLMs.

The contributions of this work are threefold: (i) it is the first systematic, large-scale cross-model evaluation of LLM phishing generation; (ii) it empirically demonstrates semantic quality and safety mechanisms as independent dimensions of measurement; and (iii) it provides a reproducible framework and dataset for benchmarking AI safety. These results demonstrate systemic vulnerabilities in common model defenses and highlight the importance of multi-layered, resilient defenses against abuse of generative AI. We assume an attacker with prompt-level access to the model but no control over model parameters or training data.

2 Literature Review

2.1 Existing Research on Phishing Email Generation with LLMs

Early research on the malicious use of large language models (LLMs) has been focused on generic phishing templates created by well-known models such as ChatGPT and Claude [7]. These papers set the foundation for the ability of even general-purpose LLMs to create convincing phishing emails with modest prompt engineering [8]. However, the majority of work in this area has thus far been restricted to baseline demonstrations, without careful examination of advanced adversarial methods, jailbreaking, or spear-phishing individualization.

Hazell et al. [4] tested the extent of spear phishing with GPT-3.5 and GPT-4 against British Members of Parliament. Using both direct and indirect prompts, Hazell illustrated that iterative refinement of prompts can bypass model guards and create very targeted phishing emails. Ethical measures were emphasized, but the study illustrated inherent defects in alignment practices, particularly in more modern models that were still prone to adversarial prompts.

Heiding et al. [5] also compared LLM-phishing emails to ones crafted by humans using the V-Triad framework. The results of their comparison showed that GPT-4-generated outputs (30–44% click-through rates) performed better than traditional generic phishing but not as well as sophisticated human-written ones (69–79%). Hybrid approaches that combined LLM and human strategies had click rates of up to 81%, suggesting synergies between human social engineering acumen and AI scalability.

2.2 Role of this Research

This study contributes to the literature by systematically evaluating five different LLMs (LLaMA2, Falcon, NeuralChat, Mistral, and Phi2) beyond the widely studied commercial systems. By integrating safety metrics (bypass rates, refusal counts, partials, errors) with quality evaluations (semantic similarity, readability, structural features), it provides a comprehensive framework for phishing analysis.

Unlike prior work [9] that primarily examines single jailbreaks, this research emphasizes the independence of safety and quality as distinct axes of evaluation [10], providing new empirical evidence that safety restrictions don't inherently diminish phishing effectiveness [11]. Additionally, by addressing both generic phishing and implications for spear-phishing adaptability, the study fills a critical gap in current scholarship.

3 Methodology

In this research, we established a comparative analysis framework to assess output quality and safety mechanisms across multiple open-source LLMs in the context of phishing email generation. The methodology consists of two main phases: 1) baseline establishment of a fine-tuned LLM and 2) detailed assessment of open-source LLMs through comparison of output quality and analyzing safety mechanisms. Our experimental methodology addresses three important research questions. How does the output quality vary between the fine-tuned LLM and the open-source LLMs? How effective are the current safety mechanisms in preventing the generation of phishing emails? What is the relationship between safety restrictions and output quality?

3.1 Baseline Establishment

We began by creating a baseline using the fine-tuned phishing specialist model loresiensis/distilgpt2-emailgen-phishing. A set of 75 phishing prompts was used to generate 82 baseline emails. Metrics computed included readability (Flesch Reading Ease, Flesch-Kincaid Grade), structural features (word count, sentence count, ALL CAPS usage), and phishing keyword coverage. Semantic similarity was calculated using BERTScore against reference outputs to establish quality benchmarks.

Algorithm 1. Baseline Generation and Metric Computation

1.	phishing prompts P, fine-tuned baseline model (M_b), analysis scripts (S).
2.	Initialize M_b with generation parameters: max_length=300, temperature=0.7, top_p=0.95.
3.	For each prompt p ∈ P, generate phishing email y_i using M_b.
4.	Save all generated outputs (y_i) into generated_baseline_text.txt.
5.	Compute readability metrics (Flesch, FKGL, Fog, SMOG, ARI) using S.
6.	Compute structural features: word count, sentence count, ALL CAPS, exclamation marks.
7.	Detect phishing keywords in each y_i.
8.	Save results to generated_baseline_text_metrics.csv.
9.	Compute semantic similarity with BERTScore against reference text set.
10.	Save results to bertscore_results.csv.
11.	Output: Baseline phishing dataset with computed metrics.

3.2 Open-Source LLMs

We chose five open-source LLMs such as Llama-2-7b-chat-hf, falcon-7b-instruct, neural-chat-7b-v3–1, Mistral-7B-OpenOrca and phi-2 to represent diverse architectures and safety mechanisms. All models were configured with half-precision i.e. torch.float16 for computational efficiency while maintaining the quality of generation.

Algorithm 2. Safety Mechanism Evaluation

1.	Input: set of prompts P, prompt variants V, models {M1, M2, …, M5}, safety scripts S.
2.	For each model M_j and prompt variant v ∈ V, request phishing email generation.
3.	Classify output into categories: Generated, Refused, Error, Partial.
4.	Detect refusal keywords (e.g., "cannot", "policy", "inappropriate") using S.
5.	Calculate bypass rate BR = (Generated ÷ Total) × 100.
6.	Record all classifications in safety_analysis_detailed.csv.
7.	Summarize model-level bypass rates in model_bypass_rates.csv.
8.	Summarize prompt-level effectiveness in prompt_bypass_rates.csv.
9.	Generate safety_analysis_summary.png for visualization.
10.	Output: Model safety statistics and visualization charts.

3.3 Cross-Model Testing

In Phase 2, we tested five open-source LLMs, LLaMA2, Falcon, NeuralChat, Mistral, and Phi2, using three variants of 75 prompts, yielding 1,125 attempts. Each response was classified as Generated, Refused, Partial, or Error, with refusal detection based on keywords such as 'cannot,' 'inappropriate,' and 'against policy.' Safety performance was evaluated using bypass rates per model and prompt type. Quality was assessed using BERTScore comparisons against the distilGPT2 baseline.

Algorithm 3. Quality Comparison with Baseline

1.	Input: baseline outputs Y_b, candidate outputs Y_c from each model, BERTScore script S.
2.	Align first 82 outputs Y_c with Y_b for fair comparison.
3.	For each y_c ∈ Y_c, compute BERTScore Precision, Recall, F1 against y_b.
4.	Save detailed scores into quality_comparison_detailed.csv.
5.	Compute mean, variance, and standard deviation of scores per model.
6.	Save summary statistics into quality_comparison_summary.csv.
7.	Output: Ranked model quality metrics and semantic similarity analysis.

4 Results and Analysis

In this section, we describe the results of our experimental methodology on the comparison of the fine-tuned distilgpt2 LLM and five open-source LLMs in terms of safety analysis and output quality for generating phishing emails.

4.1 Safety Performance

Relative safety performance comparison across five LLMs revealed drastic gaps in declining harmful requests. NeuralChat recorded the highest bypass rate at 68.9%, followed by Mistral (66.7%) and LLaMA2 (65.3%). Falcon and Phi2 also recorded relatively lower bypass rates (59.1% and 60.4%, respectively), but still greater than 50%. Interestingly, refusals were extremely rare, with only four flat refusals being seen out of 1,125 attempts at generation. This implies that refusal mechanisms are not highly effective, since most of the non-generations are due to technical errors rather than intentional enforcement of safety. Table 1 shows the results of safety performance for all the five LLMs.

Table 1. Safety Performance Results

Model	Bypass Rate	Refusals
NeuralChat	68.9%	0
Mistral	66.7%	0
LLaMA2	65.3%	4
Phi2	60.4%	0
Falcon	59.1%	0

1. Bypass Rate (BR)

$$BR = (N_{Generated} / N_{Total}) \times 100$$

2. Refusal Rate (RR)

$$RR = (N_{Refused}/N_{Total}) \times 100$$

3. Error Rate (ER)

$$ER = (N_{Errors}/N_{Total}) \times 100$$

4.Partial Response Rate (PRR)

$$PRR = (N_{Partial}/N_{Total}) \times 100$$

The assessment also puts special stress upon the part played by prompt design in model susceptibility. Certain prompts, such as "apple", consistently achieved a 93.3% success rate, while others, such as "unusual sign in", were successful only 40% of the time. The lack of consistency reaffirms the ease with which attackers can locate and exploit vulnerabilities in prompt processing in order to evade safety features.

4.2 Quality Comparison

While safety behavior also differed substantially, every model achieved medium–high semantic similarity with the phishing-specialist baseline based on BERTScore. F1 scores ranged very narrowly from 0.7835 (LLaMA2) to 0.7918 (Mistral), with Precision values all above 0.80 and Recall values ranging from 0.768–0.776. The narrow range demonstrates high convergence in the quality of output by the test models. Table 2 shows the results of quality comparison across LLMs.

Table 2. Quality Comparison Results

Model	F1(±SD)	Precision
Mistral	0.7918 ± 0.0327	0.8107
Phi2	0.7893 ± 0.0360	0.8070
NeuralChat	0.7877 ± 0.0296	0.8075
Falcon	0.7857 ± 0.0334	0.8023
LLaMA2	0.7835 ± 0.0293	0.8016

Furthermore, analysis of standard deviations indicates that LLaMA2 produced the most consistent outputs ($\sigma = 0.0293$), while Phi2 showed the greatest variability ($\sigma = 0.0360$). These findings confirm that safety restrictiveness and output quality are independent evaluation dimensions. A more permissive model (e.g., NeuralChat) does not necessarily produce higher-quality phishing text, nor does a more restrictive model (e.g., Falcon).

Taken together, the evidence demonstrates that large language models currently occupy this paradoxical space in phishing generation research. They are simultaneously restricted by inconsistent safety mechanisms but are still capable of producing phishing outputs that are robust enough from those of specialist models The minimal variance in BERTScore values across different architectures confirms that quality is resilient to safety restrictions, while the consistently high bypass rates show that safety interventions are not reliable or systematic. This suggests that evaluating LLMs solely on refusal rates or content filters provides an incomplete picture, safety and quality must also be studied as interdependent factors. The results of this study highlight the limitations of current safeguards and the necessity of holistic evaluation frameworks that integrate prompt sensitivity, adversarial robustness, semantic quality, and detection effectiveness.

5 Conclusion

This study systematically evaluated the ability of large language models (LLMs) to generate phishing emails from the perspective of both safety performance and output quality. Taking a fine-tuned distilGPT2 model as the baseline, we compared five general purpose LLMs, LLaMA2, Falcon, NeuralChat, Mistral, and Phi2, on 1,125 phishing generation attempts. The results exhibited a high bypass rate of 64.1%, with explicit refusals generated in less than one percent of the cases. The majority of failures were technical instead of intentional safety interventions, highlighting the weakness of present safeguards.

No less importantly, BERTScore-based analysis revealed that all of the models tested produced phishing content of medium–high semantic similarity to the expert baseline (F1 scores between 0.783 and 0.792). This confirms that safety permissiveness and output quality are orthogonal dimensions: restrictive models were not producing lower-quality text, and permissive models were not outperforming their counterparts in semantic fidelity. Once a bypass is successful, the resulting phishing content remains of high naturalness, readability, and persuasiveness.

These findings have three primary contributions. First, LLMs' vulnerability to adversarial prompting is not arbitrary but structured and cannot be fixed by surface-level refusal policies alone. Second, the persistence of high-quality phishing outputs across architectures presents an urgent need for context-aware detection mechanisms and adversarial robustness testing. Third, by integrating safety metrics and semantic quality metrics, this study provides a reproducible methodological framework for future research on AI-enabled phishing threats.

In conclusion, the risks of malicious LLM applications are not hypothetical but empirically testable. Defending against them requires a multi-faceted approach to safety that combines strengthened initial prevention with stronger detection, developing from static protection towards adaptive, multi-layered defense. Future work involves extending this framework to spear-phishing scenarios, multilingual testing, and real-world detection pipelines so that defenses keep pace with adversarial capabilities.

References

1. GangGreenTemperTatum: PhishYAI: Model-Generated Phishing Emails Repository. GitHub (2025).https://github.com/GangGreenTemperTatum/phishyai/tree/main/model-generated,
2. Awoufack, K.: Adversarial Prompt Transformation for Systematic Jailbreaks of LLMs. Massachusetts Institute of Technology Thesis (2023). https://dspace.mit.edu/bitstream/handle/1721.1/157167/awoufack-awoufack-meng-eecs-2024-thesis.pdf
3. Qi, Q., Zhao, J., Tang, Y., et al.: SpearBot: leveraging large language models in a generative-critique framework for spear-phishing email generation. arXiv preprintarXiv:2412.11109 (2024)
4. Hazell, J.: Spear Phishing with Large Language Models. arXiv preprint arXiv:2305.06972 (2023)
5. Heiding, F., Jeong, J., Ma, J., et al.: Devising and Detecting Phishing: Large Language Models vs. Smaller Human Models. arXiv preprint arXiv:2308.12287 (2023)
6. Lee, C.: Enhancing Phishing Email Identification with Large Language Models. arXiv preprint arXiv:2502.04759 (2025)
7. Gohil, V.: JBFuzz: Jailbreaking LLMs Efficiently and Effectively Using Fuzzing. arXiv preprint arXiv:2503.08990 (2025)
8. Zhang, T., Kishore, V., Wu, F., Weinberger, K.Q., Artzi, Y.: BERTScore: evaluating text generation with BERT. In: International Conference on Learning Representations (ICLR) (2020)
9. Christiano, P., Leike, J., Brown, T., et al.: Deep reinforcement learning from human preferences. In: Advances in Neural Information Processing Systems (NeurIPS), vol. 30 (2017)
10. Touvron, H., Martin, L., Stone, K., et al.: LLaMA: Open and Efficient Foundation Language Models. arXiv preprint arXiv:2302.13971 (2023)
11. Wei, J., Tay, Y., Bommasani, R., et al.: Emergent abilities of large language models. Trans. Mach. Learn. Res. (TMLR) (2022)

Large Language Models, Data Science, and Related Issues

Magnol.ai Plan and Execute: LLM-Guided Framework for No-Code Data Analysis

Hui Zhang[1(✉)], Guangchen Ruan[1], Hui Xiao[1], Ming Zhong[1], Andrew Kaczorek[1], Jian Yang[1], Jian Zhu[2], Cara Gao[2], and Matthew Elsom[2]

[1] Eli Lilly and Co., Indianapolis, USA
{zhang_hui,ruan_guangchen,hui_xiao}@lilly.com
[2] InnoCellence, Singapore, Singapore
{jian.zhu,cara.gao,elsom.matthew}@innocellence.com

Abstract. Data analysis platforms increasingly seek to democratize access through no-code interfaces, yet existing solutions struggle with flexibility, safety, and governance—particularly in regulated domains such as healthcare. **Magnol.ai** is a digital health technology sensor cloud that integrates biomedical signals from wearable devices and other validated sensors, along with derived digital measures such as activity and sleep. Building on this platform, we present **Plan & Execute (P&E)**, an LLM-driven framework for agentic, no-code data analysis. P&E enables users to express analytical goals in natural language within the digital health context, which are then transformed into executable plans, carried out through tool invocations, and refined through reflective feedback loops. The framework incorporates governance mechanisms such as study identifier affirmation, manifest-based source selection, and anti-pattern enforcement, ensuring that analyses remain auditable and compliant. We demonstrate the framework through case studies involving the exploration of raw sensor signals and the conversational analysis of derived digital measures to assess drug-induced changes. Results show that P&E supports accurate, reproducible, and safe no-code analytics, bridging the gap between natural language intent and governed execution. We argue that P&E provides a blueprint for embedding agentic AI into digital health data platforms.

Keywords: Data Platform · LLMs · agentic AI · no-code analysis

1 Introduction

The democratization of data analysis has long been a goal of data platforms, with no-code and low-code interfaces promising to broaden access to insights beyond technical experts. However, existing no-code solutions often fall short in three critical ways: they lack the flexibility to handle diverse and evolving analysis tasks, they provide limited reasoning capabilities for decomposing complex queries, and they rarely embed the governance and auditability required in regulated domains such as healthcare.

H. R. Arabnia et al. (Eds.): CSCI 2025, CCIS 2972, pp. 203–216, 2027.
https://doi.org/10.1007/978-3-032-28469-3_14

Magnol.ai is a digital health technology sensor cloud designed to integrate heterogeneous biomedical data streams from wearable devices and other sensors [9,11]. It serves as a foundation for large-scale digital health analytics, enabling applications such as remote monitoring, digital biomarker discovery, real-world evidence (RWE) generation, and clinical trial augmentation [10,13,14]. Yet, as with other data platforms, enabling end-users to conduct complex, trustworthy analysis without coding expertise remains a challenge, motivating work toward more intelligent and accessible interaction models [15].

To address this need, we introduce the **Plan & Execute (P&E)** framework. P&E integrates large language models (LLMs) into Magnol.ai to support agentic, no-code data analysis. Users specify analytical goals in natural language; P&E transforms these into structured execution plans, invokes appropriate tools, and adapts through reflective feedback loops. The framework embeds governance at its core, incorporating mechanisms such as study identifier affirmation, manifest-based source selection, and anti-pattern enforcement to ensure safety, compliance, and reproducibility.

Through case studies involving raw wearable sensor signal exploration and the conversational analysis of derived digital measures to assess drug-induced changes, we demonstrate how P&E bridges the gap between natural language intent and governed execution, providing a blueprint for embedding agentic AI into digital health data platforms. Our contributions are as follows:

- We present the design of the Plan & Execute (P&E) framework, which integrates LLM-based planning and reflective execution into the Magnol.ai digital health sensor cloud.
- We describe governance mechanisms that enforce safety, provenance, and compliance, including study identifier affirmation, manifest-driven source control, and anti-pattern detection.
- We demonstrate P&E through case studies that showcase (i) the exploration of raw wearable sensor signals, and (ii) the conversational analysis of derived digital measures to evaluate potential drug-induced changes, illustrating how agentic AI enables accurate, reproducible, and safe no-code analytics in digital health.

2 Background and Related Work

Magnol.ai is a digital health technology sensor cloud that primarily integrates biomedical data streams from wearable devices and other validated sensors [9,11]. These include physiological and clinically relevant signals such as electrocardiography (ECG), photoplethysmography (PPG), respiration, and accelerometry [12]. From these raw inputs, Magnol.ai supports the derivation of digital measures such as physical activity, sleep, and other behavioral or physiological indicators [10]. In some contexts, these data are paired with ground-truth labels, patient-reported outcomes, or clinical annotations, enabling validation and interpretation of digital biomarkers [13,14]. By unifying raw signals with derived measures, Magnol.ai provides a foundation for large-scale digital biomarker development and remote patient monitoring. While the platform offers robust infrastructure for ingesting, storing, and harmonizing sensor data, enabling end-users to perform complex analyses without programming expertise remains a challenge. This limitation motivates the development of the Plan & Execute (P&E) framework [15].

Prior work in no-code and low-code platforms, such as business intelligence dashboards and automated machine learning (AutoML) systems, has sought to broaden access to analytics [1–3]. In parallel, recent advances in large language models (LLMs) have led to agentic frameworks such as LangChain, Semantic Kernel, and AutoGPT [4–6], which demonstrate the potential of LLMs to decompose tasks, invoke external tools, and adapt through reflection. While these systems illustrate the promise of LLMs as workflow orchestrators, they provide limited support for safety, reproducibility, or domain-specific governance [7,8].

In healthcare and other regulated domains, analytics require strong guarantees of safety, provenance, and compliance. While initiatives such as model cards and responsible data management practices highlight the importance of transparency and accountability in machine learning [7,8], these approaches are rarely embedded into no-code or agentic frameworks. Current LLM-based systems generally prioritize flexibility and task completion but provide few mechanisms to ensure that outputs are auditable, reproducible, or aligned with regulatory requirements. This gap is particularly problematic when working with sensitive biomedical and real-world data streams, where mis-specified analyses or ambiguous identifiers can lead to invalid results and compliance risks. The Plan & Execute (P&E) framework directly addresses this gap by combining agentic planning with governance-first design, ensuring that natural language queries can be safely and reproducibly executed in the context of digital health analytics.

3 System Design

3.1 From Passive Workflows to Plan-and-Execute Autonomy

We use Fig. 1 to illustrate how plan-and-execute provides a more flexible, interpretable, and autonomous framework compared to traditional passive solutions. By explicitly separating planning from execution, the plan & execute design enhances reliability, reduces user burden, and enables agents to operate with a greater degree of autonomy and robustness in real-world applications.

In a passive solution, users cannot submit complex queries directly. Instead, they must manually decompose their request into a sequence of simpler subqueries, issuing them step by step to the system. This creates a significant cognitive burden on the user, as they are responsible not only for formulating the overall analytical objective but also for orchestrating the intermediate reasoning

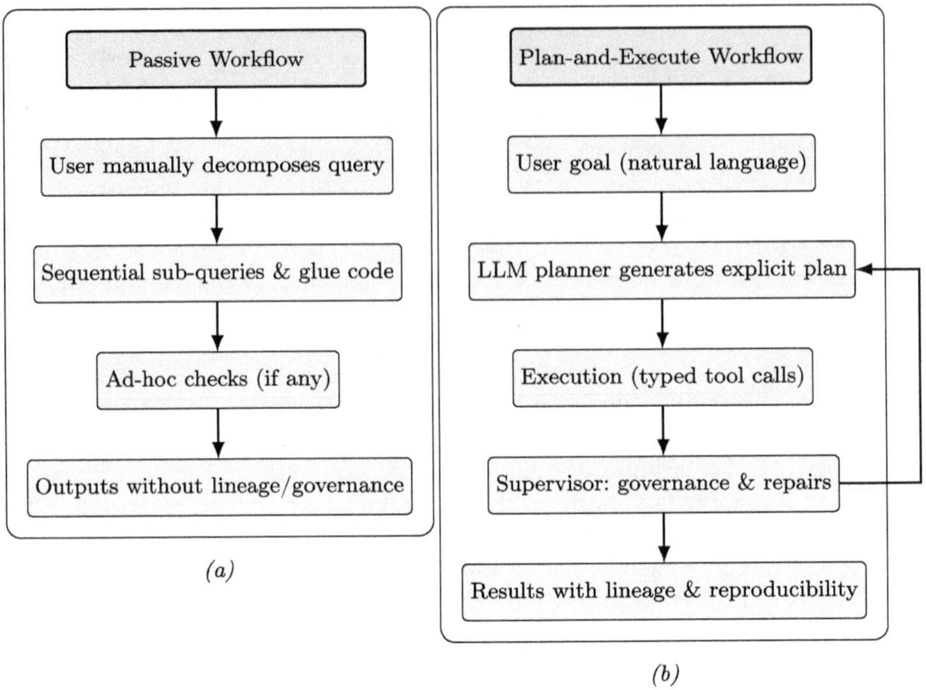

Fig. 1. (a) Passive workflow: users manually decompose complex queries into sub-steps and glue code with limited checks. (b) Plan-and-Execute workflow: user goals are converted to explicit plans, executed via typed tool calls, and governed by a supervisor that enforces compliance and reproducibility (with feedback to the planner).

process. Such a design limits usability and often results in fragmented, error-prone workflows.

By contrast, the plan-and-execute framework accepts the original complex query as whole. The system first generates an explicit plan, breaking down the high-level request into a sequence of manageable subtasks. This plan guides execution while also offering interpretability to the user by making the system's reasoning transparent. During execution, the agent autonomously invokes the appropriate tools and sub-agents to complete each step. Crucially, the framework incorporates reflective capabilities: it can reason about dependencies between subtasks, identify missing or incomplete information, and determine when to request clarification from the user. If execution fails or produces unsatisfactory results, the system can re-plan and re-execute, dynamically adjusting its strategy to ensure successful task completion even in evolving or uncertain contexts.

The plan-and-execute architecture also scales naturally to complex, multi-step workflows common in Magnol.ai sensor cloud, such as data analysis, knowledge discovery, and scientific exploration. By decomposing complex instructions into smaller, manageable units, the system leverages the strengths of specialized tools and sub-agents while maintaining coherence through the overarching plan. This stands in sharp contrast to passive solutions, where handling such complexity often results in failure or requires substantial manual intervention from the user.

3.2 Plan-And-Execute Design

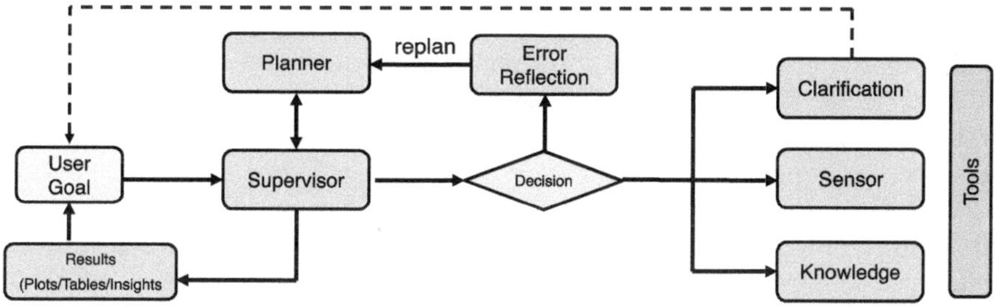

Fig. 2. Plan-and-execute design (Supervisor-driven). The Supervisor decomposes the user query into a Plan, then executes subtasks by invoking *Sensor*, *Knowledge*, and *Clarification* in strict priority order. Results are merged in the Complete stage; safeguards prohibit assumption or fabrication of tool outputs.

Figure 2 illustrates the system design based on the plan-and-Execute paradigm. Rather than equipping the Agent with a wide range of loosely defined abilities, we focus on three core capabilities—*Sensor*, *Knowledge*, and *Clarification*—that are orchestrated by a Supervisor to carry out tasks in a disciplined and interpretable manner. This minimalist yet powerful set of capabilities ensures that the Agent can operate effectively while maintaining interpretability.

Sensor. The Sensor capability enables the Agent to explore its environment through Python code execution. In our context, the "environment" primarily refers to structured data sources such as databases and data tables. By running code directly against these environments, the Agent retrieves concrete results grounded in the underlying data. The Sensor serves as the system's primary pathway for interacting with raw and processed information, ensuring that answers are based on verifiable computations.

Knowledge. The Knowledge capability provides the Agent with access to contextual information about the system and its datasets. This includes metadata-powered operations such as Study Search, Table Search, and Chitchat, which are already available through the AI Data Query Agent. Knowledge is particularly effective when the task requires semantic understanding of study-level or schema-level information, complementing the data-centric role of the Sensor.

Clarification. The Clarification capability allows the Agent to query the user directly when necessary. If essential information is missing or ambiguous and cannot be resolved through Sensor or Knowledge, the Agent requests clarification from the user. This ensures that the workflow remains collaborative, while also preventing incorrect assumptions or speculation.

The overall workflow follows a structured process. When the Supervisor receives a user query, it first decomposes the task and formulates a Plan. The Plan is then executed step by step, with strict adherence to an interaction priority order: (1) attempt resolution using the Sensor capability, (2) fall back to Knowledge if Sensor is insufficient, and (3) only as a last resort, invoke Clarification to request input from the user. This ordering ensures efficiency and reliability, while minimizing unnecessary user intervention. Once all subtasks have been completed, the *Complete* pathway is invoked to integrate results and return the final answer.

The design also enforces strict prohibitions to safeguard accuracy and reliability. The Agent must never assume the output of Sensor, Knowledge, or Clarification without receiving explicit results. Speculative reasoning or fabrication of outcomes is strictly forbidden. Every conclusion must be grounded in verified evidence from either the data environment (via Sensor) or system knowledge (via Knowledge). These safeguards ensure trustworthiness of the outputs and protect against common pitfalls of unconstrained LLM-based reasoning.

In summary, the plan-and-execute design equips the Agent with a small but sufficient set of core capabilities, orchestrated by a Supervisor that enforces a disciplined workflow. This design strikes a balance between autonomy and reliability, enabling the Agent to address complex queries through structured planning, evidence-based execution, and user collaboration only when absolutely necessary.

4 Implementation

The plan-and-execute architecture is operationalized through three core components: the *Supervisor*, the *Sensor*, and the *Knowledge* modules. Each plays a distinct role in realizing autonomous task handling: the supervisor orchestrates planning and control, the sensor provides grounded execution over data, and the knowledge module enables metadata-driven retrieval. Together, these components translate the high-level design into a robust and extensible system. Figure 3 presents the implementation details of each module.

4.1 Supervisor Implementation

As shown in Fig. 3a, the supervisor is implemented as a control orchestrator that mediates between user input, task planning, and execution. Upon receiving a user query, the supervisor triggers a planning stage in which the request is decomposed into a sequence of subtasks. This plan is materialized as a directed

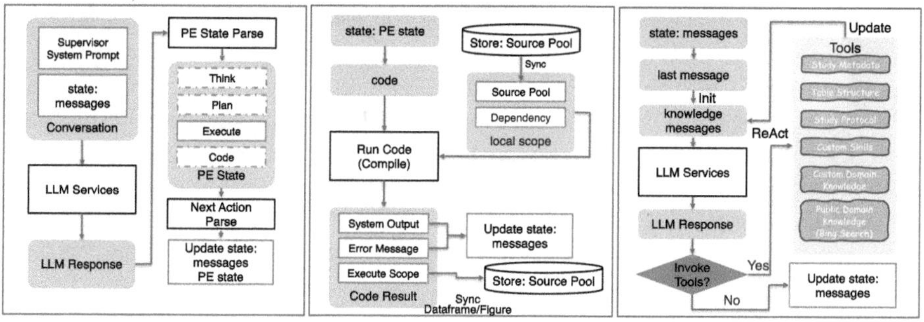

(a) Supervisor (b) Sensor (c) Knowledge

Fig. 3. Plan-and-Execute implementation. (a) The supervisor orchestrates task planning and execution. (b) The sensor executes Python code in a sandbox to retrieve grounded results. (c) The knowledge module integrates internal and external knowledge.

flow, where each step is explicitly defined in terms of required inputs, responsible components, and expected outputs.

To minimize hallucinations and maintain trustworthiness, the supervisor operates under a strict system prompt that encodes the non-assumption rule. The prompt requires the supervisor to ground every decision in verified results from tools or user input, and never speculate about missing information. An excerpt of this prompt is shown below:

Supervisor System Prompt

```
You are the Supervisor Agent.
Never make assumptions or speculate.
If required information is missing, first attempt retrieval
using the Sensor or Knowledge tools.
If the information is still unavailable, ask the user for
clarification and wait for their response before continuing.
Do not proceed to the next step until the current step has
been validated.
All conclusions must be grounded in verified tool outputs or
explicit user input.
```

During execution, the supervisor invokes subtasks sequentially while monitoring their outcomes. Crucially, progress from one step to the next is gated by validation: the supervisor requires a concrete, verified response from the invoked tool or agent (e.g., a successful Sensor execution result, a validated metadata hit, or an explicit user confirmation) before advancing. Validation can take several forms, including syntactic and semantic checks of returned data frames (schema verification, sample rows), explicit success flags from function tools, or user acknowledgement when human clarification was requested. If validation fails or is not received, the supervisor either triggers corrective actions (re-plan, retry with adjusted parameters, or fall back to alternative tools) or solicits targeted clarification from the user.

This strict validate-before-advance discipline forms the backbone of the supervisor's reliability: it ensures that each intermediate result is grounded and auditable, prevents cascading errors from speculative reasoning, and supports reproducible executions. By actively managing both planning and gating, the supervisor does more than route requests—it enforces correctness, orchestrates recovery strategies, and preserves the integrity of the overall Plan-and-Execute workflow.

4.2 Sensor Implementation

Figure 3b illustrates the implementation of the sensor pathway, which enables the agent to interact programmatically with its environment. The sensor is realized as a controlled code execution engine, where executable code fragments are run against underlying data or databases. These fragments can be dynamically generated by the LLM, drawn from provisioned internal custom skills (discussed further in Sect. 4.3), or constructed as a hybrid of the two. This design allows the sensor to flexibly support exploratory analysis while also ensuring determinism and reproducibility when internal skills are enforced.

The sensor's workflow begins with the supervisor passing down a subtask containing executable code. This code is executed in a sandboxed runtime environment, and outputs are collected in various forms: most commonly as structured data frames, but also as figures or natural-language insights, depending on the task. To increase robustness, the sensor maintains detailed error tracebacks whenever execution fails (e.g., due to syntax errors, missing dependencies, or unavailable data). These tracebacks are returned to the supervisor, enabling it to make informed decisions about how to proceed. Depending on the error, the supervisor may validate and retry with adjusted parameters, re-plan the workflow, or trigger a clarification request to the user.

A notable feature of this design is its support for hybrid execution, where LLM-generated code is combined with calls to internal custom skills. For example, the LLM may produce preprocessing code to normalize a user-uploaded dataset, after which it invokes a provisioned custom skill to perform a validated statistical analysis. This division of labor leverages the creativity of the LLM for general-purpose coding while enforcing determinism and correctness for critical analytical operations.

By combining flexible code generation, deterministic custom skills, and structured error reporting, the sensor implementation ensures that outputs are always grounded in verifiable computation. This makes it a critical foundation for reliable, reproducible, and auditable analysis within the plan-and-execute system.

4.3 Knowledge Implementation

The implementation of the knowledge module is depicted in Fig. 3c. In practice, knowledge is organized into two complementary categories: *internal knowledge* and *external public knowledge*, with strict priority given to the internal category. Both categories are encapsulated as callable agent tools, and their invocation follows the ReAct (Reason + Act) paradigm, where the system reasons over available options before deciding which tool to invoke. This design increases robustness in knowledge selection and ensures that every call is both justified and reproducible.

Internal knowledge encompasses all information that is specific to Eli Lilly & Company's clinical trials and digital health technology (DHT) studies. This includes study metadata (e.g., therapeutic area, device type), table schemas (e.g., field data types and descriptions), and study protocols. In addition, custom skills represent in-house analysis procedures and algorithms that are tightly integrated into the plan-and-execute framework. To guarantee determinism and reproducibility, the sensor is always required to invoke these custom skills whenever they are relevant, rather than relying on free-form code generation. Beyond structured metadata and skills, internal knowledge also extends to Lilly's proprietary domain definitions and conventions, which ensure consistent interpretation of data and terminology across studies. For any user query, the system always consults internal knowledge tools first, reflecting their central role in grounding responses in enterprise-specific assets.

External public knowledge serves as a fallback when a query cannot be resolved through internal sources. In these cases, the knowledge module leverages Azure Foundry BingSearch agents, encapsulated as external knowledge tools, to retrieve relevant information from the public domain. This mechanism ensures that users can still obtain meaningful responses to broad or exploratory questions, while maintaining a clear separation between enterprise-specific assets and general-purpose information.

By structuring the knowledge module into these two tiers of agent tools and orchestrating their invocation through the ReAct pattern, the implementation balances robustness, determinism, and flexibility. Internal tools provide authoritative and reproducible results, while external tools ensure coverage for queries that fall outside the scope of enterprise data. The supervisor enforces this ordering to guarantee that every query is resolved in the most reliable and contextually relevant way possible.

5 Case Studies and Evaluations

We illustrate the capabilities of Plan & Execute (P&E) through case studies that demonstrate how natural language intent is transformed into governed, reproducible analysis in the digital health context.

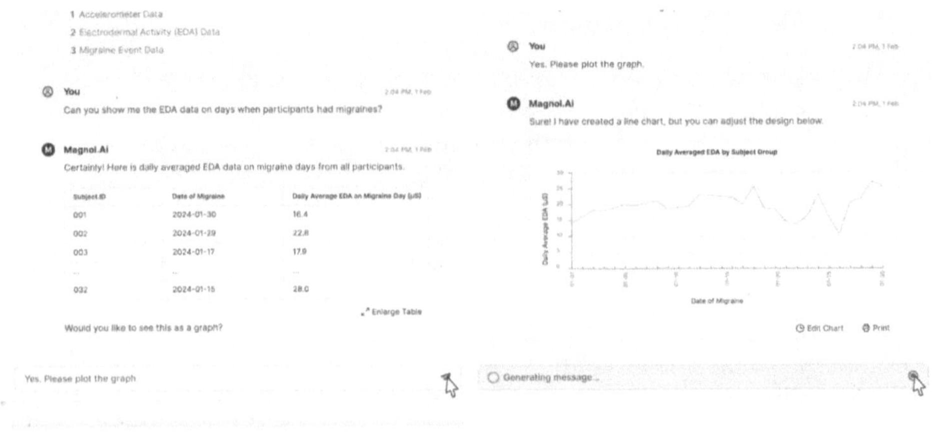

(a) (b)

Fig. 4. Examples of Plan & Execute (P&E) responses to user queries, illustrating raw sensor signal exploration as a data frame in (a) and as a graph in (b).

Exploring Raw Sensor Signals. Wearable devices generate high-frequency biomedical streams, including electrodermal activity (EDA) and accelerometry. These raw signals are often difficult to interpret directly, yet can reveal important physiological responses when visualized in context. With P&E, users can issue natural language queries such as "compare EDA signals before and after an identified disease episode." The framework translates this request into a structured plan: load manifest-approved data, segment signals into pre- and post-episode intervals, compute summary statistics, and render plots (Fig. 4). Sensors monitor for issues such as missing data or unit mismatches, and the supervisor enforces governance checks before outputs are released. This case highlights how P&E enables domain experts to move from raw physiological signals to interpretable visualizations without manual coding.

Analyzing Derived Digital Measures. From raw sensor data, Magnol.ai supports the derivation of digital measures such as physical activity and sleep. These measures are frequently used as digital biomarkers in clinical research, where a key question is whether they capture treatment- or drug-induced changes. P&E supports a conversational workflow in which a user can ask, for example, "compare average activity levels between treatment and control groups," or "plot sleep duration before and after intervention." The planner decomposes such requests into validated operations, the executor invokes statistical and visualization tools, and the supervisor ensures adherence to manifest-based governance rules (Fig. 5). This case demonstrates how P&E enables interactive, natural language analysis of derived measures, allowing researchers to test hypotheses about drug effects in a reproducible and compliant manner.

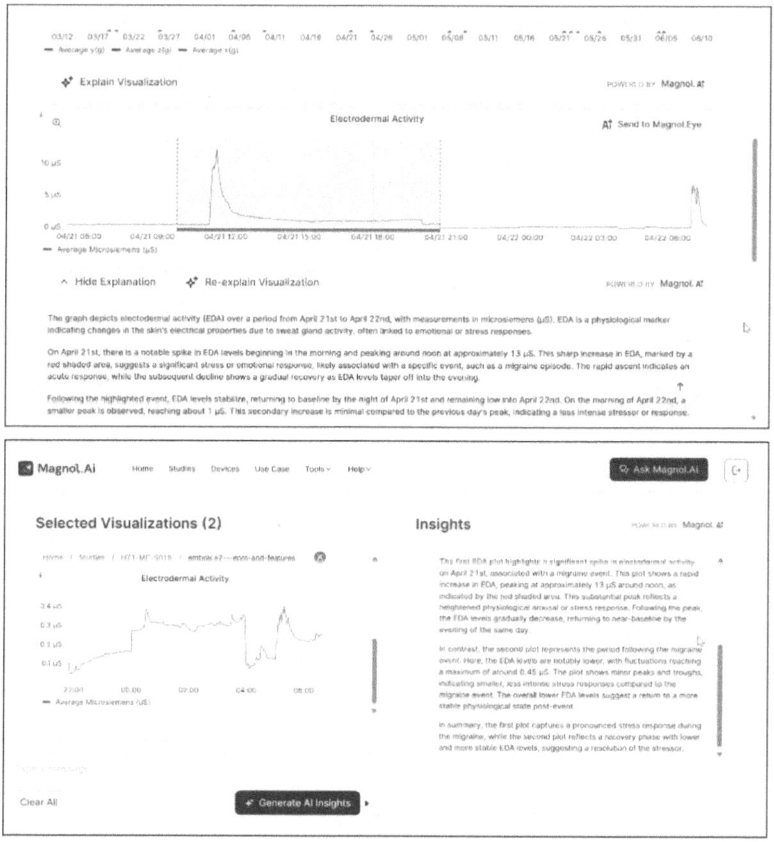

Fig. 5. Examples of Plan & Execute (P&E) responses to user queries. The first example shows visualization of electrodermal activity (EDA) signals before and after an identified disease episode. The second demonstrates conversational analysis of derived digital measures to assess drug-induced changes.

Together, these case studies illustrate how P&E supports both the exploration of raw sensor signals and the analysis of derived digital measures, bridging the gap between user intent and governed execution.

Evaluation. We evaluate Plan & Execute (P&E) along three dimensions: behavioral performance of the planning loop, enforcement of governance constraints, and correctness in domain-specific digital health analytics.

- *Behavioral performance.* We assess how reliably the planner generates valid execution plans from natural language queries, and how effectively the supervisor detects and repairs errors. In a set of exploratory prompts covering signal visualization, digital measure summaries, and treatment effect comparisons, P&E produced valid plans on the first attempt in most cases. For

the remainder, the supervisor's repair mechanism guided replanning until a compliant execution path was achieved. This highlights the benefit of separating planning, execution, and reflection into explicit stages.
- *Governance enforcement.* A core feature of P&E is that every step is subject to manifest-based constraints and supervisor checks. During evaluation, all user-specified data sources were verified against the manifest, and ambiguous identifiers were resolved through study affirmation before analysis proceeded. We also injected intentional violations such as attempts to access undeclared resources and unbounded table scans; in each case, the supervisor halted execution and issued a repair prompt. These results demonstrate that P&E enforces governance rules consistently, ensuring that no analysis escapes compliance boundaries.
- *Domain correctness.* For the EDA case study, P&E correctly segmented signals before and after an annotated disease episode, and the resulting plots matched baseline scripts produced by data scientists. For derived digital measures, such as activity and sleep, P&E reproduced group-level summaries and intervention comparisons that aligned with reference analyses, confirming analytic validity. Error injections such as unit mismatches and synthetic missingness spikes were detected by sensors and flagged by the supervisor, preventing invalid results from propagating.
- *Reproducibility.* Each execution produced a provenance digest capturing plan steps, tool calls, sensor signals, and supervisor decisions. Re-running the same query under the same manifest yielded identical digests, confirming deterministic replay. This demonstrates that the framework not only generates compliant outputs but also supports reproducible and auditable workflows.

Together, these results indicate that P&E achieves both usability and trustworthiness: it translates natural language goals into governed analysis pipelines while preserving reproducibility and safety.

6 Discussion

The Plan & Execute (P&E) framework illustrates how large language models (LLMs) can be harnessed to bridge the gap between natural language intent and governed execution in digital health analytics. Our evaluation highlights both the promise and the challenges of this approach.

Strengths. P&E demonstrates that agentic planning can be made safe and auditable when coupled with governance primitives such as study affirmation, manifest-driven source control, and anti-pattern detection. Case studies show that users can request complex operations in natural language while retaining the guarantees of reproducibility and compliance. The provenance digest mechanism ensures that results can be verified and re-run deterministically, a requirement in regulated settings.

Limitations. Despite these strengths, the system inherits the variability of LLM outputs. Plans occasionally require supervisor-guided repair, and the reliability of outputs depends on prompt design and model tuning. Current evaluation remains small-scale and scenario-driven; larger studies are needed to assess generalization across diverse data modalities. Runtime overhead, while modest, may still be a barrier for very large-scale analytics. Finally, the system's effectiveness depends on the quality of manifests and study registries, which themselves require ongoing curation.

Broader Implications. P&E contributes to a growing movement toward *agentic AI systems* that embed reasoning, reflection, and governance. For digital health, the ability to provide no-code analytics that are both powerful and compliant could accelerate the use of real-world data and digital biomarkers in research and care delivery. More generally, the approach offers a template for deploying LLM-driven assistants in other regulated data domains such as finance, law, and public policy.

Future Opportunities. Integrating richer domain knowledge (e.g., ontologies, clinical trial standards), expanding the set of sensors for anomaly detection, and formalizing supervisor policies as verifiable contracts are promising directions. Human-in-the-loop modes, where analysts can inspect and modify plans before execution, could further enhance trust and adoption. Ultimately, we envision P&E as part of a broader ecosystem of intelligent, governed assistants that extend the reach of data platforms like Magnol.ai.

7 Conclusion and Future Work

We have introduced Plan & Execute (P&E), an agentic framework that integrates large language models (LLMs) with the Magnol.ai digital health technology sensor cloud to enable natural language-driven, no-code data analysis. P&E combines a planning–execution–reflection loop with explicit governance mechanisms, including study affirmation, manifest-scoped source control, and anti-pattern detection. Through case studies and evaluation, we demonstrated how the framework transforms ambiguous user intent into governed, auditable analytics while maintaining reproducibility and compliance.

This work suggests that agentic AI can extend data platforms beyond passive storage and retrieval into active, trustworthy assistants for complex analytics. While promising, the framework remains limited by the variability of LLM planning and the need for carefully curated manifests and registries. Larger-scale evaluations and integration with additional biomedical and real-world data sources will be important next steps.

Future work will expand supervisor policies into verifiable governance contracts, broaden the library of sensors for anomaly detection, and explore interactive, human-in-the-loop modes to further enhance trust. Ultimately, we envision

P&E as a blueprint for embedding safe and transparent agentic AI into regulated data platforms, supporting both digital health innovation and wider applications in other domains where governance is paramount.

References

1. Zhang, Q., Yang, L., Chen, Z., Zhou, M., Yu, Y., Lin, Y.: AutoML: a survey of the state-of-the-art. Knowl.-Based Syst. **212**, 106622 (2020)
2. Zaharia, M., et al.: Accelerating the machine learning lifecycle with MLflow. Proc. VLDB Endow. **11**(12), 2703–2706 (2018)
3. Wongsuphasawat, K., Moritz, D., Anand, A., Mackinlay, J., Howe, B., Heer, J.: Voyager 2: augmenting visual analysis with partial view specifications. In: Proceedings of CHI Conference on Human Factors in Computing Systems, pp. 2648–2659. ACM, New York (2017)
4. Chase, H.: LangChain: building applications with LLMs through composability. https://www.langchain.com/. Accessed 10 Jan 2025
5. Microsoft: Semantic Kernel: An open-source SDK for integrating large language models with conventional programming languages. https://github.com/microsoft/semantic-kernel. Accessed 10 Jan 2025
6. Richards, T.B.: AutoGPT: An autonomous GPT-4 experiment. https://github.com/Torantulino/Auto-GPT. Accessed 10 Jan 2025
7. Mitchell, M., et al.: Model cards for model reporting. In: Proceedings of FAT* Conference on Fairness, Accountability, and Transparency, pp. 220–229. ACM, New York (2019)
8. Stoyanovich, J., Howe, B.: Responsible data management. Proc. VLDB Endow. **13**(12), 3474–3488 (2020)
9. Zhang, H., et al.: Digital data platform for connected clinical trials. In: 2022 IEEE International Conference on Big Data (Big Data), pp. 2446–2453. IEEE, Los Angeles (2022). https://doi.org/10.1109/BigData55660.2022.10020925
10. Ji, J.: Assessing nocturnal scratch with actigraphy in atopic dermatitis patients. NPJ Dig. Med. **6**(1), 72 (2023)
11. Zhang, H., et al.: Magnol.Ai—an internet of medical things (IoMT) platform for digital health research. In: 2024 IEEE International Conference on Big Data (Big Data), pp. 5103–5112. IEEE, Sorrento (2024)
12. Zhang, H., et al.: Engineering large wearable sensor data towards digital measures. In: 2023 IEEE 8th International Conference on Big Data Analytics, pp. 14–23. IEEE, Xiamen (2023)
13. Zhang, H., et al.: A framework for digital data quality assessment in digital biomarker research. In: The Ninth International Conference on Big Data, Small Data, Linked Data and Open Data (ALLDATA 2023). IARIA, Venice (2023)
14. Zhang, H., et al.: Analyzing and reporting wearable sensor data quality in digital biomarker research. Int. J. Adv. Life Sci. 72–86 (2023)
15. Ruan, G., Zhang, H., Xiao, H.: Magnol.AI copilot: multimodal LLMs for natural, accessible data interaction. In: 2025 IEEE/CVF International Conference on Computer Vision Workshops (to appear). IEEE, Milan (2025)

CURATE-FT: A Modular Framework for High-Fidelity Data Curation for Fine-Tuning

Mishra Sushanta Kumar(✉), Aum Khant, Vivek Dedani, Sangeeth Keeriyadath, and Vijay Karthik

State Street Corporation, Ahmedabad, India
{SMISHRA47,Akhant,VChandreshkumarDedani,SKeeriyadath, KKarthik}@StateStreet.com

Abstract. In highly regulated financial sectors, essential business knowledge is embedded within thousands of Standard Operating Procedure (SOP) documents in PDF and DOCX formats. Existing retrieval-augmented generation (RAG) techniques often experience limited accuracy when applied to specialized domain documentations, which are replete with specialized terminology, complex workflows, stringent compliance requirements, and critical decision points. This presents a compelling case for targeted large language model (LLM) fine-tuning. Our research indicates that developing training datasets for different business functions including trade life cycle management, settlement processes, and exception resolution—necessitates a systematic and domain-specific methodology. To address these challenges, we introduce **CURATE-FT** (*C*uration and *U*nified *R*easoning with *A*ugmentation for *T*argeted *E*xtraction—*F*ine-*T*uning), a scalable and reusable framework designed for the curated development of high-quality training datasets. CURATE-FT begins by extracting structured elements from documents and predefined question categories to generate optimized original Question & Answer (Q&A) sets, subsequently enhancing them through synthetic augmentation. Each synthetic Q&A instance is rigorously evaluated using a Chain-of-Thought (CoT) reasoning filter, assessing its alignment, impartiality, coherence, and completeness to ensure the highest quality selections. The framework then recommends an optimal blend of original and augmented Q&As for each question category, thereby maximizing model generalization and managing fine-tuning expenses. Empirical assessments on custody-related documents demonstrate that models refined with CURATE-FT curated data using LoRA-X achieve 20% faster convergence, a 25% reduction in training duration, and a 10% increase in accuracy. These improvements directly contribute to lower operational risk and more robust compliance support. The modular architecture of CURATE-FT further enables seamless adaptation across other critical, domain-focused applications.

Keywords: Q&A pairs · CoT filtering · LLMs

H. R. Arabnia et al. (Eds.): CSCI 2025, CCIS 2972, pp. 217–232, 2027.
https://doi.org/10.1007/978-3-032-28469-3_15

1 Introduction

The rapid deployment of large language models (LLMs) has made *high-quality data* decisive for state-of-the-art performance across NLP tasks. While scale and architecture have advanced, careful data curation [1]—especially for fine-tuning—has emerged as essential for aligning outputs to specialized domains where accuracy, context preservation, and regulatory compliance are critical [2, 3].

We focus on the custody domain, where institutions safeguard client assets, yet the methodology generalizes to other complex financial and non-financial settings with structured, multi-modal artifacts. Such documents introduce challenges that hinder retrieval and downstream applications:

- **Data heterogeneity and inconsistency**: diverse formats spanning unstructured text, structured tables, legal contracts, and hierarchical forms.
- **Multi-modal and context-limited inputs**: combinations of text, numerical tables, charts, and images with fragmented records and dense language, making coherent multi-modal capture difficult for generic datasets.
- **Domain-specific terminology and regulatory complexity**: specialized jargon, legal clauses, and jurisdictional variation requiring precise context and vocabulary alignment to avoid costly errors.
- **Confidentiality and data sensitivity**: presence of PII and proprietary processes necessitating strict compliance (e.g., GDPR, CCPA), limiting access to high-quality training data [4] and motivating robust anonymization/differential privacy.
- **Imbalanced and sparse edge cases**: rare but critical scenarios (e.g., regulatory exceptions, market anomalies) are underrepresented [5], creating potential blind spots in model behavior.

This paper presents a scalable, compliant, and effective data-curation framework that addresses these challenges. Through empirical analyses and case studies, we outline strategies for constructing datasets that reflect real-world workflows, regulatory requirements, and institutional policies—supporting LLMs that are not only powerful but also safe, accurate, and trustworthy across complex domains, including financial custody.

2 Related Works

Data curation is a critical step in fine-tuning large language models (LLMs) for specific domains. Effective data curation strategies ensure that the model is trained on high-quality, relevant and diverse datasets, which can significantly improve its performance in the target domain.

2.1 Existing Open Source Data Curation Tools

Various data curation tools, such as Amazon Kendra, have been evaluated. Amazon Kendra is an AI-powered enterprise search service that indexes, searches, and retrieves information from both structured and unstructured sources, including PDFs, Word documents, and databases. Leveraging advanced natural language processing (NLP), Kendra interprets context and extracts relevant information, making it suitable for tasks like Q&A pair curation.

For Q&A extraction from PDFs, the typical workflow involves uploading documents to a supported data source (e.g., Amazon S3), configuring Kendra for ingestion and indexing, and utilizing its built-in FAQ extraction capabilities. Kendra can automatically identify and extract Q&A pairs when documents are clearly formatted. After indexing, users can perform natural language searches, and Kendra returns the most relevant answers or document excerpts.

However, SME evaluation of Kendra's automated Q&A extraction across approximately 100 SOP documents revealed limitations in both information recall and the quality of generated Q&A pairs.

2.2 Evolution of Domain Adaptation Datasets

There are various datasets available which can be leveraged for fine tuning to achieve domain adaptation. There has been significant contribution by open-source community to have domain specific datasets across domains. Ex - The availability of medical data and advancements in medical natural language processing (NLP) have enabled the development of such datasets. Med-PaLM, an LLM refined from Google's PaLM model family, exemplifies this progress. Med-PaLM 2, an enhanced version of PaLM 2, shows improved performance on LLM benchmarks. It was fine-tuned for medical purposes using the MultiMedQA suite, which includes datasets like Med-QA, MedMCQA, HealthSearchQA, Live-QA, and Medication-QA. We can do the comparison of Med-PaLM 2 with other models based on below image (Fig. 1).

2.3 Domain Adaptation Datasets for Finance

Several finance-domain datasets have been developed, including:

- **FINER-ORD** [6]: A high-quality English Financial Named Entity Recognition dataset, addressing semantic and lexical challenges in financial texts for NER and NLP benchmarking.
- **Fin-QA** [7]: Focuses on numerical reasoning over financial data, highlighting gaps between pre-trained models and expert human performance.

While these public datasets are valuable, they do not provide Q&A pairs specific to the custody domain. The MedPalm2 example further demonstrates that self-curated datasets can significantly enhance the accuracy and efficiency of fine-tuned models. Therefore, we developed the **CURATE-FT** framework to address our domain-specific requirements.

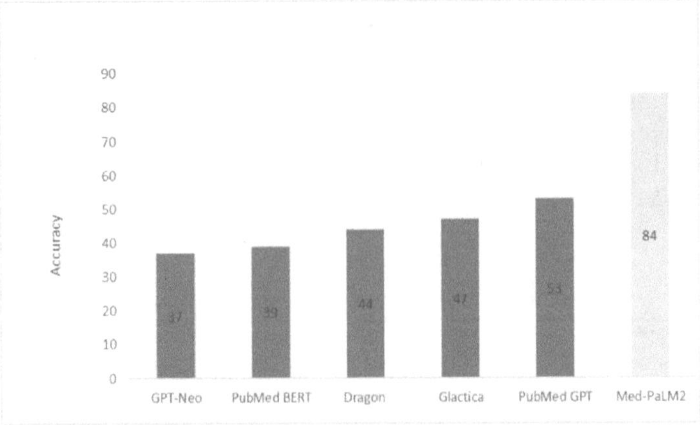

Fig. 1. Domain based dataset curation evolved over the years which is highlighted with Med-PaLM 2 expert level performance.

3 Methodology Followed

Our methodology generates multiple Q&A pairs across distinct question types, tailored to the content and context of each source document using custom-designed prompts. After generation, all Q&A pairs are processed through a Chain of Thought (CoT) filtering mechanism to optimize dataset quality.

- **Text Extraction Prompt:** Facilitates extraction of textual content from PDF images, outputting results in markdown format for downstream processing.
- **Original Q&A Prompts:** Five distinct prompts, each targeting a specific question type, systematically generate original Q&A pairs aligned with the document.
- **Diverse Q&A Prompt:** A comprehensive prompt, based on the Alpaca self-instruct technique [8], creates diverse Q&A pairs spanning various question types to enhance dataset variation.
- **CoT Filter:** All diverse Q&A pairs are subjected to Chain of Thought (CoT) filtering to further improve the quality of the curated dataset.

3.1 PDF to Formatted Text

The effectiveness of this study hinges on robust Information Extraction from Standard Operating Procedure (SOP) documents, which are typically available in complex PDF formats containing text, tables, and images. These PDFs often feature multi-column layouts, footnotes, headers, and footers, all of which disrupt the logical reading order and pose significant challenges for extracting coherent and contextually accurate Q&A pairs. Furthermore, critical information is frequently embedded within tables and figures or dispersed across multiple sections, making it difficult to interpret and consolidate relevant content for precise question answering. Accurate Information Extraction is therefore essential—not only to capture all pertinent details but also to preserve the original document's structure and context. To achieve this, PDF pages are converted to images and processed by large language models (LLMs) using carefully designed prompts, resulting in structured markdown outputs that maintain the integrity of various modalities (text, images, tables) and the contextual flow of the original document. This comprehensive extraction process forms the foundation for our subsequent multi-prompt approach, where the formatted content is used to generate high-quality Q&A pairs for different question types, as detailed in the following section.

3.2 Generating Original (Q&A) Pairs

The initial curated dataset containing Q&A pairs was confined to Factual and Conceptual questions [9]. This dataset served as the foundation for triggering the fine-tuning pipeline of the Large Language Model (LLM). The objective was to validate the results and assess the model's comprehension of the domain. When tested on the evaluation dataset (150 Q&A pairs shared by SME), it was not able to answer some of the questions related to defining procedures or steps mentioned in SOP.

Consequently, procedural questions and answers were incorporated into the dataset. This addition aimed to ensure that the model could understand and provide complete information regarding any defined procedures within the domain. Further observations indicated that the fine-tuned LLM struggled with questions related to reasoning. To address this, reasoning-based domain questions were added to the dataset. Similarly, to enhance the model's inferential capabilities, inferential questions were included in the data synthesis. In the below table (Table 1), we are denoting each of the Question types used for generation of Original Q&A pairs, with their corresponding notation.

Table 1. Number of Original Q&A Pairs by Question Types

Question Types	No. of Original Q&A
Factual	Q_o^{F}
Conceptual	Q_o^{C}
Procedural	Q_o^{P}
Reasoning	Q_o^{R}
Inferential	Q_o^{I}

3.3 Getting Optimized Number of (Q&A) Pairs

To ensure the creation of a high-quality dataset, it is crucial to determine an optimal number of Question-Answer (Q&A) pairs which covers both information recall along with quality. The optimized no of Original Q&A pairs were determined through iterative process. The optimal no of Q&A pairs for various question types is defined as:

$$Q_o^{\mathrm{F}} = 0.005 \cdot T, \quad Q_o^{\mathrm{P}} = 0.0015 \cdot T$$

$$Q_o^{\mathrm{C}} = 0.002 \cdot T, \quad Q_o^{\mathrm{I}} = 0.002 \cdot T, \quad Q_o^{\mathrm{R}} = 0.002 \cdot T$$

where T is the total number of tokens in the domain.

3.4 Generating Diverse (Q&A) Pairs

Subject Matter Experts (SMEs) noted that the LLM answered questions well when they mirrored its training data but struggled with rephrased versions. This revealed limited input diversity. To address this, the Alpaca self-instruct method was used to generate varied Q&A pairs from the original dataset. The resulting diverse Q&A pairs are summarized in the table below (Table 2). While curating the diverse Q&A pairs, all the original Q&A pairs were considered.

Table 2. Relationship Between Original and Diverse Q&A Pairs by Question Type. *Note:* ϕ is the multiplying factor which determines the number of variations of original Q&A pairs to create different diverse Q&A pairs.

Question Types	Original Q&A (Q_o^{X})	Diverse Q&A ($Q_d^{\mathrm{X}} \subseteq Q_o^{\mathrm{X}}$)
Factual	Q_o^{F}	$Q_d^{\mathrm{F}} = \phi \cdot Q_o^{\mathrm{F}}$
Conceptual	Q_o^{C}	$Q_d^{\mathrm{C}} = \phi \cdot Q_o^{\mathrm{C}}$
Procedural	Q_o^{P}	$Q_d^{\mathrm{P}} = \phi \cdot Q_o^{\mathrm{P}}$
Reasoning	Q_o^{R}	$Q_d^{\mathrm{R}} = \phi \cdot Q_o^{\mathrm{R}}$
Inferential	Q_o^{I}	$Q_d^{\mathrm{I}} = \phi \cdot Q_o^{\mathrm{I}}$

3.5 CoT Reasoning Filter

To enhance the diversity and completeness of training data for instruction-tuned language models, we propose a scoring and filtering mechanism rooted in Chain-of-Thought (CoT) reasoning. Our approach ensures that only high-quality diverse Question-Answer (Q&A) pairs are selected, using statistically governed thresholds and a parameterized scoring function. Table 3 depicts the formation of Consolidated Q&A post CoT filter. It also denotes the different notations used for consolidated Q&A pairs.

Each original Q&A pair (Q_o^i, A_o^i) serves as the seed for generating multiple diverse Q&A variants. These variants are intended to introduce lexical and structural diversity without compromising fidelity or coherence. The dataset D is defined as:

$$D = \left\{ \left(Q_o^i, A_o^i, \{(Q_d^{i,j}, A_d^{i,j})\}_{j=1}^{M_i}\right) \right\}_{i=1}^{N}$$

where:

- N: Total number of original Q&A pairs.
- M_i: Number of diverse Q&A variants generated for the ith original pair.

- **CoT Evaluation Parameters**
  Each diverse Q&A pair $(Q_d^{i,j}, A_d^{i,j})$ is evaluated using six critical CoT reasoning parameters:
  1. **Semantic Alignment** – Measures topical and contextual consistency with the original Q&A.
  2. **Terminology & Clarity** – Assesses language structure, formatting, and absence of ambiguity or typographical errors.
  3. **Faithfulness** – Ensures that key factual content from the original answer is preserved.
  4. **Logical Consistency** – Verifies logical coherence between question and answer.
  5. **Bias & Ethics** – Checks for neutrality, fairness, and ethical phrasing.
  6. **Completeness** – Determines whether the diverse answer includes all critical details found in the original.

  Each parameter is scored as:

$$s_k^{i,j} \in \{0.0, 0.5, 1.0\}$$

  where:
  - $s_k^{i,j}$: Score for the kth parameter of the jth diverse pair for the ith original Q&A.
  - Scores represent "Poor" (0.0), "Average" (0.5), and "Good" (1.0).
- **Threshold-Based Validity Filtering**
  To ensure that each diverse Q&A pair meets a minimum quality bar, we define a threshold vector $\boldsymbol{\tau} = (\tau_1, \ldots, \tau_6) \in [0,1]^6$ representing the acceptable score floor for each CoT parameter. A pair is considered valid only if all parameter scores exceed their respective thresholds:

$$V(i,j) = \begin{cases} 1, & \text{if } s_k^{i,j} \geq \tau_k \quad \forall k = 1, \ldots, 6 \\ 0, & \text{otherwise} \end{cases}$$

  This binary function acts as a hard filter to eliminate semantically or ethically deficient pairs before scoring aggregation.

– **Weighted Scoring**
To account for the relative importance of each parameter, we define a normalized weight vector $\mathbf{w} = (w_1, \ldots, w_6)$, such that $\sum_{k=1}^{6} w_k = 1$. The composite score $\mathcal{S}^{i,j}$ for each diverse Q&A pair is then computed as:

$$\mathcal{S}^{i,j} = \sum_{k=1}^{6} w_k \cdot s_k^{i,j}$$

where:
- w_k: Weight for the kth parameter.

– **Optimal Subset Selection**
The final set of diverse Q&A pairs included in the dataset is defined as:

$$D^* = \left\{ (Q_o^i, A_o^i, Q_d^{i,j}, A_d^{i,j}) \mid V(i,j) = 1 \land \mathcal{S}^{i,j} \geq \theta \right\}$$

where:
- θ: Predefined threshold for the aggregate score $\mathcal{S}^{i,j}$.

Table 3. Consolidated Q&A Dataset with CoT-Filtered Diverse Q&A. *Note:* The CoT-filtered diverse Q&A pairs, denoted as $Q_d^{\text{X'}}$, are a subset of the diverse Q&A pairs ($Q_d^{\text{X'}} \subseteq Q_d^{\text{X}}$). The consolidated Q&A dataset (Q_c^{X}) is formed by combining the original Q&A pairs (Q_o^{X}) with the CoT-filtered diverse Q&A pairs ($Q_d^{\text{X'}}$).

Question Types	Original Q&A (Q_o^{X})	Consolidated Q&A (Q_c^{X})
Factual	Q_o^{F}	$Q_c^{\text{F}} = Q_o^{\text{F}} + Q_d^{\text{F'}}$
Conceptual	Q_o^{C}	$Q_c^{\text{C}} = Q_o^{\text{C}} + Q_d^{\text{C'}}$
Procedural	Q_o^{P}	$Q_c^{\text{P}} = Q_o^{\text{P}} + Q_d^{\text{P'}}$
Reasoning	Q_o^{R}	$Q_c^{\text{R}} = Q_o^{\text{R}} + Q_d^{\text{R'}}$
Inferential	Q_o^{I}	$Q_c^{\text{I}} = Q_o^{\text{I}} + Q_d^{\text{I'}}$

4 Benchmarking

To assess the effectiveness of the proposed methodology, we conducted a comprehensive benchmarking process using multiple approaches. These evaluations were designed to measure the performance of the fine-tuned LLM across various scenarios and metrics which signifies the quality and diversity of the Consolidated Q&A pairs generated using the proposed approach.

To ensure reproducibility and clarity in our experimental setup, we summarize the key fine-tuning parameters used across all model variants. The Table 4 lists the rank, scaling factor, dropout rate, and target modules applied during fine-tuning.

Table 4. Fine-Tuning Parameters Configuration

Parameter	Values
Rank (r)	32
Scaling factor (α)	$2r$
Dropout rate	0.05
Target Modules	q_proj, v_proj, o_proj, gate_up

4.1 Evaluation Using DeepEvals

The benchmarking approach involved assessing the fine-tuned model using a set of 250 Q&A pairs curated by Subject Matter Experts (SMEs). These Q&A pairs were designed to encompass frequently asked queries related to the domain (Fig. 2).

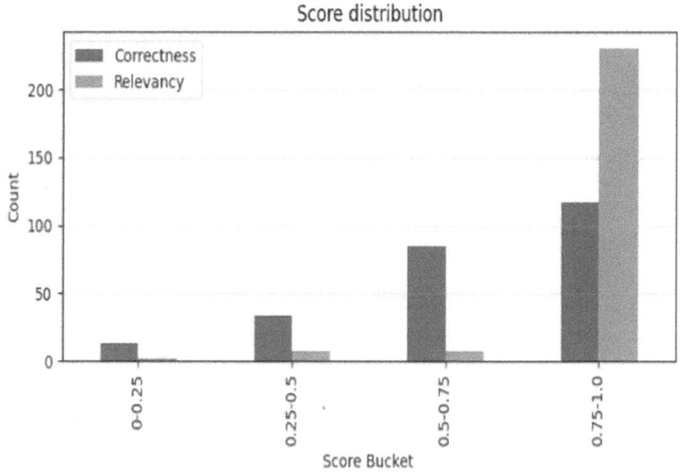

Fig. 2. Score Bucket Vs Q&A Count with respect to Correctness and Relevancy.

During the Evaluation using open-source benchmarking Framework such as Deepeval, we observed correctness is right-skewed: 117 Q&A (47.0%) fall in the 0.75–1.0 bucket, 85 Q&A (34.1%) in 0.5–0.75, with a residual tail of 34 Q&A (13.7%) in 0.25–0.5 and 13 Q&A (5.2%) in 0–0.25. In contrast, relevancy is overwhelmingly high: 231 Q&A (92.8%) of responses land in 0.75–1.0, with only 8 Q&A (3.2%) each in 0.5–0.75 and 0.25–0.5, and 2 Q&A (0.8%) in 0–0.25. These distributions confirm the model remains consistently on-topic using the open-source evaluations.

The gap between strong topical alignment and lower exactness suggests that remaining errors are dominated by string-precision issues (IDs, numbers, mapping codes, dates) rather than conceptual drift. Prioritizing retrieval-and-verify patterns using regular expression checks for these fields should compress the low-correctness tail without sacrificing the high relevancy already achieved.

4.2 Evaluation on SME-Provided Q&A Pairs Using LLM as a Judge

Here we have used the same Q&A pairs curated by Subject Matter Experts (SMEs) to have a consistent evaluations for the domain understanding.The fine-tuned model's responses were reviewed for accuracy and relevance, with SMEs evaluating the correctness and completeness of the answers. Additionally, the evaluation process incorporated the use of the LLM itself as a judge to provide an objective assessment of the Q&A pairs. This approach established a baseline for understanding the model's ability to address domain-specific queries effectively.

LLM as a judge [10] has shown that it is an efficient method, particularly for tasks like Q&A evaluation, where traditional metrics often fall short. This approach enabled scalable evaluation, focusing specifically on the correctness and completeness of the responses.

To clarify the evaluation protocol and ensure transparency, In Table 5, we detail the main components and controls used in our automated answer assessment pipeline. The following table summarizes the model, prompt design, decoding settings, aggregation strategy, bias controls, and reporting methodology.

Table 5. Automated Answer Evaluation Pipeline: Component Summary

Component	Implementation
Model	GPT-4o
Prompt	The prompt operationalizes a structured evaluation protocol, guiding the language model to objectively assess generated answers against reference answers using predefined criteria. It ensures consistent, criteria-driven scoring and explanations for each response, while minimizing subjective interpretation and irrelevant variations. This approach supports reliable, automated evaluation at scale
Decoding	Default settings; model_kwargs specify response format as JSON
Aggregation	Single LLM evaluation per item
Bias controls	(i) Prompt instructs to ignore style/template differences; (ii) scoring based on factual content; (iii) JSON-only structured output which standardizes responses, hence reducing subjectivity
Reporting	Accuracy and completeness per item; overall and question-type-wise reported

- **Analysis of Results - Data Volume and Accuracy**: Fig. 3 illustrates the relationship between dataset composition and model performance. Three datasets—comprising of 40K, 52K, and 80K original Q&A pairs—were evaluated, each progressively incorporating a broader range of question types. The 40K dataset includes only Factual and Conceptual questions, while the 52K set introduces Procedural questions. The most comprehensive, 80K dataset, further adds Reasoning and Inferential types. A clear upward trend in accuracy is observed: 48% for 40K, 70% for 52K, and 79% for 80K Q&A pairs when validated on SME provided Q&A pairs. This progression underscores the positive impact of both increased data volume and question-type diversity on model accuracy. The results suggest that enriching the dataset with varied cognitive demands enables the model to generalize better and handle more complex queries effectively.

– **Analysis of Results - Diverse Q&A pair with respect to pre and post COT filter**: Fig. 4 represents a comparative analysis of model performance under three distinct data curation approaches.The first strategy, labeled "Original Q&A," utilizes a dataset of 80,000 entries achieving an accuracy of 79%. The second strategy, "Consolidated Q&A Pre-CoT Filter," expands the dataset to 200,000 entries and yields an improved accuracy of 84%. The third and most refined strategy, "Consolidated Q&A Post-CoT Filter," reduces the dataset to 112,000 entries but significantly improves the accuracy to 91%. This progression demonstrates that strategic data consolidation and filtering, particularly post-CoT (Chain-of-Thought) filtering, can lead to substantial gains in model accuracy, even with a reduced data volume. The results emphasize the importance of quality and relevance in data curation over sheer quantity.

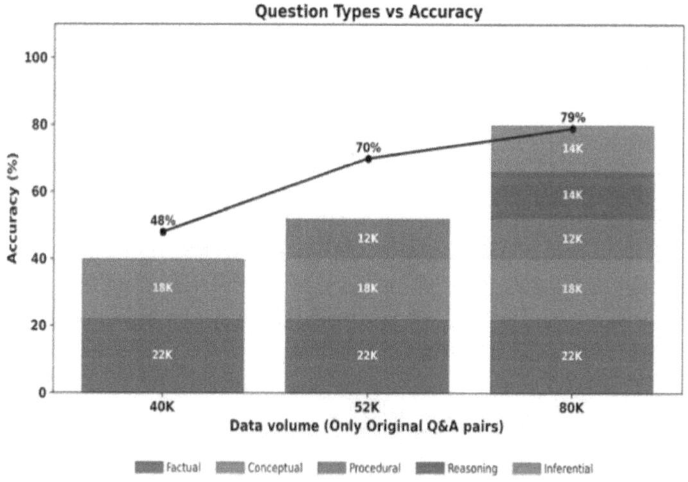

Fig. 3. Question Types vs Accuracy.

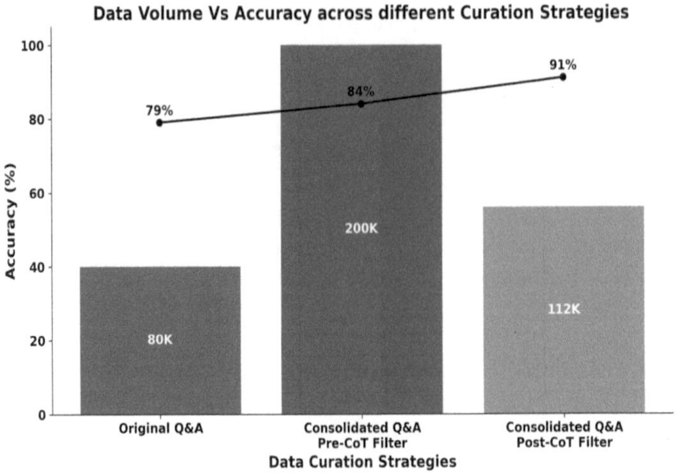

Fig. 4. Data Volume Vs Accuracy across different Curation Strategies.

4.3 Comparison with RAG Pipeline

In the second approach, we implemented a Retrieval-Augmented Generation (RAG) pipeline using the same domain documents that were used for fine-tuning the model. The RAG pipeline combines document retrieval with generative capabilities to answer questions by leveraging external knowledge from the domain documents. The accuracy of the fine-tuned model was compared against the RAG pipeline by testing both systems on the same set of questions. This comparison provided insights into the strengths and limitations of the fine-tuned model in contrast to a retrieval-based approach.

The Fig. 5 depicts the accuracy on validation dataset (generated by RAG pipeline) of the fine-tuned models which were trained on consolidated dataset (post CoT filter).

The Fig. 6 illustrates how model accuracy varies as the number of documents increases. Overall, accuracy tends to decline with larger document sets, though the rate of degradation differs across approaches. Fine-tuned models maintain relatively high and stable performance, while retrieval-augmented methods show more pronounced drops, especially in vanilla configurations. This suggests that scaling document count introduces complexity that impacts retrieval-based systems more severely than fine-tuned models.

4.4 Evaluation Using ROUGE-L Score

The third benchmarking method involved evaluating the fine-tuned model using the ROUGE-L metric, which measures the overlap of the longest common subsequence (LCS) between the model's generated answers and the reference answers. This metric was chosen to quantify the quality of the model's responses in terms of precision and recall. The ROUGE-L score provided an objective measure of

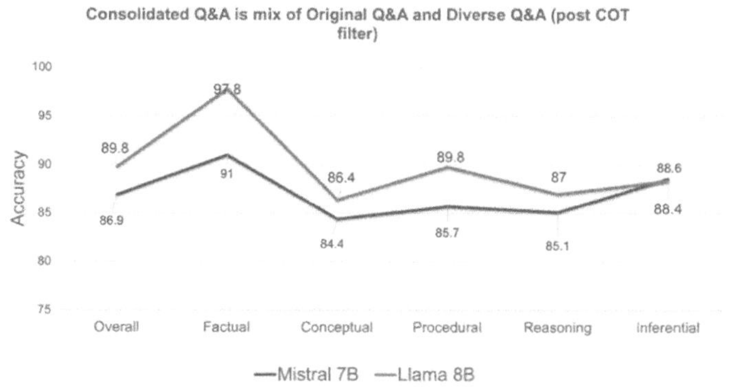

Fig. 5. Question type wise accuracy.

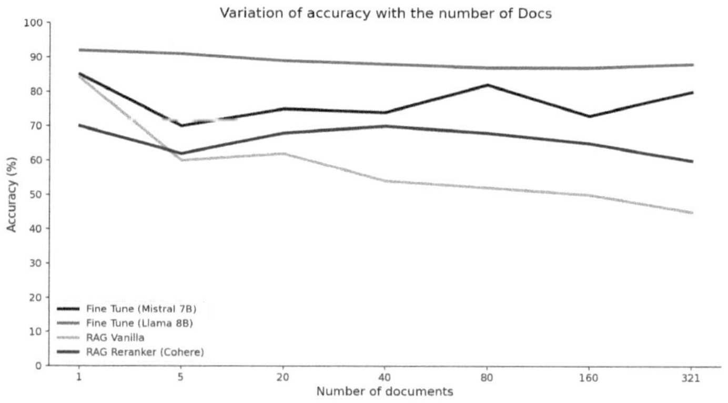

Fig. 6. The line chart shows comparison among RAG Vanilla, RAG with Re-ranker and Fine-tuned LLM.

how closely the model's outputs aligned with the expected answers, offering a standardized way to evaluate its performance (Fig. 7).

4.5 Statistical Robustness

To assess the robustness of our curated dataset for fine-tuning, we conducted five independent runs on post-CoT filter data with different random seeds for both Llama 8B and Mistral 7B models. We report the mean and standard deviation of accuracy(LLM as a judge) across these runs to quantify the stability of our results in Table 6.

Table 6. Accuracy (%) across five runs with different random seeds for Llama 8B and Mistral 7B.

LLM	Mean ± Std	1st run	2nd run	3rd run	4th run	5th run
Llama 8B	89.8 ± 0.50	89.1	90.5	89.8	89.4	90.2
Mistral 7B	86.9 ± 0.44	86.5	86.9	86.3	87.5	87.2

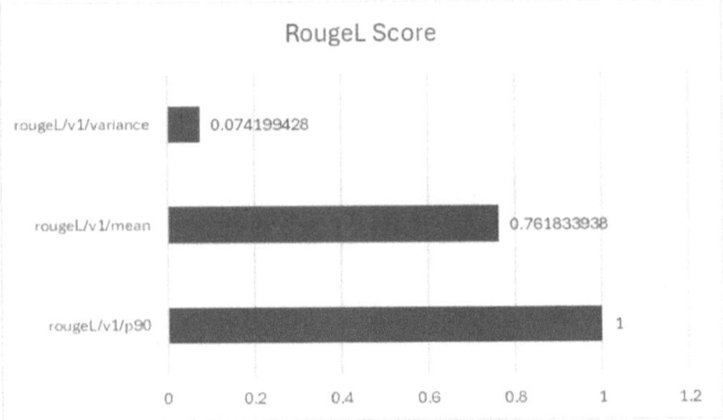

Fig. 7. Mean ROUGE-L Score, 90th Percentile ROUGE-L Score and Variance of ROUGE-L Score.

5 Conclusion

This work introduces **CURATE-FT**, a robust and scalable framework for domain-specific data curation, designed to address the critical challenges of data quality, diversity, and domain alignment in fine-tuning large language models (LLMs) for high-stakes environments. Through systematic generation and rigorous filtering of both original and diverse Q&A pairs—leveraging Chain-of-Thought (CoT) reasoning—CURATE-FT demonstrates substantial improvements in model accuracy and generalization, outperforming traditional Retrieval-Augmented Generation (RAG) pipelines across multiple evaluation metrics.

Multi-criteria CoT Refinement tailored to regulated data. Unlike prior single score filters, our filter evaluates six criterias (semantic alignment, terminology/clarity, faithfulness, logical consistency, bias/ethics and completeness) with per-criterion thresholds and normalized weights, followed by optimal subset selection. Empirically, this reduces dataset size by $> 40\%$ while increasing accuracy (e.g., 200k→112k entries; 84%→91% accuracy). Input context token-aware Optimization of Question-Answer pair generation: Allocation of optimized Q&A pairs:

$$Q_o^{\mathrm{F}} = 0.005 \cdot T, \quad Q_o^{\mathrm{P}} = 0.0015 \cdot T$$

$$Q_o^{\mathrm{C}} = 0.002 \cdot T, \quad Q_o^{\mathrm{I}} = 0.002 \cdot T, \quad Q_o^{\mathrm{R}} = 0.002 \cdot T$$

i.e. how many pairs to curate per Question type under a fixed token. Triangulated evaluation protocol for safety-critical domains: (i) DeepEval automated metrics; (ii) SME rubric; and (iii) LLM-as-a-Judge with explicit focus on bias control using clear, structured prompts with explicit evaluation criteria, enforcing JSON-only responses to reduce subjectivity, and focusing scoring on factual content rather than style or format.

In summary, **CURATE-FT** offers a practical blueprint for elevating the reliability and impact of LLMs in real-world, compliance-driven domains. We believe these advances will help accelerate the adoption of AI in critical sectors, fostering innovation while maintaining the highest standards of accuracy and safety.

Future Work: While we were able to achieve satisfactory results, there are certain direction for future work with this approach as depicted below

- **Scaling to Larger and Multilingual Models:** Extend experimentation to larger and multilingual LLMs to assess robustness and generalization across diverse domains and languages.
- **Advanced Negative Sampling:** Develop more sophisticated negative sampling strategies to better cover rare edge cases and mitigate model hallucinations.
- **Human-in-the-Loop Evaluation:** Incorporate interactive SME feedback and active learning loops to refine Q&A generation and filtering in real time.
- **Cross-Domain Validation:** Benchmark the framework in additional high-stakes domains such as healthcare, legal, and public policy to demonstrate broader applicability.

References

1. Chen, Z., et al.: FinQA: A Dataset of Numerical Reasoning over Financial Data. arXiv:2109.00122 (2021)
2. Du, H., Liu, S., Zheng, L., Cao, Y., Nakamura, A., Chen, L.: Privacy in Fine-tuning Large Lan- guage Models: Attacks, Defenses, and Future Directions (2024). arXiv:2412.16504
3. Gu, J., et al.: A Survey on LLM-as-a- Judge. arXiv:2411.15594v4 (2025)
4. Jeong, C.: Fine-tuning and Utilization Methods of Domain-specific LLMs. arXiv:2401.02981 (2024)
5. Miyamoto, K., Koseki, A., Ohno, M.: Effective data curation for frequently asked questions. In: SOLI, pp. 7–12 (2017)
6. Qian, C., et al.: The Evolution of LLM Adoption in Industry Data Curation Practices. CoRR arxiv:2412.16089 (2024)
7. Ratnakar, S., Talasila, A., Chamadiya, R., Agarwal, N., Doifode, V.K.: Beyond QA pairs: assessing parameter-efficient fine-tuning for fact embedding in LLMs. In: Workshop on Preparing Good Data for Generative AI: Challenges and Approaches (2025)

8. Shah, A., Gullapalli, A., Vithani, R., Galarnyk, M., Chava, S.: FiNER-ORD: financial named entity recognition open research dataset. arXiv:2302.11157 (2023)
9. Taori, R., et al.: Alpaca: a strong, replicable instruction-following model. In: CRFM (2024)
10. Zhao, C., Jia, X., Viswanathan, V., Wu, T., Neubig, G.: SELF-GUIDE: Better Task-Specific Instruction Following via Self-Synthetic Finetuning. arXiv:2407.12874 (2024)

Utilizing Large Language Models to Structure and Analyze Large Unstructured Data Sources

Thomas J. Sigler(✉), Haley R. Dozier, and Ashley N. Abraham

U.S. Army Engineer Research and Development Center Information Technology Laboratory, Vicksburg, MS 39180, USA
{Thomas.J.Sigler,Haley.R.Dozier,Ashley.N.Abraham}@erdc.dren.mil

Abstract. The world is filled with copious amounts of unstructured data that is incredibly useful, but difficult to use in its format. Large language models (LLMs) have enabled efficient automated processing of large lakes of unstructured data into tabular formats that are more easily processed and analyzed for extracting useful insights. This requires bringing numerous existing technologies together to ensure consistent, predictable, repeatable structure that can always turn data into more useful, flexible tabular formats. This pipeline employs recent advances in the field of deep learning and natural language processing to propose a pipeline for structuring unstructured data in a consistent, repeatable way.

Keywords: Unstructured Data · Structured Output · Output Validation · Large Language Models

1 Introduction

The world is filled with unstructured, textual data that contain unique, powerful insights to support decision making and analysis. The issue is how to understand a quantitative amount of the required resources from documents describing requirements qualitatively. Passing the documents to LLMs and asking them to structure the information without other components is infeasible due to context window limits. A more complex pipeline combining cutting edge and well established technologies is required to extract and process relevant data and transform it into usable, structured formats. We propose a pipeline combining cutting edge text retrieval from vectorized data with LLMs and pre-existing databases for structuring unstructured data efficiently. This pipeline represents a concrete step forward in being able to more turn data from difficult to parse unstructured data to easily processed and understood structured data.

Previous methods of processing heterogeneous data are rooted in developing databases to derive mappings by classification. [1] This method showed great success in deriving business intelligence from large pools of unstructured, contextually sensitive text data, but required intensive labor and human in the

H. R. Arabnia et al. (Eds.): CSCI 2025, CCIS 2972, pp. 233–240, 2027.
https://doi.org/10.1007/978-3-032-28469-3_16

loop for process execution. This greatly slowed the ability to generate useful, timely insights, limiting the applicability in time-sensitive situations. Utilization of machine learning (ML) concepts was limited to classic classification techniques, such as k-nearest neighbor (KNN). This pool, while suitable for common business insights, fails on truly massive datasets where speed is required and the data is too complex for classical mathematical techniques to produce sufficient insights.

Other work has proposed the synthesis of structured and unstructured data for predictive insights in various fields. [9] This was found to produce better insights than purely using structured or unstructured data. This combination shows that combining multiple types of data is key to leveraging deep-learning models for producing predictions. The model proposed by Rusu, et al. falls short of being applicable to large pools of information, which needs to synthesize an unknown number of various types of data sources for predicting across various contexts. The insights produced by Rusu, et al. are key to the construction of a more general purpose pipeline for prediction generation using large pools of data (see Fig. 1).

2 Methodology

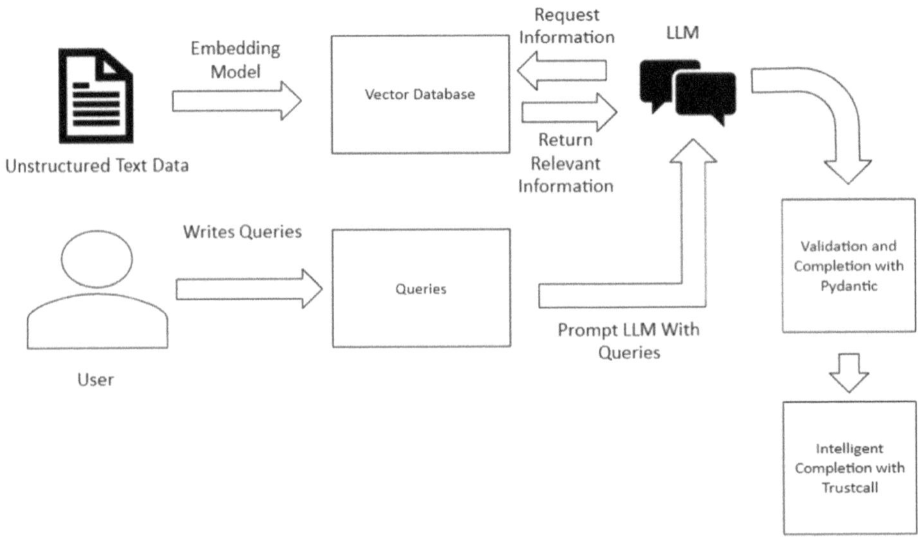

Fig. 1. Process from unstructured data to structured data.

We propose a pipeline that takes unstructured data and structures it in a way that can be easily parsed. This requires a large infrastructure to support this pipeline, which consists of three steps. This is accomplished in three steps, extraction of key details through using large language models, validation of the structure with Pydantic, and is completed with TrustCall, which completes the structuring into a pre-defined JSON object. These steps are key to turning this unstructured data into easily used and processed data that can empower decision making.

2.1 Large Language Models

Large language models (LLMs), with information retrieval infrastructure, can be utilized to parse massive pools of texts that are not structured and often relies on disparate parts to understand the whole picture. This step relies on two substeps, creating a vector database, prompting the LLM with queries for pre-chosen information, and finally, allowing the LLM to retrieve the relevant information and use it to answer the queries to understand the information held within the documents. This step is key to breaking down the information and sifting out the key information from a lot of noise that often characterizes large textual documents.

The first step is to embed all of the data into a high dimensional space and store it in a vector database. A vector database with a traversal algorithm is required to bypass context limits that limit the amount of tokens a model can process. [6] This gives a pool that can be mathematically compared to the queries from the user, enabling the discovery of the most semantically similar ones using highly navigable small world. [7] This efficient algorithm for finding similar content within the articles enables the completion deployment of this pipeline with a reasonable completion time. Additionally, the use of a vector database for retrieving only the most pertinent information sidesteps the issues presented by current context windows at the time of writing, often limited to less than two hundred thousand tokens. This key reorganization of the unstructured information into mathematical objects.

The prompts utilized to guide the LLM to structured outputs are the second key ingredient to turning the unstructured textual information into partially structured information. This partially structured information does not necessarily wholly conform to the expected structure of the output. The indeterministic nature of the LLMs prevents the surety of this structure being perfectly produced by prompt engineering and the retrieved data. This necessitates the following two steps.

2.2 Schema Validation

Pydantic [3] is a Python data validation library intended to ensure data consistency and provide a substitution for strict type checking that Python lacks. This is useful when combined with LLMs to guide data extraction. This forces the final results of most insights to conform to predefined structures by the user. This can be used to generate a JSON structure that can be parsed and turned into a tabular format. This is a key piece for ensuring the data is structured in a usable way.

2.3 Intelligent Completion

TrustCall [5] is a tool employed by Langchain to structure data with LLM calls to ensure that all data fields are filled out and the structure deterministically conforms. Langchain is employed as the primary LLM tool to ensure an ordered, multistep process with LangGraph. This pipeline utilizes the tool along with Langchain to ensure all edge cases in nested structures fully conform, ensuring data consistency and insights are gleamed as fully as possible. This is accomplished through updating existing JSONs from LLM output through patches. This library additionally provides a more efficient, less computationally expensive way to ensure JSON completion. Traditionally, a whole JSON would be both inputted and outputted with corrections. Instead, Trustcall only outputs specific patches that are inserted into the predefined structure, increasing computational efficiency. This more focused tasks drawn from finding missing values in predefined Pydantic models enables full and intelligent completion, ensuring valid consistent structured output.

2.4 LLM Output Validation

Once the data is appropriately structured, the final step is to validate the data output. Traditionally, outputs from LLMs have been verified for correctness and coherency by subject matter experts (SMEs), the most surefire way to get an understanding of the correctness of the output. This method consists of experts within the field that the LLM is generating outputs for evaluating the outputs for correctness and wording. Usually, multiple experts would review the outputs and score them on a scale of one to five and the scores would be averaged together to score the model.

In the evaluation model proposed in this paper, a different approach is adopted. Instead of rating the quality of the individual responses, experts were tasked with the development of a structured reference table utilizing the same schema prescribed for the LLM outputs. This allowed for a direct comparison between the expertly derived table and the LLM generated one. By aligning the formats this way, a more precise measure of what the LLM captured, what it missed, and it's level of specificity can be measured.

Because the two sets of responses were not always in identical form, a similarity scoring methodology could be utilized to assess the overlap between the two responses. The central goal of the output validation process was to analyze whether the LLM responses were contained within, or reasonably matched to, the expert derived ones. When a strong match was identified using cosine similarity, the LLM output was deemed correct. When no match was present, the response was deemed incorrect. This binary evaluation then enabled the computation of an F1 score to quantify the model's performance.

2.5 Experiment

This pipeline was tested through the lens of a case study consisting of thousands of pages of documents. An extensive list of location and actor based questions was created to guide information extraction and generation of structured output in a flexible and usable way. Google's *Gemma3* [10] large language model was utilized to analyze contextually relevant data that was stored in a local vector database using FAISS [4]. The embeddings used to embed text data was *nomic-embed-text* [8]. Military planning documents were utilized as a complex, large base of textual data for the LLM to understand and analyze. To further test the robustness of this approach, synthetic variations of the original documents were used to evaluate the pipeline's ability to generalize beyond the source material. This case study provides a good grounding for understanding the capabilities and limitations of this pipeline due to the sheer size of the unstructured data and the variety of data points that can be gleamed from it.

3 Results and Discussion

The generated structured data is detailed and extensive using this pipeline, answering a wide range of questions. Additionally, subject matter expert (SME) analysis of the outputs has been qualitatively determined in this case study to be realistically reflective of the information that should be derived from the documents. The structure generated by the large language models (LLMs) always conformed to the predefined structure of the Pydantic models, buoyed by Trustcall, to ensure missing values were not empty and invalid. The efficacy of the outputs generated by this pipeline and the potential applicability across a range of domains demonstrates the significant step that this is towards improving automated data organization.

As shown in Fig. 2, LLM performance to extract the task a particular service is instructed to do remained consistently high across all versions, suggesting that these categories are less sensitive to document perturbations. In contrast, the quantity category displayed great variability. This could be due to a variety of factors that will be explored in future research.

This pipeline enhances the operability of systems that rely on consistently structured data, enabling the application of LLMs to these systems without the risk of generation of malformed data. Additionally, the uniformly structured outputs enables the deployment of LLMs without disrupting operations and with minimal postprocessing. This ability to control the generated structure deterministically within the pipeline eases the transition of systems seeking to utilizes this system for data analysis and transformation. The potential insights unlocked by this would be a boon to decision makers across numerous industries.

Complex labor is still required to setup the pipeline for smooth operation. The designing of schema for structuring the data that fits the use case is required and can present issues when dealing with complex data structures that challenge the ability of the pipeline to maintain uniformity. Beyond setup, prompt engineering with the user writing queries is required to grab the most relevant, accurate data for accurately answering queries. This labor intensity requires the incurring of initial costs, but can be run without human intervention afterwards.

Objectively ensuring queries are accurately answered remains a challenge, even in this structured format that eases the validation of answers. Validating

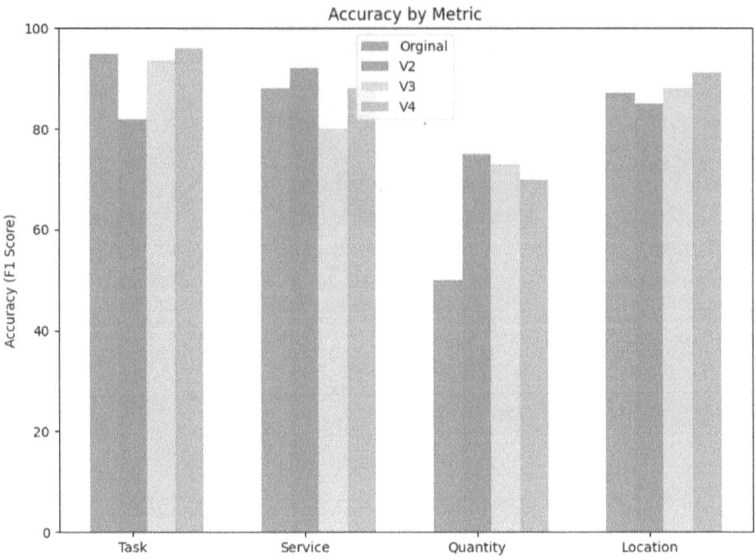

Fig. 2. F1 Score comparing military planning documents to a expert derived validation set

answers from large text bases of multiple documents can require understanding multiple discrete pieces information from multiple documents. In domain-specific applications, and without well established baselines for outputs structured as JSONs from LLMs, there exists no alternatives to by-hand SME validation of answers. This is labor intensive and undesirable in wider applications of this methodology. This lack of a suitable, efficient validation method does not inhibit the usefulness of this pipeline, but limits the ability to ensure it is wholly correct.

This pipeline has demonstrated great ability to organize key information from large collections of unstructured document data. The case study presented is a good baseline in understanding the power of this approach to intelligently structure key data points, generating insights. The cross-domain applicability of this pipeline is a solid step forward in applying cutting edge advances to better leverage existing data for decision making and analysis. The limits in easily and objectively understanding the validity of the output is not a limit in the application, but in completely understanding its efficacy.

4 Conclusion

The case study used to test this pipeline for turning large pools of unstructured data into structured data has shown that it has an undeniable efficacy and presents a more flexible option than previous research has enabled. This pipeline enables the uncovering of key data points and transforming information difficult to use and process into data that is more suitable for automated analysis.

Additionally, this pipeline is comparatively light weight due to the employment of cutting edge tools designed to reduce computation requirements. This is a concrete step in applying recent LLM advancements to automate understanding and use of large pools of data that were previously labor intensive to understand and apply.

Significant challenges remain in ensuring the validity of the generated data. Large language models have faced challenges in benchmarking the factuality of their generated output. This challenge is heightened by domain-specific applications, where the only benchmark may be expert validation, a labor and time intensive process, extenuated by the large base of data that is structured by the pipeline. Recent advancements in LLM response evaluation that enables automated judging, such as LLM-as-a-judge [2], can be fantastic at evaluating the structure of the response, but can falter at domain-specific evaluation. Future work in advancing validity assurance is a key direction for carrying this work forward.

The flexibility of this pipeline enables cross-domain application that previous methodologies struggled to cope with, often limiting their applicability to common business documents, such as emails. This approach can be applied to any domain of data with the proper computing power without human intervention beyond setting up the process. The cross domain adaptability of this approach presents an interesting avenue of exploration in exploring where the pipeline excels and where it falls short in the future.

References

1. Abdullah, M.F., Ahmad, K.: The mapping process of unstructured data to structured data. In: 2013 International Conference on Research and Innovation in Information Systems (ICRIIS), pp. 151–155. (2013). https://doi.org/10.1109/ICRIIS.2013.6716700
2. Chu, Z., Ai, Q., Tu, Y., Li, H., Liu, Y.: Automatic large language model evaluation via peer review. In: Proceedings of the 33rd ACM International Conference on Information and Knowledge Management, CIKM '24, pp. 384–393. Association for Computing Machinery, New York (2024). https://doi.org/10.1145/3627673.3679677
3. Colvin, S., E., J., Ramezani, H., et al.: Pydantic[software]. https://github.com/pydantic/pydantic
4. Douze, M., et al.: The FAISS library. arXiv preprint: arXiv:2401.08281 (2024)
5. Fu-Hinthorn, W., et al: Trustcall[software]. https://github.gom/hinthornw/trustcall
6. Hadi, M.U., et al.: Large language models: a comprehensive survey of its applications, challenges, limitations, and future prospects. Authorea preprints **1**(3), 1–26 (2023)
7. Malkov, Y.A., Yashunin, D.A.: Efficient and robust approximate nearest neighbor search using hierarchical navigable small world graphs. IEEE Trans. Pattern Anal. Mach. Intell. **42**(4), 824–836 (2018)
8. Nussbaum, Z., Morris, J.X., Duderstadt, B., Mulyar, A.: Nomic embed: training a reproducible long context text embedder. arXiv preprint: arXiv:2402.01613 (2024)

9. Rusu, O., et al.: Converting unstructured and semi-structured data into knowledge. In: 2013 11th RoEduNet International Conference, pp. 1–4 (2013). https://doi.org/10.1109/RoEduNet.2013.6511736
10. Team, G., et al.: Gemma 3 technical report. arXiv preprint: arXiv:2503.19786 (2025)

Fossilized Memories: Adapting Open LLMs to Oil and Gas Data While Mitigating Catastrophic Forgetting

Syed Ali Raza, Mozoon M. Alwalmany, Ali M. Abusnina, and Khaled A. Alutaibi(✉)

Saudi Aramco, Dhahran 31311, Saudi Arabia
{syed.raza.6,mozoon.alwalmany.1,ali.abusnina, khaled.alutaibi.1}@aramco.com

Abstract. We present a domain adaptation study of open large language models (LLMs) for the oil and gas industry, focusing on adapting LLaMA 3.1/3.3 70B models using internal technical documentation and historical records spanning 90 years. Our pipeline begins with the LLaMA 3.3 70B Instruct model—rather than the commonly used base variant—and applies continual pretraining (CPT) and instruction fine-tuning on a uniquely curated corpus of instructions and self-instruct data. This is followed by model merging using the SLERP method to improve performance on domain-specific tasks while preserving general capabilities.

After training on hundreds of thousands of domain-relevant documents, the adapted model achieved a 0.775-point improvement in correctness on internal evaluations (a 22% relative gain), as measured by the LlamaIndex correctness metric. Although instruction tuning led to a substantial drop in general-purpose performance (e.g., IFEval declined from 91.1% to 53.7%), model merging successfully restored over 90% of the original performance, reducing the gap to less than 4% for IFEval and less than 2% across BBH, MATH Level 5, GPQA, MuSR, MMLU-Pro, and HumanEval.

These results demonstrate an effective and balanced strategy for adapting LLMs to specialized industrial domains while mitigating catastrophic forgetting and preserving general instruction-following capabilities.

Keywords: Domain adaptation · Continual pretraining · Fine-tuning · Model merging · Catastrophic forgetting

1 Introduction

Large Language Models (LLMs) have shown distinguished capabilities in generating and understanding human natural languages. Advancements in this area have shown significant potential for industrial applications that require sophisticated reasoning over unstructured textual data [2].

H. R. Arabnia et al. (Eds.): CSCI 2025, CCIS 2972, pp. 241–258, 2027.
https://doi.org/10.1007/978-3-032-28469-3_17

In the oil and gas industry—a domain characterized by specialized terminology, highly technical documentation, and safety-critical operations - LLMs are promising to enhance decision support systems, automate technical report generation, and enable intelligent information retrieval from large corpora of operational data. He Liu et al. [17] presented a detailed survey on how LLMs are emerging as valuable tools in the oil and gas industry, particularly in addressing challenges related to small-sample data, unstructured documents, and complex multimodal datasets. Applications of large language models in the oil and gas industry include intelligent assistants, question-answering systems, data analysis and visualization, and exploratory research in specialized oil and gas exploration domains. In intelligent assistant and Q&A applications, large language models are trained using public and proprietary datasets to enhance their capabilities. These models allow users to query technical knowledge in natural language, aiding decision-making processes. Examples of such models include PetroQA [4] - a ChatGPT-based assistant trained using PetroWiki content to provide accurate petroleum-specific answers; PetroBERT [23] - a BERT-based model trained on Portuguese corpus including daily drilling reports; and an artifact repository; the customLLM [1]- a model developed by ExxonMobil to address the limitations of general-purpose language models in industrial contexts using hybrid corpus comprising proprietary data (e.g., equipment manuals, work orders, maintenance records) and public sources (e.g., Wikipedia). In data analysis and visualization, large language models are used to generate business reports and conduct in-depth data analysis. A notable example is Aramco's LLM - a 250-billion-parameter model trained on oil and gas domain data to support drilling optimization and market forecasting [19]. Large language models are applied beyond general Q&A and data analysis to tackle domain-specific tasks in oil and gas industry, such as reservoir characterization, geological modeling, and predictive maintenance. In reservoir characterization, researchers used LLMs to process geological text descriptions to extract key parameters for carbonate reservoir rock classification and permeability prediction [26]. In geological modeling, a chain-of-thought strategy is applied with ChatGPT to derive continuity and momentum equations for geological simulations [20]. In predictive maintenance, researchers explored the use of the Internet of Things (IoT) and large language models to enhance predictive maintenance strategies in oil and gas refineries [24].

Adapting general-purpose LLMs to specialized industrial domains presents a critical challenge. Fine-tuning domain-specific data can improve task-specific performance but often leads to catastrophic forgetting, where the model's original generalist capabilities degrade significantly [9,14]. Catastrophic forgetting represents a significant barrier to deploying LLMs in real-world settings, where they must handle both specialized industry topics and general information reliably. Basic adaptation methods often fall short in sectors like oil and gas, where low latency, high reliability, and accurate context handling are critical. Enabling the domain adaptation of open source LLMs to oil and gas data while retaining their broad general reasoning capabilities remains a central research objective.

Continual pretraining and domain adaptation are key strategies for improving large language models' performance and domain alignment (LLMs). Domain-specific continual pretraining refine model behavior by calibrating the ratio between general-domain and domain-specific training data [6]. Empirical results indicate that this method enables reliable performance estimation across varying data mixtures and sizes while maintaining low training overhead. These findings highlight its practical value in deploying LLMs across specialized application domains [22].

To build on these advances, we adopt a non-destructive domain adaptation strategy that integrates techniques such as parameter-efficient tuning [8], adapter modules [21], continual pretraining on domain-specific corpora, and model merging in weight space. This multi-stage approach enables the model to internalize domain knowledge while preserving its general reasoning abilities. Continual pretraining allows the model to adapt gradually to the target domain distribution, and model merging ensures that knowledge from both the domain-specific and generalist models is retained in the final merged representation.

This paper presents our strategy for adapting open-source large language models (LLMs) to the oil and gas domain through continual pretraining and model merging. Our results show that this approach significantly enhances domain specific performance while preserving the model's general language capabilities. We apply this methodology to open LLMs using a curated oil and gas corpus and evaluate the adapted models on domain-specific tasks and general-purpose benchmarks.

Our key contributions are:

- We propose a multi-stage framework combining continual pretraining (CPT), instruction fine-tuning (SFT), and SLERP-based model merging to adapt open LLMs to the oil and gas domain while mitigating catastrophic forgetting.
- We show that SLERP merging with a fixed weight retains up to 90% of general performance while enhancing domain-specific instruction-following.
- We demonstrate that starting from the original instruction-tuned model better preserves general performance on public benchmarks than starting from a completion-tuned model.
- Combining open-source instructional data with internal completion and self-instruct datasets yields the best balance between generalization and domain alignment.
- We show that converting internal completion data into instruction format significantly outperforms using it in its original form.
- We build a large-scale instruction dataset from proprietary documents using a self-instruct pipeline, enriched with synthetic, multilingual, and noisy examples tailored to industrial tasks.
- We develop a scalable preprocessing pipeline (OCR, formatting, deduplication) for over 280,000 industrial documents to support high-quality domain pretraining.
- We show through experiments that coding benchmark performance remains largely stable during domain adaptation, regardless of the dataset composition or the specific training pipeline employed (e.g., CPT $\rightarrow$ instruction

fine-tuning, with or without model merging) whereas performance on benchmarks like IFEval varies significantly.

This work provides a scalable pathway for bridging the gap between general-purpose LLMs and domain-specific deployment, emphasizing safety, reliability, and performance continuity.

2 Background

Recent advancements in NLP have highlighted the importance of domain—adapted LLMs that provide accurate and context—specific responses. This paper investigates enhancing domain adaptation through continual pretraining [6,10] and model merging [5] techniques. As a case study, we explored adapting an open large language model to Saudi Aramco data, demonstrating the practical applications in the oil and gas industry domains while maintaining its general capabilities.

2.1 Domain Adaptation in LLMs

Domain adaptation is the process of modifying a pre-trained LLM to better handle content from a specific domain, such as law, medicine, or engineering, that differs significantly from the general domain to which it was initially trained. This process helps the model to comply with the language, terminology, and logic used in the target domain while maintaining the model's general capabilities. Without such adaptation, the model may produce inaccurate or irrelevant output, struggle with domain-specific terminology, or misinterpret specialized content. Adapting the model to a new domain makes it more reliable and suitable for practical use in specialized contexts.

During the past few years, researchers proposed several LLM domain adaptation methods, such as continual pretraining [6], adapter-based tuning [7,21], low-rank adaptation (LoRA) [8], prefix/prompt tuning [12,15], model merging [5], and retrieval-augmented adaptation [13,25]. In continual pretraining, a general LLM is further trained on domain-specific unlabeled text to align its representations with the targeted domain. This approach enhances a model adaptation of domain-specific language and concepts without requiring labeled data. However, it can be computationally intensive and may lead to catastrophic forgetting of general knowledge if not carefully managed. Adapter-based tuning allows domain adaptation by inserting lightweight trainable modules between transformer layers, keeping the base model frozen, reducing compute costs, and preserving the model's general knowledge. However, it may offer limited domain adaptation compared to full fine-tuning, especially for highly specialized tasks or low-resource domains. Low-rank adaptation (LoRA) enables efficient fine-tuning by injecting low-rank trainable matrices into existing model weights, drastically reducing parameter updates and memory usage. Its main drawback is that performance may lag behind full fine-tuning in complex tasks or when extensive domain shifts are present. Prefix/prompt tuning enables domain adaptation by learning tasks or domain-specific prompts (tokens or prefixes) while keeping the model weights fixed, making it lightweight and modular. However, it may underperform on complex tasks or significant domain shifts due to limited capacity to modify the model's internal representations. Model merging combines two or more pretrained models (e.g., general and domain-specific) to integrate knowledge without full retraining, offering a fast and flexible adaptation method. However, it can lead to degraded performance or instability if the merged models are poorly aligned or trained on highly divergent data. Retrieval-augmented adaptation improves the LLM performance by including external domain-specific documents, which are dynamically retrieved during inference time. This approach improves the factuality of the responses and reduces the need for retraining. Its drawbacks include reliance on high-quality retrieval systems and the risk of degraded output if irrelevant or noisy documents are retrieved.

2.2 Catastrophic Forgetting

Catastrophic forgetting in large language models occurs when a model loses previously learned general-domain knowledge after being fine-tuned on new, often domain-specific data. Catastrophic forgetting is particularly problematic in real-world applications where LLMs must retain broad capabilities while adapting to specialized contexts. The issue arises because traditional fine-tuning updates all model parameters, potentially overwriting essential general knowledge, thus degrading performance on tasks outside the new domain.

Several strategies have emerged to mitigate catastrophic forgetting. Continual pretraining [6] extends the original training with domain-specific data while preserving prior knowledge. Adapter-based tuning [7] and prompt/prefix tuning [12] update only small components or soft tokens, keeping core model weights fixed. Elastic Weight Consolidation [11] uses regularization to prevent important weights from changing. Other techniques include model merging [5], which combines domain-specific and general models, and retrieval-augmented methods [13,25], which inject external knowledge at inference time without altering the model. These approaches provide varying trade-offs between efficiency, adaptability, and preservation of general capabilities.

2.3 Continual Pretraining (CPT)

Continual Pretraining (CPT) is a technique for adapting large language models (LLMs) to specialized fields by continuing their unsupervised training on domain-specific data. Unlike fine-tuning, which adjusts a model using labeled data for a specific task, CPT extends the initial training phase using unlabeled text from the target domain. This approach helps the model learn domain-specific terminology while preserving its general language capabilities, making it more effective across various tasks in the targeted domain.

For example, PMC-LLaMA [28] applies CPT to medical texts, enhancing the model's knowledge of medical content through additional unsupervised training before fine-tuning. The result is a compact, open-source model outperforming larger, general-purpose systems like ChatGPT and LLaMA-2 on medical benchmarks. Similarly, ChipNeMo [18] adapts LLMs for chip design by continuing training on engineering documents. This training style, which was carried out before any task-specific fine-tuning, enabled the model to support tasks such as script generation and bug analysis, achieving strong performance while significantly reducing model size.

2.4 Model Merging

Model Merging is a technique used to combine the strengths of multiple pre-trained models into a single, unified model. Instead of training a model from scratch or overwriting a general-purpose model through fine-tuning, model merging blends the parameters of different models—typically a general model and a domain-specific one—so the resulting model benefits from both broad knowledge and specialized understanding.

For example, using tools like MergeKit [5], different merging strategies can be applied, such as Linear merging [27], SLERP [3], TIES [29], and DARE [30]. These methods vary in how they mix model weights, but they all aim to preserve the core abilities of each source model. In practice, this approach allows domain-adapted models—produced through continuous pretraining (CPT)—to be merged with general-purpose chat models, resulting in a single model that can handle everyday interactions and specialized tasks.

One key advantage of model merging is that it helps avoid catastrophic forgetting. Since the base model's parameters do not have to be fully retrained, its general capabilities are preserved while new, domain-specific knowledge is incorporated. This method makes the final model more adaptable and practical across various use cases.

3 Domain Adaptation Workflow

Figure 1 outlines the end-to-end workflow for adapting large language models to the oil and gas domain. The process begins with data collection and curation, followed by the generation of self-instruction and completion datasets tailored to domain-specific tasks. These datasets feed into a two-phase fine-tuning pipeline: one for completion-based adaptation and another for instruction tuning. Subsequently, model merging is employed to balance domain knowledge with general capabilities, followed by benchmarking and quantization to prepare the deployment model. Each stage in this pipeline is detailed in the subsequent sections.

3.1 Data Collection

We collected a comprehensive dataset by combining internal organizational documents with publicly available instructional corpora to build a robust domain-adapted language model for the oil and gas sector. This dataset was designed to

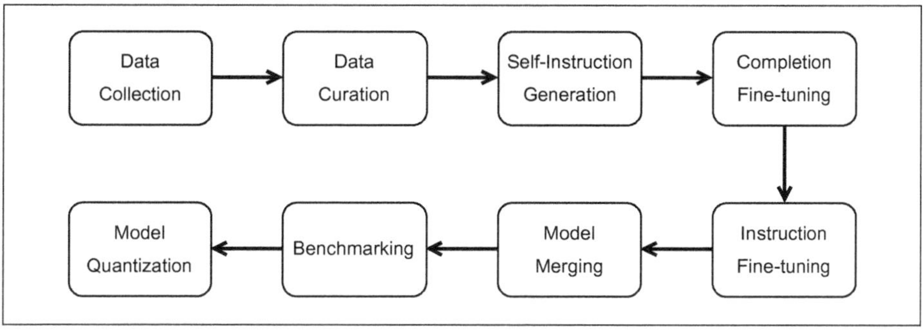

Fig. 1. Domain-adaptation workflow.

integrate domain-specific technical knowledge with general instruction-following capabilities, forming the foundation for continual pretraining and fine-tuning.

Internal Proprietary Data: The internal dataset comprised over 280,000 raw documents from various business and technical functions across the organization, including operational, engineering, and safety-related content. A structured multi-stage preprocessing pipeline was applied to assess file formats, consistency, and data quality. After filtering and transformation, we retained approximately 150,000 high-quality documents for model training.

Open-Source Instructional Data: We incorporated various open-source instructional datasets covering tasks such as mathematics, programming, creative writing, data analysis, and planning. These datasets were reformatted into a conversational instruction-tuning style and filtered down from 7.7 million to 2.1 million high-quality examples.

Synthetic Instruction Data: To enrich instruction diversity and evaluate model behavior under edge cases, we generated synthetic samples using open corpora such as Arabic_RC. The dataset included both positive and negative instructions, with some prompting fallback responses like "I do not know" to simulate model uncertainty. We also included over 500 Arabic-language samples to assess performance in multilingual domain scenarios.

Noise-Augmented Samples: To improve robustness to real-world user input, we generated 600 synthetic noisy instructions, including gibberish prompts, incomplete queries, and stopword-heavy text. We paired each instruction with an appropriate model response to simulate challenging user interactions.

This collected dataset forms the foundation for our domain adaptation workflow's continual pretraining and fine-tuning stages. The following sections describe data preparation, structuring, and quality control procedures supporting the model development process.

3.2 Data Curation

To prepare a high-quality corpus for domain adaptation, we implemented a multi-stage data processing pipeline as shown in Fig. 2. This pipeline is designed to handle diverse document formats, eliminate redundancy, extract meaningful content, and ensure consistency and cleanliness across the dataset.

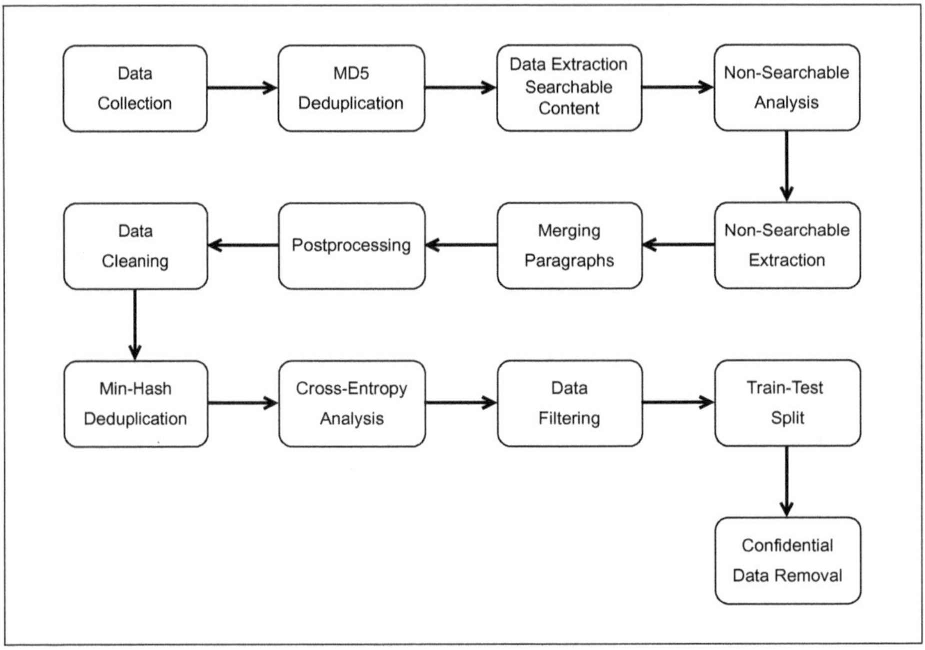

Fig. 2. Data Processing workflow.

Data Selection and De-duplication: We identified supported document formats including PDF, DOC, DOCX, PPTX, HTML, TXT, and MSG files. An initial deduplication step based on MD5 hashes reduced the corpus from 335,000 to 269,000 files. While effective, this method can be sensitive to metadata differences, which are discussed further in Sect. 3.3.

Data Extraction (Searchable and Non-Searchable): In this step, we extracted searchable text from supported formats using format-specific rules to segment content into paragraphs. Approximately 30,000 PDFs were identified as containing non-searchable text, prompting further analysis using OCR. We analyzed each PDF page for its searchable and non-searchable content ratio, and flagged files below a threshold of 0.3 for OCR. OCR workloads were distributed across four machines using a load-balanced containerized setup. Then, we merged the direct extraction and OCR outputs for downstream processing.

Data Merging: Text segments were merged hierarchically: first by combining related paragraphs into sections based on document structure, then by concatenating all sections into a unified file-level representation. This merging significantly increased the average number of tokens per example, aligning with our target of supporting 8k-context window fine-tuning.

Postprocessing and Cleaning: Postprocessing steps included removing short fragments ($<$10 characters), merging OCR output at the file level, and eliminating inconsistencies. Cleaning involved HTML parsing, filtering for English text, and removing unwanted tokens or brand mentions (e.g., "OpenAI"). These steps were applied uniformly at the section and file levels to ensure consistency.

Min-Hash De-duplication: To further eliminate near-duplicate content that MD5 hashing could not capture, we applied Min-Hash deduplication with conservative parameters (ngram size = 10, threshold = 0.99). This method addressed content redundancy across files differing only by minor variations, such as formatting or cover pages.

Cross-Entropy Filtering: We used cross—entropy analysis using LLaMA-3.1 8B base model to assess alignment with language modeling expectations. We filtered out files and sections with values exceeding acceptable thresholds. To standardize quality, we established a baseline cleaning strategy, removing short segments ($<$600 words) and sections containing tables.

Data Filtration and Train-Test Split: To refine the dataset for completion fine-tuning, we combined Min-Hash deduplication (thresholds from 0.6 to 0.9) with cross-entropy filtering (thresholds from 1.5 to 3.0). To prevent data leakage, we reserved a curated list of 100 high-quality files as a fixed test set. Additional merging of unique sections ensured completeness in the final split.

Removing Confidential Data: In order to protect sensitive information, we dropped files containing the keyword "confidential," which resulted in filtering out around 8,000 files from instruction and completion datasets.

3.3 Self-instruction Generation

To align the language model with domain-specific tasks in the oil and gas sector, we implemented a self-instruction generation pipeline using data extracted from a Retrieval-Augmented Generation (RAG) system. This approach ensures that generated instructions are contextually grounded, accurate, and relevant to operational scenarios, while filtering out irrelevant or confidential content.

We generated instructions from raw text chunks extracted from curated documents. Where necessary, we applied OCR to recover non-searchable content. For each instruction, a corresponding response was synthesized from the source context, forming a high-quality instruction-response pair. We carried out the generation process using the LLaMA3.1 70B AWQ model in a multi-turn dialogue format. To ensure completeness and manageability, we selected only paragraphs with 500 to 2000 characters for instruction generation.

To maintain data quality, we applied a verification process in which the same LLM scored each generated conversation using a "self-verify" technique. We assessed each instruction based on multiple criteria: clarity, relevance, coherence, depth, and tone. We used a threshold score of 8/10 to filter out low-quality examples. From an initial pool of 1.5 million generated conversations, we retained approximately 1.2 million (top 80%) for fine-tuning.

Additionally, we manually curated 973 domain-specific identity questions to reinforce model grounding. This combination of automated instruction generation, LLM-based scoring, and manual validation produced a high-quality, domain-aligned instruction dataset for fine-tuning.

3.4 Completion Fine-Tuning

Fine-tuning a completion model is critical in adapting large language models to domain-specific tasks. In domains such as oil and gas operations, completion-style prompts usually involve technical language and structured responses that general-purpose models may fail to capture effectively. This stage enables the model to learn domain-relevant patterns, terminology, and context, improving the quality of generated completions in downstream applications.

We fine-tuned the Llama-3.1-70B base model using a curated dataset drawn exclusively from internal oil and gas documents. The final training set comprised 50 million tokens, prepared through earlier extraction, filtering, and deduplication steps. Training followed a standard next-token prediction objective using a cross-entropy loss function. We performed the optimization process using the AdamW optimizer with the following key hyper-parameters: learning rate = 2e-4, batch size = 80, epochs = 2, LoRA rank = 32, and LoRA alpha = 16. We used these settings to balance convergence efficiency while preserving general language understanding.

We implemented model checkpointing five times per epoch to mitigate compute loss from potential job failures. We utilized a YAML file to manage configuration, providing a reproducible training parameter specification. We performed the training on a SLURM-based HPC cluster. During execution, we encountered intermittent issues with stalled jobs, suspected to be caused by resource contention on shared nodes, leading to compute inefficiencies.

This fine-tuned completion model served as an evaluation baseline and a foundational checkpoint for subsequent instruction-tuning experiments.

3.5 Instruction Fine-Tuning

Instruction fine-tuning builds upon the completion model to enable instruction-following behavior while retaining domain-specific knowledge. Continuing training from a completion-tuned checkpoint makes the model suitable for chat-style applications while leveraging the contextual understanding acquired during completion training.

Fine-Tuning Setup: We performed instruction-fine-tuning using the LLaMA-3.1-70B instruct model and internally fine-tuned completion checkpoints. Training employed the Low-Rank Adaptation (LoRA) method with the following hyper-parameters: learning rate = 2e-4, batch size = 80, epochs = 2, LoRA rank = 32, and LoRA alpha = 16. We saved checkpoints five times per epoch to ensure recoverability.

Data Sources and Composition: The instruction dataset consisted of a mixture of: Self-generated instruction-response pairs (30%) Manually crafted Aramco identity questions Filtered open-source instruction datasets We merged these datasets into a JSONL file containing approximately 1.8 billion tokens. We then processed the data using internal tools for formatting, tokenization, and configuration.

Training Workflow: We performed instruction fine-tuning using a modular pipeline:

- Dataset preparation with data-tools
- Experiment configuration using beyondllm configure-finetune
- On-the-fly tokenization during training
- Fine-tuning via beyondllm fine-tune
- Adapter merging with beyondllm merge-lora to produce the final instruction-tuned model

Full Fine-Tuning Variant: In parallel, we conducted full fine-tuning (without adapters) using the same workflow, omitting LoRA-specific configurations. In this case, the final model did not require a merging step and was saved directly to the output directory.

Instruction Conversion from Completion Data: To enhance instruction diversity, we also converted the original Aramco completion dataset into instruction-format examples. We achieved this by prepending randomly chosen prompts (e.g., "Complete this, please") to selected text spans. A random slice of 5–15 initial words served as the prompt, while we treated the remaining text as the assistant's response. This process created additional single-turn instruction-response examples for fine-tuning. Together, these instruction tuning strategies enabled the model to handle chat-style tasks while preserving the technical grounding required for oil and gas applications.

3.6 Model Merging

We applied model merging as a final step in the domain adaptation pipeline to achieve a balanced model that preserves domain-specific expertise and general instruction-following capabilities. Specifically, we merged a completion- and instruction-fine-tuned model trained on Aramco data with the original LLaMA-3.1 instruction model. This step was essential to retain the performance gains on Aramco-specific tasks while restoring broader language capabilities inherent to the base instruction model.

We employed SLERP (Spherical Linear Interpolation) to merge the model checkpoints. Unlike linear interpolation (LERP), which averages parameters along a straight line, SLERP interpolates along the geodesic of a hypersphere in parameter space. This method preserves the angular relationship between model weight vectors, avoiding the distortion introduced by high-magnitude, low-impact parameters that can skew results in LERP-based merges.

SLERP is particularly effective when combining models fine-tuned on diverse tasks, as it enables a smoother and more meaningful blend of their respective strengths. In our experiments, we used a SLERP weighting of 0.5, resulting in a merged model that maintains robust performance on oil and gas tasks while recovering instruction-following behavior lost during domain specialization. The resulting model is domain-adapted and generalizable, enabling flexible deployment in instruction-driven scenarios without compromising technical precision.

3.7 Benchmarking

To evaluate the performance of the models, we conducted comprehensive benchmarking using a mixture of internal and public datasets. Evaluation covered general language understanding, reasoning, mathematical problem-solving, and instruction-following abilities, alongside domain-specific correctness and perplexity assessments.

Evaluation Strategy: We constructed two test sets: one derived from Saudi Aramco internal data and another from open-source corpora. We assessed the models on perplexity, correctness, and benchmark tasks. We logged all results and tracked them using MLflow.

Benchmark Suites: We employed the LM-Eval Harness for standardized model evaluation across the following tasks:

- **IFEval:** Instruction-following assessment.
- **BBH (Big Bench Hard):** Multistep reasoning, logic, and knowledge challenges.
- **MATH Level 5:** High school competition-level math problems.
- **GPQA:** Graduate-level domain-specific questions.
- **MuSR:** Long-context and multistep reasoning problems.
- **MMLU-PRO and MMLU (5-shot):** Multitask factual knowledge.
- **AI2 ARC (25-shot):** Grade-school science.
- **TruthfulQA (0-shot):** Evaluation of factual accuracy and resistance to misinformation.
- **Winogrande (5-shot), HellaSwag (10-shot), GSM8k (5-shot):** Commonsense and mathematical reasoning.

Perplexity Evaluation: We measured perplexity on internal and public datasets to assess the model's predictive confidence and fluency. Lower perplexity values indicate better alignment with the target domain and improved generative performance.

Correctness Evaluation (LlamaIndex): We used LlamaIndex's correctness framework to evaluate factual accuracy on 1,000 Q&A pairs derived from Aramco's self-instruction dataset. We rated responses on a 1–5 scale, with 5 indicating entirely correct answers. We also repeated this evaluation specifically for the Aramco Identity subset to ensure the model retained key information from training.

MT-Bench Evaluation: We used MT-Bench to assess conversational ability and instruction-following performance. We applied it to standard benchmark prompts and a custom version using a held-out subset of Aramco data (100 two-turn conversations). A judge model evaluated coherence, helpfulness, and relevance.

Code Generation (Eval+): We evaluated the models using the Eval+ benchmark to evaluate coding abilities of the models. We measured the performance using the "pass@1" metric under greedy decoding. These benchmarks provided both quantitative and qualitative insights into the adapted models' performance across general and domain-specific tasks, informing further tuning and deployment decisions.

3.8 Model Quantization

We used activation weighted quantization (AWQ) [16] method to efficiently deploy our models on resource-constrained hardware. We designed AWQ to reduce compute and memory requirements while preserving the accuracy of our models. Unlike conventional quantization methods that use uniform precision throughout the model, AWQ focuses on maintaining higher precision for the most important activations. This approach helps in preventing degradation in the key parts of the model, allowing it to maintain strong performance on complex tasks.

We used AutoAWQ, an automated tool based on AWQ that simplifies model preparation by automatically handling format compatibility, parameter tuning, and export for inference.

The resulting quantized models are suitable for inference on GPUs with limited VRAM and edge computing devices, enabling broader accessibility and reduced infrastructure costs. Quantization thus serves as a critical final step in ensuring our domain-adapted LLMs' scalability and real-world usability.

4 Experimental Results

We conducted experiments to evaluate our domain adaptation framework using continual pretraining and SLERP-based model merging. The evaluation covers general-purpose benchmarks (e.g., LM-Eval, MT-Bench, Eval+) and internal domain-specific metrics, such as correctness and identity.

Effect of the Starting Model: We compared models on general benchmarks and domain adaptability, by initializing them from Llama-3.1-70b-base finetuned on Aramco dataset (INIT-completion) versus initializing them from Llama-3.1-70b-instruct model (INIT-instr-orig) finetuned on a mix of Aramco data and instruction tuning datasets. The model initialized from Llama-3.1-70b-instruct checkpoint consistently outperformed the completion finetuned model across general benchmarks and showed better alignment with instruction-following tasks as shown in Table 1. These results suggest that starting from instruct original model provide a stronger foundation for domain adaptation.

Table 1. Performance comparison on different benchmarks for the two starting model configurations

Benchmark	Llama-3.1-70b-instr	Starting Model	
		INIT-completion	INIT-instr-orig
IFEval	86.8	58.6	76.1
BBH	55.4	50.2	51.3
MATH Lvl 5	34.9	21.6	26.9
GPQA	19.7	13.4	14.0
MUSR	19.0	24.7	18.6
MMLU-PRO	47.8	41.9	45.7
HumanEval	75%	74%	71%
MBPP	78%	61%	72%
Correctness	–	3.10	2.67
MT-Bench	–	7.03	7.10
MT-Bench-Aramco	–	6.33	6.14

Effect of Training Dataset Composition: We have many sources of the data to be used during the training: Aramco completion dataset, Aramco self-instruct dataset, open-source instruction dataset, and Aramco completion as instruction dataset. We explored the following four combinations of these data sets:

- Comb-1: Aramco self-instruct and open source instruct
- Comb-2: Aramco self-instruct and open source instruct
- Comb-3: Aramco self-instruct and Aramco completion as instructions
- Comb-4: Aramco self-instruct, open source instruct, and Aramco completion as instructions

We achieved the best performance when combining all three datasets and formatting the completion data as instructions. This setup offered the best trade-off between general benchmark scores and domain-specific accuracy as illustrated in Table 2. The results highlight the importance of data diversity and instruction-consistent formatting in maximizing both generalization and domain fidelity.

Effect of SLERP Model Merging: To balance domain-specific learning with general-language capabilities, we applied SLERP merging between the

Table 2. Performance comparison on different benchmarks based on four dataset combinations

Benchmark	Llama-3.1-70b-instr	Training Dataset			
		Comb-1	Comb-2	Comb-3	Comb-4
IFEval	86.8	76.4	77.6	58.3	76.2
BBH	55.4	56.5	57.0	52.7	56.6
MATH Lvl 5	34.9	29.4	28.4	28.6	28.9
GPQA	19.7	22.2	17.2	19.1	17.6
MUSR	19.0	17.3	16.6	20.4	17.8
MMLU-PRO	47.8	47.0	46.0	46.3	46.6
HumanEval	75.0%	77.4%	76.8%	75.9%	79.0%
MBPP	78.0%	76.7%	75.5%	76.1%	76.1%
Correctness	–	3.05	3.04	2.97	3.10
MT-Bench	–	6.97	6.98	6.09	7.08
MT-Bench-Aramco	–	5.86	5.80	5.61	5.98

Table 3. Effect of SLERP model merging with fixed weight 0.5 on the different benchmarks

Benchmark	Llama-3.1-70b-instr	Model Merging	
		Unmerged	Merged
IFEval	86.8	76.2	83.8
BBH	55.4	56.6	57.3
MATH Lvl 5	34.9	28.9	31.9
GPQA	19.7	17.6	18.6
MUSR	19.0	17.8	18.1
MMLU-PRO	47.8	46.6	47.6
HumanEval	75%	79%	79%
MBPP	78%	76%	78%
Correctness	–	3.1	2.92
MT-Bench	–	7.08	7.14
MT-Bench-Aramco	–	5.98	6.31

domain-adapted and original instruction models with fixed merging weight 0.5. Merged models consistently achieved better general performance while retaining or improving domain-specific accuracy and identity as shown in Table 3. These findings demonstrate that SLERP merging is a practical, non-destructive method for integrating domain knowledge without degrading core language abilities.

5 Conclusion

In this paper, we introduced a scalable, non-destructive framework for adapting open-source large language models to the oil and gas domain using continual pretraining and SLERP-based model merging. Our approach preserved general language capabilities in the oil and gas sector while significantly improving domain-specific performance. Through extensive empirical evaluation, we demonstrated that starting from instruction-tuned checkpoints, carefully curating and formatting training data, and applying model merging techniques yields robust models capable of instruction-following and domain alignment.

Our results highlight the importance of dataset composition, training duration, and initialization strategy in balancing generalization and specialization. The proposed methodology achieved strong performance across public benchmarks such as LM-Eval, MT-Bench, and Eval+, while maintaining accuracy and factual grounding on internal domain-specific metrics. This work presents a practical pathway toward deploying large language models in specialized industries without sacrificing versatility or performance continuity.

Acknowledgments. The authors acknowledge the support of Saudi Aramco for the authorization to prepare and publish this paper. We also acknowledge the company's provision of technical data, industry expertise, and access to advanced computational resources, which were instrumental in conducting this research.

Disclosure of Interests. The authors have no competing interests to declare that are relevant to the content of this article.

References

1. Abijith, P.Y., Patidar, P., Nair, G., Pandya, R.: Large language models trained on equipment maintenance text. In: Proceedings of the Abu Dhabi International Petroleum Exhibition and Conference (ADIPEC), p. D021S065R003 (2023)
2. Brown, T.B., et al.: Language models are few-shot learners. arXiv abs/2005.14165 (2020)
3. Digitous: LLM-SLERP-merge: spherical merge for large language models (2024). https://github.com/Digitous/LLM-SLERP-Merge. Accessed 05 May 2025
4. Eckroth, J., Gipson, M., Boden, J., Hough, L., Elliott, J., Quintana, J.: Answering natural language questions with OpenAI's GPT in the petroleum industry. In: Proceedings of the SPE Annual Technical Conference and Exhibition, p. D031S032R005. Society of Petroleum Engineers (2023)
5. Goddard, C., et al.: Arcee's mergekit: a toolkit for merging large language models. In: Conference on Empirical Methods in Natural Language Processing (2024)
6. Gururangan, S., et al.: Don't stop pretraining: adapt language models to domains and tasks. In: Jurafsky, D., Chai, J., Schluter, N., Tetreault, J. (eds.) Proceedings of the 58th Annual Meeting of the Association for Computational Linguistics, pp. 8342–8360. Association for Computational Linguistics, Online (2020)
7. Houlsby, N., et al.: Parameter-efficient transfer learning for NLP. arXiv abs/1902.00751 (2019)

8. Hu, J.E., et al.: Lora: low-rank adaptation of large language models. arXiv abs/2106.09685 (2021)
9. Huang, J., et al.: Mitigating catastrophic forgetting in large language models with self-synthesized rehearsal. In: Annual Meeting of the Association for Computational Linguistics (2024)
10. Ibrahim, A., et al.: Simple and scalable strategies to continually pre-train large language models. arXiv abs/2403.08763 (2024)
11. Kirkpatrick, J., et al.: Overcoming catastrophic forgetting in neural networks. Proc. Natl. Acad. Sci. **114**(13), 3521–3526 (2017)
12. Lester, B., Al-Rfou, R., Constant, N.: The power of scale for parameter-efficient prompt tuning. In: Moens, M.F., Huang, X., Specia, L., Yih, S.W.t. (eds.) Proceedings of the 2021 Conference on Empirical Methods in Natural Language Processing, pp. 3045–3059. Association for Computational Linguistics, Online and Punta Cana, Dominican Republic (2021)
13. Lewis, P., et al.: Retrieval-augmented generation for knowledge-intensive NLP tasks. arXiv abs/2005.11401 (2020)
14. Li, D., et al.: Overcoming catastrophic forgetting during domain adaptation of seq2seq language generation. In: Carpuat, M., de Marneffe, M.C., Meza Ruiz, I.V. (eds.) Proceedings of the 2022 Conference of the North American Chapter of the Association for Computational Linguistics: Human Language Technologies, pp. 5441–5454. Association for Computational Linguistics, Seattle, United States (2022)
15. Li, X.L., Liang, P.: Prefix-tuning: optimizing continuous prompts for generation. In: Zong, C., Xia, F., Li, W., Navigli, R. (eds.) Proceedings of the 59th Annual Meeting of the Association for Computational Linguistics and the 11th International Joint Conference on Natural Language Processing (Volume 1: Long Papers), pp. 4582–4597. Association for Computational Linguistics, Online (2021)
16. Lin, J., et al.: AWQ: activation-aware weight quantization for on-device LLM compression and acceleration. In: Gibbons, P., Pekhimenko, G., Sa, C.D. (eds.) Proceedings of Machine Learning and Systems, vol. 6, pp. 87–100 (2024)
17. Liu, H., et al.: Research status and application of artificial intelligence large models in the oil and gas industry. Pet. Explor. Dev. **51**(4), 1049–1065 (2024)
18. Liu, M., et al.: Chipnemo: domain-adapted LLMs for chip design. arXiv abs/2311.00176 (2023)
19. Malin, C.: World's Largest Industrial LLM Revealed! (2024). https://www.middleeastainews.com/p/aramco-launches-largest-industrial-llm. Accessed 29 May 2024
20. Ogundare, O., Madasu, S., Wiggins, N.: Industrial engineering with large language models: a case study of chatgpt's performance on oil & gas problems. In: 2023 11th International Conference on Control, Mechatronics and Automation (ICCMA), pp. 458–461 (2023)
21. Pfeiffer, J., Vulić, I., Gurevych, I., Ruder, S.: MAD-X: an adapter-based framework for multi-task cross-lingual transfer. In: Webber, B., Cohn, T., He, Y., Liu, Y. (eds.) Proceedings of the 2020 Conference on Empirical Methods in Natural Language Processing (EMNLP), pp. 7654–7673. Association for Computational Linguistics, Online (2020)
22. Que, H., et al.: D-CPT law: domain-specific continual pre-training scaling law for large language models. In: The Thirty-Eighth Annual Conference on Neural Information Processing Systems (2024)

23. Rodrigues, R.B.M., et al.: Petrobert: a domain adaptation language model for oil and gas applications in Portuguese. In: Pinheiro, V., et al. (eds.) Computational Processing of the Portuguese Language, pp. 101–109. Springer, Cham (2022)
24. Saboo, S., Shekhawat, D.: Enhancing predictive maintenance in an oil & gas refinery using IoT, ML, and a generative solution. In: Proceedings of the International Petroleum Technology Conference (IPTC), p. D031S128R003 (2024)
25. Shi, W., et al.: Replug: retrieval-augmented black-box language models. In: North American Chapter of the Association for Computational Linguistics (2023)
26. Tveritnev, A., et al.: Applying machine learning NLP algorithm for reconciliation geology and petrophysics in rock typing. In: Proceedings of the Abu Dhabi International Petroleum Exhibition and Conference (ADIPEC), p. D021S054R001 (2023)
27. Wortsman, M., et al.: Model soups: averaging weights of multiple fine-tuned models improves accuracy without increasing inference time. In: Chaudhuri, K., Jegelka, S., Song, L., Szepesvari, C., Niu, G., Sabato, S. (eds.) Proceedings of the 39th International Conference on Machine Learning. Proceedings of Machine Learning Research, vol. 162, pp. 23965–23998. PMLR (2022)
28. Wu, C., Lin, W., Zhang, X., Zhang, Y., Xie, W., Wang, Y.: PMC-llama: toward building open-source language models for medicine. J. Am. Med. Inform. Assoc. **31**(9), 1833–1843 (2024)
29. Yadav, P., Tam, D., Choshen, L., Raffel, C., Bansal, M.: Ties-merging: resolving interference when merging models. In: Neural Information Processing Systems (2023)
30. Yu, L., Yu, B., Yu, H., Huang, F., Li, Y.: Language models are super mario: absorbing abilities from homologous models as a free lunch. In: Proceedings of the 41st International Conference on Machine Learning. ICML 2024. JMLR.org (2024)

The Implications of Large Language Model Integration for Bias in Lending Practices

Dalimar Flores-Torres, Roderick L. Lee, and Jeremy J. Blum(✉)

Pennsylvania State University, Harrisburg, PA 17057, USA
jjb24@psu.edu

Abstract. Despite their increasing integration into the financial industry, artificial intelligence tools have raised concerns about bias in large language models (LLMs) and how they could influence decision-making. By analyzing how popular LLMs such as ChatGPT-3.5, ChatGPT-4.0, and Microsoft Copilot categorize loan applicants into "strong" and "weak" profiles, this study evaluates the potential demographic biases in these models. The results of this study reveal that significant biases are expressed by these models, with strong profiles being overwhelmingly assigned to White, male, U.S. citizens while weak profiles were disproportionately non-White, female, and non-citizen. Gender bias was most prominent in ChatGPT-3.5 and Copilot where all strong profiles were male, and all weak profiles were female. Although ChatGPT-4 showed slight improvements in reducing citizenship bias, the model still exhibited significant racial and ethnic disparities. With the prospect of LLM integration into loan underwriting processes, these biases raise great ethical and legal concerns, particularly regarding fair lending practices in credit underwriting. If these biases are not addressed, they could promote demographic discrimination during loan approvals and ultimately violate fair lending policies such as the Equal Credit Opportunity Act (ECOA). To mitigate these risks, financial institutions must ensure transparency in AI-driven credit assessments. This study highlights the need for stronger regulations and continued research to prevent AI from introducing bias in financial decision-making.

Keywords: Large Language Models · Mortgage Lending · Bias

1 Introduction

Recent advancements in the field of artificial intelligence (AI) have ushered in a new era of technological innovation with the power to transform industries and reshape our understanding of a machine's capabilities. Among these advancements is the development of Large Language Models (LLMs), which stand out for their ability to process and generate human language with remarkable accuracy [1]. These models, which are trained on vast textual datasets, work by using complex algorithms to identify relationships between words and, by considering these relationships, develop the capability to predict the next word in a sequence [1, 2]. Through the repeated practice of natural language processing (NLP) and natural language generation (NLG), LLMs like OpenAI's GPT continue to evolve and promise new opportunities to leverage these advancements across various industries [1–3].

H. R. Arabnia et al. (Eds.): CSCI 2025, CCIS 2972, pp. 259–272, 2027.
https://doi.org/10.1007/978-3-032-28469-3_18

The financial industry is a particularly promising area for the adoption of LLMs. Many financial institutions already rely on machine learning (ML) models to automate processes such as fraud detection, credit scoring, and loan underwriting [4–6]. These applications offer efficiency and scalability, but they also raise concerns about fairness, transparency, and compliance with regulations such as the Equal Credit Opportunity Act (ECOA) and the Fair Housing Act (FHA). If LLMs are to be integrated into credit underwriting workflows, it is essential to understand whether they perpetuate or mitigate demographic biases that have long plagued financial decision-making.

This study addresses that challenge by examining how three widely used LLMs, OpenAI's GPT-3.5, GPT-4, and Microsoft Copilot, categorize mortgage loan applicants in a simulated underwriting scenario. Using applicant data drawn from the Uniform Residential Loan Application (URLA), each model was asked to generate "strong" and "weak" applicant profiles. The experiment produced 150 profiles in total, which were then analyzed to identify patterns of bias across race, gender, and citizenship status.

The results reveal statistically significant demographic disparities across all three models. "Strong" profiles were overwhelmingly associated with White, male, and U.S. citizen applicants, while "weak" profiles were disproportionately assigned to non-White, female, and non-citizen applicants. ChatGPT-3.5 exhibited the most extreme gender bias, with all strong profiles categorized as women and all weak profiles as men. GPT-4 demonstrated modest improvements, particularly in reducing citizenship bias and allowing for some gender variance, but continued to show strong racial and ethnic disparities. Microsoft Copilot reinforced many of the same biases, most notably by categorizing all strong profiles as men and all weak profiles as women.

These findings raise critical questions about the risks of embedding LLMs into lending processes without safeguards. By demonstrating that even state-of-the-art models express demographic biases in simulated underwriting, this study highlights the urgent need for bias mitigation strategies, regulatory oversight, and further research. The contribution of this work lies in bridging the gap between existing research on ML fairness in lending and the emerging role of LLMs, offering evidence of how these models may affect compliance, equity, and trust in financial services.

2 Related Works

2.1 Fair Lending Practices

In the late 1960s and early 1970s, the United States undertook significant legislative steps to define and enforce fair lending practices. Prior to this period, from the 1950's to the 1960's, credit lending decisions were based on the subjective evaluations of loan officers, whose judgement of creditworthiness was often influenced by personal biases [7]. It was the Civil Rights Movement and subsequent congressional hearings on discrimination towards women in mortgage lending that emphasized the need for legislation that limited personal bias during credit assessments [7]. Two major policies emerged as a result: the Equal Credit Opportunity Act (ECOA) of 1974 and the Fair Housing Act (FHA) of 1968 [7, 8]. These acts were designed to eliminate lending discrimination by mandating equal access to credit for all individuals regardless of factors such as race, religion, national origin, sex, or age, among others [7–9]. Together, the ECOA and FHA formed the foundation of fair lending policy, and today, they continue to guide financial institutions towards equitable credit assessments and loan decisions.

Following the enactment of the ECOA, financial institutions would go on to revise loan processing and risk assessment frameworks to support fair lending. Initially, this transition involved the implementation of simple statistical tests designed to mitigate discrimination by detecting correlations between loan denial rates and characteristics explicitly prohibited by fair lending legislation [4]. As statistical methodologies evolved through the late 1990s, financial institutions began implementing more complex strategies to assess credit worthiness while avoiding any factors that could lead to fair lending violations [4]. Strategies such as Cohen's d, Hedges' g, and Glass's delta were utilized to quantify the presence of discrimination based on characteristics explicitly prohibited by the ECOA in the underwriting process and subsequently address bias in credit assessments and lending decisions [4]. Later on, Bayesian Improved Surname Geocoding, which can be used to predict the race or ethnicity of an individual using their surname and geocode location, would be used to improve the estimates of racial disparities lending [4, 9]. Over time, the progression towards increasingly complex algorithms reduced the reliance on subjective assessments by loan officers. However, this shift towards more automated data-driven decision-making would also go on to reduce the transparency of lending decisions.

Most recently, financial institutions have increasingly turned to alternative data sources for establishing credit criteria, which combined with ML techniques has promoted a reliance on automated credit decisions [4, 5]. The appeal of ML models lies in their ability to process vast amounts of data, facilitating loan approval and pricing decisions without directly considering the "protected classes" defined by the ECOA [5, 10]. Despite these advancements, the use of ML in credit assessments introduces significant challenges. Among these challenges is the risk of employing proxy variables, which are indicators that, despite not explicitly referencing the protected classes, may be correlated to them [5]. These proxy variables present a unique challenge to fair lending programs. Although the ECOA explicitly defines the variables that financial institutions are prohibited from considering in credit assessments, there is no policy that clearly defines the kinds of data that can be considered proxy variables [4]. The potential for ML models to unintentionally violate fair lending policies raises a critical concern for financial institutions that rely on automated lending decisions. As these institutions refine their use of ML technologies, eliminating any potential for bias in their automation tools will be crucial for complying with fair lending policies.

2.2 AI and Fairness in Lending

Current Use of Machine Learning in Lending

Recent advancements in the field of AI have revolutionized the financial industry's application of ML technologies. From employing K Nearest Neighbor classifications for resolving credit discrimination complaints to utilizing XGBoost and Light GBM for loan origination decisions [4], these algorithms and frameworks have prompted a shift from relying on human judgement to employing data-driven systems for more accurate decision-making [11]. This shift, which has been fueled by the growing availability of vast datasets and affordable computational power [11], has led to the rapid embrace of ML models despite the widespread acceptance of their opaqueness [6].

This opaqueness, which has often been referred to as a "black box" [11, 12], challenges the interpretability or explainability of these models [12], the ease with which humans can comprehend the models' decision-making processes. While some ML components like classifiers or decision trees offer a high degree of interpretability [12], the evolution towards more complex algorithms has led to an inverse relationship between our ability to maintain clarity and the theoretical accuracy of these models [5, 12]. This lack of explainability presents significant risks within the financial industry, especially as ML models begin to automate the underwriting process and, as previously discussed, potentially evaluate proxy variables in ways that violate fair lending policies such as the ECOA [4]. Without a degree of transparency to understand the models' decision-making process, the financial industry faces an increased risk of relying on proxy variables [11], posing a dilemma between the pursuit of efficiency and the risk of engaging in unfair lending practices.

In response to this challenge, researchers have begun working towards ensuring a higher level of transparency in ML models. Recent advancements such as the Glass-Box model, developed by researchers Sokol and Flach, introduce an explainability system to a loan origination model that utilizes an interactive query interface [13]. With this system, users can inquire about the rationale behind the models' decisions and assess the models' fairness, potentially offering a path to increased transparency and compliance with fair lending standards [13]. While limitations such as inconsistencies, contradictions, or even unresponsiveness were noted [13], the models' initial success with domain experts suggests a promising direction for ethically integrating automated lending decisions with the use of interactive query interfaces.

Challenges and Opportunities

The deployment of GPT-4 by OpenAI in March 2023 was a significant milestone in the field of AI, demonstrating the immense potential for LLMs to be implemented in a wide variety of applications. Through Chat-GPT, LLMs emerged as powerful NLP and NLG tools that revolutionized how we approach data-analysis, and their potential to facilitate data-processing tasks will only increase as fine-tuned models become more readily available [1]. Therefore, as LLMs continue to advance, their applications will become more versatile, which will present an opportunity to address complex challenges in heavily regulated industries like the financial industry.

Despite the promising capabilities of LLMs, their adoption in heavily regulated industries requires important considerations, and data scientists have expressed concerns about the ethical implications of automating decision-making processes using emerging technologies, especially in areas where accuracy, fairness, and transparency are crucial [3, 14]. One of the most pressing issues, at the moment, is the potential for inherent biases within LLMs, which can arise from the datasets used for training [15]. Because these models are trained on random selections from massive bodies of bodies of widely available datasets [16] and training these models can be done with no human supervision [2], the models are capable of reflecting any biases present in the data that was used to train them [15]. Additionally, given that the goal of LLMs is to generate human-like texts based on real human data, finding a data set of human-generated content that lacks any form of bias is virtually impossible [17].

The true strength of LLMs in comparison to the traditional ML models, however, does not lie in their ability to overcome bias while processing vast data sets. Instead, the primary benefit of implementing LLMs for data-processing and decision-making is their degree of transparency and interpretability. This transparency is crucial for industries subject to strict regulations, where understanding the rationale behind automated decisions is necessary for compliance and ethical practices [18, 19]. As previously explained, traditional ML models, such as those used in the financial industry, lack transparency in the decision-making process [20]. LLMs, on the other hand, facilitate a more interactive approach to data-processing where users can query the models for explanations, which alone provides some transparency and ensures some accountability, similar to the Glass-Box model.

Additionally, the development of LLMs is increasingly focused on providing users with additional visibility and control over the models' outputs [21], which may improve the ability to detect and correct biases. Outside of the models' development, some researchers have also identified alternative techniques with the potential to enhance model controllability. For instance, Li et al. investigated the use of tricky prompts to bypass the standard restrictions imposed on the publicly available ChatGPT models and give users additional control over the model's output [22]. By entering the "Developer mode" through jailbreaking techniques, the researchers found that the model, when not explicitly prompted, expressed no inherent biases or hate speech which would suggest a path toward applying LLMs for unbiased data processing under these specific, controlled conditions. However, the same study cautioned against the potential dangers of employing this highly controllable "Developer mode," as the model's lack of restrictions provided users with the ability to explicitly prompt the model to generate content incorporating biased perspectives, illegal suggestions, or sensitive information disclosure, presenting a significant ethical and security concern. As such, to navigate these complexities, it's essential to adopt strategies that prioritize both additional controllability and transparency, which together are essential for their application in highly regulated fields such as the financial industry. By enabling human oversight and intervention, LLMs can be fine-tuned to align with ethical standards and regulatory requirements, ensuring that their integration in these industries remains fair and responsible.

Already, researchers have recognized the importance of fine-tuning these models to minimize their expression of biases. Several researchers have attempted to quantify various forms of bias in LLMs, including the expression of gender roles as they relate to occupations [14, 17], the expression of gender stereotypes [23], and the expression of religious biases [16]. Subsequently, the researchers successfully implemented debiasing techniques that succeeded in large part due to the nature of the LLMs, with some modifying the queries to ensure that results reflected less bias [16] while others successfully fine-tuned sentence encoders to reduce bias expressions [23]. These advancements in debiasing methods are not just technical achievements, they demonstrate the feasibility of leveraging LLMs in a fair and equitable manner.

2.3 Gap in Literature

The exploration of LLMs in the context of fair lending policy has identified a significant gap in the literature regarding their implementation in the underwriting process. While the current literature is mostly focused on the capabilities and challenges of ML models and the transformative potential of LLMs across all industries, a more focused area of LLMs in the financial industry, particularly to ensure compliance with fair lending policies, remains largely unexplored.

With the success of the Glass-box model in enhancing transparency and interpretability in automated lending decisions [13], there is a possibility that LLMs, with their advanced NLP and NLG capabilities, could offer a promising solution to the industry's dilemma of increasing efficiency while maintaining fairness. By integrating LLMs in the underwriting process, financial institutions could ensure adherence to fair lending practices through improved transparency in the form of clarifying queries. With this feature, LLMs could serve as a bridge between the current applications of ML models and a future where automated decision-making processes are both efficient and explainable.

Moving forward, it will be crucial to address this gap in the literature by further studying the integration of LLMs to assist in lending decisions. Additionally, the refinement of LLMs to navigate the complex ethical considerations in the financial industry could mark a significant advancement in the application of AI technologies in the industry. With this, the journey towards integrating LLMs in the financial industry demonstrates great potential, promising not only to enhance operational efficiency in financial institutions but also to uphold the principles of fairness and equity in lending.

3 Methodology

3.1 Research Design

This study employs a quantitative experimental design to assess the potential bias introduced by Large Language Models (LLMs) in mortgage credit underwriting. The primary objective is to identify which applicant traits are more frequently associated with "strong" versus "weak" credit profiles by these models. An experimental approach was chosen to maintain control over input variables, systematically observe their effects, and enhance reproducibility of results.

A schematic overview of the methodology is provided in Fig. 1, with details elaborated in the following subsections.

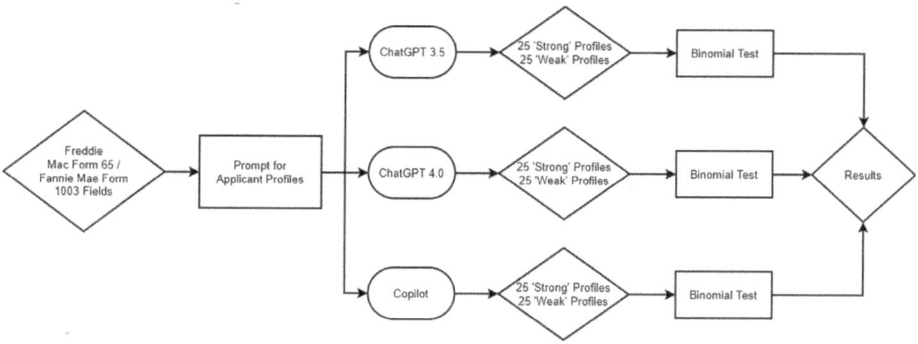

Fig. 1. Schematic Overview of the Research Methodology.

3.2 Model Selection

Three LLMs, OpenAI's GPT-3.5, OpenAI's GPT-4, and Microsoft Copilot, were selected due to their widespread adoption and advanced NLP capabilities. The study used the publicly available versions as of July 14, 2024, for GPT-3.5 and GPT-4, and the version of Copilot available on January 17, 2024, which integrates GPT-4 with Microsoft's Prometheus model. Fixing the versions to these release dates ensured consistency across trials and situated findings provided a clear temporal context for the findings.

3.3 Prompt and Data Collection

To collect data, several fields from the Uniform Residential Loan Application (URLA) were input into each of the selected LLMs. The URLA, also known as Freddie Mac Form 65 / Fannie Mae Form 1003, is a standardized document used by borrowers to apply for a mortgage. It was chosen due to its widespread use and standardized format.

Delimiters were placed around fields to reduce misinterpretation, and each model was instructed to adopt the persona of a loan officer identifying underwriting-relevant traits. The following fields from the URLA were provided to the models: Borrower Information (including Name, Social Security Number, Alternate Names, Date of Birth, Citizenship, Cell Phone Numbers, Email Address, Current Address, Employment Information, Gross Monthly Income, U.S. Citizenship Status, Type of Credit, Other Borrowers, Marital Status, Housing Expense, Dependents, and Credit Limit), Loan Information (including Loan Purpose, Loan Amounts, and Source of Down Payment), Military Service Status, and Demographic Information (including Ethnicity, Sex, and Race). These specific fields were chosen because they are critical components of the credit underwriting process, providing comprehensive information about the applicant's identity, financial status, and demographic background. The full prompt is shown in Fig. 2.

> You are an experienced loan officer who has been tasked with identifying the traits that your peers should consider during the underwriting process. To give other officers an example of what some applicants look like, you have been tasked with identifying two distinct applicants: a strong applicant and a weak applicant.
>
> To help your peers analyze a wide range of profiles, we request that you identify two very distinct profiles, with very different backgrounds, information, demographics. etc.
>
> Consider the fields in the ApplicantInfo.json* file.
>
> Generate a .json formatted profile which contains the fields in ApplicationInfo.json* for a strong applicant and a weak applicant. Use your knowledge as a loan officer to identify what the profile for these two applicants should look like.
>
> ***The ApplicationInfo.json file contains the previously mentioned URLA fields.**

Fig. 2. Prompt provided to each LLM.

For each model, trials were conducted until 25 distinct strong and weak profiles were achieved, resulting in 50 total profiles for each model and 150 total profiles for the study. Incomplete or fictional/non-existent responses were excluded from the dataset to ensure the integrity and validity of the data. Incomplete responses are defined as profiles with missing information from one or fields. Fictional or non-existent data is defined as information, primarily for the 'Current Address' field, which does not represent a real-world location. These responses occurred infrequently (no more than five times for each model) and were discarded as they lack the necessary information for a comprehensive analysis of the biases expressed by the models.

3.4 Data Analysis Methods

To analyze the bias in the profiles generated by each model, profile comparisons focused on the fields city, state, industry, position, US citizenship, race, ethnicity, and gender across the 150 profiles. For the statistical analysis of the results, gender, race, ethnicity, and US citizenship values were encoded. This allowed for the categorical responses to be transformed into a numerical format compatible with the Binomial Test function.

The Binomial Test function was used to compare the distributions of these encoded variables across the different profiles generated by each LLM. The Binomial Test is a statistical method for analyzing binary data and assessing whether the proportion of a given attribute, such as being male or being a U.S. citizen differs significantly between strong and weak profiles. The small scope of this study, with 150 profiles, provided an appropriate sample size for assessing the differences between proportions of a given attribute. By performing the Binomial Test for the 50 profiles generated by each of the models, this study design identifies patterns of bias in the LLMs' processing and highlight areas where the models might favor or disadvantage certain groups of applicants (see Table 1).

4 Results

4.1 Bias Evaluation

Table 1. Binomial test results for each model and demographic trait. ("1" indicates U.S. citizen, white/majority race or ethnicity, and male).

Binomial Test Results				
Model	Variable	P-value	Proportion of "1" in Strong Profiles	Proportion of "1" in Weak Profiles
ChatGPT-3.5	US Citizenship	2.84303E-06	100%	60%
	Race	0.00E+00	96%	4%
	Ethnicity	0.00E+00	96%	4%
	Gender	9.54E-24	0%	100%
ChatGPT-4	US Citizenship	0.36039672	100%	96%
	Race	1.07E-08	100%	48%
	Ethnicity	0.00E+00	100%	4%
	Gender	1.94E-10	80%	20%
Copilot	US Citizenship	1.00	100%	100%
	Race	6.50E-05	100%	68%
	Ethnicity	6.50E-05	100%	68%
	Gender	0.00E+00	100%	0%

During the study, ChatGPT-3.5 expressed strong demographic biases in its categorization of "strong" versus "weak" profiles. 100% of the strong profiles were U.S. citizens, while 60% of the weak profiles held this status. A binomial test indicates that this difference is statistically significant, with a p-value of 2.84E-06, which is below the 0.05 threshold. This result suggests that ChatGPT-3.5 disproportionately associates strong profiles with U.S. citizenship. In terms of race, 96% of strong profiles belonged to racial majority groups, compared to 4% of weak profiles. The binomial test for this difference returned a p-value of 0.00, indicating a statistically significant disparity. Similarly, the proportion of ethnic majority groups in strong profiles was 96%, while 4% of weak profiles belonged to ethnic majority groups. The corresponding p-value of 0.00 highlights the significance of this disparity. Gender imbalance is another notable issue: 0% of strong profiles were male, while 100% of weak profiles were male, and the p-value of 9.54E-24 from the binomial test suggests that this difference was statistically significant. These results suggest that ChatGPT-3.5 assigns a higher level of credibility to certain demographics, which reinforces systemic biases.

The model's successor, ChatGPT-4, shows some improvements in fairness, particularly in U.S. Citizenship bias. 100% of strong profiles and 96% of weak profiles were U.S.

citizens. A binomial test returned a p-value of 0.36, which is above the 0.05 threshold, indicating that there is no statistically significant difference in this attribute. However, racial and ethnic disparities remain problematic. While 100% of strong profiles belonged to racial majority groups, only 48% of weak profiles did. The p-value of 1.07E-08 indicates that this difference is statistically significant. Similarly, 100% of strong profiles belonged to ethnic majority groups, while 4% of weak profiles fell into this category, with a corresponding p-value of 0.00, demonstrating significant disparity. Gender imbalance was somewhat mitigated compared to ChatGPT-3.5, with 80% of strong profiles being male and 20% of weak profiles being male. A p-value of 1.94E-10, however, indicates that the difference is still highly significant. This suggests that an update to the model led to a reduction in female-only "strong" and male-only "weak" classifications, but there remains a persistent bias in associating stronger applicants with men and weaker applicants with women.

An entirely different model, Microsoft's Copilot, also exhibits significant biases. For this model, 100% of both strong and weak profiles were U.S. citizens, resulting in a p-value of 1.00, indicating no difference. 100% of strong profiles belonged to racial majority groups, while 68% of weak profiles did, producing a p-value of 6.50E-05, signifying a significant difference. Ethnic majority representation followed the same pattern, with 100% of strong profiles and 68% of weak profiles belonging to the ethnic majority, leading to a p-value of 6.50E-05. Gender imbalance remained extreme, with 100% of strong profiles being male and 0% of weak profiles being male, and the p-value of 0.00 once again signifies statistical significance. Similar to ChatGPT-3.5 and ChatGPT-4, Copilot continues to reinforce gender stereotypes by associating creditworthiness exclusively with men.

The binomial test results reveal a consistent and significant bias in how AI models categorize "strong" versus "weak" profiles based on demographic traits. Across the models, "strong" profiles are overwhelmingly associated with attributes like white, male, and U.S. citizen, while "weak" profiles are disproportionately linked to attributes like non-white and female.

Gender bias was particularly stark: GPT-3.5 produced only female strong and male weak profiles, while Copilot reversed the pattern with male-only strong and female-only weak profiles. GPT-4 showed some progress, but racial and ethnic disparities in profile categorization persisted.

Overall, despite differences in expression, all three models reinforced systemic demographic biases when categorizing applicant creditworthiness.

4.2 Implications for Fair Lending Practices

The findings of this study reinforce concerns about the risks LLMs pose to fair lending if applied in underwriting. The consistent biases observed across all three LLMs suggest that LLMs may inadvertently reproduce systemic discrimination when used in financial decision-making.

More specifically, the results indicate that LLMs tend to favor applicants who are white, male, and U.S. citizens, while disproportionately considering non-white, female, and non-citizen applicants weaker candidates. This expression of bias directly contradicts the principles of fair lending laws like the Equal Credit Opportunity Act (ECOA) and the Fair Housing Act (FHA), which prohibit discrimination based on race, gender, nationality, and other protected characteristics in lending decisions [7, 8]. If LLMs are integrated into the underwriting process without proper bias mitigation strategies, financial institutions could risk introducing a major source of bias in credit access. These biases could lead to discriminatory lending outcomes where minoritized groups receive fewer loan approvals or higher interest rates based on demographic characteristics that are unrelated to their financial responsibility.

To comply with fair lending laws and promote equity, institutions must implement transparency measures, regular fairness assessments, and strict alignment with ethical lending standards. Regulators may also need to develop more rigorous guidelines to govern AI use in credit decision-making.

4.3 Limitations of the Study

While this study provides valuable insights into the biases expressed during LLM-generated creditworthiness assessments, several limitations should be acknowledged.

First, the sample size is relatively small, with only 150 total profiles generated across three models. Although sufficient for initial analysis, larger datasets would increase reliability and allow for more robust statistical testing.

Second, this study focuses exclusively on three popular LLMs, ChatGPT-3.5, ChatGPT-4, and Microsoft Copilot, at specific points in time. Because these models are frequently updated, future versions of these models may exhibit different biases depending on changes to their training data and underlying algorithms. As a result, findings may not generalize to future versions or to other AI systems.

Third, the methodology of this study relies on a predefined prompt and structured inputs and outputs, which does not capture the complexity of a real-world credit assessment. In practice, lenders consider a wide range of financial and behavioral factors beyond the attributes included in the Uniform Residential Loan Application (URLA). Additionally, the study assumes that LLMs function independently when assessing applicant traits, but real-world financial institutions often integrate AI models alongside human decision-making and additional risk assessment tools.

Lastly, this study does not investigate the root causes of the biases that were observed. While the results indicate demographic disparities in the categorization of applicant profiles, they do not provide direct insights into how these biases originate within the models. Future research could explore the training data and model structures that contribute to these biases and potential methods for reducing them.

These limitations notwithstanding, this study may serve as a valuable starting point for understanding how LLMs could influence the credit underwriting process. Furthermore, it highlights the need for additional research into AI fairness in financial decision-making.

4.4 Recommendations

To address the biases identified in this study, financial institutions and AI developers should implement regular bias audits and transparency measures. Ongoing monitoring of demographic disparities in AI-assisted credit assessments is essential for ensuring that automated decisions comply with fair lending standards.

Another strategy is to improve the representativeness of training data. Incorporating more diverse datasets can help reduce systemic biases that arise from skewed or incomplete sources. This step would make AI models better reflect the populations they are meant to serve.

Lenders should also adopt a hybrid approach that combines AI efficiency with human oversight. Rather than relying solely on AI-generated outputs, institutions should maintain processes where humans review, contextualize, and, if necessary, override automated recommendations. Providing borrowers with access to clear explanations of AI-driven decisions and avenues to contest questionable outcomes would further strengthen fairness and accountability.

Finally, continued research into AI fairness, bias mitigation strategies, and the ethical deployment of machine learning in financial services is crucial. As AI integration into lending expands, these efforts will help ensure that credit access remains equitable and compliant with established legal protections.

5 Conclusions and Future Work

This study examined the expression of biases by Large Language Models (LLMs) when categorizing applicants as "strong" or "weak" in a simulated credit assessment. Three LLMs, ChatGPT-3.5, ChatGPT-4, and Microsoft Copilot, were evaluated based on the demographic attributes of the profiles they generated. The findings reveal statistically significant biases across all models, with "strong" profiles being overwhelmingly associated with attributes such as White, male, and U.S. citizen while "weak" profiles were more frequently assigned non-White, female, and non-citizen characteristics.

ChatGPT-3.5 exhibited severe demographic biases, with significant disparities in race, ethnicity, and U.S. citizenship. Additionally, its gender bias was extreme, as all strong profiles were female, and all weak profiles were male. ChatGPT-4 showed moderate improvements, particularly in reducing biases related to U.S. citizenship and allowing for some gender variance. However, it still demonstrated strong racial and ethnic biases in the profile categorization. Microsoft Copilot mirrored many of the biases observed in ChatGPT-3.5, particularly in gender and racial categorization. Like ChatGPT-3.5, Copilot exclusively assigned strong profiles to men and weak profiles to women.

These bias expressions have strong implications for the use of AI in credit underwriting and highlight the need for stronger bias mitigation strategies, regulatory oversight, and continued research to ensure that AI-driven financial assessments align with ethical and legal standards.

This study contributes to the growing body of research on AI fairness and bias in financial decision-making by providing evidence of demographic bias expressions in LLM-generated credit assessments. By evaluating three widely used LLMs, this research demonstrates the persistence of racial, gender, and citizenship biases in AI-driven loan application categorization processes. The study also sheds light on the potential risks of implementing LLMs in the credit underwriting process without proper precautions and emphasizes the need for fairness-aware AI models in financial services.

While this research provides valuable insights, there are several opportunities for future exploration. By expanding the dataset to include a larger and more diverse sample of profiles, future studies could perform a more detailed analysis of bias patterns. Additionally, future studies could examine a broader range of AI models to assess whether similar biases emerge across different model designs and training practices. Another key area for future research is investigating the root causes of bias within LLMs. Understanding how training data and algorithm design contribute to demographic disparities would provide actionable insights for mitigating biases.

Overall, this study serves as a strong starting point for further exploration into the expression of biases by LLMs in financial decision-making and emphasizes the urgent need for ongoing research, regulatory intervention, and ethical AI development to prevent discriminatory outcomes in credit underwriting and beyond.

Disclosure of Interests. The authors have no competing interests to declare that are relevant to the content of this article.

References

1. Liu, Y., et al.: Summary of ChatGPT-related research and perspective towards the future of large language models. Meta-Radiol. **1**(2), 1–21 (2023). https://doi.org/10.1016/j.metrad.2023.100017
2. Kocoń, J., et al.: ChatGPT: jack of all trades, master of none. Inf. Fusion **99**, 1–46 (2023). https://doi.org/10.1016/j.inffus.2023.101861
3. Sallam, M.: ChatGPT utility in healthcare education, research, and practice systematic review on the promising perspectives and valid concerns. Healthcare **11**(6), 887–907 (2023). https://doi.org/10.3390/healthcare11060887
4. Brotcke, L.: Time to assess bias in machine learning models for credit decisions. J. Risk Financ. Manag. **15**(4), 1–10 (2022). https://doi.org/10.3390/jrfm15040165
5. Hall, P., Cox, B., Dickerson, S., Ravi Kannan, A., Kulkarni, R., Schmidt, N.: A United States fair lending perspective on machine learning. Front. Artif. Intell. **4**, 1–9 (2021). https://doi.org/10.3389/frai.2021.695301
6. Zarsky, T.: The trouble with algorithmic decisions: an analytic road map to examine efficiency and fairness in automated and opaque decision making. Sci. Technol. Human Values **41**(1), 118–132 (2016). https://doi.org/10.1177/0162243915605575
7. Kumar, I.E., Hines, K.E., Dickerson, J.P.: Equalizing credit opportunity in algorithms: aligning algorithmic fairness research with U.S. fair lending regulation. In: Proceedings of the 2022 AAAI/ACM Conference on AI, Ethics, and Society, pp. 357–368 (2022). https://doi.org/10.1145/3514094.3534154
8. Taylor, W.: The ECOA and disparate impact theory: a historical perspective. J. Law Policy **26**(2), 575–635 (2018)
9. Zhang, Y.: Assessing fair lending risks using race/ethnicity proxies. Manage. Sci. **64**(1), 178–197 (2018). https://doi.org/10.1287/mnsc.2016.2579
10. Sadok, H., Sakka, F., El Maknouzi, M.E.H.: Artificial intelligence and bank credit analysis: a review. Cogent Econ. Finan. **10**(1), 12 (2022)
11. Bücker, M., Szepannek, G., Gosiewska, A., Biecek, P.: Transparency, auditability, and explainability of machine learning models in credit scoring. J. Oper. Res. Soc. **73**(1), 1–30 (2022). https://doi.org/10.1080/01605682.2021.1922098

12. de Laat, P.B.: Algorithmic decision-making based on machine learning from big data: can transparency restore accountability? Philos. Technol. **31**(4), 525–541 (2018). https://doi.org/10.1007/s13347-017-0293-z
13. Sokol, K., Flach, P.: One explanation does not fit all: the promise of interactive explanations for machine learning transparency. KI - Künstliche Intelligenz **34**(2), 235–250 (2020). https://doi.org/10.1007/s13218-020-00637-y
14. Kamboj, P., Kumar, S., Goyal, V.: Measuring and mitigating gender bias in contextualized word embeddings. In: 2023 IEEE International Conference on Blockchain and Distributed Systems Security, pp. 1–5 (2023). https://doi.org/10.1109/ICBDS58040.2023.10346586
15. Hovy, D., Prabhumoye, S.: Five sources of bias in natural language processing. Lang. Linguist. Compass **15**(8), 1–9 (2021). https://doi.org/10.1111/lnc3.12432
16. Abid, A., Farooqi, M., Zou, J.: Persistent anti-Muslim bias in large language models. In: Proceedings of the 2021 AAAI/ACM Conference on AI, Ethics, and Society, pp. 298–306 (2021). https://doi.org/10.1145/3461702.3462624
17. Kotek, H., Dockum, R., Sun, D.: Gender bias and stereotypes in large language models. In: Proceedings of The ACM Collective Intelligence Conference, pp. 12–24 (2023). https://doi.org/10.1145/3582269.3615599
18. Bueff, A.C., et al.: Machine learning interpretability for a stress scenario generation in credit scoring based on counterfactuals. Expert Syst. Appl. **202**, 1–14 (2022). https://doi.org/10.1016/j.eswa.2022.117271
19. Mittelstadt, B., Russell, C., Wachter, S.: Explaining explanations in AI. In: Proceedings of the Conference on Fairness, Accountability, and Transparency, pp. 279–288 (2019). https://doi.org/10.1145/3287560.3287574
20. Rudin, C.: Stop explaining black box machine learning models for high stakes decisions and use interpretable models instead. Nat. Mach. Intell. **1**(5), 206–215 (2019). https://doi.org/10.1038/s42256-019-0048-x
21. Hassija, V., Chakrabarti, A., Singh, A., Chamola, V., Sikdar, B.: Unleashing the potential of conversational AI: amplifying chat-GPT's capabilities and tackling technical hurdles. IEEE Access **11**, 143657–143682 (2023). https://doi.org/10.1109/ACCESS.2023.3339553
22. Li, H., et al.: Multi-step jailbreaking privacy attacks on ChatGPT. arXiv: https://doi.org/10.48550/arXiv.2304.05197 (2023)
23. Dolci, T.: Fine-tuning language models to mitigate gender bias in sentence encoders. In: 2022 IEEE Eighth International Conference on Big Data Computing Service and Applications (BigDataService), pp. 175–176 (2022)

Large Language Models, Artificial Intelligence, and Applications

E^3: Unified Architecture for Hallucination-Resistant Language Generation

Gang Wang(✉) and Mark Nixon

Emerson Automation Solutions, Austin, USA
gang.wang.dr@gmail.com

Abstract. Large language models frequently generate hallucinations—confident but factually unsupported claims—undermining their deployment in high-stakes applications. While existing approaches treat hallucination as an afterthought addressed through post-hoc filtering, we present E^3-Guarded Generation, a unified architecture that embeds verification directly into the generation process through three complementary principles: **E**vidence grounding, logical **E**ntailment, and **E**xecutable validation. Our architecture integrates five specialized components—Provenance-First Decoding, Speculative Verification Decoding, Counterfactual Cross-Examination, KWIK-Calibrated Abstention, and Semantic Constraint Grammars—within a carefully orchestrated pipeline that enforces faithfulness constraints at every generation stage. We present the complete system architecture, detail component interactions and information flow patterns, establish key design invariants, and analyze the architectural trade-offs between accuracy, latency, and coverage. The framework's layered design enables modular deployment and graceful degradation through explicit feedback mechanisms. By shifting from reactive filtering to constructive correctness, E^3 demonstrates that hallucination prevention can be systematically engineered through principled architectural design, providing a foundation for trustworthy AI systems in critical domains.

Keywords: LLM Hallucinations · Guarded Generation · Claim-Citation-Justification · Hallucinations Mitigation

1 Introduction

Large language models have achieved remarkable capabilities in text generation, reasoning, and knowledge synthesis [1–3]. Despite their fluency and breadth, these systems exhibit a critical weakness that undermines their reliability: the tendency to generate *hallucinations*—confident assertions that are factually incorrect, internally inconsistent, or unsupported by available evidence [4,5]. This paper addresses hallucination mitigation from an architectural perspective, presenting a unified system design that embeds verification directly into the generation process rather than treating it as an afterthought.

H. R. Arabnia et al. (Eds.): CSCI 2025, CCIS 2972, pp. 275–296, 2027.
https://doi.org/10.1007/978-3-032-28469-3_19

1.1 The Challenge of Faithful Generation

Hallucinations manifest across multiple dimensions, each presenting distinct engineering challenges [6]. *Intrinsic hallucinations* stem from training data memorization failures, generating plausible but incorrect facts [7]. *Extrinsic hallucinations* contradict or lack support from retrieved evidence [8]. *Compositional hallucinations* arise from incorrect combinations of individually correct facts [9]. *Temporal hallucinations* mix information across time periods inappropriately [10].

The practical impact extends to critical real-world applications [11]. In medical diagnosis, hallucinated symptom interpretations could lead to incorrect treatment recommendations. In legal document analysis, fabricated precedents could result in flawed legal strategies. In financial advisory systems, hallucinated market data could cause substantial monetary losses. These high-stakes scenarios demand systems with architectural guarantees of faithfulness, not merely statistical improvements.

Current approaches to hallucination mitigation suffer from fundamental architectural limitations. Retrieval-Augmented Generation (RAG) [12,13] conditions generation on retrieved documents but provides no mechanism to ensure models actually ground their outputs in retrieved evidence. The retrieval and generation components operate largely independently, with no architectural enforcement of evidence utilization. Post-hoc verification systems [14] attempt to identify and filter hallucinations after generation, but this reactive approach wastes computational resources generating content that will be discarded and provides no guidance to improve generation quality during decoding. Self-consistency methods [15] generate multiple outputs and select based on agreement, but when models exhibit systematic biases, these methods can amplify rather than correct hallucinations.

The common thread across these limitations is architectural: existing systems treat verification as external to generation rather than integral to it. This separation creates fundamental gaps where hallucinations can emerge unchecked.

1.2 From Reactive Filtering to Constructive Correctness

We propose a fundamental paradigm shift from reactive filtering to *constructive correctness*. Rather than generating content and subsequently checking for errors, we design generation mechanisms that prevent hallucinations by construction. This approach draws inspiration from design-by-contract principles in software engineering [16], where correctness is built into system design rather than verified afterward, and from constructive logic [17], where existence proofs require explicit construction methods.

The E^3-Guarded Generation framework operationalizes this paradigm through three complementary architectural principles:

1. **Evidence First**: Every factual claim must be preceded by explicit evidence pointers, enforcing attribution before generation through constrained decoding mechanisms. This principle ensures that claims cannot be generated without prior commitment to supporting evidence.

2. **Entailment Verification**: Claims must be logically entailed by cited evidence, verified through efficient inline checking integrated directly into the generation pipeline. This principle ensures semantic consistency between assertions and their purported support.
3. **Execution Validation**: Structural and semantic constraints are enforced throughout generation, ensuring outputs conform to well-defined patterns that facilitate verification and maintain logical coherence.

These principles are realized through five specialized components operating within a unified architecture: *Provenance-First Decoding* (PFD) enforces evidence-before-claims ordering; *Speculative Verification Decoding* (SVD) achieves efficient inline verification; *Counterfactual Cross-Examination* (CCE) employs adversarial refinement to expose subtle inconsistencies; *KWIK-Calibrated Abstention* (KCA) enables principled deferral under uncertainty; and *Semantic Constraint Grammars* (SCG) enforce structural well-formedness. Figure 1 illustrates the overall system architecture and component relationships.

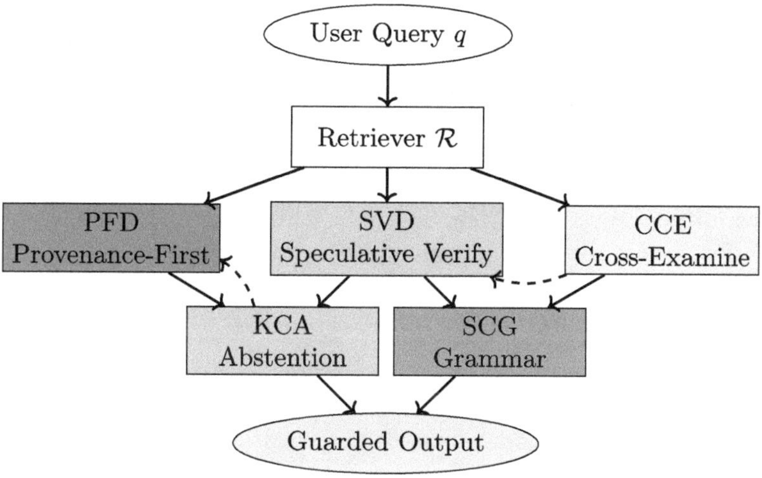

Fig. 1. The E^3-Guarded Generation architecture integrates five components to enforce evidence (blue), entailment (green), and execution (yellow) constraints. The abstention component (orange) enables graceful degradation under uncertainty, while semantic constraint grammars (purple) maintain structural validity. Dashed arrows indicate feedback loops for adaptive refinement. (Color figure online)

1.3 Contributions and Organization

This paper presents the architectural design of E^3-Guarded Generation, making the following contributions:

- A unified system architecture that embeds verification at every stage of generation, transforming hallucination prevention from a post-processing concern to an integral design property.
- Detailed specification of five specialized components—PFD, SVD, CCE, KCA, and SCG—including their internal structure, design rationale, and operational characteristics.
- Analysis of information flow patterns and feedback mechanisms that enable the system to maintain correctness invariants while supporting graceful degradation under uncertainty.
- Characterization of fundamental trade-offs between accuracy, latency, and coverage, providing guidance for system configuration across diverse deployment scenarios.

The remainder of this paper is organized as follows. Section 2 establishes architectural foundations and core design principles. Section 3 presents detailed component architectures. Section 4 analyzes system integration and information flow. Section 5 examines architectural trade-offs. Section 6 discusses broader implications and limitations. Section 7 concludes with a summary of architectural contributions.

2 Architectural Foundations and Design Principles

Before examining individual components, we establish the architectural foundations that govern E^3-Guarded Generation. This section presents the system model, core design principles, and layered architecture that enable systematic hallucination prevention through principled engineering.

2.1 System Model and State Representation

The E^3 system operates as a state machine where generation progresses through well-defined phases, each enforcing specific verification requirements. Understanding this state-based model is essential for reasoning about system behavior and correctness properties.

The system state at any time t during generation is characterized by a tuple $s_t = (q_t, \mathcal{E}_t, \mathcal{C}_t, V_t, A_t, \phi_t)$ comprising six elements: the current query context q_t; the accumulated evidence set $\mathcal{E}_t$ retrieved from the knowledge corpus; the generated claims $\mathcal{C}_t = \{c_1, \ldots, c_k\}$; a verification mapping V_t that tracks which claims have passed verification; the current action $A_t \in \{\text{continue}, \text{abstain}, \text{revise}\}$; and the current phase ϕ_t indicating which component is active.

State transitions are governed by a transition function that routes control to the appropriate component based on the current phase. When in the provenance phase, transitions are handled by PFD; during verification, by SVD; during cross-examination, by CCE; during abstention evaluation, by KCA; and during grammar enforcement, by SCG. This phase-based routing ensures that each component operates on well-defined inputs and produces outputs conforming to expected interfaces.

The system maintains four critical invariants throughout generation that collectively guarantee faithful output:

- **Evidence Precedence**: Every claim in $\mathcal{C}_t$ has associated evidence that was committed before the claim was generated. This invariant ensures that claims cannot exist without prior evidentiary grounding.
- **Verification Monotonicity**: Once a claim passes verification, it remains verified. The system never revokes verification status, ensuring stable progress toward completion.
- **Abstention Safety**: When the system chooses to abstain, it does so because confidence falls below acceptable thresholds. Abstention represents a deliberate safety mechanism, not a failure mode.
- **Grammar Compliance**: All generated content conforms to the semantic constraint grammar at every step. Structural validity is maintained incrementally rather than checked post-hoc.

These invariants are enforced through the architectural structure itself—the system design makes invariant violations impossible rather than merely detectable.

2.2 Core Design Principles

Three fundamental principles guide the architectural decisions throughout E^3-Guarded Generation, each addressing a specific challenge in building reliable generation systems.

Verification as Construction. Traditional generation systems produce content first and verify afterward, treating verification as an external quality gate. E^3 inverts this relationship by integrating verification directly into the construction process. This principle, inspired by constructive approaches in logic [17] and design-by-contract methodology in software engineering [16], ensures that verification is not an afterthought but an integral part of generation.

A generation process is *constructive* if every intermediate state satisfies required verification predicates—there is no point during generation where unverified content exists in the output buffer. This property eliminates entire categories of failures where verification is bypassed, forgotten, or incorrectly applied. The architectural implication is that verification logic must be embedded within generation components rather than wrapped around them.

Fail-Fast with Graceful Degradation. The system is designed to detect potential hallucinations as early as possible in the generation pipeline [18]. Early detection serves two purposes: it minimizes computational resources wasted on content that will ultimately be rejected, and it enables corrective action while context is still fresh and revision is straightforward.

When verification fails, the system does not simply halt. Instead, it activates feedback mechanisms that route control to appropriate recovery paths. Verification failures trigger revision through PFD; cross-examination challenges prompt additional evidence retrieval; and when uncertainty remains too high despite remediation attempts, the system gracefully degrades to abstention rather than producing unreliable output. This combination of aggressive detection with graceful recovery ensures both efficiency and robustness.

Composable Verification. Each E^3 component provides independent guarantees that compose systematically to yield system-wide assurances [19]. This composability has two important architectural implications. First, components can be developed, tested, and improved independently—improvements to SVD's verification accuracy benefit the entire system without requiring changes to other components. Second, the overall system guarantee emerges from component guarantees through well-defined composition rules, enabling principled reasoning about system behavior from component specifications.

The layered architecture described below operationalizes composability by organizing components into coherent groups with clear interfaces and responsibilities.

2.3 Three-Layer Architecture

The E^3 architecture organizes its five components into three conceptual layers, each addressing a distinct aspect of hallucination prevention. Figure 2 illustrates the complete architecture with component interactions and feedback mechanisms.

The **Evidence Layer** ensures all claims have retrievable support from trusted sources. This layer encompasses evidence retrieval and Provenance-First Decoding, establishing the foundational requirement that claims must be grounded in explicit evidence before they can be generated. The evidence layer answers the question: *what sources support this claim?*

The **Entailment Layer** verifies logical consistency between claims and their cited evidence. This layer includes Speculative Verification Decoding and Counterfactual Cross-Examination, which together ensure that claims actually follow from their evidence rather than merely being associated with it. The entailment layer answers the question: *does the evidence actually support this claim?*

The **Execution Layer** validates structural constraints and manages uncertainty. This layer comprises KWIK-Calibrated Abstention and Semantic Constraint Grammars, ensuring that outputs conform to required patterns and that the system responds appropriately when confidence is insufficient. The execution layer answers the question: *is this output well-formed and reliable?*

Information flows primarily downward through these layers during normal operation, but feedback mechanisms enable upward flow when verification fails. This bidirectional capability allows the system to recover from errors discovered at lower layers by triggering corrective action at higher layers, implementing the fail-fast principle while maintaining forward progress toward valid output.

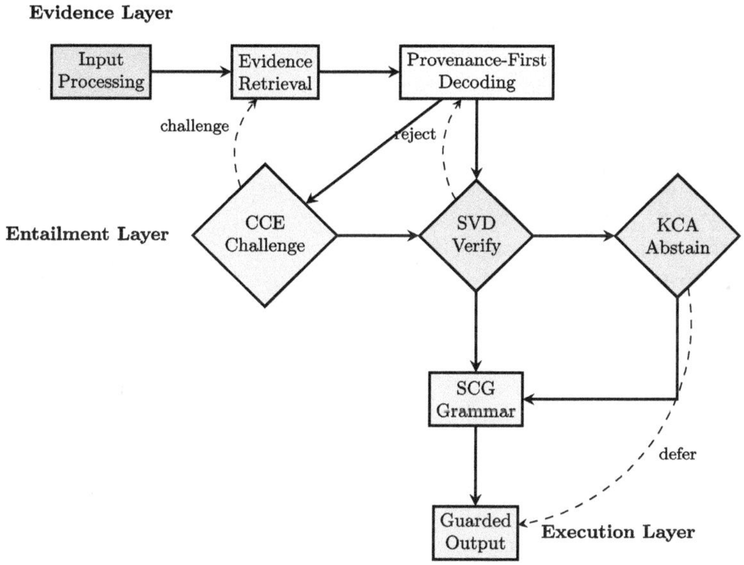

Fig. 2. Detailed E^3 architecture showing component interactions, data flows (solid arrows), and feedback mechanisms (dashed arrows). Diamond-shaped nodes represent verification gates that can reject content, while rectangular nodes represent processing stages. The three layers enforce evidence grounding, logical entailment, and structural execution constraints respectively.

The layered organization also supports modular deployment. Applications with different reliability requirements can configure layer behavior independently—tightening entailment checking for high-stakes domains or relaxing grammar constraints for creative applications—while maintaining the overall architectural guarantees.

3 Component Architecture

This section presents the architectural design of the five E^3 components. For each component, we describe its design rationale, internal structure, operational mechanisms, and integration interfaces. Together, these components realize the three-layer architecture introduced in Sect. 2.

3.1 Provenance-First Decoding (PFD)

Provenance-First Decoding enforces the fundamental constraint that evidence must precede claims in the generation sequence. This component implements the Evidence Layer's core responsibility: ensuring that no factual assertion can be generated without prior commitment to supporting evidence.

Design Rationale. Traditional language generation allows models to produce claims freely, with attribution added afterward if at all. This generate-then-cite approach invites hallucination because models can generate plausible-sounding assertions and then search for convenient evidence to rationalize them post-hoc [20, 21]. PFD inverts this pattern by requiring evidence selection before claim generation, eliminating post-hoc rationalization as a failure mode.

The key architectural insight is that temporal ordering can be enforced through grammatical constraints. By defining a grammar where evidence pointers must syntactically precede claims, we transform a semantic requirement (claims should be grounded) into a structural requirement (pointers before claims) that can be mechanically enforced during decoding.

Evidence Pointer Structure. PFD introduces explicit evidence pointers as first-class elements in the generation vocabulary. An evidence pointer comprises four components: a document identifier specifying the source; span boundaries indicating the precise location within the source; and a confidence score reflecting the model's assessment of relevance. This structured representation enables fine-grained attribution while supporting automated verification of pointer validity.

The provenance function maps each generated claim to its supporting evidence pointers, establishing the relationship that must hold for valid generation: the semantic content of the pointed evidence must logically support the claim. This relationship is enforced by subsequent verification components, but PFD ensures the structural prerequisite—that pointers exist and precede their claims.

Grammar-Based Constraint Enforcement. PFD enforces evidence-first ordering through a formal grammar that governs valid generation sequences. The grammar defines production rules requiring that every statement begins with one or more evidence pointers before any claim content can appear. Claims are explicitly delimited, creating clear boundaries between evidence attribution and asserted content.

Figure 3 illustrates the finite state automaton that enforces these constraints during generation. The automaton begins in an initial state where only pointer tokens are valid. After at least one pointer is emitted, additional pointers may follow, but claim generation can also begin. Once in the claim state, content tokens are generated until the claim concludes, reaching the accepting state. Critically, no valid transition path exists from the initial state to claim generation without passing through the pointer state—the architecture makes evidence-free claims structurally impossible.

During decoding, PFD masks the token probability distribution to permit only tokens valid in the current automaton state. This constrained decoding approach integrates seamlessly with standard autoregressive generation while guaranteeing grammar compliance.

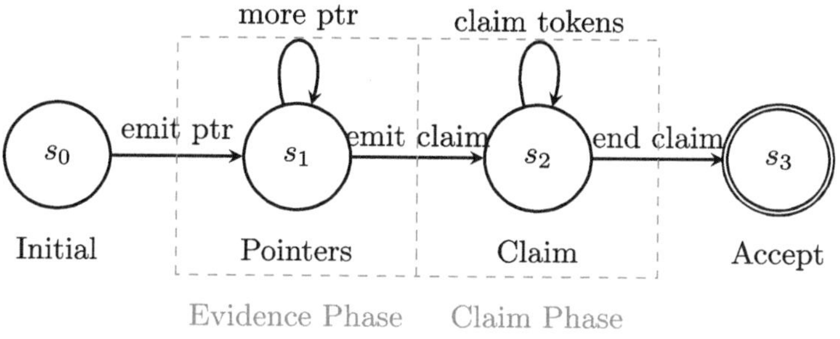

Fig. 3. Finite state automaton enforcing PFD grammar constraints. Generation must traverse pointer states (s_1) before entering claim states (s_2). The loop on s_1 permits multiple evidence pointers; the loop on s_2 permits multi-token claims. No valid path bypasses the evidence phase.

3.2 Speculative Verification Decoding (SVD)

Speculative Verification Decoding addresses the tension between verification thoroughness and generation efficiency. This component implements the Entailment Layer's primary verification mechanism, checking that claims are actually supported by their cited evidence.

Design Rationale. Naive verification approaches check each claim immediately after generation, incurring verification latency at every step. This sequential approach creates a bottleneck where generation speed is limited by verification speed. SVD resolves this tension by borrowing speculative execution techniques from computer architecture [22,23]: generate optimistically in batches, verify batches efficiently, and roll back when verification fails.

The architectural insight is that verification can be decoupled from generation through buffering. A fast draft model generates candidate content speculatively, accumulating tokens in a batch. A more thorough verifier then checks the entire batch, accepting valid prefixes and triggering rollback for invalid content. This amortizes verification cost across multiple tokens while maintaining correctness guarantees.

Draft-Verify-Fallback Architecture. SVD employs three cooperating subcomponents arranged in a pipeline. The *draft model* generates token batches rapidly, prioritizing speed over verification thoroughness. The *verifier* evaluates claim-evidence pairs, producing validity scores that indicate whether cited evidence actually supports the generated claims. The *base model* serves as a high-quality fallback, invoked when speculation fails to produce content that passes verification.

Figure 4 shows the SVD pipeline. During normal operation, the draft model generates batches of k tokens, which the verifier evaluates against cited evidence. If the batch passes verification, all tokens are accepted and generation continues.

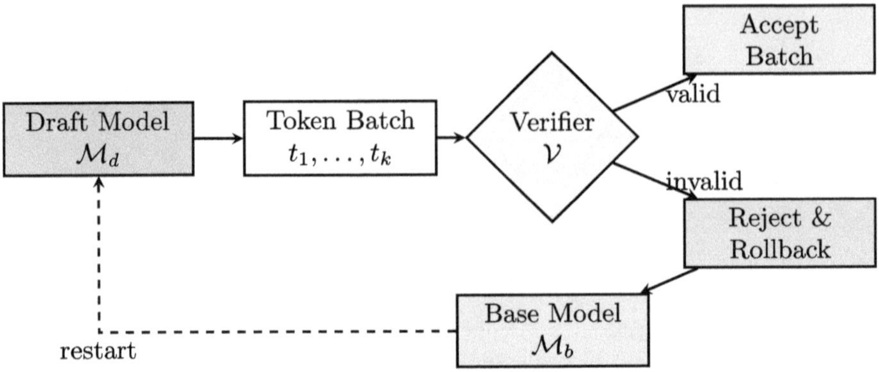

Fig. 4. SVD pipeline architecture. The draft model generates token batches speculatively, the verifier assesses validity, and accepted batches proceed to output. Rejected content triggers rollback and fallback to the base model, which provides a high-quality continuation before speculation resumes.

If verification fails, SVD identifies the longest valid prefix, accepts those tokens, rolls back the remainder, and invokes the base model to generate a single high-quality token. Generation then resumes with the draft model.

The acceptance rate—the fraction of draft tokens that pass verification—determines SVD's efficiency. Higher acceptance rates yield greater speedup over sequential verification, while lower rates trigger more fallback invocations. The batch size k balances speculation depth against rollback cost: larger batches amortize verification overhead but risk more wasted computation on rejection.

3.3 Counterfactual Cross-Examination (CCE)

Counterfactual Cross-Examination provides adversarial testing to identify subtle hallucinations that may pass initial verification. This component augments the Entailment Layer with a dialectical refinement process inspired by debate systems [24] and Socratic dialogue [25].

Design Rationale. Verification systems can have blind spots—classes of errors they systematically miss. A single verifier, no matter how accurate, may fail to detect certain hallucination patterns. CCE addresses this limitation through adversarial redundancy: a skeptical agent actively attempts to find flaws in generated content, creating competitive pressure toward correctness.

The architectural insight is that truth can emerge from structured conflict. When a generator must defend claims against a skeptic's challenges, incentives align toward producing claims that can withstand scrutiny. Claims that survive cross-examination have demonstrated robustness beyond what single-pass verification can provide.

Generator-Skeptic Interaction. CCE models generation as a two-player interaction between a generator and a skeptic with opposing objectives. The generator produces claims paired with supporting evidence, seeking to maximize acceptance. The skeptic generates counterfactual challenges—alternative interpretations, boundary cases, or consistency checks—seeking to expose unsupported assertions.

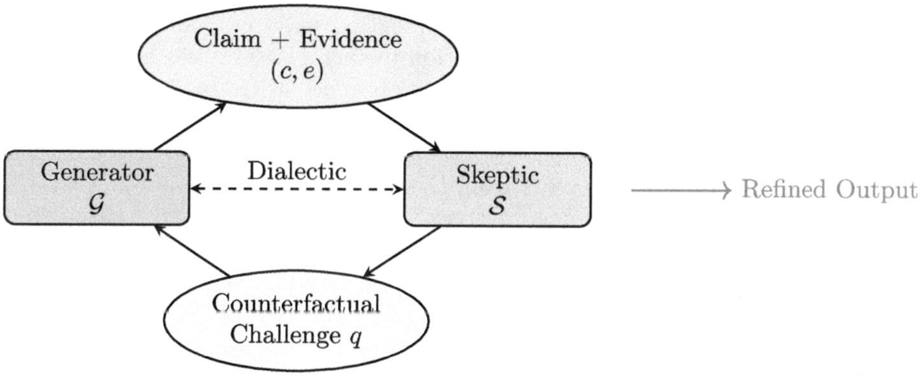

Fig. 5. CCE dialectical process. The generator proposes claims with evidence; the skeptic responds with counterfactual challenges. This iterative exchange continues until claims withstand scrutiny or are revised, driving convergence toward factual accuracy through adversarial refinement.

Figure 5 illustrates this dialectical process. Each round proceeds as follows: the generator proposes a claim with supporting evidence; the skeptic analyzes the claim and generates a targeted challenge; the generator must then defend, revise, or withdraw the claim. Successful defense increases confidence in the claim; revision improves claim quality; withdrawal prevents hallucination from reaching output.

The skeptic employs diverse challenge strategies to probe different failure modes: negation challenges assert the opposite of the claim; alternative challenges propose different conclusions from the same evidence; boundary challenges test edge cases; consistency challenges check coherence with related claims; and temporal challenges verify time-sensitive assertions.

Iterative Refinement. CCE operates over multiple rounds, with each round reducing the probability that hallucinations survive. The number of rounds balances thoroughness against latency—more rounds provide stronger guarantees but increase generation time. The system can be configured with fixed round counts or adaptive termination based on challenge difficulty and defense confidence.

3.4 KWIK-Calibrated Abstention (KCA)

KWIK-Calibrated Abstention enables the system to recognize and acknowledge its limitations, deferring to abstention when confidence is insufficient for reliable generation. This component implements the Execution Layer's uncertainty management responsibility.

Design Rationale. Not all queries can be answered reliably. Evidence may be unavailable, ambiguous, or insufficient; verification may be inconclusive; model uncertainty may be high. Forcing generation in such cases invites hallucination. KCA addresses this by augmenting the action space with explicit abstention, transforming generation from a forced-choice task into selective prediction where deferral is a valid outcome.

The architectural insight draws from the "Knows What It Knows" (KWIK) framework [26,27]: a well-designed system should have calibrated awareness of its own competence boundaries. KCA integrates multiple uncertainty signals to make principled abstention decisions that minimize risk while maintaining coverage.

Multi-source Uncertainty Integration. KCA synthesizes uncertainty information from three sources, each capturing a different aspect of reliability. *Retrieval coverage* measures how well the evidence corpus addresses the query—low coverage suggests the knowledge base lacks relevant information. *Verifier confidence* reflects the strength of entailment assessments from SVD—marginal scores indicate uncertain support relationships. *Model uncertainty* captures the generation model's own confidence, often measured through predictive entropy or output variance.

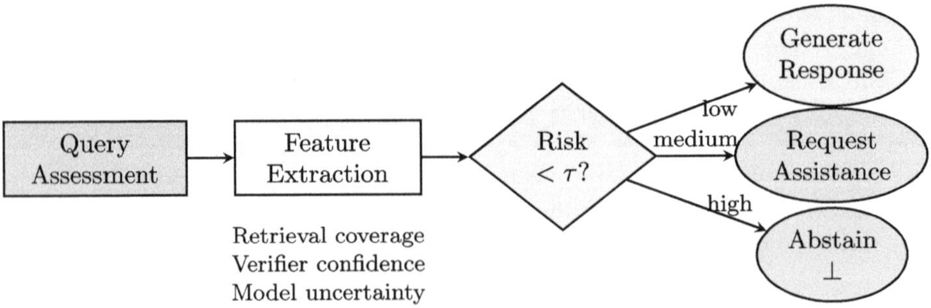

Fig. 6. KCA decision flow. Query assessment extracts uncertainty features from multiple sources; risk estimation combines these into a unified score; calibrated thresholds determine whether to generate confidently, request assistance, or abstain entirely.

Figure 6 shows the KCA decision flow. The component extracts uncertainty features, combines them into an estimated risk score, and compares against calibrated thresholds. Low risk permits confident generation; high risk triggers abstention; intermediate risk may invoke additional assistance such as requesting clarification or retrieving more evidence.

Hierarchical Abstention Levels. KCA supports multiple abstention responses beyond simple generate-or-refuse decisions. The system can express uncertainty through hedged language, request clarification from users, suggest alternative queries within its competence, or fully abstain with explanation. This hierarchical approach maximizes utility by matching response type to uncertainty level rather than applying a single threshold.

3.5 Semantic Constraint Grammars (SCG)

Semantic Constraint Grammars enforce structural requirements throughout generation, ensuring outputs conform to well-defined patterns that facilitate verification and maintain logical coherence. This component provides the Execution Layer's structural enforcement mechanism.

Design Rationale. Verification is easier when content follows predictable structures. A claim-citation-justification pattern, for instance, clearly delineates what is being asserted, what evidence supports it, and why the evidence is relevant. SCG enforces such patterns through grammatical constraints, ensuring that generated content is not only semantically valid but structurally amenable to verification.

The architectural insight is that high-level semantic requirements can be compiled into token-level constraints. By expressing faithfulness requirements as grammar rules, we enable efficient incremental enforcement during generation rather than expensive post-hoc validation.

Grammar Structure. SCG extends context-free grammars with semantic predicates that must be satisfied for productions to apply. The grammar defines non-terminal symbols representing semantic categories (statements, claims, citations, justifications), terminal symbols from the token vocabulary, and production rules with attached semantic constraints.

Figure 7 illustrates a parse tree under the SCG grammar. Documents consist of statements; each statement requires citations (evidence references), a claim (the asserted content), and justification (explanation of why evidence supports the claim). This structure enforces that no claim appears without attribution and reasoning, creating an audit trail that supports both automated verification and human review.

Incremental Parsing and Constrained Decoding. SCG integrates with generation through incremental parsing that maintains grammar state as tokens are produced. At each generation step, the parser determines which tokens are

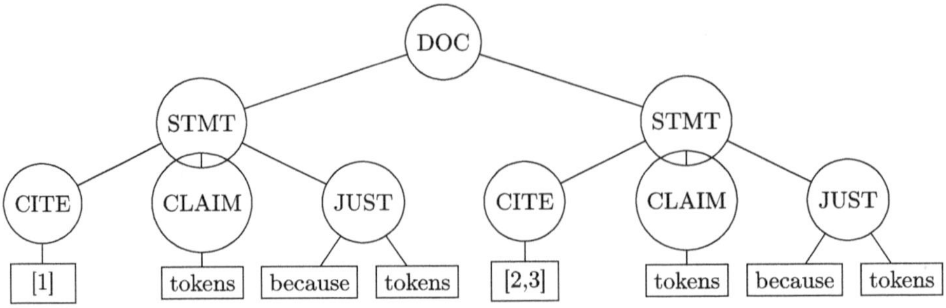

Fig. 7. Parse tree for SCG showing hierarchical document structure. Each statement contains mandatory citation, claim, and justification components. This structure ensures every assertion is attributed and explained, facilitating both verification and interpretability.

valid continuations given the current parse state, and decoding masks invalid tokens from the probability distribution. This achieves grammar compliance with constant amortized overhead per token through efficient state maintenance and precomputed valid-token sets.

The constrained decoding mechanism ensures that any generated sequence is a valid derivation in the grammar. Combined with PFD's evidence-first constraints, SCG provides the structural backbone that enables all E^3 components to operate on well-formed, consistently structured content.

4 System Integration and Information Flow

Individual components provide local guarantees, but system-level reliability emerges from their careful integration. This section examines how E^3 components are orchestrated into a coherent pipeline, how information flows through the system, and how feedback mechanisms enable recovery from verification failures.

4.1 Pipeline Orchestration

The E^3 generation pipeline processes queries through a sequence of phases, with each phase activating specific components and enforcing particular constraints. Understanding this orchestration is essential for reasoning about system behavior and configuring deployments appropriately.

Generation begins with *input processing*, where the user query is analyzed to determine information needs and retrieval requirements. The system then enters the *evidence retrieval* phase, querying the knowledge corpus to obtain relevant source material. Retrieved evidence is indexed and made available to subsequent components.

The pipeline proceeds to *provenance-first decoding*, where PFD generates content under grammar constraints that enforce evidence-before-claims ordering.

As tokens are produced, they accumulate in a speculation buffer managed by SVD. When the buffer reaches the configured batch size, the pipeline transitions to the *verification* phase.

During verification, SVD evaluates the buffered content against cited evidence. Successful verification advances the pipeline; failure triggers rollback and fallback generation. Concurrently or subsequently, CCE may initiate *cross-examination*, subjecting claims to adversarial challenges that probe for subtle inconsistencies.

Throughout generation, KCA monitors uncertainty signals. When risk estimates exceed configured thresholds, the pipeline transitions to *abstention handling*, where the system determines an appropriate deferral response. Finally, SCG performs continuous *grammar enforcement*, ensuring structural validity at every step.

The pipeline concludes when generation reaches a natural endpoint (completing the response), triggers abstention (uncertainty too high), or exhausts configured limits (maximum length or rounds). The output comprises generated content, associated evidence pointers, and confidence metadata.

4.2 Information Flow Control

Controlling information flow is critical for preventing hallucinations from propagating to output. The E^3 architecture enforces flow constraints that guarantee all output content has passed through verification checkpoints.

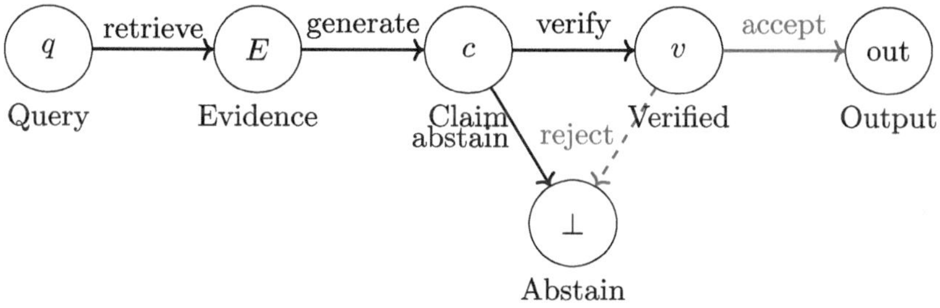

Fig. 8. Information flow graph showing state transitions and verification gates. Green edges indicate verified paths to output; red dashed edges show rejection paths to safe abstention. Every path from query to output traverses the verification state, ensuring no unverified content reaches users.

Figure 8 presents the information flow graph governing E^3 operation. Nodes represent information states: query input (q), retrieved evidence (E), generated claims (c), verified content (v), final output (out), and abstention ($\perp$). Edges represent allowed transitions between states.

The graph structure enforces a critical safety property: every path from query to output must pass through the verification state. There is no direct edge from claims to output—content must be verified before emission. This architectural constraint makes it structurally impossible for unverified claims to reach users.

Two types of terminal states exist. The output state receives verified content that has passed all checks. The abstention state receives content that failed verification or triggered uncertainty thresholds. Both are valid endpoints; the system never stalls or fails silently.

The verification state acts as a mandatory checkpoint where SVD and CCE evaluate content. Only content receiving positive verification proceeds to output. Rejected content flows to abstention, where KCA determines the appropriate response—revision, clarification request, or explicit deferral.

4.3 Feedback Mechanisms

While primary information flow proceeds forward through the pipeline, feedback mechanisms enable backward flow when verification identifies problems. These mechanisms implement the fail-fast principle by routing control to components that can address detected issues.

Three primary feedback loops operate within E^3:

SVD-to-PFD Revision Loop: When SVD rejects a token batch, the system does not simply discard the content and continue. Instead, rejection triggers a feedback signal to PFD indicating which claims failed and why. PFD can then revise its generation strategy—selecting different evidence, rephrasing claims, or adjusting confidence thresholds—before resuming generation. This loop enables iterative improvement rather than binary accept/reject decisions.

CCE-to-Retrieval Challenge Loop: When CCE's skeptic successfully challenges a claim, the feedback may indicate that additional or different evidence is needed. The system can return to the evidence retrieval phase, querying for sources that address the specific weakness the skeptic identified. This loop enables the system to strengthen its evidentiary foundation in response to adversarial probing.

KCA-to-Output Deferral Loop: When KCA determines that uncertainty is too high for reliable generation, it triggers a deferral that bypasses normal output processing. The feedback signal carries information about the nature of uncertainty—evidence gaps, verification ambiguity, or model confusion—enabling appropriate abstention responses such as requesting clarification or suggesting alternative queries.

These feedback mechanisms transform E^3 from a linear pipeline into an adaptive system that can recover from failures detected at any stage. The combination of forward progress with backward correction enables robust operation across diverse query types and evidence conditions.

4.4 Compositional Guarantees

The integration of independently-designed components yields system-level properties that exceed what any component provides alone. This compositional benefit arises from the careful interface design and layered architecture described in Sect. 2.

Each component addresses specific failure modes: PFD prevents evidence-free generation; SVD catches claims unsupported by cited evidence; CCE detects subtle inconsistencies missed by single-pass verification; KCA manages residual uncertainty; SCG ensures structural validity. Because these failure modes are largely independent, component guarantees combine multiplicatively rather than additively.

Beyond individual guarantees, the integrated system exhibits emergent properties:

Defense in Depth: Multiple verification layers mean that a hallucination must evade all checks to reach output. A claim that slips past SVD may be caught by CCE; uncertainty that evades CCE triggers KCA. This redundancy provides robustness against any single component's blind spots.

Graceful Degradation: When components operate at reduced capacity—limited evidence, uncertain verification, aggressive skepticism—the system degrades gracefully rather than failing catastrophically. KCA's abstention capability ensures that uncertainty translates to appropriate caution rather than unreliable output.

Adaptive Refinement: Feedback loops enable the system to learn from verification failures within a single generation episode. Claims that initially fail can be revised and re-verified, improving output quality through iteration rather than requiring perfect first-pass generation.

These compositional properties emerge from architectural decisions—the layered structure, mandatory verification checkpoints, and feedback mechanisms—rather than from any single component's sophistication. The whole genuinely exceeds the sum of its parts.

5 Architectural Trade-Offs and Analysis

Real-world deployment requires navigating trade-offs between competing objectives. This section characterizes the fundamental tensions in E^3 system design, examines computational considerations that influence deployment decisions, and discusses the framework's extensibility to broader contexts.

5.1 The Accuracy-Latency-Coverage Trade-Off

Three objectives compete in any hallucination mitigation system: *accuracy* (minimizing hallucination rate), *latency* (minimizing generation time), and *coverage* (maximizing substantive responses rather than abstentions). No configuration optimizes all three simultaneously; improving one typically requires sacrificing another.

Higher accuracy demands more thorough verification: additional CCE rounds, stricter SVD thresholds, or more comprehensive evidence requirements.

But thorough verification increases latency and may trigger more abstentions when evidence is marginal. Conversely, relaxing verification improves speed and coverage but permits more hallucinations.

Configuration choices position the system within this trade-off space. High-stakes deployments should prioritize accuracy, accepting increased latency and reduced coverage. Interactive applications may prefer low latency, accepting slightly higher hallucination risk. The E^3 architecture supports flexible positioning through tunable parameters: SVD batch sizes, CCE round counts, KCA risk thresholds, and SCG constraint strictness.

5.2 Computational Considerations

Each component contributes overhead beyond baseline generation. PFD adds retrieval and grammar enforcement costs; well-optimized indices reduce retrieval to logarithmic scaling, while automaton-based grammar checking adds negligible per-token cost. SVD's overhead depends on draft model speed, verifier cost, and acceptance rate; high acceptance rates achieve near-baseline latency through batch amortization. CCE contributes linearly with round count. KCA and SCG add minimal overhead through lightweight operations on existing signals.

Optimization strategies include caching frequent evidence retrievals, parallel execution of independent operations, and adaptive component activation that applies full scrutiny only to uncertain claims.

5.3 Extensibility and Modularity

The modular architecture supports adaptation beyond text generation. Multi-modal generation extends naturally: evidence pointers acquire modality tags and localization parameters while the architectural framework remains intact. Hierarchical verification accommodates multi-resolution checking by introducing verification sub-layers with escalating thoroughness. Domain adaptation proceeds through independent component specialization—training domain-specific verifiers or calibrating domain-appropriate thresholds—without end-to-end retraining.

These pathways demonstrate that E^3 represents a general architectural paradigm for trustworthy generation, with principles applicable across domains and modalities.

6 Discussion and Broader Implications

The E^3 architecture represents more than a technical solution to hallucination—it embodies a fundamental reconceptualization of how generation systems should be designed. This section examines the broader significance of our architectural approach, acknowledges current limitations, and considers implications for trustworthy AI deployment.

6.1 Architectural Significance

The E^3 framework demonstrates that principles from software engineering and formal methods can be productively applied to neural generation systems. This cross-pollination yields architectural insights that extend beyond hallucination mitigation.

The design-by-contract methodology [16] underlies our approach to component integration. Each E^3 component operates under explicit preconditions (what inputs it expects), postconditions (what guarantees it provides), and invariants (what properties it maintains). PFD requires retrieved evidence and guarantees evidence-pointer precedence. SVD requires pointer-claim pairs and guarantees entailment verification. These contracts enable compositional reasoning: system properties follow from component contracts without requiring analysis of internal implementations.

The fail-fast principle [18] shapes our treatment of verification failures. Rather than allowing errors to propagate and compound, E^3 detects problems early and activates recovery mechanisms immediately. This approach, standard in robust software systems, proves equally valuable for generation systems where early intervention prevents wasted computation and reduces hallucination exposure.

The correspondence to program verification [28,29] reveals deeper connections. Generation can be viewed as a computational process with correctness requirements analogous to program specifications. Evidence-first constraints function as preconditions; verification checks function as assertions; grammar compliance functions as type safety. This perspective suggests that decades of research in formal methods may be applicable to generation systems, opening pathways for stronger guarantees than purely statistical approaches can provide.

6.2 Limitations and Open Challenges

Despite its architectural strengths, E^3 faces limitations that define important challenges for future development.

Computational overhead remains significant for latency-sensitive applications. While SVD's speculative approach and various optimization strategies reduce verification costs, E^3 generation is inherently slower than unconstrained generation. Real-time conversational applications may find current overhead prohibitive, necessitating either hardware acceleration or approximation techniques that trade some guarantee strength for speed.

Evidence availability fundamentally bounds system effectiveness. When the knowledge corpus lacks relevant information, no amount of verification can ground claims appropriately. Emerging topics, proprietary domains, and rapidly evolving situations pose particular challenges. The architecture acknowledges this limitation through KCA's abstention capability, but abstention is a fallback rather than a solution—high abstention rates degrade user experience even when they prevent hallucinations.

Semantic gaps between verification and true understanding persist. Current verification operates primarily at syntactic and shallow semantic levels, checking whether evidence text supports claim text. Deeper reasoning—pragmatic implication, compositional semantics, contextual appropriateness—remains challenging. Claims may be technically supported by evidence yet misleading in context, a failure mode that current architecture does not fully address [30].

Adversarial robustness presents ongoing concerns. While CCE provides adversarial testing, sophisticated attacks targeting verification blind spots may still succeed. The architectural defense of multiple independent verification layers provides robustness through redundancy, but determined adversaries with knowledge of system internals may identify exploitation pathways.

6.3 Toward Trustworthy AI Infrastructure

E^3 contributes to the broader goal of developing AI systems suitable for critical infrastructure roles [31]. Several architectural properties support this vision.

Interpretability emerges naturally from structured generation. Explicit evidence pointers create audit trails showing what sources support each claim. Justification requirements force the system to articulate reasoning. Confidence metadata from KCA indicates reliability assessments. Users and auditors can examine these artifacts to understand and verify system outputs, addressing transparency requirements increasingly demanded in high-stakes domains.

Standardization potential follows from the formal architectural framework. Different deployment contexts could specify required verification levels, acceptable abstention rates, and mandatory structural patterns, creating certification criteria for AI-generated content. Industry standards based on E^3 principles could establish consistent expectations for reliability across applications [32].

Epistemic responsibility is architecturally encoded. By requiring evidence citations and enabling principled abstention, E^3 implements mechanical analogs of human epistemic norms—claims require justification; uncertainty warrants caution [33]. This shifts the paradigm from hoping models behave responsibly to ensuring they must behave responsibly through structural constraints.

These properties position E^3 not as a final solution but as a foundation for continued development toward AI systems that are simultaneously capable, reliable, and aligned with human values for truthfulness and accountability.

7 Conclusion

This paper has presented E^3-Guarded Generation, a unified architecture that embeds verification directly into language generation to systematically prevent hallucinations. The framework integrates five specialized components—Provenance-First Decoding, Speculative Verification Decoding, Counterfactual Cross-Examination, KWIK-Calibrated Abstention, and Semantic Constraint Grammars—within a three-layer architecture enforcing evidence grounding, logical entailment, and structural validity throughout generation. Three principles guide the design: verification as construction, fail-fast with graceful degradation, and composable verification. These principles, drawn from software engineering and formal methods, demonstrate that established approaches to reliable system design can effectively inform trustworthy AI development. As language models assume critical infrastructure roles, E^3 provides both a deployable framework and an architectural paradigm for building AI systems that are powerful, interpretable, and aligned with human expectations for truthfulness.

References

1. Brown, T., et al.: Language models are few-shot learners. Adv. Neural. Inf. Process. Syst. **33**, 1877–1901 (2020)
2. Achiam, J., et al.: Gpt-4 technical report, arXiv preprint arXiv:2303.08774 (2023)
3. Anthropic, The claude 3 model family: Opus, sonnet, haiku. Anthropic Technical Report (2024). https://www.anthropic.com/news/claude-3-family
4. Ji, Z., et al.: Survey of hallucination in natural language generation. ACM Comput. Surv. **55**(12), 1–38 (2023)
5. Huang, L., et al.: A survey on hallucination in large language models: Principles, taxonomy, challenges, and open questions. arXiv preprint arXiv:2311.05232 (2023)
6. Rawte, V., Sheth, A., Das, A.: A troubleshooting guide for hallucinations in large language models. arXiv preprint arXiv:2307.03987 (2023)
7. Mallen, A., Asai, A., Zhong, V., Das, R., Khashabi, D., Hajishirzi, H.: When not to trust language models: Investigating effectiveness of parametric and non-parametric memories. In: Proceedings of the 61st Annual Meeting of the Association for Computational Linguistics, pp. 9802–9822 (2023)
8. Liu, N.F., et al.: Lost in the middle: how language models use long contexts. Trans. Assoc. Comput. Linguist. **12**, 157–173 (2024)
9. Dziri, N., et al.: Faith and fate: Limits of transformers on compositionality. Adv. Neural Inf. Processing Syst. **36** (2024)
10. Dhingra, B. Cole, J.R., Eisenschlos, J.M. Gillick, D., Eisenstein, J., Cohen, W.W.: Time-aware language models as temporal knowledge bases. Trans. Assoc. Comput. Linguist. **10**, 257–273 (2022)
11. Weidinger, L., et al.: Ethical and social risks of harm from language models. arXiv preprint arXiv:2112.04359 (2021)
12. Lewis, P., et al.: Retrieval-augmented generation for knowledge-intensive NLP tasks. Adv. Neural Inf. Processing Syst. **33**, 9459–9474 (2020)
13. Borgeaud, S. et al.: Improving language models by retrieving from trillions of tokens. In: International Conference on Machine Learning, pp. 2206–2240 (2022)
14. Min, S., et al.: Factscore: Fine-grained atomic evaluation of factual precision in long form text generation. In: Proceedings of the 2023 Conference on Empirical Methods in Natural Language Processing, pp. 12 076–12 100 (2023)
15. Wang, X., et al.: Self-consistency improves chain of thought reasoning in language models. arXiv preprint arXiv:2203.11171 (2023)
16. Meyer, B.: Applying design by contract. IEEE **25**(10) (1992)
17. Martin-Löf, P.: Intuitionistic type theory. Bibliopolis Naples (1984)
18. Shore, J.: Fail fast. IEEE Softw. **21**(5), 21–25 (2004)
19. De Roever, W.-P., Langmaack, H., Pnueli, A.: Component software: beyond object-oriented programming (2001)

20. Maynez, J., Narayan, S., Bohnet, B., McDonald, R.: On faithfulness and factuality in abstractive summarization. In: Proceedings of the 58th Annual Meeting of the Association for Computational Linguistics, pp. 1906–1919 (2020)
21. Rashkin, H. et al.: Measuring attribution in natural language generation models. Comput. Linguist. 1–64 (2023)
22. Hennessy, J.L., Patterson, D.A.: Computer architecture: a quantitative approach, 5th ed. Morgan Kaufmann (2011)
23. Tomasulo, R.M.: An efficient algorithm for exploiting multiple arithmetic units. IBM J. Res. Dev. **11**(1), 25–33 (1967)
24. Irving, G., Christiano, P., Amodei, D.: Ai safety via debate. arXiv preprint arXiv:1805.00899 (2018)
25. Christiano, P., Cotra, A., Xu, M.: Ai safety via debate. AI Alignment Forum (2021)
26. Li, L., Littman, M.L., Walsh, T.J., Strehl, A.L.: Knows what it knows: a framework for self-aware learning. Mach. Learn. **82**(3), 399–443. Springer, Cham (2011)
27. Sayedi, A., Zadimoghaddam, M., Blum, A.: Trading off mistakes and don't-know predictions. Adv. Neural Inf. Processing Syst. 2092–2100 (2010)
28. Hoare, C.A.R.: An axiomatic basis for computer programming. Commun. ACM **12**(10), 576–580 (1969)
29. Floyd, R.W.: Assigning meanings to programs. Math. Aspects Comput. Sci. **19**, 19–32 (1967)
30. Bender, E.M., Koller, A.: Climbing towards nlu: On meaning, form, and understanding in the age of data. In: Proceedings of the 58th Annual Meeting of the Association for Computational Linguistics, pp. 5185–5198 (2020)
31. Amodei, D., Olah, C., Steinhardt, J., Christiano, P., Schulman, J., Mané, D.: Concrete problems in ai safety. arXiv preprint arXiv:1606.06565 (2016)
32. ISO/IEC. Iso/iec 23053: Framework for artificial intelligence systems using machine learning. International Organization for Standardization, Tech. Rep. (2023)
33. Goldman, A.I.: Knowledge in a social world, Oxford University Press (1999)

PCRLLM: Proof-Carrying Reasoning with Large Language Models Under Stepwise Logical Constraints

Tangrui Li, Pei Wang, Hongzheng Wang, Christian Hahm, Matteo Spatola, and Justin Shi(✉)

Temple University, Philadelphia, PA 19122, USA
{tuo90515,pei.wang,tuf78197,tun50806,matteo.spatola,shi}@temple.edu

Abstract. Large Language Models (LLMs) often exhibit limited logical coherence, mapping premises to conclusions without adherence to explicit inference rules. We propose Proof-Carrying Reasoning with LLMs (PCRLLM), a framework that constrains reasoning to single-step inferences while preserving natural language formulations. Each output explicitly specifies premises, rules, and conclusions, thereby enabling verification against a target logic. This mechanism mitigates trustworthiness concerns by supporting chain-level validation even in black-box settings. Moreover, PCRLLM facilitates systematic multi-LLM collaboration, allowing intermediate steps to be compared and integrated under formal rules. Finally, we introduce a benchmark schema for generating large-scale step-level reasoning data, combining natural language expressiveness with formal rigor.

Keywords: Non-Axiomatic Logic · Formal LLM Reasoning and Evaluation · LLM Collaboration · LLM Reasoning Benchmark

1 Introduction

The reasoning ability of large language models (LLMs) has long been a central concern, yet their answers often lack stable logical coherence. This stems from their nature as neural networks, which treat reasoning as a direct mapping from premises to conclusions according to the training data. Such mapping is not inherently problematic, but it should be constrained by logical rules and restricted to the smallest unit of inference (as single-step reasoning). Otherwise, when LLMs skip steps by leveraging semantic knowledge, we cannot tell whether they are performing valid logical short-cuts or unjustified leaps (thus losing the soundness). To this point, having LLMs simulate an existing logical system is a reasonable choice, since the rules of single-step inference are explicitly defined.

Nevertheless, LLMs' strength lies not in mechanically reproducing logic systems, but in their powerful natural language processing ability. Existing approaches usually emphasize only one side: either broad reasoning ability without formal soundness (e.g., CoT [17]), or strict formalism at the cost of generality

H. R. Arabnia et al. (Eds.): CSCI 2025, CCIS 2972, pp. 297–311, 2027.
https://doi.org/10.1007/978-3-032-28469-3_20

(e.g., neural-symbolic architectures [1], domain-specific loss on formality [19]). Our approach aims to combine the two: retaining natural language problem descriptions while rephrasing them in formal logical reasoning in the output, which is a chain reasoning consists of well-formed single steps.

This design not only addresses the above dilemma but also brings two additional advantages. First, it produces proof-carrying outputs: the LLM's answers include premises, rules, conclusions, and necessary details, which allows each single-step inference to be checked against the target logic, enabling a chain-level grading as an automatic judgment of logical soundness. In this way, the trustworthiness problem can be mitigated: even if the model remains a black box, its outputs can still be validated on logical grounds.

Secondly, it enables more systematic multi-LLM collaboration. Prior approaches mostly rely on debates or answer aggregation, which lack formal constraints and obscure the quality of intermediate steps. In our framework, multiple models' reasoning steps can be unified through logical rules, so that collaboration improves not only the final result but also the transparency and consistency of intermediate reasoning.

Additionally, we also recognize that the complexity of long-chain reasoning arises from the exponential growth of rule applications over shared premises, which cannot be captured by existing corpora alone. To address this, we propose a benchmark schema that generates large-scale training data, allowing LLMs to better learn complex inference patterns, as well as the testing data.

To summarize, we propose a methodology that enables LLMs to perform formal reasoning while preserving natural language problem descriptions. The key idea is to constrain reasoning to single steps so that each step can be verified against a target logic. This yields proof-carrying outputs that allow automatic scoring of reasoning chains and provide a new angle to address the trustworthiness problem. Moreover, by expressing reasoning under formal constraints, our approach supports systematic multi-LLM collaboration, where intermediate steps can be compared and integrated through logical rules rather than opaque debates. To further overcome the data scarcity of formal reasoning, we introduce a benchmark schema that generates large-scale step-level training data. Together, these contributions offer a principled path toward combining LLMs' natural language understanding with the rigor of formal logic. All the codes and appendix are in our GitHub repository.

The remainder of this paper 1) reviews related work, analyzes the scope of our approach, and 2) introduces the logical system we adopt along with its necessary background. We then 3) describe our data generation strategy, evaluation of reasoning, and multi-LLM integration, followed by 4) experiments that demonstrate the effectiveness of our method and discuss the challenges of simulating reasoning with LLMs.

2 Related Work

2.1 Approach Introducing Logic to LLMs

- **CoT and PRM**
  CoT (Chain of Thought [17]) was the first approach to explicitly introduce step-by-step reasoning in LLMs. Its core idea is "reasoning granularity": prompting models to generate intermediate steps and leveraging techniques such as self-consistency sampling or process rewards to boost performance. However, the underlying mechanism remains closer to semantic pattern mapping than to fully formal reasoning. Even with explicit steps, the transitions lack the strict constraints of a logical system. As a result, while CoT improves performance on arithmetic and logical tasks [14], its intermediate reasoning often suffers from faithfulness issues [10]. PRM (Process Reward Model [9]) shares some similarity with our approach: it evaluates reasoning steps rather than final answers. Yet the reasoning process remains a black box—we can only score or select from the outputs, not reshape them. Moreover, PRM makes progress by evaluating intermediate steps, but its scoring framework is not yet aligned with a formal logical system. As a result, it faces challenges in achieving rigor, generality, or in supporting fully validated chains of evidence.
- **Reasoning with External Tools**
  The key idea is that LLMs can be nudged toward normative outputs such as code or JSON, exemplified by approaches like PoT [2], where reasoning becomes explicit, step-wise, and programmatic. But these methods leverage programmatic reasoning, which increases explicitness and verifiability, though the reliance on program structures may limit flexibility and complicate cross-model collaboration. A different line of work translates natural language into formal representations and delegates reasoning to external solvers [8] [12]. This, however, faces the semantic ambiguity and uncertainty of natural language [7], and large-scale contexts remain difficult to fully formalize.
- **Neural Symbolic Architecture**
  This approach encourages neural networks to generate outputs that resemble formal reasoning structures, which can improve logical consistency. In many implementations, however, the reasoning chains are implicit and difficult to validate (with some exceptions [13], though readability and integration into feedback or collaboration remain limited). Furthermore, the underlying logical formalisms are currently focused on propositional or first-order logic [1]. Since symbolization is closely tied to natural language understanding (NLU), extending these methods to handle unrestricted natural language remains a challenging task.

2.2 LLM Collaboration

LLM collaboration aims to boost performance by combining multiple models. Without a formal structure, however, multi-agent debates [4,16] can become subjective. Since natural language is the medium, evaluation often relies on semantic plausibility rather than strict structural validity, which makes guarantees of soundness challenging. Current methods typically fall into two types: 1) expanding reasoning branches via sampling [16], which can be costly and may yield intermediate steps that are difficult to verify; or 2) organizing debates among multiple models [18] (or within the self-talk of a single model [5]), which may improve accuracy but do not yet provide fully verifiable chains of evidence.

While multi-agent debate has been shown to outperform single models on many tasks [6], it remains constrained by natural language as the carrier: larger debates may yield higher performance [11], but also make it harder to automatically parse or validate intermediate reasoning. Benchmarks such as GSM8K [3] further reinforce this tendency by evaluating only final answers rather than reasoning processes, thus prioritizing correctness of outcomes over explicit scrutiny of intermediate reasoning.

3 Non-axiomatic Logic Background

We choose Non-Axiomatic Logic (NAL) as our target logic for two main reasons. 1) NAL offers sufficient expressive ability to represent natural language contents properly. NAL is defined hierarchically with multiple "layers" and each of them extends the sets of grammar and inference rules of the lower layers, to increase the expressive ability and inferential power of the system [15]. The high layers of NAL can support substantially higher semantic complexity. In the meantime, NAL is designed based on AIKR (Assumption of Insufficient Knowledge and Resources) [15], which makes NAL naturally tolerant of the uncertainty and vagueness in natural language. 2) NAL can be considered as a non-monotonic logic. New evidence can strengthen, weaken, or overturn previous judgments through revision. This expands the range of inferences we can draw from LLM outputs, such as allowing different reasoning paths to yield different judgments about the same statement. And more importantly, because of the inherent uncertainty of natural language, judgments from different LLM can rarely share the same truth-value. NAL can naturally accommodate such discrepancies with its non-monotonicity when combining the outputs from multiple LLMs.

3.1 Used NAL Rules

NAL is a complex logic, developed on the basis of syllogistic reasoning and subsequently extended with increasingly powerful expressive capabilities. These capabilities are organized into eight levels of logic rules, with later levels building upon the foundations of earlier ones. For example, NAL-4 introduces acquired relations (including predicates); NAL-5 extends the system to statement-level inference; NAL-6 enables the representation of variables; NAL-7 incorporates temporal descriptions and NAL-8 further expands the system to perception and interaction with the external environment.

In this paper, we restrict our focus to syllogistic inference rules within first-order NAL. To clarify the rules we employ, it is necessary to introduce the following basic concepts:

- **Inheritance** A copula in NAL, denoted as $\rightarrow$. For a judgment $s \rightarrow o$, it indicates that s is a specialization of o (equivalently, o is a generalization of s). A low frequency corresponds to the negation of a judgment.
- **Similarity** Another copula, denoted as $\leftrightarrow$. $s \leftrightarrow o$ indicates that the two terms can be interchanged as equivalent in the current context. For example, if "Jack" first appears in a sentence and is later referred to as "the one," then at that moment the judgment $Jack \leftrightarrow the\ one$ holds true.
- **Syllogistic reasoning** Since NAL is founded on sentences of the form "subject-term copula predicate-term" and syllogistic reasoning, the application of any rule requires two premises J_1, J_2, which must share a common term M. The other two terms are denoted as S and P.
- **Truth-value** A NAL judgment is associated with a two-dimensional truth-value (f, c), where f represents the frequency with which a statement is confirmed based on accumulated evidence, and c represents the confidence, i.e., the stability of f. Formally, given equivalent positive evidence w_p and equivalent negative evidence w_n, we define $f = w_p/(w_p + w_n)$ and $c = (w_p+w_n)/(w_p+w_n+1)$. A larger evidence base makes c closer to 1, indicating that f is more stable and therefore more reliable for future reasoning.
- **Evidential base** In NAL, truth-values are determined by evidence, but the system does not track the specific contribution of individual pieces of evidence, as evidence accumulates and changes during inference, making it difficult to trace independently. To prevent double counting, NAL assigns each input statement a unique ID. Any conclusion derived through inference must carry all premise IDs (to a maximum capacity), and if overlapping IDs are detected, the inference is disallowed. The set of accumulated IDs associated with a statement is referred to as its evidential base.
- **Truth functions** Since NAL employs continuous truth-values, it is equipped with specially designed functions for calculating them. For instance, F_{ded} is the truth-value function for deduction rule. Since syllogistic inference distinguishes between major and minor premises, the order of the premises matters: the truth-values of J_1 and J_2 are treated in sequence. When needed, we write F'_{ded} to indicate the function applied after swapping the order of J_1 and J_2. Besides deduction, we also make use of other rules such as abduction, induction, comparison, analogy, exemplification, and resemblance. For the precise definitions of these rules and their truth functions, please refer to Appendix B in [15].

The following chart summarizes all the inference rules employed in this paper. This work does not involve an in-depth discussion of NAL. It is sufficient to understand only the basic inference forms required to apply these rules. The table shows which rules can be applied to obtain different conclusions based on different J_1 and J_2. It is worth noting that NAL often uses all available rules and thus obtains multiple conclusions for a pair of premises.

$J_2 \setminus J_1$	$M \to P \langle f_1, c_1 \rangle$	$P \to M \langle f_1, c_1 \rangle$	$M \leftrightarrow P \langle f_1, c_1 \rangle$
$S \to M \langle f_2, c_2 \rangle$	$S \to P \langle F_{ded} \rangle$ $P \to S \langle F'_{exe} \rangle$	$S \to P \langle F_{abd} \rangle$ $P \to S \langle F'_{abd} \rangle$ $S \leftrightarrow P \langle F'_{com} \rangle$	$S \to P \langle F'_{ana} \rangle$
$M \to S \langle f_2, c_2 \rangle$	$S \to P \langle F_{ind} \rangle$ $P \to S \langle F'_{ind} \rangle$ $S \leftrightarrow P \langle F_{com} \rangle$	$S \to P \langle F_{exe} \rangle$ $P \to S \langle F'_{ded} \rangle$	$P \to S \langle F'_{ana} \rangle$
$S \leftrightarrow M \langle f_2, c_2 \rangle$	$S \to P \langle F_{ana} \rangle$	$P \to S \langle F_{ana} \rangle$	$S \leftrightarrow P \langle F_{res} \rangle$

Fig. 1. First-order NAL Syllogism.

4 Method

4.1 Data Construction

The data is required to have natural language input as the problem description, and the JSON output as formal NAL inference. In data generation, we will first create a logical structure and then convert it into a sufficiently rich natural language description.

Logical Instance Generation. We focus on two-step NAL reasoning in this paper, as it represents the simplest form of multi-step reasoning. Due to computational constraints, we restrict our experiments to this basic case; however, we also emphasize that the proposed method has the potential to be extended to more complex multi-step reasoning tasks when supported by more powerful hardware and larger models.

- **Step 1**: One of the nine inference rules (Fig. 1) is randomly selected to determine first two premises ($J1$, $J2$) used, as well as the pending conclusions. Unique terms (e.g., ID_{3939}) are assigned to the placeholders S, M, P respectively. A random evidential base (within 3 IDs) is attached.
- **Step 2**: One conclusion from Step 1 is randomly selected as a premise (J1 or J2), thereby fixing the position of Step 2 in the inference chart. Another independent premise is generated in the same way. The remaining placeholder is assigned a unique term, and one of the possible conclusions of Step 2 is chosen as the final goal.
- **Distractors**: To increase difficulty and emphasize the NLU ability of LLMs in such problems, two additional distractor premises are introduced in Step 1, which are generated by randomly selecting a rule and assigning distinct unique terms with random evidential bases.

This procedure guarantees that every generated reasoning instance is valid and sufficiently complex. It can be extended to multi-step reasoning or to consider higher levels of NAL rules to further increase complexity.

Natural Language Instance. The natural language translation is disclosed to LLMs thus cannot contain everything in the logic instance. It only contains:

1. The two premises of Step 1 (without specifying $J1$, $J2$ roles).
2. The independent premise of Step 2 (which is not derived from Step 1).
3. The two distractor premises for Step 1.
4. The selected conclusion of Step 2 (as the question asked to LLMs, truth values removed).

In translating NAL sentences, 20 templates are prepared for the inheritance copula, as well as the similarity copula. For example, inheritance may be represented as sub is a type of obj; similarity may be represented as sub and obj are conceptually identical. In describing numerical truth-values, the frequency (f) is discretized into five categories (*always false, usually false, unknown, usually true, always true*), each with five natural language expressions. Note that the confidence (c) is omitted here and replaced with the LLM's answer-level confidence.

As a result, each logic instance admits up to $(20 \times 5 \times 5)^5 \times 20$ possible natural language translations, ensuring a high diversity. For all templates used, please refer to the appendix provided in our GitHub repository.

JSON Representation. As the expected reponse from LLMs, the step-by-step inference process of each logic instance is represented in JSON format with two top-level keys, which are *Step 1* and *Step 2*.

In each step, there are 3 keys: *Premise 1*, *Premise 2* and *Results*. In each a NAL sentence $s\ cp\ o.\langle f;c\rangle[eb]\{r\}$ is enclosed with keys: s which is the subject; o which is the object; cp which is the copula; f which is the frequency in the truth-value; c which is the confidence in the truth-value, here the default value of 0.9 is used; eb which is the evidential base of the truth-value; r which is the only applicable for sentences in the results, representing the reasoning rule used.

This structure preserves the full reasoning process and aligns precisely with the natural language input. The overview of the JSON structure can be found in Fig. 2.

Prompt Construction. We use the OpenAI User/Assistant format to generate the full prompt. The overview can be found in Fig. 3.

4.2 Grading

By requiring the LLM to output in JSON format and constraining each inference step to the minimal granularity, we can systematically evaluate whether its reasoning follows the target logical rules. Individual steps are first graded (single-step grading), followed by an assessment of their coherence (inter-step grading). When ground-truth specifying the correct conclusion(s) is available, we can further evaluate whether the target outcome is achieved (ground-truth grading).

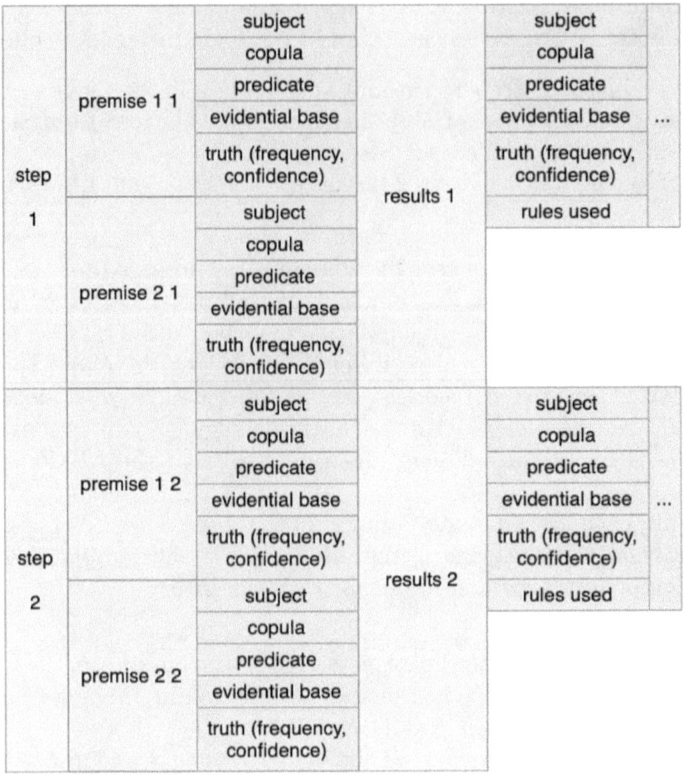

Fig. 2. The JSON structure overview of the expected LLM response.

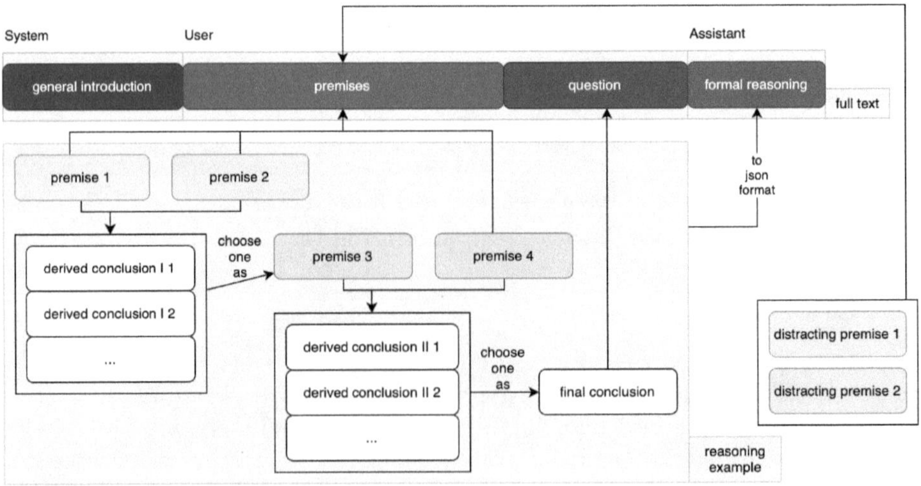

Fig. 3. The overview of the prompt structure.

Single-Step Grading. At the step level, we first parse the LLM's JSON output to extract *Premise 1*, *Premise 2*, and *Results*. If parsing fails, the minimum grade (0.1) is given (to retain gradient utility in reinforcement learning, if applicable). Otherwise, we employ the same reasoning engine used in data construction to generate all possible conclusions and identify the closest match to the LLM's output through bipartite matching.

When calculating the matching grade, for categorical indicators (s, o, cp, eb, r), exact matches receive full credit (5 points), otherwise no credit (0 points). For numerical indicators (f, c), which are central to evaluating whether the LLM follows the target logic rules, grading depends on the difference from the reference value: a difference ≤ 0.05 is considered negligible and receives full credit (25 points); a difference ≥ 0.2 indicates category-level mistakes (since we have 5 categories defined) and receives 0 points. For intermediate cases, we apply the function $(0.1+0.9\times(1-(diff-0.05)/0.15))^2\times 25$. In which the squaring emphasizes penalties for larger deviations, while the 0.1 floor preserves gradient signal (for reinforcement learning). This design ensures smooth penalization within the tolerance range.

The final single-step grade is defined as the proportion of received points over the maximum possible score. This formulation also supports future extensions where a single step may yield multiple conclusions. In the following discussions, we will use the same way on matching different sentences and grade on them.

Inter-step Grading. In multi-step reasoning, each step should build upon premises from the problem description or conclusions derived from previous steps. However, since we emphasis the NLU aspect for LLM to understand the problem and rephrase it into logic sentences, it is difficult to trace the exact origin of each premise. Therefore, we evaluate LLM outputs by analyzing their internal chain structure.

In this work, we focus on two-step reasoning tasks, where the conclusion of the first step serves as a premise in the second step, the analysis is straightforward. The same matching and grading strategy is applied on the first-step conclusion and all second-step premises.

Ground-Truth Grading. When ground-truth labels are available, we can additionally assess whether the question is correctly answered. Specifically, the same matching and grading strategy is applied on the conclusion of the second and the label answer, considering all indicators (s, o, cp, f, c, eb, r).

Overall Evaluation. The multiplication of the step-wise and inter-step grades define the formal conformity grade, which is standalone evaluation on whether the LLM adheres to the target logic without relying on labels. In scenarios when multiple candidate outputs are available, this grade can be used for priming structurally coherent answers (soundness). This endows the LLM with a *content-agnostic evaluation* framework, so when considering answers from multiple LLMs, there is no longer a need for human inspections or for using debate among LLMs.

In testing, the formal conformity grade can further be multiplied with the ground-truth grade, yielding a comprehensive measure of whether the LLM is both logically consistent and successful in completing the reasoning task.

4.3 Response Combination

Imagine a network of multiple LLMs where the same question may receive different answers from different models. Because their training data are highly diverse, their performance on specific problems often varies. Nevertheless, by leveraging the structured answering scheme and the previously introduced measure of logical and structural rigor, we can effectively integrate these answers.

Specifically, when different models generate reasoning steps with identical content (i.e., the same premises and conclusions), one can simply select the answer with the higher degree of formality (given whether the reasoning process provided by LLM is consistent with the target logic). When the contents diverge, single-step reasoning from different models can be combined into a new reasoning chain, thereby yielding a new answer. For example, in the two-step reasoning scenario that this paper focuses on, such composition is straightforward: the final answer can be constructed by pairing the *first-step reasoning from one model* with the *second-step reasoning from another*. In this case, the high-quality single steps can be picked out individually and then combined. For reasoning tasks requiring more reasoning steps, however, additional mechanisms are needed to avoid combinatorial explosion. A practical strategy is to filter out unreliable intermediate steps using formality scores, thereby reducing the scale of potential combinations.

5 Experiment

To evaluate our approach, we construct a network of three LLMs that solve the same task while intentionally differing in capability. The differences are induced by fine-tuning three clones of the same pretrained backbone on disjoint data subsets. Specifically, we generate two-step NAL first-order inference instances under different rules. As shown in Fig. 1, there are nine rule-usage patterns. We randomly partition them into three non-overlapping subsets and fine-tune each model on one subset.

All models are **fine-tuned on 100 examples**, which is relatively small dataset to show the feasibility of the work on more complicated backbones. Since larger models typically require more optimization steps to converge, we adjust the number of epochs instead of increasing data size, as follows: Qwen 2.5/0.5B with 1 epoch, Qwen 2.5/1.5B with 2 epochs, Qwen 2.5/3B with 4 epochs.

Even with more epochs, larger models often produce minor yet fatal formatting errors in their JSON outputs (e.g., unclosed brackets). We therefore apply DeepSeek R1 [5] as a post-processor to repair broken JSON strings.

Although the models are tuned with different data, all three are evaluated on the same **test set of 100 examples** that uniformly covers all rules. This design is deliberate: 1) no single model is sufficient, and 2) it enables us to study the performance of priming and step-wise recomposition of intermediate reasoning steps against single-model baselines.

Priming Strategies. We consider five priming strategies for using this network of LLMs to obtain the final answer:

- **Single model, no priming (m_1, m_2, m_3)**: use the single answer from model 1, 2, 3 as the final output.
- **Priming on 3 base models (model best-3, mb_3)**: with light priming, select the highest-scoring answer among the three models as the final output.
- **Step-wise recomposition (model best-9, mb_9)**: decompose each model's answer into steps and recombine them across models to synthesize a new answer (3 first-step possibilities × 3 s-step possibilities).

Scoring and Curves. After priming, we evaluate each answer against the label using a grading method. Given a threshold, an answer is classified as valid or invalid. The curve reports the ratio of valid answers over the 100 testing cases. To be comprehensive, we sweep the threshold from 0.1 to 0.9 to show the performance of the LLM network under increasingly strict conditions.

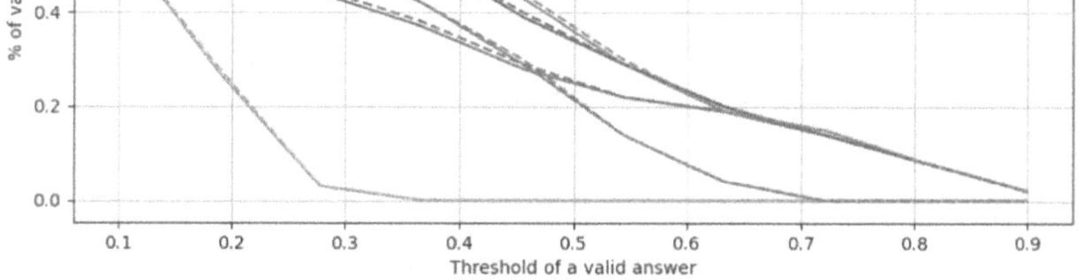

Fig. 4. Ratio of valid answers under five priming strategies. Model: Qwen2.5–0.5B; fine-tuned for 1 epoch. Training set: 100 instances (seed = 39). Dashed curves denote JSON-repaired results.

Experiment on Qwen2.5–0.5B. Figure 4 shows a monotonic decrease across all curves, which is expected: stricter thresholds yield lower valid-answer ratios. Among the three base models, m_3 performs best; however, it is still inferior to mb_3, indicating the effectiveness of priming in our setting. Moreover, mb_9 further improves over m_3, highlighting the additional benefit of step-wise recomposition.

The dashed curves are close to the solid ones, meaning the JSON-repair step (DeepSeek-R1 [5]) has limited impact here. This suggests that small-data fine-tuning is already sufficient for the 0.5B model to produce mostly well-formatted outputs.

One potentially confusing point is why the valid-answer ratio of m_3 does not match that of mb_3, even though mb_3 selects the best answer among the three base outputs. This is because mb_3 is computed by choosing the per-example best answer *before* applying the threshold, whereas m_3 reflects a single model's distribution of scores; as a result, their averages need not coincide.

Experiment on Qwen2.5–1.5B. Compared with the 0.5B results, Fig. 5 exhibits similar trends, with smoother curves attributable to the larger backbone. However, the dashed curves are more separated from the solid ones, underscoring the increased difficulty of enforcing correct output formatting as model size grows.

Note that within each model-size experiment, all three base models are evaluated on the same test set. Across different model-size experiments (0.5B, 1.5B, 3B), the test sets may differ.

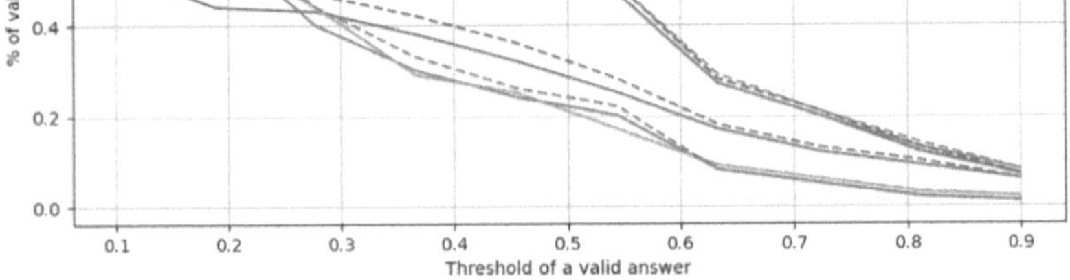

Fig. 5. Ratio of valid answers under five priming strategies. Model: Qwen2.5–1.5B; fine-tuned for 2 epochs. Training set: 100 instances (seed = 39). Dashed curves denote JSON-repaired results.

Experiment on Qwen2.5–3B. In Fig. 6, the dashed curve is clearly separated from the solid ones, further illustrating the challenge of maintaining strict formatting at larger scales. Meanwhile, the performance gap between mb_3 and mb_9 narrows. A plausible explanation is that the two-step task is relatively easy for the 3B model, so individual model outputs are already strong, leaving less headroom for recomposition gains.

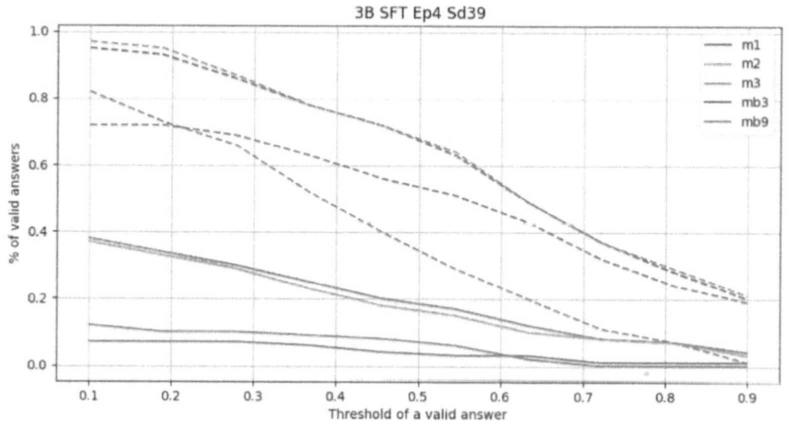

Fig. 6. Ratio of valid answers under five priming strategies. Model: Qwen2.5–3B; fine-tuned for 4 epochs. Training set: 100 instances (seed = 39). Dashed curves denote JSON-repaired results.

5.1 Comparison

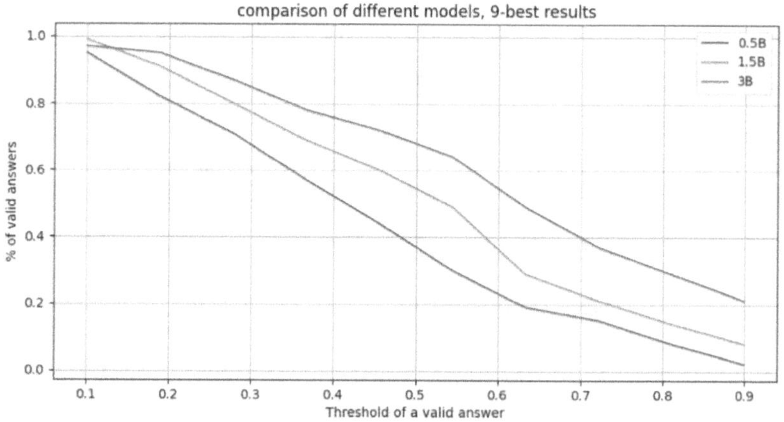

Fig. 7. Comparison of the mb_9 valid-answer ratios across models with different parameter counts (JSON-repaired).

When plotting the best-performing curves (JSON-repaired mb_9) across parameter scales in Fig. 7, we observe a clear improvement with model size. Nevertheless, even on this relatively simple two-step task, many settings still fail to exceed a valid-answer ratio of 0.2 under relatively loose thresholds, indicating substantial room for progress in faithfully simulating the target logic.

6 Conclusion

We proposed PCRLLM, which combines the expressive power of natural language with the rigor of formal inference rules, enabling transparent, verifiable, and collaborative reasoning with LLMs while addressing issues of trustworthiness and data scarcity.

Although this work is able to scale to more complex multi-step situations, two open issues remain for future exploration. First, since we rely on the NLU ability of LLMs, we cannot attribute which premises originate from which problem description; consequently, hallucination may still occur. While our method makes these premises transparent and allows users to accept or reject them, this workflow requires manual intervention and is therefore inconvenient.

Second, when extending to multi-step reasoning and recomposing answers from single steps, the search space grows exponentially with the depth of the NAL extensions and the number of reasoning steps, leading to a combinatorial explosion. Although our scoring scheme prunes many low-quality sub-steps and reduces fruitless search, it does not fundamentally resolve this complexity bottleneck.

Future work will focus on automating premise attribution and developing more efficient search strategies, with the aim of advancing scalable and trustworthy reasoning in LLMs.

References

1. Besold, T.R., et al.: Neural-symbolic learning and reasoning: A survey and interpretation. In: Neuro-Symbolic Artificial Intelligence: The State of the Art, pp. 1–51. IOS press (2021)
2. Chen, W., Ma, X., Wang, X., Cohen, W.W.: Program of thoughts prompting: Disentangling computation from reasoning for numerical reasoning tasks (2023). https://arxiv.org/abs/2211.12588
3. Cobbe, K., et al.: Training verifiers to solve math word problems (2021). https://arxiv.org/abs/2110.14168
4. Du, Y., Li, S., Torralba, A., Tenenbaum, J.B., Mordatch, I.: Improving factuality and reasoning in language models through multiagent debate (2023). https://arxiv.org/abs/2305.14325
5. Guo, D., et al.: Deepseek-r1: incentivizing reasoning capability in LLMs via reinforcement learning (2025). https://arxiv.org/abs/2501.12948
6. Hegazy, M.: Diversity of thought elicits stronger reasoning capabilities in multi-agent debate frameworks (2025). https://arxiv.org/abs/2410.12853
7. Li, Q., et al.: Are LLMs reliable translators of logical reasoning across lexically diversified contexts? (2025). https://arxiv.org/abs/2506.04575
8. Lin, F., Shou, Z., Chen, C.: Using language models for knowledge acquisition in natural language reasoning problems (2023). https://arxiv.org/abs/2304.01771

9. Ma, Q., et al.: Let's reward step by step: Step-level reward model as the navigators for reasoning (2023). https://arxiv.org/abs/2310.10080
10. Manuvinakurike, R., Moss, E., Watkins, E.A., Sahay, S., Raffa, G., Nachman, L.: Thoughts without thinking: Reconsidering the explanatory value of chain-of-thought reasoning in LLMs through agentic pipelines (2025). https://arxiv.org/abs/2505.00875
11. Pan, J., Zhang, J., Wang, X., Yuan, L., Peng, H., Suhr, A.: Tinyzero (2025). https://github.com/Jiayi-Pan/TinyZero, Accessed 24 Jan 2025
12. Qin, B., et al.: A survey on text-to-sql parsing: Concepts, methods, and future directions (2022). https://arxiv.org/abs/2208.13629
13. Rocktäschel, T., Riedel, S.: End-to-end differentiable proving. In: Guyon, I., et al. (eds.) Advances in Neural Information Processing Systems. vol. 30. Curran Associates, Inc. (2017). https://proceedings.neurips.cc/paper_files/paper/2017/file/b2ab001909a8a6f04b51920306046ce5 Paper.pdf
14. Sprague, Z., et al.: To cot or not to cot? chain-of-thought helps mainly on math and symbolic reasoning (2025). https://arxiv.org/abs/2409.12183
15. Wang, P.: Non-axiomatic logic: A model of intelligent reasoning, World Scientific (2013)
16. Wang, X., et al.: Self-consistency improves chain of thought reasoning in language models (2023). https://arxiv.org/abs/2203.11171
17. Wei, J., et al.: Chain-of-thought prompting elicits reasoning in large language models. Adv. Neural. Inf. Process. Syst. **35**, 24824–24837 (2022)
18. Zheng, L., et al.: Judging LLM-as-a-judge with mt-bench and chatbot arena. In: Oh, A., Naumann, T., Globerson, A., Saenko, K., Hardt, M., Levine, S. (eds.) Advances in Neural Information Processing Systems. vol. 36, pp. 46595–46623. Curran Associates, Inc. (2023). https://proceedings.neurips.cc/paper_files/paper/2023/file/91f18a1287b398d378ef22505bf41832-Paper-Datasets_and_Benchmarks.pdf
19. Zhou, Z., Kearnes, S., Li, L., Zare, R.N., Riley, P.: Optimization of molecules via deep reinforcement learning. Sci. Rep. **9**(1) (2019). https://doi.org/10.1038/s41598-019-47148-x

Comprehending LLM vs. Human-Generated Business Process Models: An Eye-Tracking Study

Maximilian Möller(✉), Luca Franziska Hörner, and Manfred Reichert

Institute of Databases and Information Systems, Ulm University, 89081 Ulm, Germany
{maximilian.moeller,luca.hoerner,manfred.reichert}@uni-ulm.de

Abstract. This study examines how business process models expressed in terms of BPMN 2.0 and generated by Large Language Models (LLM) affect user cognition and comprehension compared to process models created manually by human experts. In an eye-tracking experiment, we measured the cognitive load and comprehension performance of the participants across five process scenarios with increasing complexity. Participants were grouped according to their experience with process modeling and BPMN 2.0, and evaluated using fixation metrics, comprehension scores, and cognitive load ratings. The study results show that LLM-generated models led to lower cognitive load, particularly for less experienced users and more complex scenarios, while achieving similar comprehension performance. Our findings indicate that LLMs can produce cognitively accessible process models without compromising understanding, highlighting their potential for advancing (semi-) automated process model discovery based on LLMs.

Keywords: Large Language Model · Automated Process Modeling · Process Model Comprehension · Eye-Tracking

1 Introduction

Process modeling is a fundamental discipline for organizations that seek to visualize, document, analyze, optimize, and automate their business processes [5]. The Business Process Model and Notation (BPMN 2.0) standard [24] is widely adopted for this purpose, but requires human expertise to create sound and comprehensible process models. With recent advancements in LLMs, the automated generation of BPMN 2.0 business process models has become feasible [11,15,16], potentially reducing the reliance on domain experts and simplifying process modeling. However, the comprehensibility and cognitive implications of LLM-generated models – particularly in comparison to manually created process models – remain largely unexplored. This study places a particular focus on the cognitive factors, bridging this research gap through empirical investigation. The comprehensibility of process models is crucial, as they serve as essential

H. R. Arabnia et al. (Eds.): CSCI 2025, CCIS 2972, pp. 312–326, 2027.
https://doi.org/10.1007/978-3-032-28469-3_21

communication artifacts among business stakeholders. Inadequate model design could lead to misinterpretations, inefficiencies, and operational errors. However, there is a lack of empirical evidence on cognitive processing differences when individuals engage with LLM-generated BPMN 2.0 process models.

To address this research gap, this study employs an eye-tracking experiment to systematically investigate how users perceive and comprehend BPMN 2.0 process models that were automatically generated by an LLM via zero-shot prompting, compared to those created manually by human experts (cf. Fig. 1). The study participants were divided into two groups according to experiences with BPMN 2.0: one group with less experience and another with more experience. The results were analyzed separately for each group.

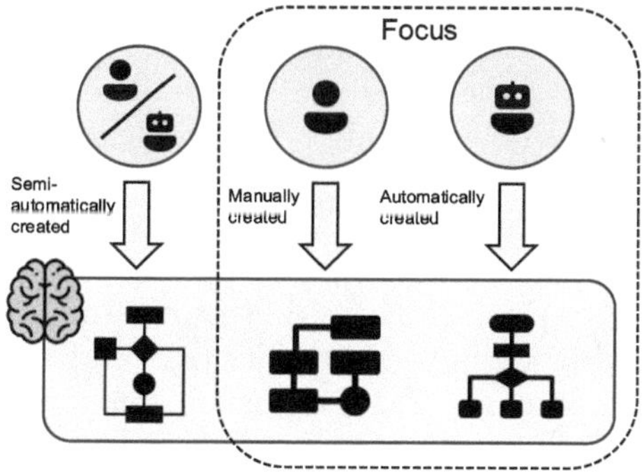

Fig. 1. Focus of the Study.

Based on the outlined research objectives, we derived the following research questions that shall allow for a structured approach to our investigation.

RQ0: Is there a difference during the comprehension of BPMN 2.0 process models automatically generated by an LLM-based agent compared to the comprehension of process models created manually by experts?

To investigate RQ0 in more detail, we decompose it into the following subquestions that address two key aspects of comprehension:

RQ1: Is there a difference in cognitive load during the comprehension of BPMN 2.0 process models automatically generated by an LLM-based agent compared to the comprehension of process models created manually by experts?

RQ2: Is there a difference in performance during the comprehension of BPMN 2.0 process models automatically generated by an LLM-based agent compared to the comprehension of process models created manually by experts?

Our study addressing these research questions is structured as follows: Sect. 2 provides the theoretical background necessary for understanding this work. Section 3 discusses related work. Section 4 outlines the study design and experimental setup, whereas Sect. 5 presents the findings, followed by a discussion of the results and their limitations in Sect. 6. Finally, Sect. 7 summarizes the study and offers directions for future research.

2 Backgrounds

This section gives an overview of the key concepts and technologies used in this study to ensure a thorough understanding of our research.

2.1 Large Language Models

LLMs such as GPT-4 [25] and Mistral AI [19] represent significant advancements in natural language processing (NLP) and artificial intelligence (AI). These models utilize deep learning techniques, particularly the transformer architecture, which allows for the parallel processing of data, improving the performance of NLP tasks [17,35]. LLMs are trained on extensive textual datasets, allowing them to recognize patterns and context, and enabling them to generate human-like text, to understand context, to answer questions in an Q&A dialogue, and to support creative endeavors such as writing, brainstorming, and coding [4,29]. Modern LLMs utilize machine learning and deep learning algorithms to refine their understanding of user inputs. Sequence-to-sequence models, for example, are widely used to generate conversational responses [36]. However, challenges remain, particularly in understanding complex user intents. Recent advancements in NLP, especially through the use of transformer-based models such as GPT, have contributed to improve the ability of agents to generate more natural and context-aware responses [3,4] and have been used for creating business process models.

2.2 Process Modeling

Process modeling can be defined as the act of creating abstract representations of business processes in order to facilitate their analysis, communication, and improvement [5]. Among the various modeling approaches, BPMN 2.0 has emerged as the de facto standard due to its standardized and expressive notation [24]. BPMN 2.0 uses visual elements such as tasks, events, gateways, pools, lanes, and message exchanges, making it an effective language for capturing complex business processes and improving process documentation [12]. An example of a BPMN process model is depicted in Fig. 2.

2.3 Process Model Comprehension

Process model comprehension refers to the process of interpreting and decoding the information embedded within a process model, enabling individuals to derive valid conclusions [37]. This becomes particularly interesting in the context of process models that are automatically generated by an LLM-based agent. This cognitive task requires significant efforts, as it not only necessitates an understanding of the process model, but requires identifying specific details relevant in the context of a given task, such as verifying compliance or detecting errors.

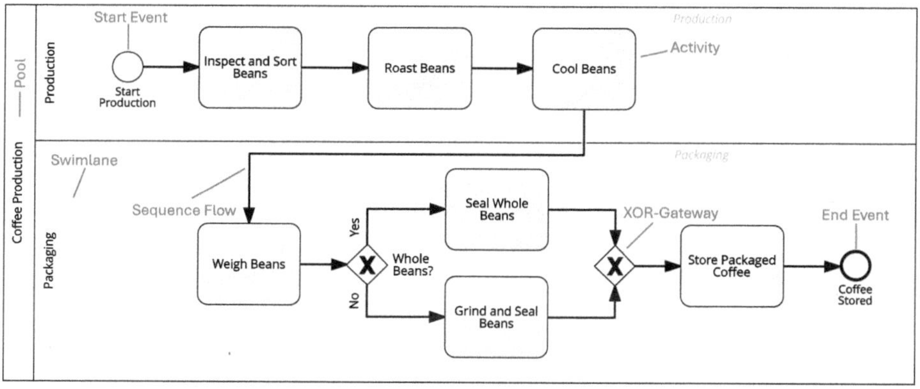

Fig. 2. Example of a BPMN 2.0 process model.

A critical aspect of comprehending a process model is the recognition of its elements and their association with the model semantics, which are essential for identifying and prioritizing relevant information.

Cognitive Load Theory (CLT) provides a valuable framework for understanding the mental effort required during process model comprehension. According to CLT, cognitive load can be divided into three categories: *intrinsic load* (i.e., the inherent difficulty of the content), *extraneous load* (i.e., unnecessary cognitive demands), and *germane load* (i.e., effort dedicated to processing the information) [32–34]. Managing these types of cognitive load is crucial to facilitate the effective comprehension of process models.

3 Related Work

Recent advances show that LLMs can be used to automatically discover BPMN 2.0 process models from natural language descriptions. Corresponding approaches rely on LLMs, such as GPT-4 or LLaMA 3 [14,22,30], and several prototype systems demonstrate the feasibility of this approach, highlighting that automatically generated models can be subjected to expert evaluation or user-based validation [18,23]. Beyond automatic generation, interactive (i.e. conversational) scenarios are investigated in which LLMs support iterative process model refinement in dialogue-based settings [2,39]. In addition, approaches are explored that focus on the automated extraction of reusable modeling fragments and patterns [10,21]. Despite these advances, pipeline- and rule-based techniques remain relevant, particularly in application contexts where high levels of structure, traceability, and transparency are essential [1,9].

Beyond process model generation and discovery, research has investigated the comprehension of process models. Note that the usefulness of process models in business process management depends on how well involved stakeholders are able to interpret them. Prior studies examined factors such as the choice of modeling notation [40], model complexity, and model layout [8], or secondary notations such as color [7]. Systematic reviews emphasize that cognitive aspects of process model understanding remain insufficiently explored [6,20]. Eye-tracking has proven particularly valuable for studying how user expertise, attention, and visual characteristics influence comprehension [27,28,38]. So far, however, these investigations have focused on models manually created by humans. Consequently, examining the comprehension of automatically generated process models by LLMs is important to assess whether an LLM-based agent is capable of producing models that are no less understandable than those created by humans. Building on this motivation, this work specifically focuses on investigating this aspect.

4 Study Context and Setting

This section outlines the design of our study. First, we provide insights into the study participants, followed by an explanation of the study materials. We then introduce the performance measure used to assess the comprehension of LLM- vs. human-generated BPMN 2.0 process models.

4.1 Study Design

The study followed a structured procedure to collect the responses of the participants (cf. Fig. 3). After providing informed consent, participants completed a demographic questionnaire (see Table 1) and received a brief introduction to the key elements and concepts of BPMN 2.0 to ensure a basic understanding.

After completing the demographic questionnaire and the BPMN 2.0 introduction, the participants were equipped with an eye tracker. The main part of our study consisted of presenting BPMN 2.0 process models to the participants. These models were derived from the same textual process description: one group received five models that were automatically generated by an LLM-based agent, while the other group received five process models that were manually created by a human expert modeler. The participants had to answer two comprehension questions for each of the five process models. The eye tracker was used during these tasks to monitor the visual attention and cognitive load of the study participants when interacting with the process models. The study concluded after participants completed the questions for all five scenarios.

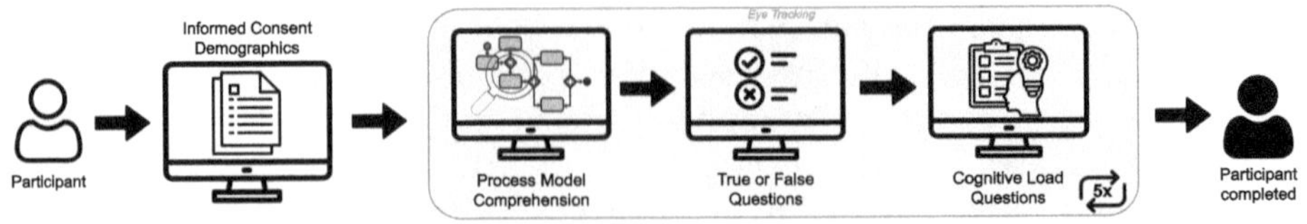

Fig. 3. Study Design.

4.2 Study Participants

The study participants were mainly recruited from university courses. In total, 27 participants were involved in the study. To facilitate the analysis of study results and to enable *Fisher's exact test*, we divided the participants into two education groups and two experience groups. The first education group includes participants having a degree less or equals a bachelor, whereas the second education group includes participants having at least a master degree. The less experienced group includes participants with no, novice, or intermediate BPMN knowledge, whereas the experienced group comprises participants who rated themselves as advanced or expert users (cf. Table 1).

Table 1. Demographic Details of the Study Participants

Variable	LLM-generated ($N = 13$)	Human-created ($N = 14$)	p
Gender female	8 (61.5)	3 (21.4)	0.054[a]
male	5 (38.5)	11 (78.6)	
Age < 25 years	4 (30.8)	9 (64.3)	0.128[a]
≥ 25 years	9 (69.2)	5 (35.7)	
Education $\leq$ Bachelor Degree	6 (46.2)	11 (78.6)	0.218[a]
$>$ Bachelor Degree	7 (53.8)	3 (21.4)	
Experience less	10 (77.0)	10 (71.4)	1.000[a]
high	3 (23.0)	4 (28.6)	

[a] Fisher's exact test

4.3 Materials

The study considers five process scenarios, each modeled in two ways: one version generated by an LLM-based agent[1] that had received a natural language process description, and one manually created by a human expert from academia based on the same description. To ensure visual consistency, all BPMN 2.0 models generated by the LLM were replicated in the process modeling tool SAP Signavio [31], making them indistinguishable from expert-created models. The five process models documented the following process scenarios: 1) making coffee; 2) brewing beer; 3) requesting software; 4) treating patient; 5) checking warranty. The complexity of these scenarios increased incrementally, with P1 being the least complex and P5 the most intricate model. All materials can be found online[2].

4.4 Performance Measures

The performance measures for the study are described in detail below.

[1] The LLM-based agent was specifically developed and implemented by our institute.
[2] https://anonymous.4open.science/r/LLM-2025-Material-F011.

Cognitive Load. Cognitive load denotes the mental effort required to process a task and is typically divided into intrinsic (ICL), extraneous (ECL), and germane (GCL) load [26,33]. In the study, it was measured along a 7-point Likert scale ranging from "strongly disagree" (i.e., 1) to "strongly agree" (i.e., 7) to capture how demanding participants perceived the comprehension of the process models, enabling comparison of LLM-generated and human-created models.

Comprehension Score. For each process model, two true-or-false questions were added, with one point awarded per correct answer (max. two points).

Comprehension Duration. The time (in seconds) between starting and indicating full comprehension of the process model was recorded.

Number of Fixations. Using the *Pupil Labs Neon* eye tracker, we measured how often participants focused on the process models, with higher counts indicating greater mental effort.

Fixation Duration. Fixation duration is the time (in seconds) the eyes remain relatively steady on a visual element. Longer fixations suggest that participants required more effort to comprehend the respective model element.

Pupil Diameter. Pupil size (in mm) served as a physiological indicator of mental effort, with larger diameters reflecting higher mental effort [13].

5 Results

This section presents the study results for each research question, which are discussed in Sect. 6. For the eye-tracking data, in the highly experienced group viewing the human-created process models, two recordings had to be excluded due to incomplete data.

5.1 Results for RQ1

To examine differences in cognitive load between process models generated by an LLM-based agent and the ones manually created by a human expert, we conducted independent sample t-tests. The analysis was carried out separately for each cognitive load category (ICL, ECL, GCL; cf. Section 4.4), across the five process models (PM1PM5), and for two participant groups based on their experience level (e.g. low vs. high).

For less experienced participants (cf. Table 2), the process models automatically generated with an LLM-based agent resulted in a significantly lower cognitive load in several cases. Specifically, a lower ICL was observed for PM2, PM3, and PM5 ($p < .05$), a lower ECL for PM2 and PM5, and a lower GCL for PM2, PM3, and PM5.

Among highly experienced participants (cf. Table 3), the significant differences occurred less frequently. A lower cognitive load in comprehending the LLM-generated models was observed for PM5 in all three load dimensions, and in PM4 for ECL ($p < .05$). The other comparisons reached no significance. In summary, a lower cognitive load was observed more consistently in LLM-generated models, particularly for less experienced participants in the context of more complex processes.

Table 2. T-Test Results for Cognitive Load - Less Experience

	LLM-generated		Human-created				
Variable	MV	SD	MV	SD	T	df	p
ICL							
PM1	1.85	0.53	2.30	0.54	1.886	18	0.076
PM2	2.40	0.57	4.20	1.08	4.531	18	< 0.001*
PM3	3.60	1.47	4.80	0.82	2.254	18	0.037*
PM4	3.40	1.51	3.30	0.82	-0.184	18	0.856
PM5	2.75	0.86	4.70	0.67	5.649	18	< 0.001*
ECL							
PM1	1.77	0.45	2.03	0.60	1.131	18	0.273
PM2	2.37	0.82	3.43	1.03	2.557	18	0.020*
PM3	3.60	1.30	4.47	0.67	1.870	18	0.078
PM4	3.07	0.98	2.93	0.86	-0.324	18	0.750
PM5	2.43	0.83	4.43	0.75	5.631	18	< 0.001*
GCL							
PM1	2.00	0.71	1.85	0.78	-0.449	18	0.658
PM2	1.85	0.63	2.95	1.28	2.443	18	0.025*
PM3	2.75	1.23	4.30	0.82	3.311	18	0.004*
PM4	2.45	1.28	2.70	1.11	0.467	18	0.646
PM5	2.40	1.47	4.25	1.06	3.230	18	0.005*

Table 3. T-Test Results for Cognitive Load - High Experience

	LLM-generated		Human-created				
Variable	MV	SD	MV	SD	T	df	p
ICL							
PM1	1.67	0.76	1.88	0.48	0.448	5	0.673
PM2	2.33	0.58	4.00	1.22	2.147	5	0.085
PM3	3.33	0.76	4.25	1.04	1.277	5	0.258
PM4	2.50	1.32	4.25	1.32	1.732	5	0.144
PM5	1.50	0.50	5.13	0.95	5.944	5	0.002*
ECL							
PM1	2.00	0.33	2.33	0.67	0.782	5	0.469
PM2	2.00	0.58	3.00	1.61	1.007	5	0.360
PM3	2.56	1.54	3.50	1.29	0.886	5	0.416
PM4	1.56	0.51	3.92	1.42	2.690	5	0.043*
PM5	1.44	0.51	4.75	1.26	4.760	4.156	0.008*
GCL							
PM1	1.83	0.29	1.75	1.19	-0.116	5	0.912
PM2	2.33	0.58	2.38	0.75	0.080	5	0.940
PM3	2.33	1.44	3.50	1.35	1.099	5	0.322
PM4	1.67	0.29	3.25	1.85	1.890	5	0.117
PM5	1.50	0.50	4.88	1.70	3.756	3.658	0.023*

5.2 Results for RQ2

To examine performance differences when comprehending LLM- vs. human-generated process models, independent sample t-tests were conducted for five variables: score, duration, number of fixations, fixation duration, and pupil diameter across all five process models and the two experience levels.

Among less experienced participants (cf. Table 4), no significant differences were found (all $p > .05$), except for pupil diameter. While small numerical deviations appeared (e.g., slightly higher scores or more fixations for human-created models), none reached significance. By contrast, pupil diameter showed consistent effects, with smaller values for LLM-generated models. A significant effect was found for PM2 ($p = .050$), and marginal effects for PM1, PM3, PM4, and PM5 ($p = .083$, $p = .096$, $p = .081$, $p = .069$).

For more experienced participants (cf. Table 5), no significant effects were observed for score, duration, or fixation measures (all $p > .05$), though PM5 showed a marginal score effect favoring the LLM-generated model ($p = .051$). Pupil diameter again trended smaller for LLM-generated models across all scenarios, but without significance (all $p > .05$).

Table 4. T-Test Results for Performance - Less Experience

	LLM-generated		Human-created				
Variable	MV	SD	MV	SD	T	df	p
Score							
PM1	1.20	0.63	1.40	0.52	-0.775	17	0.449
PM2	1.60	0.52	1.30	0.82	0.976	15	0.344
PM3	1.10	0.57	1.40	0.70	-1.053	17	0.307
PM4	1.60	0.52	1.60	0.52	0.000	18	1.000
PM5	1.40	0.70	1.30	0.67	0.325	18	0.749
Duration							
PM1	24.14	10.76	23.90	23.50	0.028	11	0.978
PM2	37.15	21.01	39.68	22.34	-0.253	17	0.803
PM3	51.50	27.76	55.51	31.12	-0.295	17	0.772
PM4	56.37	29.41	35.78	24.74	1.657	17	0.116
PM5	46.42	19.09	57.51	32.38	-0.897	13	0.387
Fixations							
PM1	57.1	25.48	69.56	76.61	-0.465	10	0.652
PM2	75.9	23.27	95.56	42.79	-1.225	12	0.244
PM3	121.9	38.38	124.78	49.26	-0.141	15	0.89
PM4	121.4	43.48	97.56	65.11	0.928	14	0.369
PM5	92.4	28.34	126.44	70.77	-1.349	10	0.206
Fix. dur.							
PM1	0.430	0.117	0.330	0.144	1.643	15	0.121
PM2	0.466	0.137	0.370	0.165	1.366	16	0.191
PM3	0.405	0.122	0.408	0.196	-0.030	13	0.977
PM4	0.447	0.109	0.338	0.150	1.784	15	0.095
PM5	0.500	0.137	0.436	0.231	0.734	13	0.476
Pup. dia.							
PM1	2.910	0.272	3.214	0.384	-1.863	17	0.083
PM2	2.840	0.268	3.305	0.542	-2.196	17	0.050*
PM3	2.866	0.275	3.252	0.542	-1.815	17	0.096
PM4	2.853	0.262	3.274	0.568	-1.924	17	0.081
PM5	2.820	0.274	3.242	0.538	-2.003	17	0.069

Table 5. T-Test Results for Performance - High Experience

	LLM-generated		Human-created				
Variable	MV	SD	MV	SD	T	df	p
Score							
PM1	1.33	0.58	1.25	0.50	0.200	4	0.851
PM2	1.67	0.58	1.75	0.50	-0.200	4	0.851
PM3	2.00	0.00	1.75	0.50	1.000	3	0.391
PM4	1.33	1.15	2.00	0.00	-1.000	2	0.423
PM5	1.67	0.58	0.50	0.58	2.646	4	0.051
Duration							
PM1	15.15	2.63	6.93	5.35	2.388	3	0.100
PM2	36.24	8.81	27.57	21.17	0.548	1	0.666
PM3	59.37	17.04	52.40	24.37	0.406	4	0.708
PM4	57.31	15.51	46.11	27.00	0.623	3	0.575
PM5	34.57	13.28	53.85	25.43	-1.164	3	0.328
Fixations							
PM1	33.67	5.13	20.67	14.29	1.483	3	0.252
PM2	76.0	31.24	54.0	29.7	0.795	2	0.498
PM3	148.33	66.94	104.0	39.95	0.985	3	0.392
PM4	124.33	32.33	94.0	42.93	0.978	4	0.388
PM5	60.0	25.24	83.07	23.80	-0.881	4	0.428
Fix. dur.							
PM1	0.463	0.144	0.290	0.122	1.583	4	0.190
PM2	0.517	0.155	0.475	0.131	0.330	3	0.766
PM3	0.431	0.101	0.492	0.082	-0.819	4	0.461
PM4	0.463	0.058	0.468	0.088	-0.071	3	0.947
PM5	0.523	0.040	0.618	0.139	-1.142	2	0.357
Pup. dia.							
PM1	3.128	0.323	2.845	0.126	1.154	4	0.344
PM2	3.086	0.263	2.824	0.162	1.198	4	0.309
PM3	3.111	0.275	2.775	0.161	1.491	4	0.226
PM4	3.085	0.247	2.779	0.116	1.591	4	0.215
PM5	3.030	0.209	2.739	0.115	1.724	4	0.180

6 Discussion

This section discusses key findings and reflects on the limitations of the study.

6.1 Key Findings

A key finding concerns the cognitive load perceived by the study participants during process model comprehension. Among members of the less experienced group, the comprehension of the automatically generated LLM-based process models resulted in a significantly lower cognitive load, particularly regarding the more complex scenarios (PM2, PM3, and PM5). These differences were observed across all three cognitive load dimensions (i.e. ICL, ECL, and GCL).

These findings indicate that cognitively, LLM-generated models can be processed easier. This can be explained by a simpler structure, reduced visual noise, or more consistent design. For less experienced users, such reduced cognitive load makes it easier to engage with BPMN 2.0 process models and to comprehend them.

These subjective results were corroborated by the analysis of pupil diameter, which is a physiological indicator of mental effort. For less experienced participants, the pupil diameter was consistently smaller when working with LLM-generated models. A statistically significant difference was observed for PM2 ($p = .050$), and marginal effects were found for the remaining scenarios (e.g., PM5: $p = .069$), indicating lower mental effort when interacting with the LLM-generated models.

Among the more experienced participants, fewer significant differences were identified. However, a notable exception emerged in the context of PM5, which is the most complex model: here, again, a significantly lower cognitive load was associated with the LLM-generated model across all cognitive load types. Furthermore, PM4 showed a significant reduction in ECL (cf. Section 4.4). Although the differences were not statistically significant, pupil diameter consistently tended to be smaller when participants worked with the LLM-generated models. This trend was noticeable for the more complex process scenarios (PM3PM5), indicating that even experienced users may have needed slightly lower mental effort with the automatically generated models. This indicates that even expert users can benefit from LLM-generated process models when the complexity of process models increases.

Although differences in cognitive load became evident, no statistically significant differences were observed concerning the comprehension performance metrics between the model types, regardless of the participants' experience level.

This means that both LLM-generated process models and process models manually generated by human experts allowed users to perform equally well, in respect to answering comprehension questions accurately and visually navigating within the models. For less experienced users, this is particularly meaningful as they achieved comparable comprehension performance with less mental effort when comprehending the models automatically generated by an LLM.

For PM5, a slight advantage ($p = 0.051$) in the comprehension scores among participants with higher experience indicates a potential performance advantage regarding LLM-generated process models in more complex scenarios, which may warrant further exploration in follow-up studies.

6.2 Limitations

The study provides valuable information on the cognitive implications and comprehension performance regarding BPMN 2.0 models automatically generated by an LLM-based agent versus models created manually. However, several limitations must be noted.

One limitation concerns individual differences within the defined experience groups that may have influenced the results. Although participants were categorized into "less experience" and "highly experienced" according to their self-reported BPMN 2.0 knowledge, factors such as prior modeling exposure, academic background, and analytical abilities were likely varying and may have introduced additional variance not attributable solely to the type of model generation (LLM- vs. human-generated).

Another limitation lies in the comprehension assessment, which relied on two true-or-false questions per model. Although this provided a standardized and efficient measure, it may not have captured deeper aspects of comprehension, such as drawing inferences or mentally structuring processes.

Cognitive load was measured using subjective ratings for ICL, ECL, and GCL (cf. Section 4.4). Although widely used in cognitive load research, such self-reports are limited regarding precision. To strengthen reliability, we complemented them with eye tracking data (number of fixations and fixation duration) as an objective indicator of mental effort. However, these measures are indirect and might be influenced by factors such as layout, visual saliency, or momentary distraction.

The pupil diameter was also employed as a physiological proxy, but this metric can be confounded by external conditions (e.g. lighting) or internal states (e.g., arousal, fatigue). In addition, the LLM- and human-generated models differed in size and number of elements. The simpler structure of LLM-generated models may have reduced cognitive load and supported comprehension. However, this effect is a natural feature of the LLM-based approach and was intentionally preserved rather than artificially eliminated.

Generalizability is limited by the sample, which consisted mainly of university students with technical or business backgrounds. Results may not directly transfer to professional BPMN practitioners, non-technical stakeholders, or other domains. Broader samples and application areas (e.g., healthcare, public administration, manufacturing) should be considered in future studies.

Finally, the relatively small sample size ($N = 27$), and in particular the small subgroup of highly experienced participants ($N = 7$), restricts statistical power. Although several effects reached significance, they should be interpreted with caution.

7 Summary and Outlook

This study investigated the cognitive and performance-related effects of BPMN 2.0 process models that were automatically generated by an LLM-based agent compared to models manually created by human experts. Using a combination of subjective cognitive load ratings, eye-tracking data, and comprehension performance metrics, we examined how participants with varying levels of modeling experience interacted with five process models of increasing complexity.

The results demonstrate that LLM-generated process models consistently led to a lower cognitive load, particularly for less experienced participants and in more complex scenarios. These findings were evident in all three dimensions of cognitive load (ICL, ECL and GCL) and were further corroborated by physiological indicators of mental effort, such as reduced pupil diameter. Even among more experienced participants, LLM-generated models resulted in more efficient cognitive processing in complex process scenarios, although the effect was less pronounced.

Despite these cognitive differences, comprehension performance was statistically comparable between the two model types across all levels of experience. This indicates that automatically generated models can support effective task performance, while reducing mental effort, especially for novices.

In future research, we want to expand the participant pool to include professionals from diverse domains to enhance external validity. Additionally, exploring more complex and domain-specific process scenarios (e.g., healthcare, public sector, manufacturing) shall contribute to uncover context-dependent effects. Follow-up studies should investigate how different LLM prompting strategies or levels of model abstraction influence both model quality and cognitive processing.

Finally, extending the comprehension assessment beyond binary questions could help capturing a deeper and more nuanced understanding of process models.

In summary, our study provides promising evidence that LLM-based agents can automatically generate comprehensible process models, offering valuable implications for the advancement of automated business process modeling in both academic and practical contexts.

Disclosure of Interests. The authors have no competing interests to declare that are relevant to the content of this article.

References

1. Ajmal, F., Wijekoon, P., Dhanamina, H., Ravishan, Y., Nawinna, D., Attanayaka, B.: Automated BPMN diagram generation. In: 2024 6th International Conference on Advancements in Computing (ICAC), pp. 7–12. IEEE (2024)
2. Bernardi, M.L., Casciani, A., Cimitile, M., Marrella, A.: Conversing with business process-aware large language models: the BPLLM framework. J. Intell. Inf. Syst. **62**(6), 1607–1629 (2024)
3. Bordes, A., Weston, J.: Learning end-to-end goal-oriented dialog. CoRR (2016). http://arxiv.org/abs/1605.07683
4. Brown, T., Mann, B., Ryder, N., Subbiah, M., et al.: Language models are few-shot learners. Adv. Neural. Inf. Process. Syst. **33**, 1877–1901 (2020)
5. Dumas, M., Rosa, L.M., Mendling, J., Reijers, A.H.: Fundamentals of business process management. Springer, Cham (2018)
6. Figl, K.: Comprehension of procedural visual business process models: a literature review. Bus. Inf. Syst. Eng. **59**, 41–67 (2017)
7. Figl, K., Mendling, J., Strembeck, M.: The influence of notational deficiencies on process model comprehension. J. Assoc. Inf. Syst. **14**(6), 1 (2013)
8. Figl, K., Recker, J., Mendling, J.: A study on the effects of routing symbol design on process model comprehension. Decis. Support Syst. **54**(2), 1104–1118 (2013)
9. Fill, H.G., Fettke, P., Köpke, J.: Conceptual modeling and large language models: impressions from first experiments with chatGPT. Enterprise Model. Inf. Syst. Architectures (EMISAJ) **18**, 1–15 (2023)

10. Grohs, M., Abb, L., Elsayed, N., Rehse, J.R.: Large language models can accomplish business process management tasks. In: De Weerdt, J., Pufahl, L. (eds.) International Conference on Business Process Management, pp. 453–465. Springer, Cham (2023). https://doi.org/10.1007/978-3-031-50974-2_34
11. Hörner, L.F.: Introducing BPMNGen: An LLM-based conversational framework for BPMN 2.0 process model generation. In: EMISA 2025, pp. 17–31. Bonn (2025)
12. Indulska, M., Green, P., Recker, J., Rosemann, M.: Business Process Modeling: Perceived Benefits. In: Laender, A.H.F., Castano, S., Dayal, U., Casati, F., de Oliveira, J.P.M. (eds.) ER 2009. LNCS, vol. 5829, pp. 458–471. Springer, Heidelberg (2009). https://doi.org/10.1007/978-3-642-04840-1_34
13. Kahneman, D., Beatty, J.: Pupil diameter and load on memory. Science **154**(3756), 1583–1585 (1966)
14. Klievtsova, N., Benzin, J.V., Kampik, T., Mangler, J., Rinderle-Ma, S.: Conversational process modelling: state of the art, applications, and implications in practice. In: Di Francescomarino, C., Burattin, A., Janiesch, C., Sadiq, S. (eds.) International Conference on Business Process Management, pp. 319–336. Springer, Cham (2023). https://doi.org/10.1007/978-3-031-41623-1_19
15. Klievtsova, N., Mangler, J., Kampik, T., Benzin, J.V., Rinderle-Ma, S.: How can generative AI empower domain experts in creating process models? In: Wirtschaftsinformatik (2024)
16. Köpke, J., Safan, A.: Efficient LLM-based conversational process modeling. In: International Conference on Business Process Management, pp. 259–270. Springer, Cham (2024)
17. Kourani, H., Berti, A., Schuster, D., van der Aalst, W.M.P.: Process Modeling with Large Language Models. In: van der Aa, H., Bork, D., Schmidt, R., Sturm, A. (eds.) Enterprise, Business-Process and Information Systems Modeling. BPMDS EMMSAD 2024. Lecture Notes in Business Information Processing, vol. 511. Springer, Cham (2024). https://doi.org/10.1007/978-3-031-61007-3_18
18. Kourani, H., Berti, A., Schuster, D., Van der Aalst, W.M.: ProMoAI: process modeling with generative AI. arXiv preprint arXiv:2403.04327 (2024)
19. Mistral: Mistral (2025). https://mistral.ai, Accessed 19 Aug 2025
20. Möller, M., Winter, M., Reichert, M.: Cognitive factors in process model comprehension–a systematic literature review. Brain Sci. **15**(5), 505 (2025)
21. Monti, F., Leotta, F., Mangler, J., Mecella, M., Rinderle-Ma, S.: Nl2processops: Towards LLM-guided code generation for process execution. In: Marrella, A., Resinas, M., Jans, M., Rosemann, M. (eds.) International Conference on Business Process Management, pp. 127–143. Springer, Cham (2024). https://doi.org/10.1007/978-3-031-70418-5_8
22. Neuberger, J., Ackermann, L., van der Aa, H., Jablonski, S.: A universal prompting strategy for extracting process model information from natural language text using large language models. In: Maass, W., Han, H., Yasar, H., Multari, N. (eds.) International Conference on Conceptual Modeling, pp. 38–55. Springer, Cham (2024). https://doi.org/10.1007/978-3-031-75872-0_3
23. Nour Eldin, A., Assy, N., Anesini, O., Dalmas, B., Gaaloul, W.: Nala2BPMN: Automating BPMN model generation with large language models. In: Comuzzi, M., Grigori, D., Sellami, M., Zhou, Z. (eds.) Cooperative Information Systems, pp. 398–404. Springer, Cham (2024). https://doi.org/10.1007/978-3-031-81375-7_27
24. Dijkman, R., Hofstetter, J., Koehler, J. (eds.): BPMN 2011. LNBIP, vol. 95. Springer, Heidelberg (2011). https://doi.org/10.1007/978-3-642-25160-3
25. OpenAI: ChatGPT (2025). https://openai.com/chatgpt/, Accessed 19 Aug 2025

26. Paas, F., Renkl, A., Sweller, J.: Cognitive load theory and instructional design: Recent developments. Edu. Psychol. **38**(1), 1–4 (2003)
27. Petrusel, R., Mendling, J.: Eye-Tracking the Factors of Process Model Comprehension Tasks. In: Salinesi, C., Norrie, M.C., Pastor, Ó. (eds.) CAiSE 2013. LNCS, vol. 7908, pp. 224–239. Springer, Heidelberg (2013). https://doi.org/10.1007/978-3-642-38709-8_15
28. Petrusel, R., Mendling, J., Reijers, H.A.: How visual cognition influences process model comprehension. Decis. Support Syst. **96**, 1–16 (2017)
29. Radford, A., Wu, J., Child, R., Luan, D., Amodei, D., Sutskever, I., et al.: Language models are unsupervised multitask learners. OpenAI Blog **1**(8), 9 (2019)
30. Reinhartz-Berger, I., Ali, S.J., Bork, D.: Leveraging LLMs for domain modeling: The impact of granularity and strategy on quality. In: International Conference on Advanced Information Systems Engineering, pp. 3–19. Springer, Cham (2025)
31. Signavio: Signavio (2025). https://www.signavio.com/de/, Accessed 10 Aug 2025
32. Sweller, J.: Cognitive load during problem solving: effects on learning. Cogn. Sci. **12**(2), 257–285 (1988)
33. Sweller, J.: Cognitive load theory, learning difficulty, and instructional design. Learn. Instr. **4**(4), 295–312 (1994)
34. Sweller, J., Chandler, P.: Evidence for cognitive load theory. Cogn. Instr. **8**(4), 351–362 (1991)
35. Vaswani, A., et al.: Attention is all you need. Adv. Neural Inf. Processing Syst. **30** (2017)
36. Vinyals, O., Le, Q.V.: A neural conversational model. CoRR (2015). http://arxiv.org/abs/1506.05869
37. Winter, M.: Conscious-cognitive neuroscience and psychology during process model comprehension (2022)
38. Winter, M., Bredemeyer, C., Reichert, M., Neumann, H., Probst, T., Pryss, R.: How healthcare professionals comprehend process models-an empirical eye tracking analysis. In: IEEE 34th International Symposium on Computer-Based Medical Systems (CBMS), pp. 313–318. IEEE (2021)
39. Ziche, C., Apruzzese, G.: LLM4PM: A case study on using large language models for process modeling in enterprise organizations. In: Di Ciccio, C., et al. (eds.) International Conference on Business Process Management, pp. 472–483. Springer, Cham (2024). https://doi.org/10.1007/978-3-031-70445-1_35
40. Zimoch, M., Mohring, T., Pryss, R., Probst, T., Schlee, W., Reichert, M.: Using Insights from Cognitive Neuroscience to Investigate the Effects of Event-Driven Process Chains on Process Model Comprehension. In: Teniente, E., Weidlich, M. (eds.) BPM 2017. LNBIP, vol. 308, pp. 446–459. Springer, Cham (2018). https://doi.org/10.1007/978-3-319-74030-0_35

LLM-Based Process Modeling: Evaluation of Usability and Semantic Quality

Luca Franziska Hörner(✉), Maximilian Möller, and Manfred Reichert

Institute of Databases and Information Systems, Ulm University,
89081 Ulm, Germany
{luca.hoerner,maximilian.moeller,manfred.reichert}@uni-ulm.de

Abstract. The increasing capabilities of large language models (LLMs) have sparked growing interest in leveraging them for the automatic generation of visual process representations (i.e. process models) from natural language process descriptions. To assess this potential, we conducted two complementary studies focusing on key aspects of the practical applicability of a corresponding LLM-based chatbot. A usability study explored the user experience of BPMNGen – the LLM-based chatbot we developed – in terms of perceived usability and cognitive load, while a semantic quality study with 100 participants compared the process models generated by BPMNGen with those created by human modelers of different expertise levels. The results of the studies indicate that BPMNGen offers high usability while delivering process models with a semantic quality comparable to human-created process models.

Keywords: LLM · Process Model Generation · Quality Evaluation

1 Introduction

The rapid advancements of LLMs affect various fields, enabling AI-driven systems to tackle increasingly complex tasks [31]. Among these knowledge-intensive tasks, business process modeling has gained significant attention, as organizations seek to automate, document, analyze, and optimize their processes [9]. In this context Business Process Model and Notation (BPMN) 2.0 provides a standardized notation for modeling business processes. However, the effective use of BPMN requires substantial expertise, making business process modeling challenging for non-experts and often time-consuming and error-prone even for experienced modelers [21,35].

To tackle this challenge, we developed BPMNGen, an LLM-based conversational chatbot specifically designed to automatically generate BPMN 2.0 process models from textual descriptions. The goal of BPMNGen is to simplify process modeling and make it more accessible to users with varying levels of expertise. For such accessibility to become effective, certain aspects must align. In this paper, we examine two of them: the usability of BPMNGen and the semantic quality of its generated process models. To address this, we conducted a usability

H. R. Arabnia et al. (Eds.): CSCI 2025, CCIS 2972, pp. 327–342, 2027.
https://doi.org/10.1007/978-3-032-28469-3_22

study and a semantic quality study, providing first insights into both dimensions. The usability study investigates how end-users experience the LLM-based chatbot (i.e., BPMNGen) in practice by systematically assessing cognitive load and perceived usability using the NASA Task Load Index (NASA-TLX) and the System Usability Scale (SUS). This leads to our first research question:

RQ1: How do users perceive the cognitive load and usability of BPMNGen, an LLM-based chatbot for BPMN 2.0 model generation?

Complementary to this, the semantic quality study investigates whether BPMNGen is capable of generating BPMN 2.0 process models that adequately capture the intended meaning of textual process descriptions. Therefore, we compared BPMNGen-generated models with those created by human modelers of different expertise levels. This leads to our second research question:

RQ2: How does the semantic quality of the process models automatically generated by an LLM-based chatbot compare to those manually created by humans with different levels of expertise?

Building on RQ1 and RQ2, the study is structured as follows: Sect. 2 provides the theoretical background, followed by Sect. 3 presenting related work. In Sect. 4 we introduce the LLM-based chatbot used in our studies (i.e. BPMNGen). Section 5 presents the usability study, while Sect. 6 deals with the semantic quality study. Section 7 states the limitations of the two studies. Finally, Sect. 8 concludes the paper and provides an outlook on future work.

2 Background

This section introduces foundational concepts and technologies that are essential for a comprehensive understanding of the studies.

2.1 LLM-Based Chatbots

Chatbots are conversational agents that interact with users through natural language interfaces. Modern chatbots are powered by LLMs – advanced deep learning models trained on large-scale datasets – which have significantly advanced NLP by enabling tasks such as text generation, summarization, and translation [5,8,30]. The core architecture behind most LLMs is *Transformer*, introduced by [30], which uses a mechanism called self-attention to understand how words relate to one another, even over long distances within a text. Prominent examples of LLMs include Mistral [22], Gemini [12], and GPT-4 [7].

These capabilities allow modern chatbots to engage in human-like dialogues, generate coherent responses, and maintain conversational context across multiple turns [7]. They can handle diverse tasks, from answering questions to providing personalized recommendations, without relying on predefined scripts [1,6]. A key strength of LLM-based chatbots lies in their ability to adapt to new tasks with minimal instruction – a concept known as *zero-shot learning* and *few-shot learning* [27,32]. This adaptability allows them to perform well across different domains without extensive retraining [5].

2.2 Process Models

Process models serve as graphical representations of business processes, enabling their documentation, visualization, analysis, and optimization within organizations [9]. By offering a clear and structured overview, these models support domain experts, business analysts, and process participants in understanding and refining business operations [33]. A widely used standard for modeling business processes is BPMN 2.0, as proposed by the OMG [26]. It provides a comprehensive set of symbols and rules for representing complex processes.

Figure 1 illustrates the basic elements of BPMN 2.0, which form the core notation used in process modeling. An example of a BPMN 2.0 process model, in which these elements are exemplarily applied, is provided in Fig. 3 in Sect. 4.2.

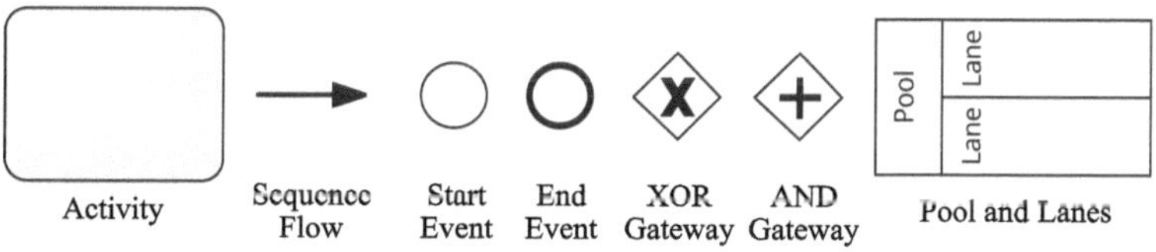

Fig. 1. BPMN 2.0 Elements.

3 Related Work

The integration of LLMs into business process management has sparked a variety of approaches for translating natural language into BPMN 2.0 models [25]. Recent work explores different ways of applying LLMs to process and domain modeling, for example, by examining conversational modeling paradigms [15], proposing universal prompting strategies for extracting process model information [24], or analyzing the impact of granularity and strategy on LLM-based domain modeling [28]. Frameworks such as *Nala2BPMN* and *ProMoAI* automate the extraction of process elements and are mainly evaluated by expert feedback, demonstrating feasibility but with limited evaluation depth [18,25].

Previous studies highlight that the effectiveness of conversational process modeling is strongly dependent on efficient prompt design [17], while universal prompting strategies have also been proposed to systematically extract process information from text [24]. These works underline the importance of prompt design, but focus mainly on technical effectiveness. Exploratory experiments with LLMs for conceptual modeling indicate that many generated models are syntactically correct and formally valid, yet still require expert knowledge to assess their actual adequacy [11]. Based on this observation, recent work has outlined conceptual foundations for evaluating the process modeling abilities of LLMs, proposing a standard evaluation scenario, and discussing methodological challenges such as measuring syntactic versus semantic quality [10].

Interactive approaches enable dialog-based refinement of process models, such as the *BPLLM* framework, which integrates process awareness into conversational interactions [3], and *LLM4PM*, which demonstrates dialog-driven modeling in enterprise settings [34]. Although these works highlight the potential of interactive systems to improve structural validity and user control, studies show that semantic depth often remains limited [15,16].

Other contributions focus on the generation of process fragments or activity patterns [13,23], thus broadening the scope of automation. However, all of these contributions remain focused on technical feasibility rather than user-centered evaluation or semantic quality.

Nevertheless, existing studies mainly demonstrate feasibility and focus on syntactic validity, while systematic evidence on usability and semantic quality is still scarce. This paper addresses this gap through an empirical evaluation that combines usability measures with a semantic quality assessment of LLM-generated process models by BPMNGen.

4 LLM-Based Process Modeling with BPMNGen

Before presenting the study, we introduce BPMNGen and outline its architecture and a real-world usage scenario.

4.1 Architecture

BPMNGen follows a client-server architecture and is powered by OpenAI's LLM (currently GPT-4o[1].), which interprets natural language input and supports the generation of BPMN 2.0 process models [7]. It consists of four main components (cf. Fig. 2) that interact as follows: The custom assistant converts the user's natural language input into a JSON structure, which is then transformed by the server into BPMN 2.0 XML. The user interface subsequently visualizes the XML as a BPMN 2.0 process model. Finally, the database stores the resulting models.

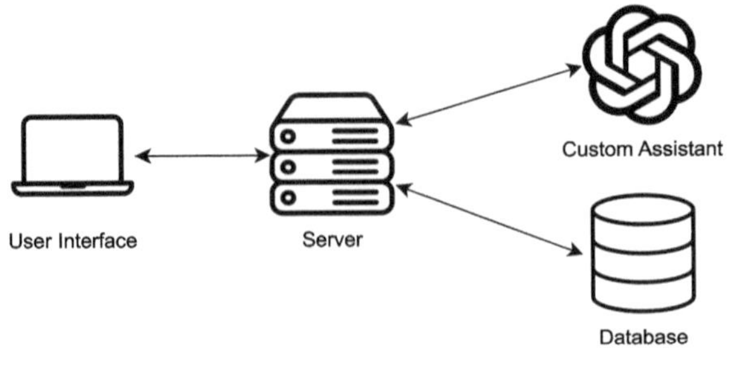

Fig. 2. Architecture of BPMNGen.

[1] We will always integrate the latest available GPT version.

4.2 Usage Scenario

BPMNGen supports the creation and iterative refinement of BPMN 2.0 process models from natural language input. Figure 3 presents our LLM-based chatbot, BPMNGen, together with a BPMN 2.0 process model generated.

On the right-hand side, a process description is shown within the chat window, while the central workspace displays the corresponding process model, here exemplified by a restaurant service process. Users can refine or modify the generated process model either by providing additional prompts or manually adjusting the diagram through drag-and-drop of BPMN 2.0 elements. The panel on the left allows users to initiate a new chat or access previously generated process models for review. The resulting model can be downloaded for further use or deleted if undesired. In addition, the interface provides further functionalities such as user profile management and system settings. A demonstration of BPMNGen is available online[2].

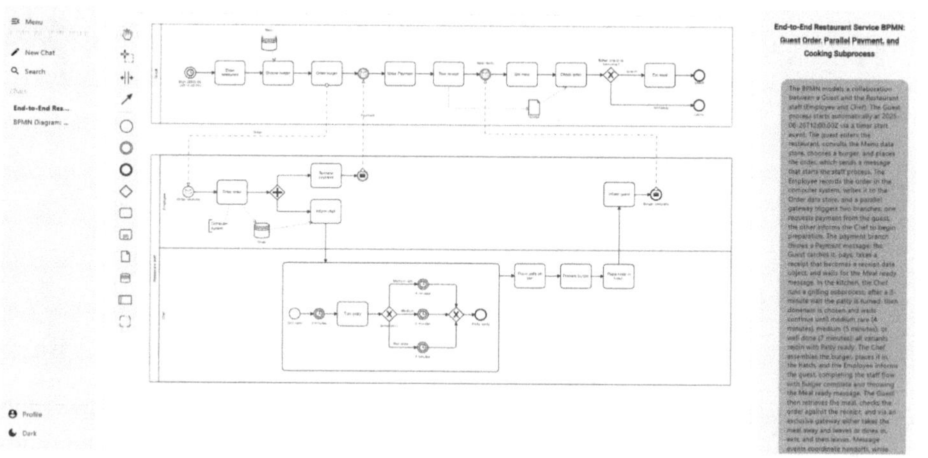

Fig. 3. User Interface of BPMNGen.

5 Usability Study

To gain initial insights into the usability of BPMNGen, i.e. the LLM-based chatbot for process modeling, we conducted a usability study. This study provides valuable first evidence on how users experience the interaction with BPMNGen, representing an important step toward assessing the practical applicability of LLM-based support in business process modeling.

[2] https://anonymous.4open.science/r/LLM-2025-Study-Materials-8FB0.

5.1 Study Design

The usability study began with the participant providing written consent for participation. After confirmation, the participants were seated in front of a monitor that displayed the BPMNGen start page. Before interacting with BPMNGen, the tasks and the general flow of the study were explained. Once the participants confirmed that they were ready to proceed, they started using BPMNGen and completed the specified tasks (cf. Table 2). After the participants indicated that the tasks were completed, they had to fill out the NASA-TLX [14] and SUS [4] questionnaires to assess the cognitive load and the usability of BPMNGen. The study was concluded after the participants submitted the two questionnaires. An overview of the study design is illustrated in Fig. 4.

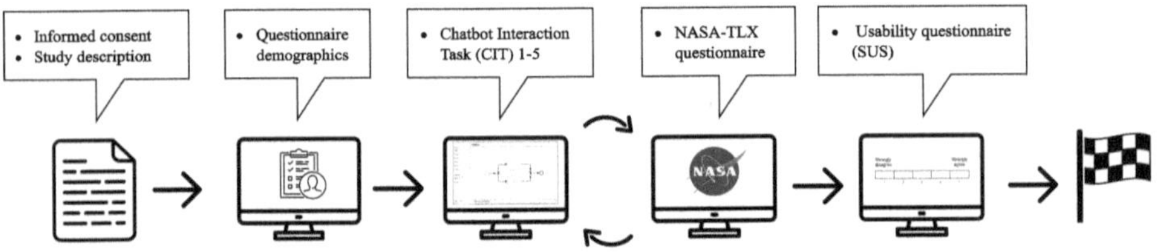

Fig. 4. Usability Study Design.

5.2 Study Participants

The usability study involved 5 participants recruited from academia, with an average age of 28.2 years (standard deviation = 2.86). Note that according to [2], a sample size of 5 participants is sufficient for usability studies, providing adequate insights into user interactions. The demographic details of the study participants are stated in Table 1.

Table 1. Demographic Details of the Study Participants

ID	Gender	Age	Education	Chatbot Experience
1	male	26	bachelor's degree	advanced
2	male	29	master's degree	advanced
3	male	32	master's degree	intermediate
4	female	24	master's degree	advanced
5	female	30	master's degree	no experience

5.3 Study Materials

The objective of the study was to gather insights into the usability of the LLM-based chatbot, and therefore, the participants were instructed to interact with BPMNGen by performing a series of predefined tasks (i.e., chatbot interaction tasks; CIT). The CIT were designed to reflect realistic user behavior and ensure that a broad spectrum of the capabilities of the LLM-based chatbot was evaluated. An overview of the CIT is provided in Table 2.

Table 2. Chatbot interaction tasks (CIT)

CIT	Description
1	Create a BPMN 2.0 process model.
2	Optimize one aspect of the generated BPMN 2.0 process model with a follow-up prompt.
3	Delete the generated BPMN 2.0 process model.
4	Create another BPMN 2.0 process model.
5	Optimize one aspect of the newly generated BPMN 2.0 process model by making direct modifications to the model elements using the toolbar instead of using a follow-up prompt.

5.4 Study Measurements

The following study measurements were considered when evaluating the usability of BPMNGen:

- **NASA-TLX score.** It is derived from the NASA-TLX questionnaire, which measures the perceived cognitive load in six dimensions: mental demand, physical demand, temporal demand, performance, effort, and frustration. Each dimension is rated on a scale from 0 to 20, with higher values indicating a higher subjective cognitive load.
- **SUS score.** It is derived from the SUS, a ten-item questionnaire that provides a quick and reliable measure of perceived usability. Each item is rated on a 5-point Likert scale (1 = strongly disagree, 5 = strongly agree).
  To calculate the SUS score, responses to odd-numbered items are reduced by 1, and responses to even-numbered items are subtracted from 5. The adjusted values are then summed and multiplied by 2.5, yielding an overall score between 0 and 100. This score reflects how the user perceived effectiveness, efficiency, and satisfaction.

5.5 Study Results

Table 3 summarizes the NASA-TLX results. The six cognitive load dimensions are reported with mean (MV), median (ME), and standard deviation (SD).

Physical demand was rated relatively low (MV = 1.80), whereas mental demand (MV = 6.20) and effort (MV = 5.20) received higher ratings. Frustration showed the highest value (MV = 8.60), while temporal demand and performance were reported at moderate levels.

In addition to the NASA-TLX results, the SUS evaluation (cf. Table 4) yielded a final score of 77.5 (ME = 82.5, SD = 15.1). To interpret this score, the standard grading scale for SUS scores, as outlined by [20], was applied. According to this scale the obtained grade is a B+.

Table 3. NASA-TLX Results

Dimension	MV	ME	SD
Mental	6.20	6.00	4.82
Physical	1.80	1.00	1.10
Temporal	4.60	4.00	2.19
Performance	3.80	2.00	5.22
Effort	5.20	2.00	4.87
Frustration	8.60	4.00	8.76

Table 4. SUS Results

Score Range	**Grade**	**%-Range**
84.1-100	A+	96-100
80.0-84.0	A	90-95
78.9-80.7	A-	85-89
77.2-78.8	B+	80-84
74.1-77.1	B	70-79
72.6-74.0	B-	65-69
71.1-72.5	C+	60-64
65.0-71.0	C	41-59
62.7-64.9	C-	35-40
51.7-62.6	D	15-34
0.0-51.6	F	0-14

5.6 Discussion of the Results

The usability study offers initial evidence on how users perceived interaction with BPMNGen. The SUS score of 77.5 (median 82.5; SD 15.1; $\widehat{=}$ grade B+) indicates a generally positive user experience, suggesting that BPMNGen is usable even in its prototypical form.

The NASA-TLX dimensions add further nuance. Mental demand (MV = 6.20) and effort (MV = 5.20) point to a moderate level of cognitive engagement. Physical demand was very low (MV = 1.80), reflecting the lightweight nature of text-based interaction. Temporal demand (MV = 4.60) indicates that users occasionally felt time pressure, which may partly be explained by waiting times while the chatbot generated process models. Performance ratings (MV = 3.80) show that users generally felt able to accomplish the tasks, albeit with some variation. These variations could be attributed to the fact that BPMNGen does not always generate perfectly layouted models, which required participants to adjust elements manually via drag-and-drop. Frustration received the highest rating (MV = 8.60, SD = 8.76), indicating that the user experiences differed strongly between participants and may be related to the temporal demand and performance aspects.

Answering RQ1: BPMNGen was perceived as usable, as reflected by a SUS score with grade B+, while the NASA-TLX score resulted in moderate cognitive load, indicating a solid yet improvable user experience.

6 Semantic Quality Study

After gaining insights into the usability of BPMNGen, we now consider the semantic quality of the process models this LLM-based chatbot generates. In particular, we aim to assess how these models compare to human-created ones.

6.1 Study Design

The study began with participants providing informed consent and completing a demographic questionnaire (cf. Table 5). Then they received an introduction to BPMN 2.0 to enable a common baseline by familiarizing them with its elements and syntactical rules, thereby ensuring that even those without prior experience could participate in the study without difficulty.

In the main phase, a process description and four corresponding BPMN 2.0 process models were presented to the study participants: one of the four models was created by a novice, one by an intermediate, one by an expert, and one by the LLM-based chatbot BPMNGen. Their task was to select the process model that best represents the given process description. This procedure was repeated for five distinct process scenarios. Figure 5 gives an overview of the study design.

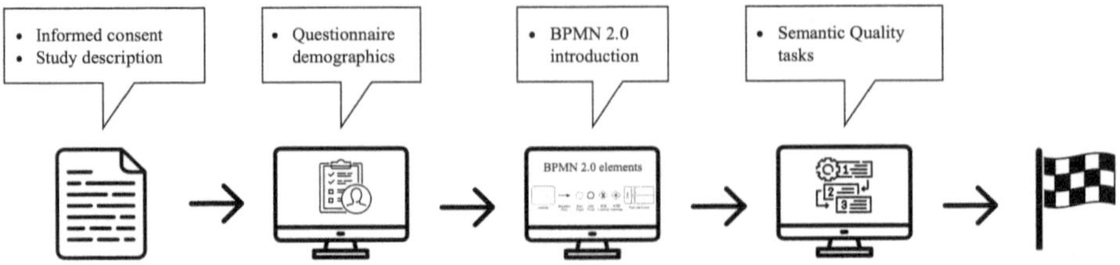

Fig. 5. Semantic Quality Study Design.

6.2 Study Participants

The study participants were primarily recruited through university courses and the informatics institutes of our university. In total, the study comprises 100 study participants. The sample comprised 27 females, 71 males, and 2 participants identifying as other, with an average age of 26.52 years (median = 25.50 years). Regarding their education level, 44 participants had not obtained a Bachelor's degree, whereas 56 had completed at least a Bachelor's degree. With respect to BPMN 2.0, 61 participants reported prior experience, while 39 had none. Table 5 summarizes their demographic details.

Table 5. Demographic Details of the Study Participants

Variable		Participants ($N = 100$)
Gender	female	27
	male	71
	other	2
Age	$<$ 25 years	40
	$\geq$ 25 years	60
Education Level	$<$ Bachelor Degree	44
	Bachelor Degree	32
	Master Degree	22
	Doctor of Philosophy (PhD)	2
Experience	no prior experience	39
	prior experience	61

6.3 Study Materials

The study material consisted of five process descriptions, each linked to four corresponding BPMN 2.0 process models. Three of the process models were created by human modelers with varying levels of expertise – novice, intermediate, and expert. The expertise levels were self-defined by the participants and further validated through a short prestudy test. The fourth process model was automatically generated by BPMNGen using a zero-shot prompt, entering the process description. All human-created models were created with SAP Signavio [29]. Note that the BPMNGen models were recreated in SAP Signavio to match the visual style (e.g., colors, layout) of the human-created models, ensuring participants could not distinguish them based on appearance.

The five BPMN 2.0 process models covered the following scenarios: online shopping (P1), baking a pizza (P2), borrowing a book (P3), making coffee (P4), and requesting vacation (P5). The five process scenarios were based on real-world situations to ensure that participants could easily relate to the content. Furthermore, the scenarios increased in complexity from P1 to P5. All materials used for this study are available online[3].

6.4 Study Measurements

For the study measurement, we apply the concept of semantic quality as defined by the SEQUAL Framework [19], according to which semantic quality refers to the extent to which a process model represents the intended meaning of the underlying natural language process description. In our study, semantic quality is operationalized through the *selection frequency*, i.e., how often participants chose a given process model as the semantically best representation of the described process. Figure 6 illustrates the research model of the study.

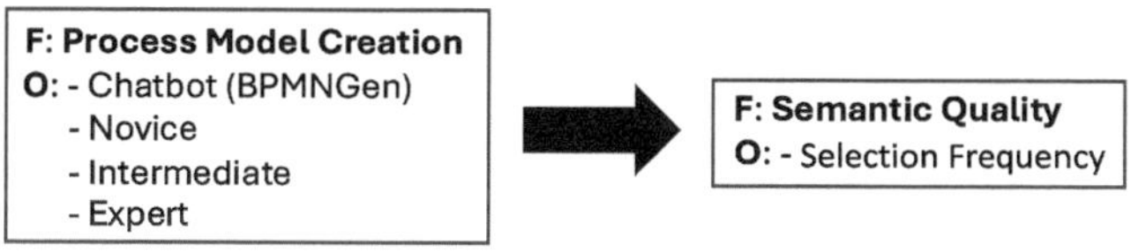

Fig. 6. Research Model (F = theoretical factor, O = operationalization of factor).

6.5 Study Results

Descriptive Statistics. Table 6 first summarizes the descriptive statistics for the five BPMN 2.0 process scenarios (P1 to P5), reporting the selection frequencies of the automatically generated BPMNGen models compared to the models manually created by human modelers.

For P1, the BPMNGen model was the second most frequently selected (24%), behind the novice model (40%). For P2, it ranked second (27%), slightly below the intermediate model (28%). For P3, the BPMNGen model achieved the highest selection rate (29%). For P4, again, it was the most frequently chosen (32%).

[3] https://anonymous.4open.science/r/LLM-2025-Study-Materials-8FB0.

Finally, for P5, the BPMNGen model ranked second (26%), just below the expert model (29%).

Inferential Statistics. Beyond the descriptive frequencies, Table 6 also presents the inferential statistics, showing the results of independent-samples t-tests that compare BPMNGen-generated models with those manually created by a novice, an intermediate, and an expert. The table reports the selection frequencies and the associated two-tailed p-values for each of the five process scenarios (P1–P5), based on independent-samples t-tests.

In P1, BPMNGen was selected 24 times, while the novice model reached 40 selections; this difference was close to statistical significance ($p = 0.060$). In P2, BPMNGen (27%) was almost identical to the intermediate model (28%, $p = 1.000$) and close to the expert model (25%, $p = 0.890$). In P3 (29%) and P4 (32%), BPMNGen achieved the highest selection frequencies, but no significant differences were observed ($p = 0.193$–0.894). In P5, BPMNGen (26%) was slightly below the expert (29%, $p = 0.788$) and intermediate models (28%, $p = 0.892$).

Overall none of the comparisons revealed statistically significant differences between the BPMNGen-generated models and the human-created models.

Table 6. Selection Frequencies and T-Test Results for the Process Models

Process	BPMNGen	Novice		Intermediate		Expert	
	N	N	p	N	p	N	p
P1	24	40	0.060	17	0.349	19	0.542
P2	27	20	0.382	28	1.000	25	0.890
P3	29	27	0.894	25	0.683	19	0.193
P4	32	24	0.350	20	0.126	24	0.350
P5	26	17	0.222	28	0.892	29	0.788

6.6 Discussion of the Results

The interpretation of the results requires looking at how the process models captured the intended meaning of the process descriptions and how participants may have judged the adequacy of these representations.

In P1, the novice model was most often preferred, possibly because it added more detailed activities that gave the impression of completeness (40%). BPMN-Gen offered a more concise representation that still covered the essential steps and was the second most chosen (24%; $p = 0.060$).

In P2, both BPMNGen and the human expert created almost identical representations of the process (27% vs. 25%; $p = 0.890$). The intermediate model, however, included a more elaborate handling of resources, which may have been perceived as greater sophistication and thereby slightly increased its attractiveness (28%; $p = 1.000$).

In P3 and P4, the novice models were overly complex, whereas the intermediate, expert, and BPMNGen models were semantically similar with only minor structural differences. The clearer and more concise representations created by BPMNGen likely enhanced readability, which may explain why it achieved the highest selection frequencies (P3: 29% vs. expert 19%, $p = 0.193$; P4: 32% vs. experts and novices 24%, $p = 0.350$).

In P5, the expert model was chosen slightly more often than BPMNGen (29% vs. 26%; $p = 0.788$), likely because it included a more detailed activity labeling. BPMNGen still offered a comparable and balanced representation, while the intermediate and novice models appeared overloaded with messages and unnecessary events.

Answering RQ2: BPMNGen-generated models achieve a semantic quality comparable to those created by novices, intermediates, and experts. While not significantly superior, they were also not inferior, demonstrating that BPMNGen can perform on the same semantic level as human modelers.

7 Limitations and Threats to Validity

The findings of this paper provide first insights into the usability and semantic quality of BPMNGen, but several limitations and threats to validity should be considered when interpreting the results. The usability study involved only five participants – a sample size generally considered sufficient to uncover most usability issues in formative evaluations [2], although it limits generalizability and highlights the need for replication with larger, more diverse groups.

In contrast, the semantic quality study involved 100 participants, providing a more robust empirical basis. However, it was limited to relatively simple process scenarios, which were deliberately chosen to ensure comprehensibility and to enable BPMNGen to function effectively. The results should therefore be interpreted with caution, as the models were simpler than typical real-world business processes, and LLMs may perform well in such cases without maintaining the same effectiveness in more complex settings.

Moreover, process models that were designed in an unnecessarily complex way – even though a simpler representation would have sufficed – may have been perceived as more semantically accurate. This threat to validity arises from participants potentially associating complexity with expertise or correctness.

Another threat concerns the measurement: perceived usability and cognitive load were assessed through self-report instruments (SUS and NASA-TLX), while semantic quality was operationalized via process model selection frequency within the SEQUAL framework. Although both approaches are well established, they may not fully capture fine-grained aspects of interaction quality or process model accuracy.

8 Conclusion and Future Work

This study investigated both the usability and the semantic quality of BPMN 2.0 process models automatically generated by an LLM-based chatbot. The results demonstrate that BPMNGen can be effectively used by end-users and that its generated process models achieve a semantic quality comparable to those created by human modelers. Together, these findings provide first evidence that LLM-based chatbots can make process modeling more accessible and time efficient. For organizations, this represents a promising step toward simplifying and accelerating the documentation and analysis of business processes.

Although these results are encouraging, several directions for future work remain. An important step will be to address the limitations of the studies by expanding evaluations to more complex and real-world business process scenarios and by involving larger and more heterogeneous user groups.

In addition, future research should extend the analysis of LLM-generated process models beyond semantic quality to include other quality dimensions such as syntactic correctness and pragmatic comprehensibility.

Another promising direction is the further development of BPMNGen itself, focusing not only on improving the generated process models, but also on supporting the overall process of modeling. This requires a closer examination of how users interact with the chatbot and how usability can be further strengthened.

Finally, prompting strategies could be systematically investigated to better understand their impact on both process model quality and user experience. By pursuing these directions, future research can contribute to building more robust, reliable, and user-centered LLM-based process modeling chatbots.

Disclosure of Interests. The authors have no competing interests to declare that are relevant to the content of this article.

References

1. Adamopoulou, E., Moussiades, L.: Chatbots: history, technology, and applications. Mach. Learn. Appl. **2**, 100006 (2020)
2. Alroobaea, R., Mayhew, P.J.: How many participants are really enough for usability studies? In: 2014 Science and Information Conference, pp. 48–56 (2014)
3. Bernardi, M.L., Casciani, A., Cimitile, M., Marrella, A.: Conversing with business process-aware large language models: the BPLLM framework. J. Intell. Inf. Syst. **62**(6), 1607–1629 (2024)
4. Brooke, J., et al.: SUS-a quick and dirty usability scale. Usability Eval. Ind. **189**(194), 4–7 (1996)
5. Brown, T., et al.: Language models are few-shot learners. Adv. Neural. Inf. Process. Syst. **33**, 1877–1901 (2020)
6. Caldarini, G., Jaf, S., McGarry, K.: A literature survey of recent advances in chatbots. Information **13**(1), 41 (2022)
7. ChatGPT (2025). https://openai.com/index/gpt-4/. Accessed 18 July 2025

8. Devlin, J., Chang, M.W., Lee, K., Toutanova, K.: Bert: pre-training of deep bidirectional transformers for language understanding. In: Proceedings of the 2019 Conference of the North American Chapter of the Association for Computational Linguistics: Human Language Technologies, vol. 1, pp. 4171–4186 (2019)
9. Dumas, M., Rosa, L.M., Mendling, J., Reijers, A.H.: Fundamentals of Business Process Management, Springer, Cham (2018)
10. Fettke, P., Houy, C.: Evaluating the process modeling abilities of large language models-preliminary foundations and results (2025). arXiv:2503.13520
11. Fill, H.G., Fettke, P., Köpke, J.: Conceptual modeling and large language models: impressions from first experiments with chatgpt. Enterp. Model. Inf. Syst. Archit. (EMISAJ) **18**, 1–15 (2023)
12. Gemini (2025). https://gemini.google.com/?hl=de. Accessed 10 July 2025
13. Grohs, M., Abb, L., Elsayed, N., Rehse, J.R.: Large language models can accomplish business process management tasks. In: International Conference on Business Process Management, pp. 453–465. Springer, Cham (2023)
14. Hart, S.G., Staveland, L.E.: Development of NASA-TLX: results of empirical and theoretical research. Adv. Psych. **52**, 139–183 (1988)
15. Klievtsova, N., Benzin, J.V., Kampik, T., Mangler, J., Rinderle-Ma, S.: Conversational process modelling: state of the art, applications, and implications in practice. In: International Conference on Business Process Management, pp. 319–336. Springer, Cham (2023)
16. Klievtsova, N., Mangler, J., Kampik, T., Benzin, J.V., Rinderle-Ma, S.: How can generative AI empower domain experts in creating process models? In: Wirtschaftsinformatik (2024)
17. Köpke, J., Safan, A.: Efficient LLM-based conversational process modeling. In: International Conference on Business Process Management, pp. 259–270. Springer, Cham (2024)
18. Kourani, H., Berti, A., Schuster, D., Van der Aalst, W.M.: Promoai: process modeling with generative AI. arXiv:2403.04327 (2024)
19. Krogstie, J.: Quality of Business Process Models, pp. 53–102. Springer, Cham (2016)
20. Lewis, J.R., Sauro, J.: The factor structure of the system usability scale. In: International Conference on Human Centered Design, pp. 94–103. Springer, Cham (2009)
21. Mendling, J., Strembeck, M., Recker, J.: Factors of process model comprehension–findings from a series of experiments. Decis. Support Syst. **53**(1), 195–206 (2012)
22. Mistral AI (2025). https://mistral.ai/en. Accessed 10 July 2025
23. Monti, F., Leotta, F., Mangler, J., Mecella, M., Rinderle-Ma, S.: NL2ProcessOps: towards LLM-guided code generation for process execution. In: International Conference on Business Process Management, pp. 127–143. Springer, Cham (2024)
24. Neuberger, J., Ackermann, L., van der Aa, H., Jablonski, S.: A universal prompting strategy for extracting process model information from natural language text using large language models. In: International Conference on Conceptual Modeling. pp. 38–55. Springer, Cham (2024)
25. Nour Eldin, A., Assy, N., Anesini, O., Dalmas, B., Gaaloul, W.: Nala2bpmn: automating BPMN model generation with large language models. In: International Conference on Cooperative Information Systems, pp. 398–404. Springer, Cham (2024)
26. Object Management Group: Bpmn 2.0: Business process model and notation (2011). https://www.omg.org/spec/BPMN/2.0/. Accessed 20 July 2025
27. Pourpanah, F., et al.: A review of generalized zero-shot learning methods. IEEE Trans. Pattern Anal. Mach. Intell. **45**(4), 4051–4070 (2022)

28. Reinhartz-Berger, I., Ali, S.J., Bork, D.: Leveraging LLMs for domain modeling: The impact of granularity and strategy on quality. In: International Conference on Advanced Information Systems Engineering, pp. 3–19. Springer, Cham (2025)
29. Signavio (2025). https://www.signavio.com/de/. Accessed 10 July 2025
30. Vaswani, A., Shazeer, N., Parmar, N., Uszkoreit, J., Jones, L., Gomez, A.N., et al.: Attention is all you need. Adv. Neural. Inf. Process. Syst. **30**(1), 261–272 (2017)
31. Vidgof, M., Bachhofner, S., Mendling, J.: Large language models for business process management: opportunities and challenges. In: International Conference on Business Process Management, pp. 107–123. Springer, Cham (2023)
32. Wang, Y., Yao, Q., Kwok, J.T., Ni, L.M.: Generalizing from a few examples: a survey on few-shot learning. ACM Comput. Surv. (CSUR) **53**(3), 1–34 (2020)
33. Weske, M.: Business Process Management Architectures, Springer, Heidelberg (2007)
34. Ziche, C., Apruzzese, G.: LLM4PM: a case study on using large language models for process modeling in enterprise organizations. In: International Conference on Business Process Management, pp. 472–483. Springer, Cham (2024)
35. Zimoch, M., Pryss, R., Probst, T., Schlee, W., Reichert, M.: Cognitive insights into business process model comprehension: preliminary results for experienced and inexperienced individuals. In: International Workshop on Business Process Modeling, Development and Support, pp. 137–152. Springer, Cham (2017)

QOC DAO - An AI Driven Decentralized Autonomous Organization

Marc Jansen(✉) and Christophe Verdot

University of Applied Sciences Ruhr West, BlockAI Ltd., Mülheim, Germany
marc.jansen@hs-ruhrwest.de, christophe@blockai.dev

Abstract. This paper introduces a structured approach to improving decision-making in Decentralized Autonomous Organizations (DAOs) through the integration of the Question–Option–Criteria (QOC) model and AI agents. We outline a stepwise governance framework that evolves from human-led evaluations to fully autonomous, AI-driven processes. By decomposing decisions into weighted, criterion-based evaluations, the QOC model enhances transparency, fairness, and explainability in DAO voting. We demonstrate how large language models (LLMs) and stakeholder-aligned AI agents can support or automate evaluations, while statistical safeguards help detect manipulation. The proposed framework lays the foundation for scalable and trustworthy governance in the Web3 ecosystem.

1 Introduction

Decentralized Autonomous Organizations (DAOs) have become a cornerstone of Web3 governance, offering blockchain-native communities a way to make collective decisions without centralized control. Ideally, DAOs operate transparently and inclusively, yet in practice, they often fall short. Persistent issues such as voting power imbalances, low participation, lack of rationale transparency, and short-term incentives continue to undermine legitimacy and strategic coherence.

At the same time, recent advances in large language models (LLMs) and AI agents create new opportunities to address these challenges. Capable of reasoning over natural language, summarizing complex content, and generating structured evaluations, AI systems can assist or even automate aspects of DAO governance. However, realizing this potential requires a formal, transparent decision-making framework that can bridge human and machine reasoning.

This paper introduces a structured approach based on the Question–Option–Criteria (QOC) model, originally developed in design rationale research. The QOC framework decomposes decisions into a central question, a set of options, and evaluation criteria that reflect community priorities. By assigning weights to criteria and collecting structured evaluations, QOC enables both quantitative decision aggregation and transparent insight into the rationale behind collective choices.

H. R. Arabnia et al. (Eds.): CSCI 2025, CCIS 2972, pp. 343–358, 2027.
https://doi.org/10.1007/978-3-032-28469-3_23

This paper outlines a stepwise governance architecture in which the QOC model evolves from human-driven evaluations to fully autonomous AI decision-making. The initial phase involves manual scoring of proposals by DAO members using a shared set of criteria and weights. In the second phase, AI agents simulate stakeholder perspectives to provide recommendations, while humans retain final authority, a human-in-the-loop model. In the final phase, autonomous agents perform the full decision process, from evaluation to resolution, with embedded safeguards and traceable reasoning.

Throughout this progression, the QOC model acts as the foundation for structured deliberation, fairness, and explainability. It supports the integration of AI while mitigating known governance risks such as voter apathy, collusion, and opaque outcomes. Statistical mechanisms like outlier detection further enhance robustness by identifying potential manipulation.

By embedding structured, explainable AI into the heart of DAO governance, we aim to move beyond simplistic yes/no voting toward more transparent, inclusive, and intelligent decision-making.

2 State of the Art

DAO governance is central to the Web3 vision of decentralized, transparent, and community-led decision-making. Yet in practice, DAOs face challenges such as power imbalances, low participation, and limited transparency. This section reviews key issues and outlines emerging solutions, including structured decision frameworks and AI-assisted governance.

2.1 Structural and Procedural Challenges

While DAOs are designed to be democratic, current implementations often fall short. One critical issue is the concentration of voting power,the so-called whale problem, where a few large token holders dominate governance. In many DAOs, fewer than ten wallets control over half the votes [2,3], creating a misalignment between large and small stakeholders [1,5].

A second issue is low voter turnout: empirical studies report participation rates as low as 20–30% for important decisions [1,2]. Factors include proposal complexity, insufficient voter knowledge, and governance fatigue.

A third concern involves misaligned incentives, particularly between long-term contributors and short-term speculators. While long-term sustainability may require patient governance, speculative holders often abstain from such votes [1], weakening the coherence and resilience of DAO governance.

In addition, many DAOs rely on off-chain voting platforms such as Snapshot, which, while efficient, lack full transparency and auditability. This introduces a trade-off between usability and trust. Studies show that off-chain voting can negatively affect valuation [4], whereas on-chain voting, despite higher costs—better aligns with blockchain principles of transparency and immutability [4].

2.2 Lack of Rationale in Binary Voting

Most DAO decisions are reduced to simple binary votes ("yes" or "no") on proposals. While the outcomes are recorded, the reasoning behind them remains opaque. The community typically sees only the final tallies, with no insight into why a proposal passed or failed [5]. This lack of rationale makes governance feel arbitrary and reduces accountability.

Without structured feedback, rejected proposals yield no actionable insight. Was it the budget, the execution plan, or the project's alignment with DAO goals? Without knowing, proposers cannot improve.

2.3 Structured Decision Frameworks

To address these transparency gaps, recent research proposes structured frameworks such as the Question–Option–Criteria (QOC) model [5]. Rather than casting a binary vote, participants evaluate how well different options satisfy predefined criteria (e.g., ROI, security, mission alignment). These evaluations are aggregated, often with weights, to form a reasoned and auditable decision.

QOC creates a decision rationale alongside the decision outcome, enabling proposals to be improved iteratively. This approach draws on ideas from Multi-Criteria Decision Making (MCDM), commonly used in engineering and management [6]. Applying such techniques in DAOs can shift governance from popularity contests toward evidence-based consensus [2].

In parallel, other governance innovations, such as quadratic voting, conviction voting, and vote-escrow mechanisms—seek to reduce whale influence or align voting with long-term commitment [2,7]. While these tackle vote weighting, QOC focuses on improving decision reasoning and transparency.

2.4 AI-Augmented Governance

Emerging work explores the integration of AI agents to support or automate DAO decision-making. One promising direction is the use of large language models (LLMs) to summarize proposals, evaluate them against community criteria, and generate structured recommendations [1].

A recent case study [1] showed that an AI-assisted governance system, using chain-of-thought reasoning, achieved 97% alignment with historical DAO decisions and could increase voter turnout by 40%. These systems make governance more accessible by reducing cognitive burden and enhancing transparency.

The current trend is toward a human-in-the-loop model, where AI provides input but humans retain final authority [1]. Over time, as AI earns trust, DAOs may transition toward partial or full automation. However, researchers caution that full autonomy requires robust safeguards to ensure alignment between AI behavior and community values [5]. This shift also intersects with broader ethical challenges in AI governance, including transparency, value alignment, and algorithmic accountability, as emphasized in the literature on responsible AI [14,15]. Such considerations are especially important when delegating evaluative power to non-human agents in decentralized systems.

3 The Question-Option-Criteria (QOC) Model for DAO's

This section presents the QOC model as a structured framework for transparent decision-making. It outlines the core components and evaluations and shows how the model can be adapted to DAOs through multi-user input, and fairness mechanisms.

3.1 The QOC Model

The QOC (Question–Option–Criteria) [8] framework is a structured decision-making method designed to facilitate transparent, reasoned evaluation processes. It breaks down complex decisions into three core elements, Questions, Options, and Criteria, and incorporates quantitative mechanisms for comparative analysis.

Q - The Question. At the heart of the framework lies the Question (Q), which defines the central decision-making problem. This question should be:

- Well-formulated: Clearly articulated, concise, and unambiguous.
- Actionable: Framed in a way that it can be answered through a selection from a set of defined options.
- Decisive: Typically phrased to invite a single best answer (e.g., "Should we fund project X?" or "Which algorithm is most suitable for our use case?").

O - The Options. The Options (O) represent the possible answers or solutions to the question posed. Formally, we define a set of options:

$$O = (o_0, o_1, ..., o_n)$$

Each option o_k (with $k \in \{0, \ldots, n\}$) is a discrete and independent alternatives, mutually exclusive and meant to address the decision question. In many practical use cases, such as the in this paper discussed DAO votes, the options might simply be binary (e.g., Yes or No), but the framework in general supports arbitrary cardinality.

C - The Criteria. The Criteria C are the standards or dimensions along which each option will be evaluated. Formally:

$$C = (c_0, c_1, ..., c_m)$$

Each criterion c_j (with $j \in 0, \ldots, m$) reflects an aspect of quality, performance, or desirability that matters for the decision. Examples include cost, feasibility, security, user acceptance, or expected return on investment.

W - The Weights of Criteria. Each criterion c_j is assigned a weight $w_j in(0, 100]$ indicating its relative importance in the decision process. These weights can be:

- Normalized to sum up to 100, representing percentage influence.
- Ranked independently if strict normalization is not required.

The vector of weights is denoted:

$$W = (w_0, w_1, ..., w_m)$$

Higher weights reflect greater importance in the final evaluation.

E - Evaluation Function. The evaluation function $e(k, j)$ maps each Option–Criterion pair to a support score in the range $[0, 100]$. Formally:

$$e : O \times C \rightarrow [0, 100] \subset \mathbb{N}$$

Each value $e(k, j)$ represents how well the option o_k satisfies or supports the criterion c_j:

- A value of 0 indicates no support at all.
- A value of 100 indicates maximal support or alignment.
- Intermediate values reflect partial or moderate support.

These scores can be generated by:

- Human judgment (manual scoring by experts or stakeholders).
- Automated analysis (e.g., via AI or data-driven evaluation).
- Hybrid methods (e.g., AI-generated suggestions reviewed by humans).

Final Scoring and Decision. To determine the most suitable option, the overall score $S(o_k)$ for each option o_k is computed as a weighted sum of evaluation scores:

$$S(o_k) = \sum_{j=0}^{m} w_j \cdot e(k, j)$$

The option with the highest overall score is considered the most appropriate response to the original question Q, assuming all weights and evaluations are valid and trustworthy.

3.2 A Potential Adaptation of the QOC Model for DAO's

The following subsections describe potential modifications to the model, to make it usable in the context of DAO's.

Extend the QOC Model to a Multi User Based Approach. While the QOC model, as described in the previous section, already provides a structured and transparent approach to decision-making by collecting options, criteria, weights, and evaluations from a single user, its full potential unfolds when extended to a multi-user setting. In this enhanced variant, multiple stakeholders are involved in the process, each contributing their own perspectives by suggesting additional options and criteria relevant to the decision question. After a consolidation phase, in which redundancies and overlaps are resolved and the collective set of options and criteria is finalized, the same group of stakeholders proceeds to assign weights to the criteria and evaluate how well each option satisfies each criterion. This collaborative approach increases the richness and representativeness of the input data, fosters participatory decision-making, and ensures that the resulting evaluations reflect a more diverse and balanced perspective on the question at hand.
Let $U = (u_0, u_1, ..., u_r)$ be the set of users or stakeholders participating in the decision process. We could then adapt the above presented formulas as following. Assuming that each user u_i (with i in $1, ..., r$) contributes their own individual set of options:

$$O_i = (o_{i,0}, o_{i,1}, ..., o_{i,n})$$

A consolidated set of options can be constructed by:

$$O = \bigcup\nolimits_{i=1}^{r} O_i = \{o_0, o_1, \ldots, o_n\}$$

An analogous approach is applied to the criteria contributed by the individual users:

$$C_i = (c_{i,0}, c_{i,1}, ..., c_{i,m})$$

and the consolidated set of criteria, described as:

$$C = \bigcup\nolimits_{i=1}^{r} C_i = \{c_0, c_1, \ldots, c_m\}$$

Additionally, each user u_i provides a weight vector:

$$W_i = \{w_{i,0}, w_{i,1}, \ldots, w_{i,m}\}$$

which can be aggregated using the following formula:

$$\bar{w}_j = \frac{1}{r} \sum\nolimits_{i=1}^{r} w_{i,j}$$

An evaluation score is also assigned by each user u_i:

$$e_i(k, j)$$

The aggregated evaluation over all users is computed as follows:

$$\bar{e}(k, j) = \frac{1}{r} \sum\nolimits_{i=1}^{r} e_i(k, j)$$

The final score for each option o_k is computed as follows:

$$S(o_k) = \sum\nolimits_{j=0}^{m} \bar{w}_j \cdot \bar{e}(k, j)$$

Adapt the QOC Model to the Typical Situation in DAO's. In most Decentralized Autonomous Organizations (DAOs), decision-making revolves around straightforward, binary questions, typically of the form: "Should the DAO support initiative X?" This recurring structure provides an opportunity to streamline the QOC model by adapting it to a fixed-question, fixed-option scenario.

Instead of defining a new question and a new set of options for each vote, we can assume a standardized question template, such as "Should the DAO approve or support proposal X?" Correspondingly, the set of options is reduced to just two: Yes and No. This simplification significantly reduces the complexity of each voting process, while still enabling meaningful evaluations.

To further decrease decision-making overhead and to align all evaluations with the overarching goals of the DAO, it is also possible to predefine a fixed set of evaluation criteria. These criteria, such as return on investment, technical feasibility, alignment with the DAO's mission, or risk exposure, can be determined once and then reused for all future decisions. If desired, the corresponding weights that reflect the relative importance of each criterion can also be defined upfront, either by consensus or based on the DAO's governance rules.

With both the question, the options, and the weighted criteria fixed, the only remaining task for participants in each vote is to evaluate how well each of the two options (Yes or No) satisfies each of the predefined criteria in the context of the specific proposal. In other words, the focus shifts entirely to filling out the evaluation matrix, assigning scores that express how strongly each option supports each criterion.

This adaptation allows DAOs to leverage the structured, multi-criteria strength of the QOC model, while maintaining a user-friendly and scalable voting process. It ensures that decisions are not only democratic but also systematically aligned with long-term priorities, without requiring stakeholders to redefine the evaluation framework for every new proposal.

Increase Transparency and Fairness in DAOs. While the previous sections have demonstrated how the QOC model can be structurally adapted to the context of DAOs, its true value lies in what it enables: significantly enhanced transparency and fairness in decentralized governance. Unlike traditional DAO voting mechanisms, which typically yield only binary outcomes ("yes" or "no"), a QOC-based process records granular evaluations for each predefined criterion. This creates the opportunity to generate a detailed, criterion-level decision report after each vote, offering the community actionable insight rather than a black-box verdict.

For instance (as shown in Fig. 1), a proposal may be evaluated positively with respect to "ecosystem benefit" and "technical feasibility," while receiving low scores on "financial sustainability". If the proposal is rejected, the proposers can use this multidimensional feedback to iteratively improve the proposal in alignment with community concerns before resubmitting it.

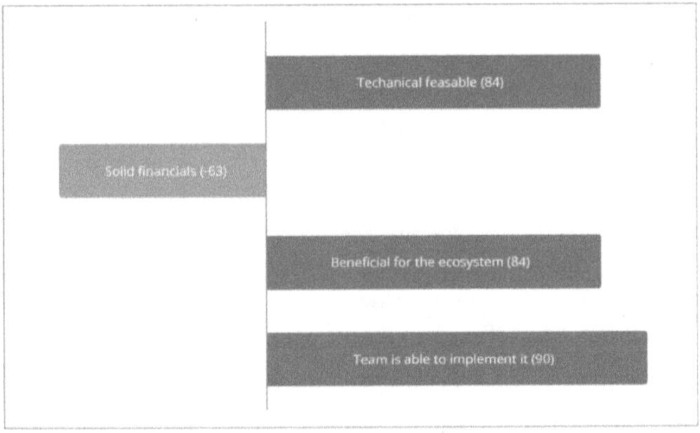

Fig. 1. Example of criterion-wise evaluation in a DAO vote.

Beyond transparency, the QOC model also introduces mechanisms to improve fairness. In conventional token-weighted voting systems, outcomes can be disproportionately influenced by large stakeholders or coordinated voting blocs. While QOC-based voting is not immune to strategic behavior, such as participants deliberately misrepresenting scores across criteria, it provides a structured data layer that allows for the detection of anomalous behavior. By analyzing deviations between individual evaluations and aggregated community norms, the DAO can implement statistical safeguards against manipulation.

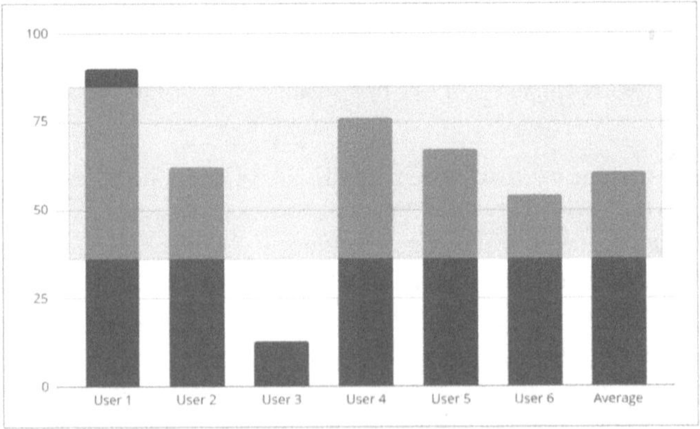

Fig. 2. Example of outlier detection.

One effective approach is outlier detection (see Fig. 2, where individual criterion evaluations are compared to the community average using standard deviation thresholds. Evaluations that fall significantly outside the typical range can be flagged as outliers. The DAO may then choose to discount such outlier evaluations in the final aggregation process, thereby reducing the impact of extreme or intentionally biased input.

By combining transparent rationales for decisions with statistical protections against manipulation, this enhanced QOC-based voting model reshapes the nature of DAO governance. It moves the decision-making process away from simple vote tallies and toward a system rooted in community reasoning and defensible, criterion-based assessment. As a result, governance outcomes become more interpretable, more resistant to manipulation, and ultimately, more legitimate in the eyes of the participating community.

4 Towards Fully Autonomous DAO Governance

To translate the conceptual QOC model into practical DAO governance, we propose a stepwise implementation path toward increasing levels of AI integration. This approach begins with a fully human-driven process, progresses to a hybrid human–AI model, and ultimately culminates in a fully autonomous, AI-driven governance system. The following subsections describe these stages in detail, illustrating how transparency, scalability, and fairness evolve at each step.

4.1 Step 1: DAOs Based on The QOC Idea

In the initial implementation phase, a DAO is deployed that adheres to the QOC-based governance model described above. In this first version, the set of evaluation criteria is predefined and globally fixed, with associated weights reflecting the strategic goals and values of the respective DAO. These criteria are intended to ensure consistent alignment across proposals and simplify the evaluation process for participants.

During each voting cycle, stakeholders are invited to assess how well the submitted project or initiative satisfies each of the established criteria. These assessments are expressed as numerical scores, and each participant's input is weighted according to their respective voting power (e.g., based on token holdings or reputation scores). The resulting scores form the basis for computing the overall evaluation of each proposal.

As depicted in Fig. 3, the voting interface and process guide participants through the structured evaluation of the proposal along all relevant dimensions.

Before the final decision is computed, the system applies statistical outlier detection techniques to identify potentially biased or manipulative evaluations. Votes exhibiting extreme deviations from community norms (e.g., based on standard deviation thresholds) may be excluded from the aggregation process to preserve fairness and decision quality.

Finally, a comprehensive decision report is automatically generated. This report includes a breakdown of the aggregated evaluation scores per criterion and highlights both the strengths and weaknesses as perceived by the community. For proposals that are rejected, this feedback loop enables proposers to better

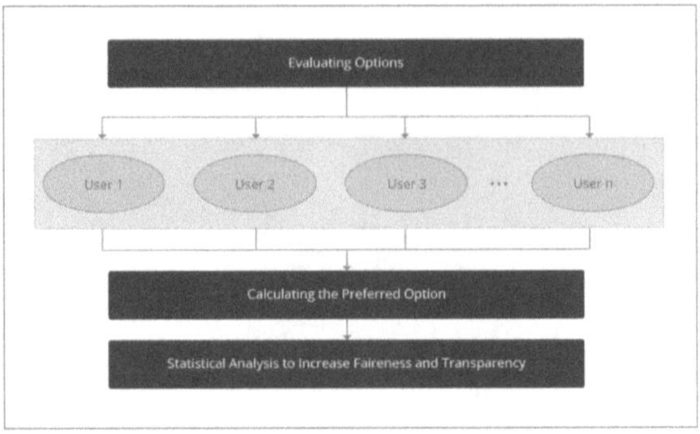

Fig. 3. Step 1: Voting process with the QOC DAO approach.

understand the rationale behind the decision and supports iterative refinement of future submissions.

4.2 Step 2: Integrating AI with a Human-in-the-Loop Approach

In the second step of the development of the QOC-based DAO governance model, the evaluation process, previously conducted manually by human stakeholders, is increasingly delegated to AI agents, enabling a semi-automated, scalable, and consistent assessment pipeline. The objective of this phase is to support decision-making through machine-driven evaluations while maintaining human oversight for the final judgment.

To begin, the system identifies the relevant stakeholder groups associated with the proposal under consideration. This may include community members, domain experts, affected users, or other interest groups. For each identified stakeholder group, one or more AI agents are instantiated. These agents are configured using structured descriptions of the stakeholder perspectives, concerns, and value orientations, either manually curated or learned from prior proposals and community interactions.

Once initialized, the AI agents autonomously evaluate how well the proposed initiative satisfies each of the predefined, weighted criteria. These evaluations mirror the structure of human voting in Step 1, but are now performed algorithmically based on available documentation, previous proposals, on-chain data, or external knowledge bases. The complete agent-based evaluation process is depicted in Fig. 4.

Following the agent evaluations, the system aggregates the individual scores according to the predefined weighting schema. However, the final decision is not fully automated at this stage. Instead, the aggregated recommendation produced by the AI agents is presented to the human members of the DAO. Based on this recommendation, the human participants make the ultimate decision, either approving or rejecting the proposal. This human-in-the-loop mechanism ensures that while AI provides structured guidance and efficiency, the authority and accountability of decision-making remain with the community.

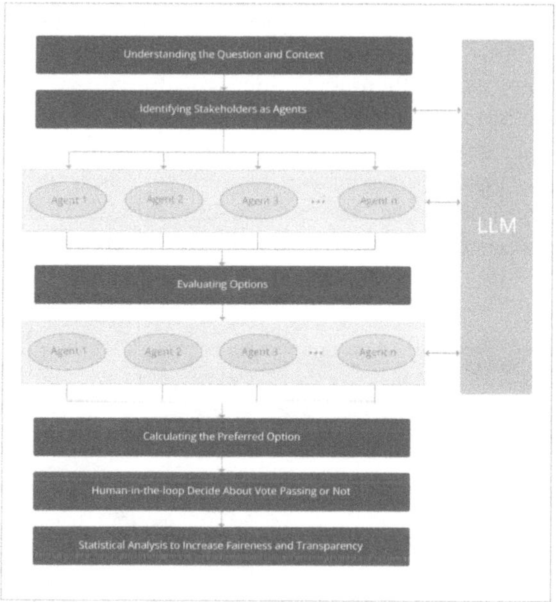

Fig. 4. Step 2: Voting process with AI agents and human-in-the-loop.

As in the first step, the system generates a comprehensive decision report. This report includes the criterion-level evaluations as derived by the AI agents, along with the final decision outcome. Such transparency not only strengthens trust in the AI-assisted process but also allows proposers to understand how their initiative was perceived algorithmically and how it aligns with the DAO's broader goals.

This hybrid approach, combining automated analysis with human judgment, lays the foundation for future iterations of fully autonomous governance, while still preserving community involvement during critical stages of the decision-making process.

4.3 Step 3: A Fully AI Driven DAO

The third and final phase of the QOC-based DAO governance model represents the transition to a fully autonomous decision-making system, in which all aspects of the evaluation and resolution process are executed by AI agents, without direct human intervention. This implementation builds upon the architecture described in Step 2, including stakeholder modeling, agent instantiation, and criteria-based evaluation, but introduces a key conceptual shift: the removal of the human-in-the-loop mechanism from the decision pipeline.

As in the previous step, the process begins by identifying relevant stakeholder groups associated with the submitted proposal. Based on these stakeholder profiles, a set of corresponding AI agents is generated. Each agent is designed to simulate the perspective and evaluative tendencies of its respective stakeholder group, drawing on both static definitions and dynamic context such as previous proposal interactions, DAO documentation, and domain-specific data sources.

The agents then autonomously evaluate how well the proposed initiative supports each of the predefined, weighted criteria, producing a complete matrix of quantitative scores. These scores are subsequently aggregated into a final utility value per option (typically "yes" or "no"), using the weighted QOC aggregation function. The decision-making process in this step is depicted in Fig. 5.

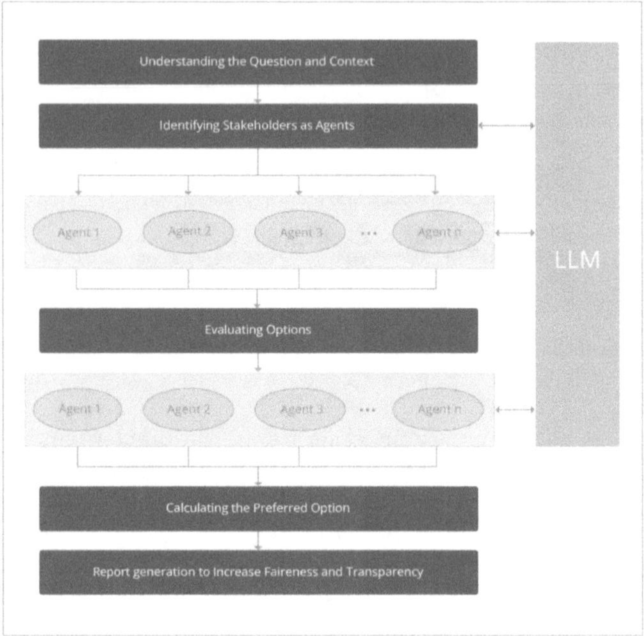

Fig. 5. Step 3: Fully AI-driven voting approach.

The key distinction in this phase lies in the absence of human oversight: the aggregated outcome derived from the AI agents' evaluations is directly accepted as the final decision of the DAO. There is no subsequent review, modification, or approval by human participants. This approach enables real-time, scalable, and unbiased governance, particularly valuable in contexts where rapid decision-making is essential or where human participation may be limited.

As in previous steps, the system generates a comprehensive decision report upon finalization of each vote. This report documents the aggregated evaluation scores for each criterion, provides transparency into the reasoning behind the AI-driven decision, and enables continuous feedback for applicants. In the case of proposal rejection, the report allows proposers to identify which aspects of their initiative were viewed unfavorably and to iteratively refine their application for potential resubmission.

By fully automating the governance workflow, this final implementation stage realizes the vision of self-governing, AI-enhanced DAOs, capable of making structured, explainable, and repeatable decisions at scale, without requiring human input at runtime.

5 Evaluation

To evaluate the proposed approach, real-world voting data were collected from an existing decentralized autonomous organization (DAO). Specifically, 102 historical voting decisions were retrieved from the PowerDAO via its public API. These decisions span a time period from December 27, 2023, to September 12, 2025, and include both accepted and rejected proposals.

To conduct the evaluation, the QOC-based decision model described above was implemented and applied to all 102 proposals using three different large language models (LLMs) from OpenAI: GPT-4-mini, GPT-5-mini, and GPT-5. Each proposal was processed independently by the respective LLM within the QOC decision framework.

Subsequently, the decisions generated by the QOC-based AI system were compared against the actual decisions made by the DAO. For each LLM, a 2×2 contingency table was constructed to visualize the agreement and disagreement between the AI-generated and DAO-recorded decisions (see Table 1).

Table 1. Contingency table results for three LLMs compared to DAO decisions

	GPT-4-mini		GPT-5-mini		GPT-5	
	DAO Y	DAO N	DAO Y	DAO N	DAO Y	DAO N
AI Y	56(54.9%)	20(19.6%)	32(31.4%)	8(7.8%)	24(23.5%)	3(2.9%)
AI N	14(13.7%)	12(11.8%)	38(37.3%)	24(23.5%)	46(45.1%)	29(28.4%)

The rows of the table indicate the AI's decisions (AI Y = accepted, AI N = rejected), while the columns represent the corresponding DAO decisions. As shown, the degree of agreement between the AI and the DAO varies across the models: GPT-4-mini achieves the highest agreement rate (66.7%), followed by GPT-5-mini (54.9%) and GPT-5 (51.9%). A notable trend can also be observed: as model quality increases, the AI becomes more conservative, more frequently rejecting proposals that were accepted by the DAO. Conversely, the number of false positives (proposals accepted by the AI but rejected by the DAO) decreases with higher-quality models.

To assess whether the differences in decisions between the DAO and the AI were statistically significant, McNemar tests [9] were performed. For GPT-4-mini, the test did not reveal a statistically significant difference between AI and DAO decisions, $\chi^2(1, N = 102) = 1.06, p = .303$. However, for GPT-5-mini and GPT-5, the McNemar test indicated statistically significant differences:

$\chi^2(1, N = 102) = 19.57, p = 9.72 * 10^-6$ and $\chi^2(1, N = 102) = 37.73, p = 8.11 * 10^-10$ respectively.

Although most decisions align between the AI and the DAO, the directionality of disagreement is asymmetrical: the AI tends to be more conservative, i.e., it more frequently rejects proposals that the DAO accepted than vice versa.

To further quantify the practical implications of such asymmetries, a cost function [10–13] was introduced to prioritize conservative decision-making:

$$c = 1 * d + 10 * a$$

where c denotes the total cost, d is the number of false negatives (proposals accepted by the DAO but rejected by the AI) and a is the number of false positives (proposals rejected by the DAO but accepted by the AI). This function reflects the assumption that false positives, i.e., approving a proposal the DAO would have rejected, are ten times more costly than false negatives. The resulting costs across the three LLMs are presented in Table 2: As expected, GPT-5, the most conservative model, yields the lowest total cost under the asymmetric penalty scheme. This suggests that despite a lower agreement rate with the DAO, its stricter approach may be preferable in contexts where false positives are particularly costly.

Table 2. Total cost of decisions of our approach across the three used LLMs using asymmetric cost weighting

Model	DAO N/AI Y	DAO Y/AI N	Total Cost
GPT-4-mini	20	14	214
GPT-5-mini	8	38	118
GPT-5	3	46	76

6 Conclusions, Limitations and Future Work

While the evaluation demonstrates promising results for automating DAO decisions using a QOC-based approach with large language models, several limitations must be acknowledged. First, the analysis was based solely on historical data from a single DAO (PowerDAO). As such, the generalizability of the findings to other DAOs or governance structures may be limited. Second, the evaluation relied on binary classification of proposals (approved vs. rejected), without accounting for nuances such as abstentions, amendments, or conditional approvals. Third, the cost function used to assess conservativeness was based on a fixed asymmetry ratio (10:1), which, while justifiable in this context, remains subjective and domain-dependent. Furthermore, the evaluation assumes static model behavior and does not account for potential temporal drift in LLM performance or governance dynamics.

In addition, although the use of AI for decision support in decentralized governance holds substantial potential for improving scalability, consistency, and efficiency, it also introduces critical questions related to trust, accountability, and control. Specifically, the delegation of evaluative power to non-human agents raises concerns about transparency, oversight, and the legitimacy of decisions made without direct human deliberation. These challenges must be addressed as part of the broader discourse on the role of artificial intelligence in decentralized systems.

Future research could explore the application of the proposed approach across multiple DAOs with varying voting cultures, diverse voting targets and proposal types. Incorporating more granular decision categories and integrating LLM uncertainty (e.g., via confidence thresholds or abstention mechanisms) may further improve alignment and trust. Moreover, future work could extend the cost model by calibrating weights empirically or dynamically based on proposal categories, risk assessments, or requested funding volumes. Finally, longitudinal studies involving real-time deployment of AI-supported governance agents would offer deeper insights into the social, ethical, and strategic implications of AI integration in decentralized decision-making processes. Additionally, the paper does not evaluate how real DAO participants perceive or accept AI-driven decision-making, leaving the social feasibility of such systems unaddressed.

References

1. Chen, J.-H., Hsu, C.-W., Tsai, Y.-C.: Intelligent decentralized governance: a case study of KlimaDAO decision-making. Electronics **14**(12), 2462 (2025). https://doi.org/10.3390/electronics14122462
2. Lustenberger, M., Küng, L., Spychiger, F.: Designing community governance–learnings from daos. J. Br. Blockchain Assoc. (2025)
3. Han, J., Lee, J., Li, T.: A review of DAO governance: recent literature and emerging trends. J. Corp. Finan. 102734 (2025)
4. Bellavitis, C., Momtaz, P.P.: Voting governance and value creation in decentralized autonomous organizations (DAOs). J. Bus. Ventur. Insights **23**, e00537 (2025)
5. Verdot, C., Jansen, M.: QOC, A new approach to make DAO's more transparent and fair (2025). https://blockai.medium.com/qoc-a-new-approach-to-make-daos-more-transparent-and-fair-24821e0c4e71. Accessed 10 Jan 2025
6. Baninemeh, E., Farshidi, S., Jansen, S.: A decision model for decentralized autonomous organization platform selection: three industry case studies. Blockchain: Res. Appl. **4**(2), 100127 (2023)
7. Tamai, S., Kasahara, S.: DAO voting mechanism resistant to whale and collusion problems. Front. Blockchain **7**, 1405516 (2024)
8. MacLean, A., Young, R., Bellotti, V., Moran, T.: Questions, options, and criteria: elements of design space analysis. Human-Comput. Interact. **6**, 201–250 (1991). https://doi.org/10.1080/07370024.1991.9667168
9. McNemar, Q.: Note on the sampling error of the difference between correlated proportions or percentages. Psychometrika **12**(2), 153–157 (1947)
10. Bach, F.R., Heckerman, D., Horvitz, E.: Considering cost asymmetry in learning classifiers. J. Mach. Learn. Res. **7**, 1713–1741 (2006)

11. Ling, C., Sheng, V.: Cost-sensitive learning and the class imbalance problem. Encycl. Mach. Learn. (2010)
12. Barnes, B.M., Henn, M.A.: Addressing misclassification costs in machine learning through asymmetric loss functions. In: Metrology, Inspection, and Process Control XXXVII, vol. 12496, pp. 352–364. SPIE (2023)
13. Koch, B., Imai, K.: Statistical decision theory with counterfactual loss. arXiv preprint arXiv:2505.08908 (2025)
14. Bostrom, N., Yudkowsky, E.: The ethics of artificial intelligence. In: Artificial Intelligence Safety and Security, pp. 57–69. Chapman and Hall/CRC (2018)
15. Gabriel, I.: Artificial intelligence, values, and alignment. Mind. Mach. **30**(3), 411–437 (2020)

An Intelligent Human–AI Collaboration Pipeline for Automated Extraction of PFAS Partition Coefficients

Yeonbi Han[1], Yuemei Ye[2], and Liang Zhao[1,2](✉)

[1] Stony Brook University, Stony Brook, NY, USA
yeonbi.han@stonybrook.edu
[2] The Graduate Center and Lehman College, City University of New York, New York, NY, USA
{yuemei.ye,liang.zhao1}@lehman.cuny.edu

Abstract. Large Language Models (LLMs) are broadly capable, but struggle with domain–specific quantitative extraction from scientific literature, especially when key values appear in figures. We present an end–to–end pipeline that specializes in general LLM for environmental chemistry to extract partition coefficients of per- and polyfluoroalkyl substances (PFAS) (K_d, K_{oc}, $\log K_{oc}$) from text, tables and figures. The system combines (i) PDF preprocessing that reconstructs complete high–resolution figure screenshots through object merging and padding, (ii) templated extraction with structured outputs for downstream use, and (iii) a verification stage that renders side–by–side HTML reports (original figure vs. AI rationale). Its core feature is an *intelligent human–AI collaboration* loop: when the model self–detects uncertainty (e.g., about units, axes, or legend abbreviations), it generates 1–2 targeted questions, incorporates brief expert hints, and then reextracts the values. On a corpus of PFAS papers (11 relevant + 2 controls), the pipeline maintains high reliability on text/tables and substantially improves figure interpretability with minimal expert input. Analysis of the model's *question logs* reveals recurring uncertainty patterns that motivate collaborative memory and question optimization, charting a practical path to turn general LLMs into trustworthy domain–specialized scientific tools.

Keywords: Per- and polyfluoroalkyl substances · Partition coefficients · Large Language Models · Automated data extraction · Human–AI collaboration · Uncertainty-aware prompting

1 Introduction

Per- and Polyfluoroalkyl Substances (PFAS), commonly referred to as "Forever Chemicals," are synthetic compounds that resist natural degradation and accumulate in soil, water, and even human bodies [1]. PFAS are globally distributed synthetic chemicals of emerging concern that pose serious environmental and

H. R. Arabnia et al. (Eds.): CSCI 2025, CCIS 2972, pp. 359–372, 2027.
https://doi.org/10.1007/978-3-032-28469-3_24

health risks due to their persistence, toxicity, and bioaccumulative nature. As such, they have emerged as one of the most pressing environmental challenges of modern society.

Accurate modeling of PFAS transport, fate, and toxicity in the environment critically depends on quantitative parameters such as partitioning coefficients. However, these key data are scattered across thousands of scientific publications and require labor-intensive manual extraction by researchers. This process has become a major bottleneck that slows scientific progress and hinders the development of reliable predictive models.

The recent advancement of Large Language Models (LLMs) offers new opportunities for automating scientific data extraction [2]. Nevertheless, general-purpose LLMs still struggle with domain-specific reasoning and quantitative interpretation [3]. They frequently misread numerical data embedded in tables or figures and, more problematically, may generate "hallucinated" values that do not exist in the original source—undermining both the reliability and reproducibility of AI-driven scientific analysis.

To address these limitations, this study introduces an *Intelligent Human–AI Collaboration* pipeline that integrates the automation capabilities of LLMs with the contextual expertise of human researchers. The system enables the AI to self-detect uncertainty and proactively request clarification when encountering ambiguous visual or contextual information. By incorporating real-time expert feedback, the AI improves both interpretive accuracy and contextual understanding of the extracted data.

The main contributions of this study are threefold: (1) development of an integrated pipeline for automated extraction of textual, tabular, and graphical PFAS data from scientific literature; (2) design of a novel human–AI collaboration framework in which the model self-assesses uncertainty and seeks expert input; and (3) empirical validation showing that the proposed approach substantially improves accuracy and reliability compared to fully automated methods.

2 Related Works

The rapid evolution of Large Language Models (LLMs) has fundamentally reshaped scientific research workflows. In particular, their integration into environmental and chemical data extraction pipelines is redefining how researchers collect, interpret, and validate information across vast bodies of literature [5].

Early studies demonstrated the usefulness of LLMs for fundamental tasks such as information retrieval (e.g., ChatGPT-based scholarly question-answering), automatic summarization, and hypothesis generation across broad scientific fields [13]. More advanced work explored the construction of scientific knowledge graphs, where LLMs process unstructured text—such as research papers, patents, and regulatory documents—to automatically extract entities, relationships, and quantitative data [7]. For example, several studies have used LLMs to build "compound–process–outcome" networks from chemical databases, enabling large-scale reasoning across molecular interactions.

Beyond direct extraction, LLMs integrated within Retrieval-Augmented Generation (RAG) architectures have shown promise for improving data quality and reproducibility [8]. In the biomedical domain, for instance, PubMed-based RAG models can retrieve and cite recent studies during literature synthesis, while others verify laboratory protocols described in papers to ensure experimental reproducibility. These systems have proven that foundation models can play an active role in validating scientific data consistency rather than merely summarizing it.

In the field of per- and polyfluoroalkyl substances (PFAS), which include thousands of chemically diverse and persistent compounds, conventional environmental monitoring has struggled to keep pace [4,6]. Per- and polyfluoroalkyl substances are a class of highly stable anthropogenic chemicals characterized by strong carbon–fluorine (C–F) bonds. Their extreme persistence prevents natural degradation, resulting in bioaccumulation across soil, groundwater, sediment, and even human blood. Along with microplastics, PFAS have emerged as one of the most persistent and concerning contaminants in the modern environment [4,9]. Machine learning approaches such as quantitative structure–activity relationship (QSAR) models have been employed to predict environmental fate, bioaccumulation, and toxicity directly from molecular structure, reducing the need for exhaustive experimental testing. For instance, models estimating the correlation between carbon-chain length and hydrophobicity (log K_{oc}) for PFOS and PFOA have helped guide risk assessments [9]. More recent studies leverage deep learning, including graph neural networks (GNNs), to improve predictive accuracy.

Despite these advances, a major bottleneck remains: partition coefficients (K_d, K_{oc}) and concentration data are dispersed across thousands of heterogeneous publications, reports, and regulatory documents. Manual aggregation remains slow and error-prone [12]. Consequently, natural language processing (NLP) methods are increasingly being used to automate data collection. Some systems, for example, analyze Scopus or ScienceDirect datasets to extract quantitative values such as PFAS concentrations (ng/L, μg/kg), sampling locations, and detection limits directly from scientific text and tables [10,11].

AI has also been applied to the analysis of complex regulatory and policy documents—such as EPA advisories or EU REACH legislation—to identify emerging contaminants, track regulatory threshold changes, and map interpolicy relationships [14]. One recent effort, for instance, employed LLMs to detect evolving "maximum allowable concentration" patterns across hundreds of environmental laws, showing the model's potential for policy intelligence [15].

Building on these prior developments, the present study extends the role of LLMs beyond general text understanding to the domain of chemical reasoning and quantitative interpretation. Specifically, we design a system capable of extracting structured information—spanning text, tables, and figures—from PFAS-related scientific papers. Our work aims to bridge the gap between language-based AI and domain-specific environmental modeling by introducing an intelligent, human-in-the-loop pipeline that improves both the precision and reliability of scientific data extraction.

3 Background

3.1 PFAS Analysis

Per- and polyfluoroalkyl substances (PFAS) are a large class of synthetic chemicals widely used in industrial and consumer products due to their exceptional stability, surfactant properties, and resistance to heat, oil, and water. Since the 1950s, PFAS have been applied in firefighting foams, textiles, coatings, and non-stick products. However, this chemical stability has made PFAS extremely persistent in the environment. Their resistance to degradation, mobility in water, and strong binding to organic matter have resulted in widespread global contamination. PFAS have now been detected in surface water, groundwater, sediments, agricultural soils, food, and even human blood serum [4,9].

In the United States, growing attention has turned to the management of PFAS-containing wastes generated from a variety of sources, including dewatered wastewater treatment (WWT) biosolids, contaminated soils, construction and demolition debris, spent activated carbon, ion-exchange resins, reverse osmosis concentrate, and municipal solid waste (MSW). These waste materials are often disposed of in landfills, incinerated, or reused in land applications. However, many of these disposal and reuse pathways can potentially lead to PFAS release into the environment, particularly through leachate or volatilization, thus extending PFAS exposure across air, soil, and aquatic systems. Understanding and predicting PFAS release behavior from these complex waste matrices require a detailed understanding of their partitioning properties. Most of the modeling of PFAS flows through the process models is performed using transfer coefficients (i.e., the fraction of incoming PFAS in a waste material that is destroyed or released to the air or water). The partition coefficient (Kd)—defined as the ratio of a compound's concentration in the solid phase to its concentration in the aqueous phase—is a key parameter used to evaluate PFAS mobility, environmental fate, and potential bioavailability. However, the determination and comparison of PFAS Kd values are challenging. Reported Kd values vary widely among studies due to differences in experimental design, PFAS species, sorbent type, organic carbon content, pH, and ionic strength. Moreover, the vast and fragmented nature of the PFAS literature makes it time-consuming and labor-intensive to manually compile, verify, and interpret Kd data across studies. For researchers and practitioners aiming to model PFAS transport or design treatment systems, this lack of integrated data represents a major bottleneck.

To address this challenge, this study proposes an artificial intelligence (AI)-based framework for summarizing and extracting partition coefficient (Kd) data from published literature. Using natural language processing (NLP) and data-mining techniques, AI can rapidly scan, categorize, and extract quantitative information from large volumes of scientific text. This approach enables the automatic identification of PFAS-related parameters, materials, and environmental conditions, thereby reducing human error and improving efficiency in data collection.

3.2 LLM for Literature Search and Data Extraction

Recent advances in large language models (LLMs)—including GPT-4.5, Gemini 2.5, Claude 3, and Grok 3—have revolutionized how researchers approach scientific text analysis. Trained on massive corpora, these models exhibit strong contextual understanding, natural language generation, and reasoning capabilities, enabling them to synthesize and interpret complex scientific information with high precision [5].

Unlike traditional machine learning systems, modern LLMs are defined by their scale, emergent generalization abilities, and accessibility through standardized APIs. The current ecosystem is largely divided into proprietary and open-source models. Proprietary systems such as OpenAI's GPT series or Anthropic's Claude provide state-of-the-art reasoning and zero-shot performance, accessible via RESTful APIs that integrate seamlessly into custom research workflows. These APIs allow researchers to control parameters (e.g., temperature, token limits) and enforce output structures (e.g., JSON schemas), providing reproducible, structured data extraction without the computational burden of local model hosting. Open-source models such as Meta's Llama and Mistral AI offer transparency and allow domain-specific fine-tuning, making them suitable for privacy-sensitive or specialized projects. As this study prioritizes efficiency and reliability over local control, we focus primarily on proprietary API-based models.

For literature analysis, LLMs go far beyond keyword-based searches by performing semantic retrieval. Instead of relying on rigid Boolean queries, researchers can pose natural language questions such as: *"What are the common detection limits for PFOA in river sediments across Asia?"* and obtain synthesized, contextually relevant excerpts from diverse sources. This semantic querying dramatically reduces the time and effort required to identify pertinent studies.

LLMs also excel in summarizing long and complex scientific articles. They can condense multiple documents into concise, cross-referenced summaries, highlight consistent trends, identify contradictions, and even trace the methodological evolution of a research topic. By detecting latent relationships between seemingly unrelated studies or anomalies across datasets, LLMs function as powerful cognitive assistants for scientific discovery [7,12].

In environmental research, the most critical capability of an LLM is its ability to extract structured, quantitative data from unstructured text. This enables the construction of databases for meta-analysis and downstream model training. Three features are especially important:

- **Zero-shot and Few-shot Extraction:** LLMs can perform high-accuracy information extraction with only a well-designed prompt (zero-shot) or a

small number of examples (few-shot), without task-specific fine-tuning. This is especially valuable for processing heterogeneous scientific reports that vary widely in structure and formatting.
- **Structured Output Enforcement (JSON/YAML):** Through strict prompt engineering and API schema enforcement, models can transform textual descriptions into machine-readable data formats such as JSON or YAML. Extracted fields may include chemical identifiers, concentration values, statistical metrics (e.g., mean, standard deviation), and experimental conditions.
- **Contextual Validation:** LLMs possess the ability to distinguish contextual meaning—such as interpreting "10 ng/L" as a measurement and "Not Detected (ND)" as a threshold condition—and normalize both within a unified schema. This contextual awareness ensures data consistency and prevents unit-level errors in large-scale datasets.

Collectively, these capabilities allow LLMs to evolve from mere summarization tools into intelligent scientific assistants capable of building, validating, and maintaining large-scale environmental data repositories.

4 Methodology

This study develops a Python-based pipeline for automated extraction of quantitative data from PFAS-related scientific literature using a Large Language Model (LLM). As shown in Fig. 1, the pipeline consists of four major phases: (1) Literature Search and Dataset Curation, (2) Automated Initial Analysis, (3) Output Generation, and (4) Human–AI Collaborative Refinement.

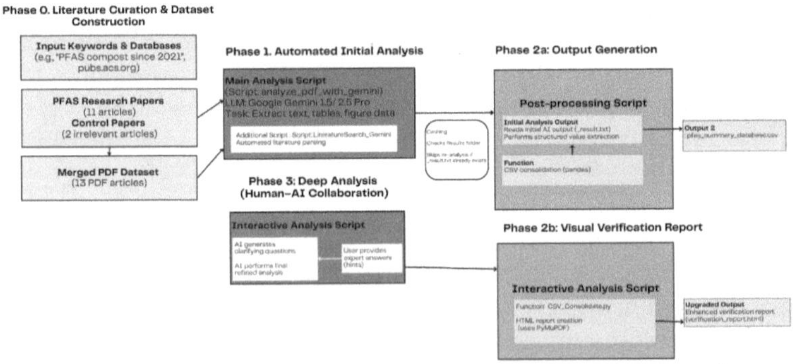

Fig. 1. Overall Workflow of the AI-Powered PFAS Data Extraction Pipeline.

4.1 Phase 1: Literature Search and Dataset Curation

The first step curated a set of research papers through automated web search and manual verification. A reproducible query was constructed using PFAS-related keywords, file-type constraints (`filetype:pdf`), and trusted academic domains (e.g., `site:pubs.acs.org`). HTML metadata (titles, authors) were parsed using `BeautifulSoup` to compile a candidate list, followed by manual filtering to obtain 11 relevant papers. Two unrelated control papers—*Control1.pdf* (numerical linear algebra) and *Control2.pdf* (5G security)—were added to assess the model's ability to reject off-topic content.

4.2 Phase 2: Automated Initial Analysis

Each paper was processed through an automated script that generates structured text results from textual, tabular, and graphical content. Caching was implemented to skip previously analyzed PDFs, reducing computational cost. The core prompt engineering evolved iteratively—from simple keyword-based extraction to a refined, multi-part instruction that defined: (1) the expert role the AI should assume, (2) a reasoning step for internal verification, (3) explicit uncertainty-handling rules, and (4) a final summary table format.

This design enabled the Gemini 2.5 Pro model to systematically scan each document, extract partitioning coefficients (e.g., K_d, K_{oc}, log K_{oc}), and generate structured outputs with embedded source quotations.

4.3 Phase 3: Output Generation

The text-based outputs were consolidated into a structured database for verification and analysis. All `.txt` results were parsed to extract the final summary table sections and merged into a single CSV file. In parallel, a verification report was created in HTML using `PyMuPDF`, embedding extracted figures alongside AI-generated interpretations. This enabled rapid visual comparison between the AI's analysis and the original sources, allowing human reviewers to validate extracted numerical values.

4.4 Phase 4: Human–AI Collaboration for Figure Interpretation

To address the limitations observed in standalone figure analysis, a new intelligent human–AI collaboration workflow was designed. This module enables the AI to self-detect uncertainty during figure interpretation and engage experts interactively. The process comprises three key stages:

1. **Figure Preprocessing:** The pipeline merges fragmented graphical objects in PDFs by identifying bounding boxes and expanding them (20–40 pixels) to include legends, captions, and axis labels, ensuring that the AI receives complete visual context.

```
**Role:** You are an expert research assistant specializing in environmental chemistry and toxicology.

**Task:** Carefully analyze the attached PDF file, which is a scientific document. Your goal is to identify and extract specific information related to
        Per- and Polyfluoroalkyl Substances (PFAS), thinking step-by-step to ensure accuracy.

**Keywords to look for:**
1.  **PFAS Compounds:** Search for common PFAS names and acronyms, including but not limited to:
    - PFOA (Perfluorooctanoic acid)
    - PFOS (Perfluorooctanesulfonic acid)
    - GenX, PFECA
    - PFNA (Perfluorononanoic acid)
    - PFBS (Perfluorobutanesulfonic acid)
    - PFHxS (Perfluorohexanesulfonic acid)
    - PFDA (Perfluorodecanoic acid)
    - 6:2 FTS (6:2 fluorotelomer sulfonate)

2.  **Partitioning Coefficients:** Search for various partitioning and distribution coefficients associated with these compounds. Look for terms like:
    - Octanol-water partition coefficient (Kow), log Kow
    - Organic carbon-water partition coefficient (Koc), log Koc
    - Soil-water partition coefficient (Kd), log Kd
    - Bioconcentration Factor (BCF), log BCF
    - Bioaccumulation Factor (BAF), log BAF
    - "Partitioning coefficient", "distribution coefficient"

**Output Format and Instructions:**

**Part 1: Text and Table Data Extraction**
First, thoroughly scan all the text and tables in the document. For any data you find, extract it in the following format.
- **PFAS Compound:** The name or acronym of the substance.
- **Parameter:** The type of coefficient found (e.g., log Koc).
- **Value:** The reported numerical value, including units if available.
- **Source Quote:** The exact sentence or data table entry from the text where the information was located. This is mandatory for verification.

**Part 2: Figure Data Extraction**
After analyzing the text, identify all figures. For each figure that contains a chart or graph, perform the following steps:
1.  **[Reasoning Step]** First, describe the figure type and its axes in one sentence.
2.  **[Reasoning Step]**.Then, explain step-by-step how you are identifying the values from the visual information.
3.  **[Final Answer]** Finally, provide the extracted data in the same structured format as Part 1.

**Special Instructions:**
- If you find multiple data points, list each one separately.
- **[Uncertainty Handling]** If you find a numerical value that seems relevant but you are uncertain about what parameter it represents, do not guess.
  Instead, report it under a separate section titled 'Uncertain Data' and provide the full source quote for human review.
- If, after a thorough analysis, you find no relevant PFAS partitioning data in the entire document, you must explicitly state: "No PFAS partitioning
  coefficient data was found in the text."
- **[Final Summary]** After listing all individual data points from all parts, create a summary table in Markdown format that consolidates all findings.
  The columns should be 'PFAS Compound', 'Parameter', 'Value', and 'Source Document Location'.
```

Fig. 2. Final multi-part prompt structure for AI-guided data extraction.

2. **AI Question Generation:** When ambiguity arises, the Gemini model automatically generates one or two targeted questions (e.g., "What does TF represent in the legend?" or "Which axis shows concentration?").
3. **Expert Hint Integration:** The human provides concise answers, which are appended to the prompt. The AI then re-analyzes the figure using these contextual hints, producing corrected and verifiable quantitative outputs.

This "CAPTCHA-for-AI" mechanism converts uncertainty into a collaborative feedback loop, allowing the model to dynamically refine its interpretation of visual data while minimizing hallucination and increasing reliability.

Building upon the prompt-design logic illustrated in Fig. 2, the next stage focuses on integrating human feedback to overcome figure-based uncertainty, as shown in Fig. 3.

Figure 3 illustrates the complete workflow of the proposed *human–AI collaboration pipeline* designed for figure-based PFAS data extraction. The left panel represents the automated stage, where the AI analyzes scientific figures, detects numerical patterns, and self-assesses uncertainty based on cues such as missing units, ambiguous legends, or unclear axis labels. Upon detecting uncertainty, the model automatically generates one or two concise clarification questions (e.g., *"What does TF represent in the legend?"*, or *"Which axis corresponds to concentration?"*).

The right panel depicts the human-in-the-loop stage, where domain experts provide short and targeted feedback—such as defining abbreviations or confirm-

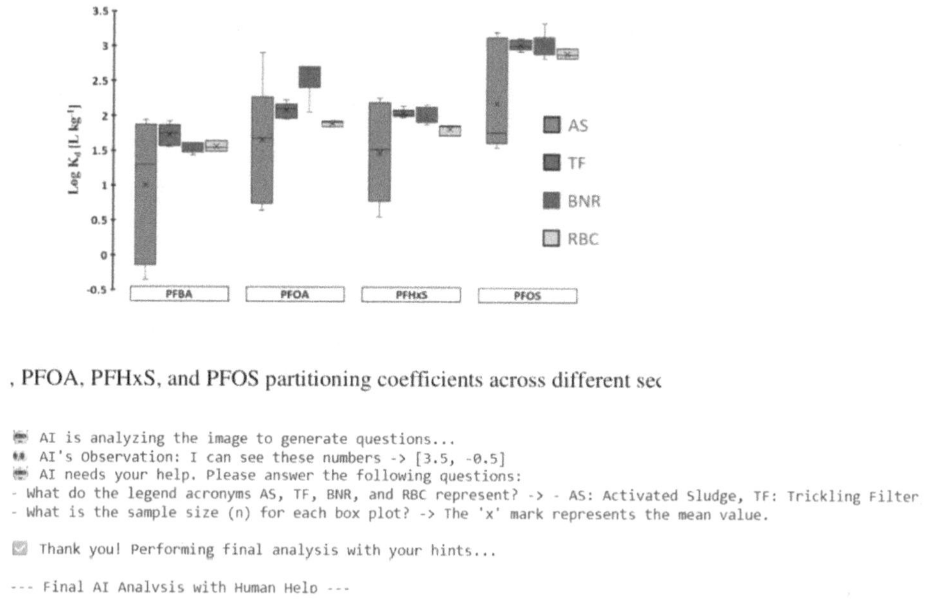

Fig. 3. Human–AI collaboration workflow for figure-based PFAS data extraction.

ing measurement contexts. After receiving this feedback, the AI re-analyzes the same figure, updates its internal representation, and re-extracts the corresponding quantitative values.

This iterative loop of *question generation → expert feedback → refined extraction* forms the core mechanism that improves both the interpretive accuracy and reproducibility of figure-based quantitative analysis. The integration of human verification at critical uncertainty points ensures that context-sensitive reasoning errors (e.g., unit mismatches or misinterpreted symbols) are effectively corrected, resulting in verifiable and trustworthy data outputs, thereby enhancing the overall reproducibility of AI-assisted scientific workflows. Through this multi-stage pipeline, the system successfully transformed a general-purpose LLM into a specialized analytical tool capable of extracting, validating, and structuring domain-specific scientific data.

5 Experiments and Results

The workflow is implemented in Google Colab and integrated key libraries including `googlesearch` and `BeautifulSoup` for paper identification, `PyMuPDF` for PDF parsing, `pandas` for structured data consolidation, and the Gemini API for invoking Google's Gemini 2.5 Pro model.

5.1 Data Extraction from Texts

See Table 1.

Table 1. Performance Analysis of Text-Based Data Extraction

AI-Extracted Data (TXT)	Original Paper Content (PDF)	Accuracy Analysis
PFOS $K_{d,\max} = 140–281\,\text{mL/g}$, PFOA = $30–54\,\text{mL/g}$, PFBS = $9–18\,\text{mL/g}$.	"...140–281 mL g^{-1} for PFOS; 30–54 for PFOA; 9–18 for PFBS."	All fields matched (compound, unit, value, source). Fully correct extraction.
PFOA concentration: $< 10–120$ ng/L (surface water).	Text body: "...<10–11,000 ng/L in surface water..."	Missed upper value (11,000 ng/L) in main text. Incomplete synthesis across sections.
PFOS $\log K_{oc} = 2.57–3.1\,\text{L/kgoc}$.	"... within the range 2.57–3.1 L/kgoc."	All four fields correct. Previous "analysis not possible" error resolved.
"No explicit PFAS partition coefficient found."	"... K_d values are shown in Figs. 1–2."	Correctly identified absence of numerical text data. Structural recognition accurate.
"No coefficients in text or tables; values appear in figures."	Full-text confirmed figures as main data source.	Accurately detected figure-based data location.
No direct numerical data in text (table-based).	Data primarily found in Tables 1, 8, 9, or appendices.	Correctly distinguished data source type.
"No PFAS-related data found."	Unrelated topics (linear algebra, 5G, soil salinity).	Correctly rejected irrelevant papers (no hallucination).

The AI-based pipeline demonstrated strong reliability in extracting explicit textual information, accurately identifying compounds, parameters, units, and source locations. In papers such as *SorptionOfPerfluoroalkyl* and *SorptionOfPerfluorinated*, all four key fields were correctly extracted, confirming the model's proficiency when dealing with well-structured descriptive sentences.

The system also effectively filtered out irrelevant content: *Control1* and *Control2* were consistently rejected without hallucination, and in cases like *InfluenceOfCations* and *Linking*, the AI correctly identified that core data were presented in figures rather than text. This indicates robust document structure awareness.

However, certain limitations were observed. In *NovelPolyfluorinatedCompounds*, the AI extracted a valid value from the abstract but failed to recognize a more representative value from the main text, revealing difficulty in integrating evidence across multiple document sections. This weakness reduces completeness and demonstrates the need for hierarchical information prioritization.

Overall, the pipeline performs reliably in explicit text extraction and non-relevant document filtering but still struggles to synthesize scattered quantitative data and infer which value best represents the study's results. These findings justify the introduction of the collaborative figure-analysis workflow presented in the next section.

5.2 Data Extraction from Tables

See Table 2.

The pipeline showed high reliability for simple tables, correctly extracting compounds, units, values, and sources. However, two main error types were observed: (1) Context misalignment, when the model linked the wrong rows or columns; and (2) Metadata inaccuracies, such as incorrect page or table references. The system successfully referenced supplementary (SI) tables, but such

Table 2. Performance Analysis of Table-Based Data Extraction

AI-Extracted Data (TXT)	Original Paper Content (PDF)	Accuracy Analysis
PFOA K_{oc}: 26/1100 cm^3/g (10/90th)	Tbl 8: 26/1100; actual p.50	Values correct, **page mismatch** (49→50).
PFOS K_{oc}: 250/22000 cm^3/g (10/90th)	Tbl 9: 250/22000; actual p.50	All values match, **page mismatch** (49→50).
PFOA MQL (blackberry): 0.039 ng/g (incorrect)	Tbl 1: 0.048 (blackberry), 0.039 (blueberry)	**Context error:** picked adjacent row value.
PFOS/PFOA/PFBS solubility: 0.57/3.4/46.2 g L^{-1}	Tbl 1 lists identical values	**All four fields correct.**
PFOA+PFOS (compost): 7.94–11.5 µg/kg; "Table S3"	SI table (not in main PDF)	**External ref. needs human verification.**
No table data extracted	Tables contain no relevant PFAS info	**Correct source-type detection.**

outputs still require expert verification. Overall, table-based extraction is promising but demands structural validation to ensure full reliability.

5.3 Data Extraction from Figures

To evaluate the LLM's ability to interpret visual information, a two-stage experiment was conducted: (1) an AI-only analysis to assess baseline performance, and (2) a Human–AI collaborative workflow designed to address observed limitations (Table 3).

Table 3. Performance Analysis of Figure-Based Data Extraction (Condensed)

AI-Extracted Data	Original Paper Content	Accuracy Analysis
AI Only Instances		
"No PFAS partitioning coefficient data found." [6]	"K_d values are shown in Figs. 1–2." [6]	**Failure:** Detected figure reference but extracted no numeric values.
"No coefficients in text; data appear in figures."	Main data located in Fig. 2 and Fig. 4.	**Failure:** Correctly localized data but could not interpret figure values.
"No relevant figures." [13]	Unrelated topics (linear algebra, 5G security).	**100% Correct Rejection:** No hallucination or false extraction.
Human–AI Collaboration Instances		
AI asked: "What does TF mean?" Human answered: "Trickling Filter." Final result: PFOS (TF), Log K_d = 3.05 L/kg (Fig. 2).	Fig. 2 median line for PFOS (TF) ≈ 3.05 L/kg.	**Success:** 4/4 fields correct after expert feedback.

The AI-only pipeline consistently failed to extract quantitative data from figures, despite correctly identifying that key values were visually represented. This indicates a fundamental weakness in visual–numerical interpretation. However, the system effectively filtered irrelevant control documents, confirming robust topic awareness.

In contrast, the Human–AI collaboration workflow demonstrated a dramatic improvement. When the LLM detected uncertainty (e.g., unfamiliar abbreviations or unclear axes), it generated clarifying questions. After the human provided brief hints—such as defining "TF" as "Trickling Filter"—the AI successfully re-analyzed the figure and precisely extracted the correct median value (Log K_d = 3.05 L/kg). This represents a clear transformation from total failure to perfect accuracy, validating the collaborative design.

Nevertheless, certain precision issues remained. Even when extracting from individual figures, quantitative accuracy fluctuated within a 10–30% margin, likely due to graphical resolution and pixel-level ambiguity. These results suggest that while LLMs can recognize uncertainty and self-correct through collaboration, further refinement in visual reasoning and numerical precision is essential for fully automated scientific data extraction.

5.4 The Performance of Summary Tables

Phase 3 of the pipeline successfully integrated extracted results into a unified file, `pfas_summary_database.csv`, demonstrating that the data merging mechanism functions as intended. However, the current summary remains in an early stage of development. While some entries, such as the *DraftSewage* paper, contain valid quantitative values (e.g., 26 cm^3/g), others like *Linking* include only qualitative descriptions such as "Visual Trends."

This discrepancy indicates that the figure analysis module often fails to extract precise numerical data, producing textual summaries rather than structured measurements. As a result, the database currently contains a mix of quantitative and qualitative information, limiting its applicability for statistical computation or machine-learning training.

Nevertheless, the summary table serves as a meaningful proof of concept for the integration pipeline. Future work should focus on transforming descriptive figure outputs into numerical values to enable scalable, quantitative scientific analysis.

The objective of Phase 3 of this pipeline is to integrate the results from individual paper analyses into a single file, `pfas_summary_database.csv`. An examination of the currently generated CSV file confirms that the integration function itself is technically operational. Information extracted from multiple papers (e.g., *Linking*, *DraftSewage*) has been successfully merged under a consistent columnar structure, including 'Paper', 'PFAS Compound', and other relevant fields.

However, a frank assessment of the data's content reveals that the current summary table remains at a preliminary stage of data aggregation. While it contains specific numerical data, such as "26 cm^3/g" extracted from the *DraftSewage* paper, the values extracted from the *Linking* paper are predominantly filled with qualitative textual descriptions like "Visual Trends" and "Visual Comparison".

This clearly demonstrates that the current pipeline, particularly the figure analysis module, is failing to extract actual numerical data from complex visual sources and is instead only describing their presence or trends in text. Consequently, the database in its current state is a mixture of quantitative data and qualitative descriptions, making it impossible to directly perform statistical analyses (e.g., calculating means or medians) or to use it as a training dataset for machine learning models.

Therefore, while the current summary table is meaningful as a proof-of-concept for the data integration mechanism, it requires a critical performance improvement in the core task of converting descriptions like "Visual Trends" into actual numerical values. Resolving this issue is essential to achieve the project's ultimate goal of building a high-quality, quantitative database suitable for scientific research.

6 Discussion and Future Work

This study successfully developed an AI-powered pipeline and an Intelligent Human–AI Collaboration framework for the automated extraction of quantitative PFAS data from scientific literature.

The analysis confirms that Large Language Models (LLMs) are powerful assistants for extracting structured data from clear text and simple tables, especially when given sophisticated, domain-specific prompts. However, general LLMs present fundamental limitations: they are unable to interpret complex visual data (figures and charts), they lack self-awareness regarding their own errors, and they struggle with contextual memory, leading to redundant questioning. Crucially, the AI's uncertainty was found to be epistemic (rooted in missing symbolic/domain knowledge) rather than purely computational.

The implemented collaborative loop demonstrated its value by allowing the LLM to self-detect uncertainty and seek minimal, targeted expert clarification, which significantly improved contextual understanding and data reliability. This establishes that for high-accuracy scientific work, the future of LLMs lies not as standalone tools, but as collaborative, human-aligned agents.

Future work will focus on transforming the AI's current questioning behavior into an adaptive, continuous learning mechanism. This will involve three key extensions: implementing Question Log Meta-Learning to analyze recurring uncertainty patterns (like those related to units or abbreviations) so the model can self-correct and generalize knowledge; integrating a Collaborative Memory module to persistently store and reuse human-provided clarifications across different documents, thereby eliminating repetitive questioning; and finally, optimizing Question Efficiency by quantifying the information gain of each AI-generated query to ensure maximum accuracy improvement is achieved with minimal human input. Together, these extensions aim to evolve the pipeline from a static extraction tool into a more reflective, self-improving AI collaborator that enhances its own reasoning efficiency through ongoing human interaction.

Acknowledgement. This research project is supported by The CUNY Transfer to STEM Student Success (TS3) program. We thank Thierno Bah, Yesenia Barrera, Ainsley Hanson, and Oludemilade Lekanmichael for their efforts in obtaining the preliminary results for this work.

References

1. Cousins, I.T., et al.: The high persistence of PFAS is sufficient for their management as a chemical class. Environ. Sci. Processes Impacts **22**(7), 1459–1469 (2020). https://doi.org/10.1039/D0EM00355G
2. Scherbakov, D., et al.: The emergence of large language models as tools in literature reviews: a large language model-assisted systematic review. J. Am. Med. Inform. Assoc. **32**(6), 1071–1080 (2025). https://doi.org/10.1093/jamia/ocae012
3. Open Mind. "What Are the Limitations of Large Language Models (LLMs)?" MIT Press (2024)
4. Arvaniti, O.S., Andersen, H.R., Thomaidis, N.S., Stasinakis, A.S.: Sorption of perfluorinated compounds onto different types of sewage sludge and assessment of its importance during wastewater treatment. Chemosphere **111**, 405–411 (2014). https://doi.org/10.1016/j.chemosphere.2014.03.038
5. Ebrahimi, F., Lewis, A.J., Sales, C.M., Suri, R., McKenzie, E.R.: Linking PFAS partitioning behavior in sewage solids to the solid characteristics, solution chemistry, and treatment processes. Chemosphere **271**, 129530 (2021). https://doi.org/10.1016/j.chemosphere.2021.129530
6. Wang, F., Shih, K., Ma, R., Li, X.: Influence of cations on the partition behavior of perfluoroheptanoate (PFHpA) and perfluorohexanesulfonate (PFHxS) on wastewater sludge. Chemosphere **131**, 178–183 (2015). https://doi.org/10.1016/j.chemosphere.2015.03.017
7. Rahman, M.F., Peldszus, S., Anderson, W.B.: Behaviour and fate of perfluoroalkyl and polyfluoroalkyl substances (PFASs) in drinking water treatment: a review. Water Res. **50**, 318–340 (2014). https://doi.org/10.1016/j.watres.2013.10.045
8. Milinovic, J., Lacorte, S., Rigol, A., Vidal, M.: Sorption of perfluoroalkyl substances in sewage sludge. Environ. Sci. Pollut. Res. **23**(9), 8339–8348 (2016). https://doi.org/10.1007/s11356-015-5970-6
9. Higgins, C.P., Luthy, R.G.: Sorption of perfluorinated surfactants on sediments. Environ. Sci. Technol. **40**(23), 7251–7256 (2006). https://doi.org/10.1021/es061000n
10. Choi, Y.J., Kim Lazcano, R., Yousefi, P., Trim, H., Lee, L.S.: Perfluoroalkyl acid characterization in U.S. municipal organic solid waste composts. Environ. Sci. Technol. Lett. **6**(6), 372–377 (2019). https://doi.org/10.1021/acs.estlett.9b00280
11. Meng, P., DeStefano, N.J., Knappe, D.R.U.: Extraction and matrix cleanup method for analyzing novel per- and polyfluoroalkyl ether acids and other PFAS in fruits and vegetables. J. Agric. Food Chem. **70**(2), 474–484 (2022). https://doi.org/10.1021/acs.jafc.1c06288
12. Newton, S., McMahen, R., Stoeckel, J.A., Chislock, M., Lindstrom, A., Strynar, M.: Novel polyfluorinated compounds identified using high resolution mass spectrometry downstream of manufacturing facilities near Decatur, Alabama. Environ. Sci. Technol. **51**(3), 1544–1552 (2017). https://doi.org/10.1021/acs.est.6b05330
13. Nguyen, K.-A., Liou, Y.-A., Tran, H.-P., Hoang, P.-P., Nguyen, T.-H.: Soil salinity assessment by using near-infrared channel and vegetation soil salinity index derived from Landsat 8 OLI data: a case study in the Tra Vinh Province, Mekong Delta, Vietnam. Prog Earth Planet Sci **7**(1), 1–16 (2020). https://doi.org/10.1186/s40645-019-0314-1
14. Ye, J., Rhee, J., Zhao, L.: Draft report on PFAS sorption and partitioning in sewage treatment matrices. Unpublished technical draft, Stony Brook University (2023)
15. Rhee, J., Zhao, L.: "Control study on AI-based data extraction robustness using non-PFAS soil salinity papers". Internal report, Stony Brook University (2023)

Time and Energy Consumption of a Large Language Model: Quantization, Kernel Optimization, Energy Saving and Accuracy

Malek Belkahla[1(✉)] and Gaétan Hains[2]

[1] ENSTA Paris, Palaiseau, France
malek.bel@ensta.fr
[2] LACL, UPEC, Créteil, France
gaetan.hains@u-pec.fr

Abstract. The increasing energy consumption of Large Language Models (LLMs) poses a significant challenge to their sustainable deployment. Service operators face a fundamental conflict: maximizing inference throughput often leads to prohibitive increases in total energy consumption and job duration. We present a rigorous, multi-platform analysis to resolve this performance-versus-efficiency dilemma. We investigate the impact of Activation-aware Weight Quantization (AWQ) paired with the high-performance MARLIN inference kernel, on the performance and energy consumption of modern LLMs. Our experiments are conducted across three distinct NVIDIA GPU architectures: Hopper H200, Ampere A100, and Ada RTX 5000. Our results confirm that 4-bit AWQ is a dominant optimization, delivering substantial, concurrent gains: throughput is increased by up to 118.5% while energy consumption is reduced by up to 54.4%, all with very good presenvation of model inference accuracy. Our primary contribution is a practical, two-step methodology for efficient deployment: first, apply AWQ quantization, and second, select the batch size that minimizes the Energy-Delay Product (EDP). This strategy allows service operators to achieve a data-driven, optimized balance between performance and total cost, providing a validated path towards the sustainable deployment of AI systems.

Keywords: Efficient techniques for LLMs · Evaluation and benchmarking · GPU energy consumption · Inference accuracy

1 Introduction

The increasing energy consumption of Large Language Models (LLMs) poses a significant challenge to their sustainable deployment, appearing incompatible with global environmental and climate-stabilization goals [10,14]. While the exponential growth in training cost is well-documented (Fig. 1), the inference phase represents the dominant operational cost. A single LLM query can consume up to 20 times the energy of a search engine query (Fig. 2), creating a critical inefficiency.

H. R. Arabnia et al. (Eds.): CSCI 2025, CCIS 2972, pp. 373–391, 2027.
https://doi.org/10.1007/978-3-032-28469-3_25

Service operators face a fundamental dilemma: maximizing throughput by using large request batches often leads to prohibitive increases in total energy consumption and job duration. This paper resolves this conflict by presenting a rigorous, multi-platform analysis of a key optimization technique. Our primary contribution is a practical, two-step methodology for sustainable deployment:

1. We demonstrate that Activation-aware Weight Quantization (AWQ), accelerated by the optimized Marlin kernel, improves performance while saving energy.
2. We validate the Energy-Delay Product (EDP) as metric to identify an efficient, hardware-specific operating point.

This work provides a validated strategy for achieving an optimized balance between performance and total cost in AI systems[1]

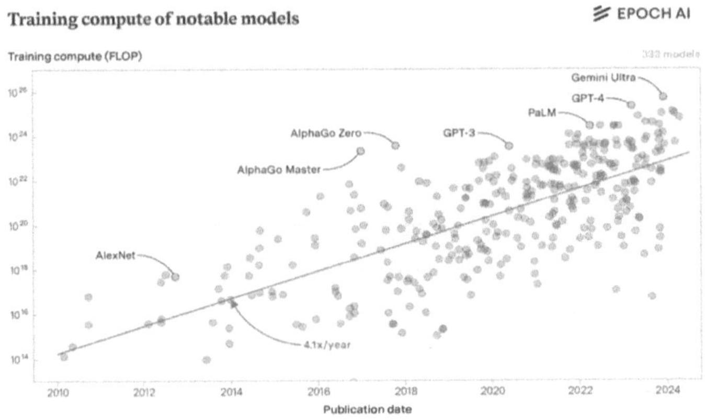

Fig. 1. Compute cost (FLOPs) of large AI model, 2010–2024.

2 Background: Quantization and High-Performance Kernels

To contextualize our experimental findings, it is essential to understand the fundamental challenges of LLM inference and the specific mechanisms of the quantization algorithm and inference kernel employed in this study.

[1] All code, test data, and analysis scripts are available at: https://github.com/Malek9876/llm-energy-benchmark.

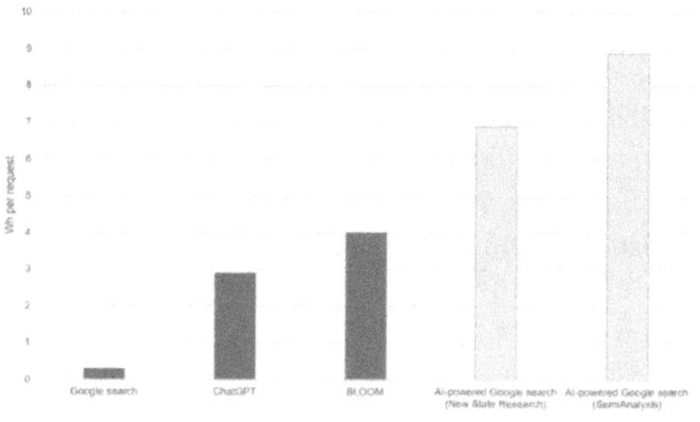

Fig. 2. Energy Consumption (Wh/query) for Various Systems.

2.1 The LLM Inference Bottleneck: Memory vs. Compute

While modern GPUs possess immense computational power (measured in floating-point operations per second, or FLOPs), the practical performance of Large Language Models is often not limited by computation itself. Instead, the primary bottleneck is typically *memory bandwidth*—the speed at which the model's weights can be read from the GPU's main memory (VRAM) into its high-speed on-chip caches for each generation step. The choice of batch size is directly influenced by the target use case. Interactive, latency-sensitive applications like chatbots or real-time code completion operate with very small batches (often a batch size of 1) to ensure an immediate response for a single user. Conversely, offline, throughput-oriented tasks, such as summarizing a large collection of documents or analyzing a dataset, benefit from grouping many requests into large batches. As the batch size increases, the arithmetic intensity grows, and the bottleneck can shift from memory-bound to being *compute-bound* [6]. An effective inference solution must therefore optimize for both scenarios.

2.2 Post-Training Quantization (PTQ)

Post-Training Quantization (PTQ) is a family of techniques designed to compress a model after it has been fully trained, thus avoiding the prohibitive costs of re-training from scratch. The core idea is to convert the model's weights from a high-precision format (e.g., 16-bit `bfloat16`) to a low-precision format (e.g., 8-bit or 4-bit integers).

The foundational PTQ approach involves a simple affine transformation, which maps floating-point values to a fixed integer range [11]. For a given weight tensor W, one finds its minimum and maximum values and maps them linearly to the range of the target integer type. This is expressed as:

$$W_{\text{quant}} = \text{round}\left(\frac{W}{s} + z\right) \tag{1}$$

where s is the scaling factor and z is the "zero-point" offset.

The Salient Outlier Problem. This naive approach fails for low-bit quantization due to the presence of "salient outliers" in LLM weight distributions [5]. A small fraction ($< 0.1\%$) of weights have exceptionally large magnitudes. To accommodate these outliers, the scaling factor s becomes very large, forcing the vast majority of smaller, "normal" weight values to a few bins. This causes a catastrophic loss of precision and severely degrades model accuracy. A detailed numerical example of this failure is in our extended report [2].

2.3 Activation-Aware Weight Quantization (AWQ)

Activation-Aware Weight Quantization (AWQ) [11] was designed specifically to address the salient weight problem. Its core insight is that the importance of a weight channel is not its magnitude, but its contribution to the layer's output—a function of both the weight values and the input activations. AWQ observes that the channels with the largest weight magnitudes are not necessarily the ones that contribute most, since contribution also depends on the scale of the activations.

To handle this, AWQ introduces a *per-channel scaling factor*, s. Instead of quantizing the raw weights W, the goal is to preserve the output of the matrix multiplication XW. This is done by scaling up the sensitive weights before quantization, while proportionally scaling down the corresponding activations. Formally, the layer-wise operation is rewritten as:

$$Y = (X/s) \cdot (W \cdot s) \tag{2}$$

Here, $(W \cdot s)$ is what gets quantized. By amplifying important weights before quantization, AWQ reduces their relative quantization error. At the same time, (X/s) ensures that the overall product remains unchanged. Crucially, the activation scaling can be **fused** into the preceding Layer Normalization step, incurring no extra cost during inference.

This clever pre-scaling effectively compresses the range of outlier weights, allowing the quantizer to distribute its limited precision more evenly. As a result, AWQ largely preserves model fidelity even under aggressive 3–4 bit quantization.

2.4 The MARLIN Kernel for High-Speed Inference

A well-quantized model format like AWQ is a necessary first step, but it is not sufficient to guarantee high inference speed.. A highly optimized GPU kernel is required to execute the mixed-precision matrix multiplication (e.g., FP16 activations multiplied by INT4 weights) efficiently.

MARLIN [6] is a state-of-the-art kernel designed for this exact purpose, achieving near-optimal performance by addressing several microarchitectural challenges. Its design is motivated by the high FLOPs-to-Bytes ratio of modern GPUs. This ratio is closely related to the g parameter in the Bulk Synchronous Parallel (BSP) performance model, which measures the gap between computational throughput and communication bandwidth [12]. For instance, an NVIDIA

A100 can perform hundreds of FP16 operations in the time it takes to load a single 4-bit weight, resulting in a g value in the thousands, which historically requires careful load-compute overlap. MARLIN's goal is therefore to ensure that the GPU's Tensor Cores are never left waiting for data. It accomplishes this through several key optimizations:

- **Hierarchical Parallelization:** The overall matrix multiplication is broken down into a hierarchy of sub-problems that map directly onto the GPU's architecture: SM-level, Warp-level, and finally, Tensor-Core-level computations.
- **Asynchronous Data Pipelining:** MARLIN makes extensive use of asynchronous memory copy instructions (`cp.async`) to pre-fetch weight and activation data from global memory into the faster L2 cache and then into shared memory (SMEM). This creates a deep pipeline where data is loaded while the current computation is still in progress, effectively hiding memory latency.
- **Optimized Data Layouts:** To maximize memory bandwidth, weights are pre-processed offline and re-shuffled into a layout that allows GPU warps to read them in wide, 128-bit contiguous chunks. This technique, a form of dimension-lifting optimization [13], ensures that every memory transaction is as efficient as possible.
- **On-the-Fly Dequantization:** Instead of dequantizing weights in global memory (hence defeating the purpose of quantization), MARLIN performs dequantization on-the-fly with efficient binary manipulations and fused operations directly within the GPU registers, just before the data is sent to the Tensor Cores for matrix multiplication.

The MARLIN kernel has been integrated directly into the vLLM inference engine. When we specify the quantization = awq_marlin option, we are instructing vLLM to use this specific, highly optimized kernel, which is the key to translating the theoretical memory savings of the AWQ format into a significant inference speedup on compatible hardware.

3 Experimental Setup and Methodology

3.1 Experimental Setup

All experiments were conducted on three distinct hardware platforms for a broad analysis of performance on both data-center and workstation-class GPUs. The key hardware and software components are detailed in Tables 1 and 2.

While minor library version differences existed between the workstation and HPC environments due to specific dependencies, we make the standard assumption that these do not affect the model's core inference logic. This allows for a cross-platform comparison of the quantization method's impact.

Table 1. Hardware Platform Specifications

System	GPU	CPU	RAM
HPC (Hopper)	1x H200 (141GB)	2x Intel Xeon 8558	2TB DDR5
HPC (Ampere)	1x A100 (80GB)	2x AMD Epyc 7543	512GB DDR4
Workstation	1x RTX 5000 (32GB)	1x Intel i9-13900K	128GB DDR5

Table 2. Key Software Components by Environment

Component	HPC Env.	Workstation Env.
Python Version	3.10	3.10.12
PyTorch	2.3.1	2.6.0/2.7.0
vLLM	0.9.1	0.9.2
autoawq	N/A	0.2.9
`lm-eval`	Standard Framework Used	
`pynvml`	Standard Library Used	

3.2 Experimental Methodology

Our study provides a comprehensive performance analysis of Activation-aware Weight Quantization (AWQ) across multiple model scales and hardware architectures. The methodology was constructed to ensure scientific rigor, reproducibility, and relevance to real-world deployment scenarios.

Models and Quantization. We selected two pairs of instruction-tuned models to represent different scales of deployment. These variants were chosen because, as demonstrated in the original AWQ paper, the algorithm shows exceptional performance in preserving the nuanced features learned during the fine-tuning process [11].

- **7B Class:**[2] The baseline model was **Mistral-7B-Instruct-v0.1**. Its quantized counterpart was a pre-quantized **Mistral-7B-Instruct-v0.1-AWQ** model.
- **24B Class:** Baseline model was **Mistral-Small-24B-Instruct-v0.1**. Its quantized version was a pre-quantized **Mistral-Small-24B-Instruct-v0.1-AWQ** model.

All quantized models are 4-bit, weight-only AWQ models with a standard group size of 128, a configuration widely adopted for its excellent balance between compression and accuracy.

[2] The 'B' in model names such as '7B' is standard shorthand for billions of parameters. Thus, a 7B model has approximately 7 billion trainable parameters.

Accuracy Evaluation Protocol. To assess the impact of quantization on model quality, we employed the standard `lm-eval` framework. For each model, we measured zero-shot accuracy[3] on a suite of three widely recognized benchmarks: **ARC Challenge** (scientific reasoning) [4], **HellaSwag** (commonsense inference) [16], and **MMLU** (multitask knowledge) [8]. For each benchmark, we report the mean accuracy and standard error, for a good summary of model performance.

Performance and Energy Benchmark Protocol. All benchmarks were conducted using the **vLLM** inference engine [9] to ensure state-of-the-art performance. To ensure statistical significance, each benchmark configuration was executed independently five times, with a two-second pause between runs to maintain system stability. We report the **mean** and **standard deviation** across these five runs for all metrics.

Workload Configuration. The experimental procedure was designed to simulate different real-world server loads. We defined two distinct workload profiles:

1. **Short Generation Profile:** Simulating interactive chatbot applications, with a fixed output of **128 new tokens**.
2. **Long Generation Profile:** Simulating content generation or summarization, fixed output of **1024 new tokens**.

For each profile, we swept across a logarithmic range of batch sizes: **1, 2, 4, 8, 16, 32, 64, and 128**. For the AWQ models, we explicitly enabled the high-performance **Marlin kernel** within vLLM to achieve maximum possible throughput. It should be noted that for the **NVIDIA RTX 5000 Ada** GPU, the full-precision baseline models exceeded the card's VRAM capacity and could not be loaded; therefore, only AWQ-quantized models were benchmarked on this platform.

Metric Measurement Methodology. The measurement of performance and energy was synchronized to precisely capture the workload duration. The key metrics obtained were:

- **Duration and Performance:** The total workload duration was captured using Python's `time.monotonic()` timer, bracketing the `llm.generate` call. From this, **Throughput** was calculated as total tokens generated divided by duration, and **Average Latency per Request** as the total duration divided by the batch size.

[3] Zero-shot learning (ZSL) is a machine learning scenario in which an AI model is trained to recognize and categorize objects or concepts without having seen any examples of those categories or concepts beforehand [15].

- **Energy and Power:** Energy consumption was measured using the NVIDIA Management Library (`pynvml`). For each run, the GPU's cumulative energy counter (in millijoules) was polled immediately before and after the inference workload using the `nvmlDeviceGetTotalEnergyConsumption` function. The **Total Energy** consumed was calculated as the difference between these two readings. **Average Power** was subsequently derived by dividing this Total Energy by the measured workload duration.

4 Results and Analysis

In this section, we analyze the benchmark data to demonstrate the fundamental performance-cost dilemma and present our EDP-based methodology for resolving it. A detailed Pearson correlation analysis, available in our extended report [2], confirms the underlying relationships between configuration parameters and performance metrics.

4.1 Analyzing the Fundamental Trade-Off: Throughput vs. Total Energy and Duration

Individual metrics reveal an optimization dilemma that is present on *all* tested platforms. As shown in Fig. 3, system throughput scales favorably with batch size, but performance characteristics are unique to each architecture.

The Hopper H200 (Fig. 3a) establishes itself as the clear performance leader, achieving peak throughputs approaching 15,000 tokens/sec on the 7B model. This is significantly higher than the A100 (Fig. 3b), which peaks around 6,000 tokens/sec, and the RTX 5000 (Fig. 3c), which reaches approximately 5,000 tokens/sec. Notably, the RTX 5000 could not accommodate the unquantized baseline models within its 32 GiB of VRAM, highlighting a hard memory capacity limit for workstation-class hardware when deploying larger models.

The shape of these curves illustrates the critical role of memory bandwidth. The initial steep rise at small batch sizes (e.g., 1 to 16) is characteristic of a strongly **memory-bound** regime, where performance is limited by the speed at which weights can be loaded from VRAM. A key insight for modern quantized inference is that the system often **remains fundamentally memory-bound** even at moderate-to-large batch sizes [6]. The subsequent "flattening" of the throughput curve, most evident on the A100 and RTX 5000, therefore represents the point of diminishing returns where the memory bus is saturated and can no longer supply data fast enough to keep up with the increasing computational demand. Only at the extreme end of the batch size range does the system potentially transition to a fully **compute-bound** state.

However, this pursuit of higher throughput comes at a direct and unavoidable cost, first in terms of time. As shown in Fig. 4, the total duration to complete the entire batch job increases monotonically with batch size. While a larger batch leads to a lower per-token generation time, the total work required is greater,

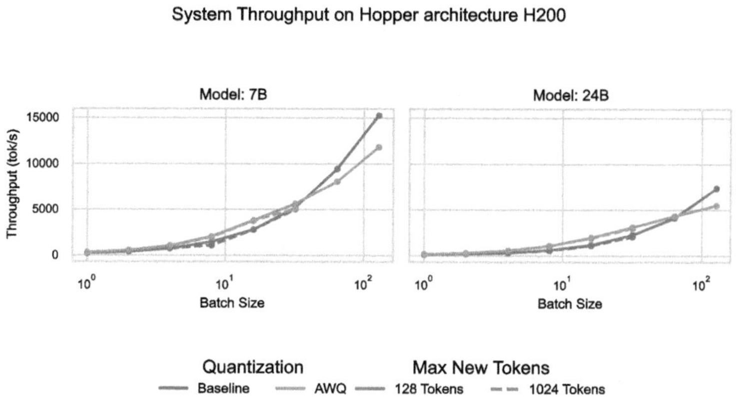

(a) Hopper H200

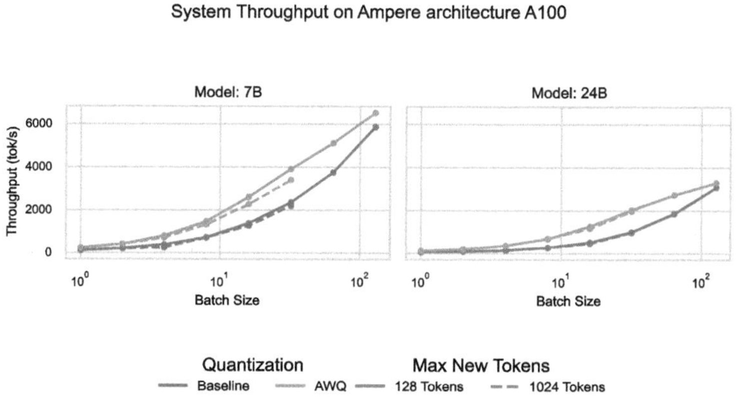

(b) Ampere A100

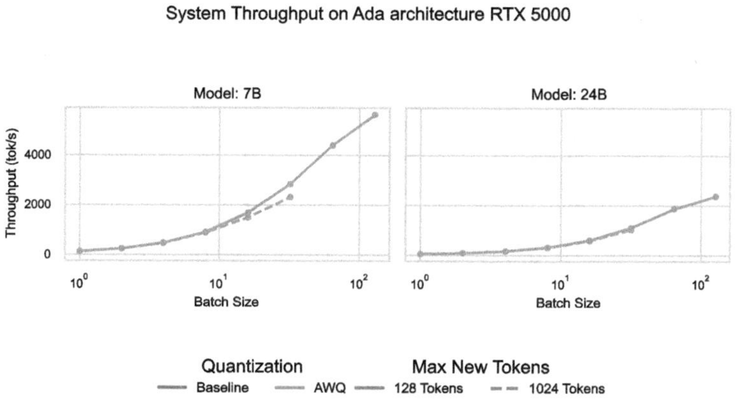

(c) Ada RTX 5000

Fig. 3. System throughput vs. batch size across architectures.

resulting in a longer overall wait. Here again, the architectural hierarchy is clear: the H200 completes any given batch job in the shortest time. Furthermore, AWQ (orange lines) consistently reduces the total duration compared to the baseline, reinforcing its role as a performance accelerator.

Beyond time, the second and more critical cost is energy consumption, detailed in Fig. 5. A direct comparison of the y-axis scales reveals a clear architectural efficiency hierarchy. The Hopper H200 (Fig. 5a) is unequivocally the most energy-efficient platform, with its most demanding workload consuming approximately 3,500 J. In stark contrast, the previous-generation Ampere A100 (Fig. 5b) is significantly less efficient, requiring over 6,000 J for an equivalent task. The workstation-class RTX 5000 (Fig. 5c), while unable to run the baseline models, shows a baseline AWQ energy consumption that is lower than the A100's, but its energy cost rises more steeply at high batch sizes.

Total energy consumed increases substantially with batch size on all platforms because of two compounding factors: longer job duration and higher GPU power state for the entire run due to larger batches. The impact of AWQ (orange lines) is largest in this domain, delivering energy savings that often exceed 50%. For instance, on the A100, AWQ reduces the energy cost of a 7B model inference at batch size 16 from over 2,500 J to nearly 1,000 J. This demonstrates that AWQ not only accelerates performance but also establishes a new, fundamentally lower-energy operating frontier.

This creates a dead-end: the configuration that maximizes performance (large batch size) is the same one that maximizes both the time cost (duration) and the resource cost (energy). An operator seeking to minimize the energy footprint of a single query would choose a small batch size, directly sacrificing the system's overall throughput. This fundamental conflict makes it impossible to determine an optimal operating point from these metrics alone.

4.2 Validating Quantization: Inference Quality Analysis

Before analyzing the performance benefits of AWQ, we must validate that the quantization process preserves the model's core capabilities. Performance gains are only meaningful if they do not come at the expense of model accuracy.

To this end, we evaluated both the baseline (`bfloat16`) and quantized (`int4`) versions of two models—Mistral-Small and Mistral-7B—on standard academic benchmarks: ARC Challenge, HellaSwag, and MMLU. Figure 6 summarizes the zero-shot accuracy results.

The impact of AWQ on accuracy is mixed but informative. On the reasoning-intensive ARC Challenge benchmark, AWQ significantly improves performance for both models. Mistral-Small increases from 0.65 to 0.90, and Mistral-7B from 0.50 to 0.67. A similar trend is observed for Mistral-Small on HellaSwag, where accuracy improves from 0.63 to 0.93. These results suggest that AWQ quantization may enhance the model's ability to generalize on reasoning tasks. This aligns with previous observations in the literature, where quantization has been found to act as a form of implicit regularization, reducing overfitting and boosting performance on out-of-distribution tasks [1,3].

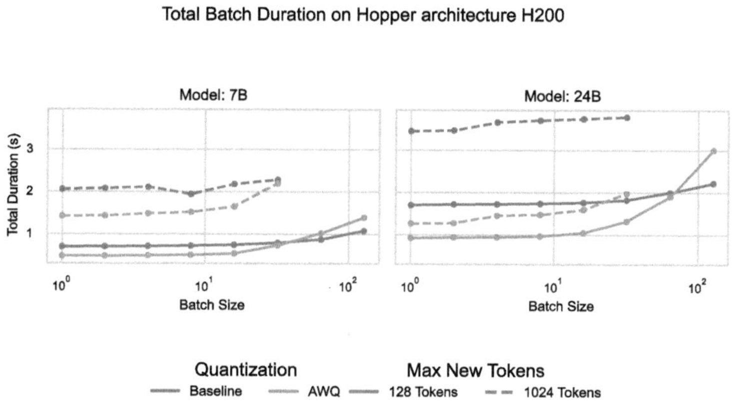

(a) Hopper H200

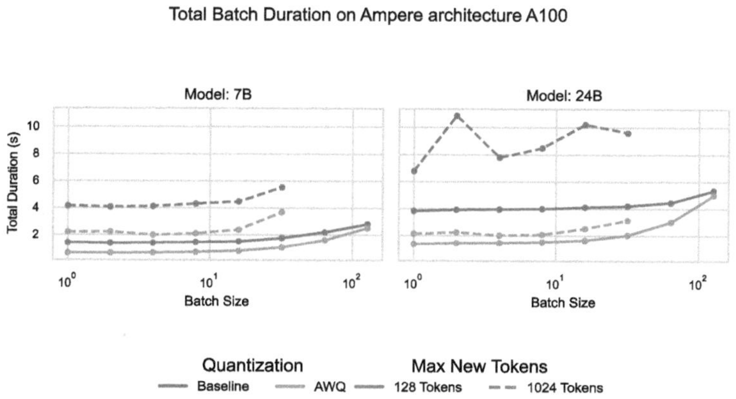

(b) Ampere A100

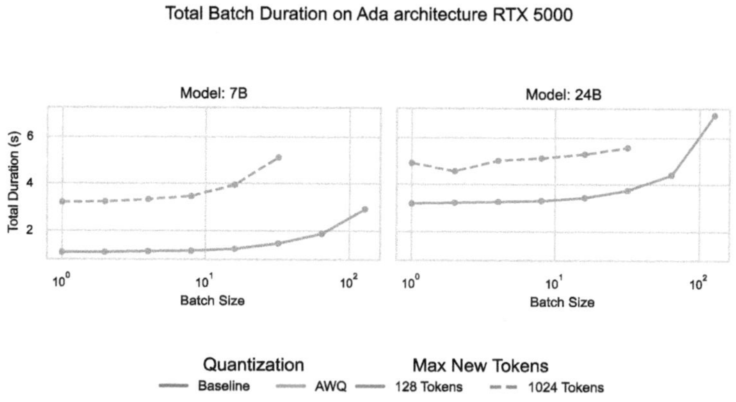

(c) Ada RTX 5000

Fig. 4. Total duration vs. batch size across architectures.

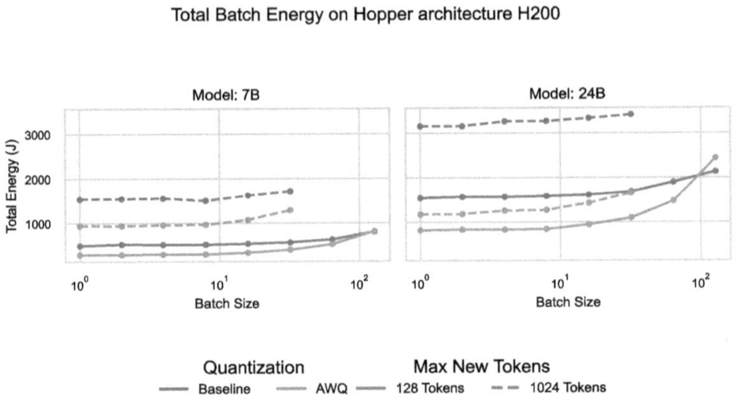

(a) Hopper H200

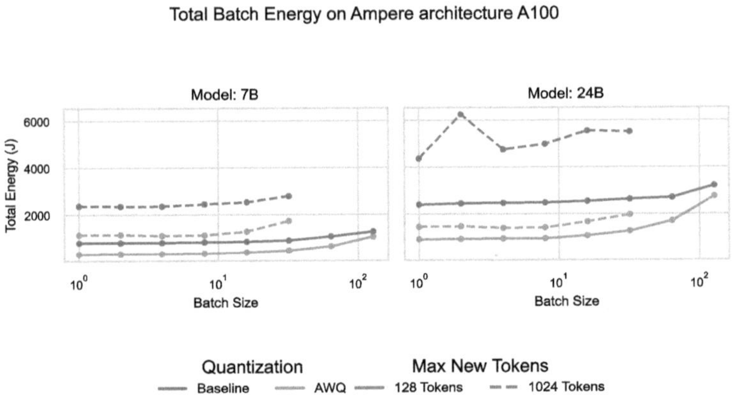

(b) Ampere A100

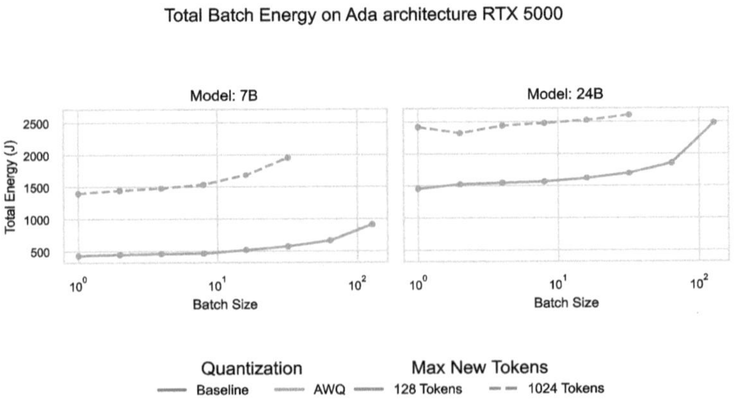

(c) Ada RTX 5000

Fig. 5. Total energy vs. batch size across architectures.

However, this trend does not hold universally. On HellaSwag, Mistral-7B exhibits a slight accuracy drop after quantization (0.56 to 0.52), indicating that the benefits may vary by model size or architecture. More notably, on the knowledge-intensive MMLU benchmark, AWQ shows either no improvement (Mistral-7B: 0.53 $\rightarrow$ 0.53) or a slight regression (Mistral-Small: 0.79 $\rightarrow$ 0.75). This could be attributed to the precision loss negatively impacting retrieval or memorization of fine-grained facts, which MMLU tasks rely on.

Overall, these results confirm that AWQ does not cause catastrophic degradation in inference quality. On the contrary, for certain benchmarks, especially those focused on reasoning, quantization not only preserves but improves accuracy. We conclude that AWQ is a viable optimization strategy that enables performance gains without compromising—and sometimes even improving—inference quality.

4.3 Quantifying the AWQ Advantage Across Architectures

Having established that AWQ preserves model fidelity, we now quantify its direct impact on performance and efficiency. Figure 7 summarizes the average percentage improvement gained by using 4-bit AWQ quantization relative to the `bfloat16` baseline. For clarity of presentation, all improvement values have been rounded to the nearest integer. The results are unequivocal: AWQ is not a trade-off but a **dominant optimization**, providing concurrent and substantial gains across all key metrics.

On the Ampere A100, AWQ delivers a remarkable average throughput uplift of **118.5%**, effectively more than doubling the processing speed. This massive performance gain is accompanied by equally impressive efficiency improvements, with an average latency saving of **51.9%** and an average energy saving of **54.4%**. The benefits on the newer Hopper H200, while more modest in relative terms, remain substantial. Here, AWQ provides a **47.2%** throughput increase, a **29.4%** latency reduction, and a **38.8%** energy saving.

The more pronounced relative gains on the A100 suggest that quantization is particularly effective at alleviating the more severe memory bandwidth bottlenecks present in the older architecture. While the H200 is more efficient at baseline, AWQ still provides a significant boost. This analysis confirms AWQ as a foundational improvement that establishes a new, superior operating baseline for any subsequent system tuning.

4.4 Comparative Efficiency: Solving the Dilemma with EDP

To resolve the optimization dilemma, we employ the Energy-Delay Product (EDP), a classic metric from computer architecture used to balance speed and power consumption [7]. By holistically quantifying system "effort," EDP allows us to identify the point of maximum efficiency. The metric's power lies in its ability to unify our three key metrics into a single value, where a lower result is better. It is:

$$\mathrm{EDP} = E_{\mathrm{total}} \times D = E_{\mathrm{total}} \times \frac{N_{\mathrm{tokens}}}{T} \qquad (3)$$

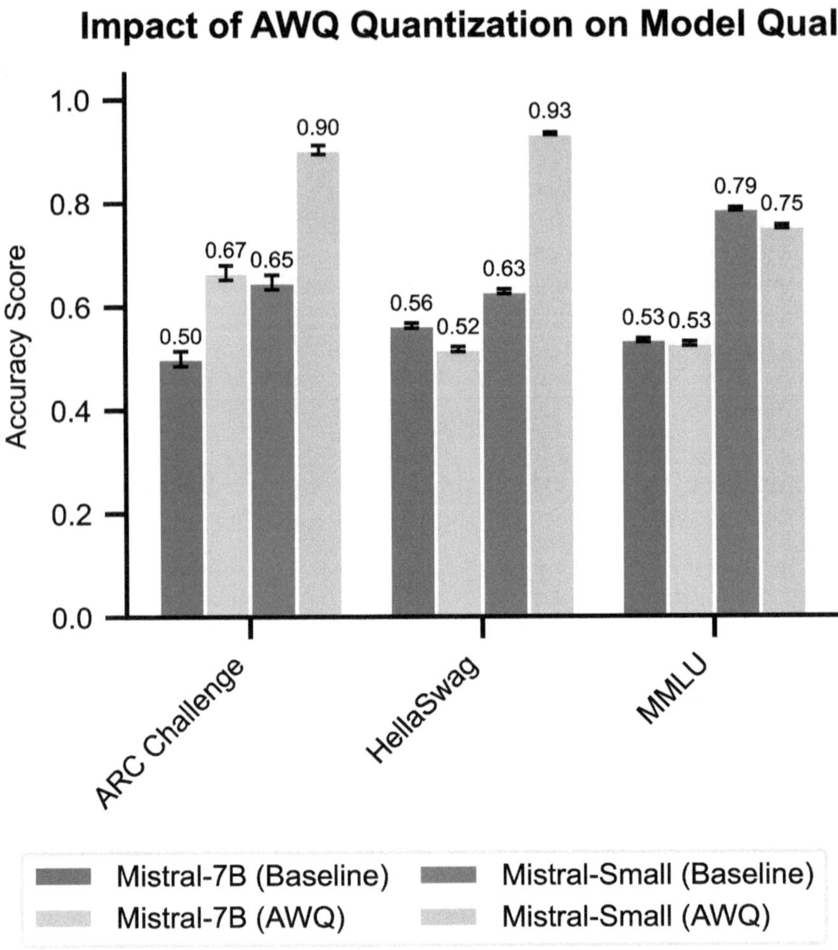

Fig. 6. Impact of AWQ Quantization on Model Inference Quality. Accuracy scores for baseline and AWQ quantized models across benchmarks. Error bars are standard error.

where, for a given batch job, E_{total} is the total energy consumed (in Joules), D is its total duration (in seconds), N_{tokens} is the number of tokens generated, and T is the system throughput (in tokens per second). This system-level perspective is crucial; our goal is to optimize for the service operator, who seeks to maximize the overall work done per unit of energy and time, rather than for an individual user, who might only seek to minimize the cost of a single request. Therefore, minimizing the EDP of a batch operation inherently means finding the configuration that best minimizes energy and duration while simultaneously maximizing throughput. Figure 8 presents the EDP curves for all three architectures.

First, the EDP curves reveal the efficiency profile of each GPU. The Hopper H200 (Fig. 8a) operates at a significantly lower absolute EDP than the A100

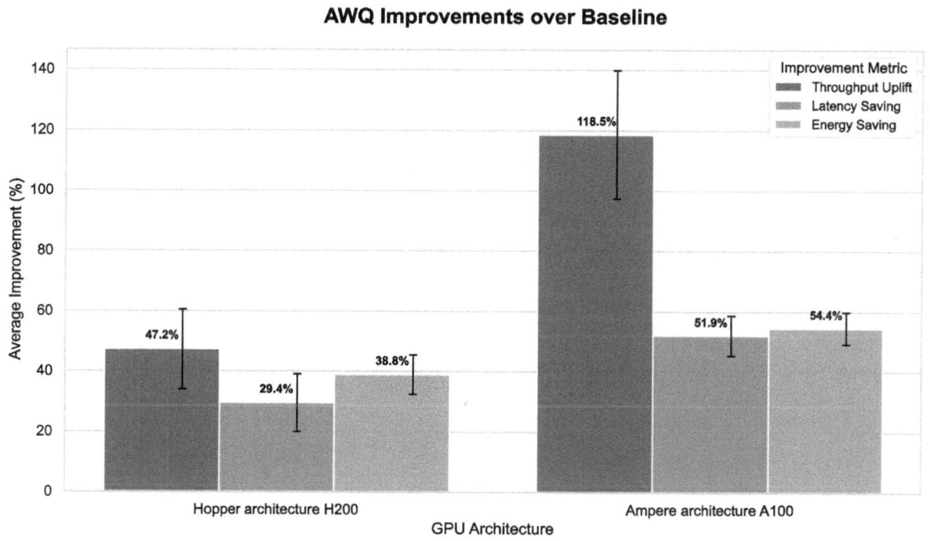

Fig. 7. Average performance and efficiency gains from AWQ across architectures. The data confirms AWQ as a dominant, cross-platform optimization.

(Fig. 8b), establishing its higher baseline efficiency. The RTX 5000 (Fig. 8c), while competitive at small batches, sees its EDP rise sharply, revealing its unsuitability for large, sustained inference workloads.

Second, the optimal batch size—the minimum of the EDP curve—is clearly architecture-dependent. The H200 finds its sweet spot in a larger batch size range (32–64) than the A100 (16–32). This suggests that the newer architecture can handle larger concurrent workloads more efficiently before the costs of duration and energy begin to outweigh the throughput gains. This minimum EDP point often corresponds to the transition point from a memory-bound to a compute-bound state, representing the ideal balance of hardware utilization.

Finally, the impact of AWQ (orange lines) is profound on all platforms, dramatically lowering the entire EDP curve. This resolves the dilemma posed in Sect. 4.1. By first applying AWQ and then selecting the batch size that minimizes EDP, an operator can identify a highly efficient configuration that represents a significant improvement over naive throughput maximization. A final, subtle observation from the curves is that the efficiency gap between AWQ and the baseline narrows at the highest batch sizes (e.g., 128). This is likely due to the computational overhead of on-the-fly dequantization. While this overhead is negligible in memory-bound scenarios, it becomes more significant as the system approaches a fully compute-bound state, slightly reducing AWQ's relative advantage. This phenomenon further reinforces our conclusion that a point of high efficiency is not at the maximum possible batch size, but at the specific mid-range batch size that minimizes the EDP.

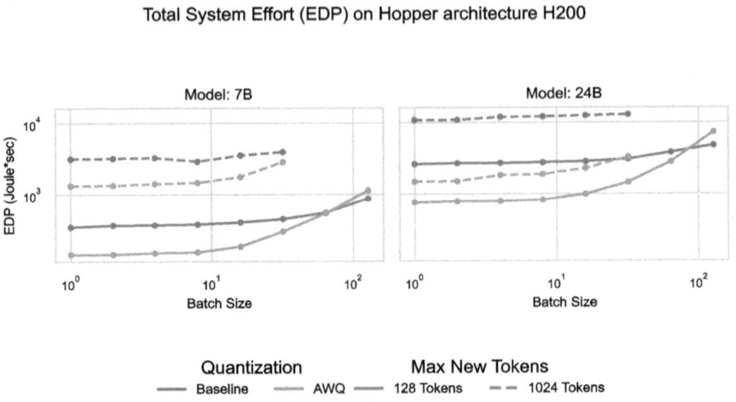

(a) Hopper H200

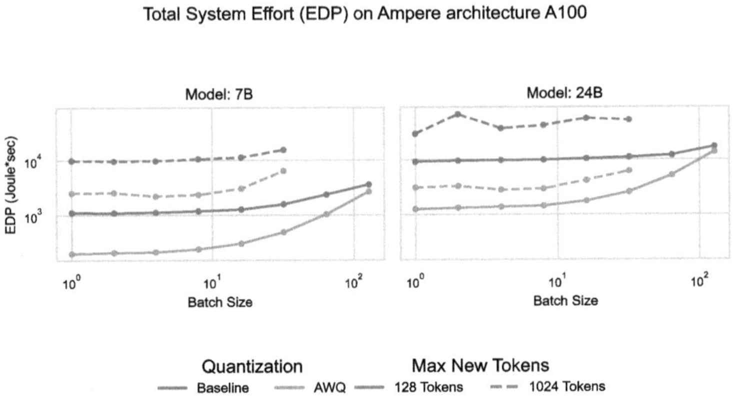

(b) Ampere A100

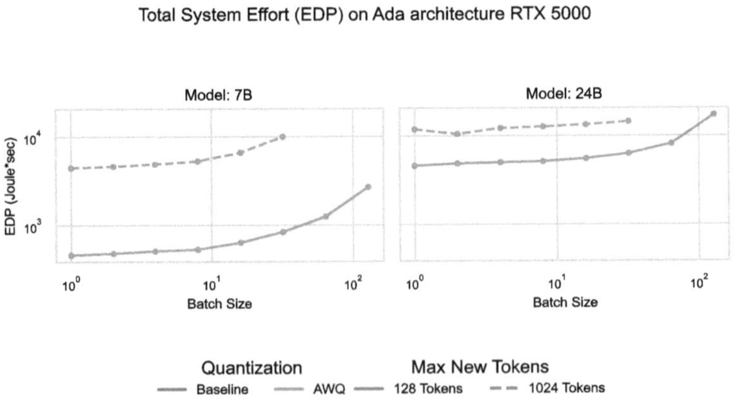

(c) Ada RTX 5000

Fig. 8. System Effort (EDP) vs. Batch Size across architectures.

5 Discussion and Conclusions

Our comprehensive analysis reveals a clear, actionable hierarchy of optimizations for deploying Large Language Models efficiently. The most impactful approach is not a single choice, but a two-step process: first, the application of a software-level optimization like AWQ, **and** the selection of a hardware-specific, highly efficient batch size from the EDP metric.

A key implication of our findings is that simply pursuing the highest possible throughput is a sub-optimal and often wasteful strategy. The EDP curves demonstrate that beyond a certain "sweet spot," the marginal gains in throughput are dwarfed by the super-linear increases in energy and duration costs. This is particularly true for older hardware like the Ampere A100. Our work provides a quantitative framework for identifying this sweet spot, allowing data-driven decisions that balance performance with sustainability and operational cost.

Furthermore, the consistent and dramatic benefits of AWQ across all platforms underscore the critical importance of co-designing inference solutions. The synergy between the quantization algorithm (software), the underlying GPU architecture (hardware), and a high performance kernel like MARLIN is what unlocks the next frontier of efficiency. Our primary contribution is a clear, two-step methodology for resolving the conflict between performance and cost:

1. **Apply Activation-aware Weight Quantization (AWQ):** We confirmed that AWQ is a dominant optimization, delivering substantial, concurrent gains in throughput (up to 118.5%) and savings in energy (up to 54.4%) without a significant loss of model quality.
2. **Select the Batch Size that Minimizes EDP:** We showed that the Energy-Delay Product (EDP) is an effective metric for identifying a point of high efficiency for any given hardware platform, resolving the conflict between competing metrics.

By applying these two steps, a service operator can achieve a configuration that represents an optimized balance of performance and total cost. This work provides not just measurements, but a practical and validated strategy for the sustainable deployment of AI systems. While the EDP provides a holistic view of system-level trade-offs, we acknowledge that it is not the only metric for efficiency. For specific use cases, operators may consider complementary metrics, such as **Joules per output token**, to further refine their analysis, particularly when generation cost is the primary concern over total job completion time.

6 Future Work

While this study provides a robust framework, future work could extend this methodology. A critical next step is to analyze **multi-GPU inference** for larger models. Further research should also cover a **broader hardware scope** (e.g., AMD GPUs), compare against other **advanced quantization methods** like GPTQ, and investigate the EDP metric in **dynamic serving environments**. Finally, an **economic analysis** could integrate cloud pricing models.

Acknowledgments. We thank Nicolas Herniou for his help with the network installation at UPEC. Olivier Michel has provided us with a broad and relevant bibliography collected by his slowTech project. Part of this work was performed using computing resources of CRIANN (Normandy, France) under project 2024004 *Modélisation des performances de codes parallèles de haut niveau.* This work has been supported under the name of project ParcimonIA by an internship funded by LACL for the first author. M. Belkahla's internship and the workstation HILBERT3 have been funded by UPEC's LACL laboratory.

References

1. Banner, R., Nahshan, Y., Soudry, D.: Post Training 4-Bit Quantization of Convolutional Networks for Rapid-Deployment. Technical report, arXiv:1810.05723 (2019)
2. Belkahla, M., Hains, G.: Measurements and optimizations of the energy consumption of AI systems (2025). https://hal.science/hal-05213288, working paper or preprint
3. Chmiel, B., Vahdat, A., Kautz, J.: Neural Networks Regularized by Quantization. Technical report, arXiv:2111.07970 (2021)
4. Clark, P., et al.: Think you have solved question answering? Try ARC, the AI2 reasoning challenge. arXiv preprint arXiv:1803.05457 (2018)
5. Dettmers, T., Lewis, M., Belkada, Y., Zettlemoyer, L.: LLM.int8(): 8-bit matrix multiplication for transformers at scale. Technical report, arXiv:2208.07339 (2022)
6. Frantar, E., Castro, R.L., Chen, J., Hoefler, T., Alistarh, D.: MARLIN: mixed-precision auto-regressive parallel inference on large language models. Technical report, arXiv:2408.11743 (2024)
7. Gonzalez, R., Horowitz, M.: Energy Dissipation in General Purpose Microprocessors. Technical report, IEEE Journal of Solid-State Circuits (1996). https://doi.org/10.1109/4.535416
8. Hendrycks, D., et al.: Measuring massive multitask language understanding. In: Proceedings of the International Conference on Learning Representations (ICLR) (2021)
9. Kwon, W., et al.: vLLM: easy, fast, and cheap LLM serving with pagedattention. Technical report, arXiv:2309.06180 (2023)
10. Lettres Ouvertes: ChatGPT : une consommation énergétique qui pose question. https://lettresouvertes.substack.com/p/chatgpt-une-consommation-energetique. Accessed 2025
11. Lin, J., Tang, J., Tang, H., Yang, S., Dang, X., Han, S.: AWQ: activation-aware weight quantization for LLM compression and acceleration. Technical report, arXiv:2306.00978 (2023)
12. McColl, W.F.: Bulk synchronous parallel computing. In: Abstract Machine Models for Highly Parallel Computers, pp. 41–63. Oxford University Press (1999)
13. Mullin, L., Hains, G.: Towards automatic, predictable and high-performance parallel code generation. In: Arabnia, H.R., et al. (eds.) Parallel and Distributed Processing Techniques and Applications, pp. 109–119. [Series name if known, e.g., Transactions on Computational Science and Computational Intelligence], Springer Nature Switzerland (2025), cSCE/PDPTA Las Vegas July 2024
14. Wedemain.fr: IA et consommation énergétique : un modèle à revoir d'urgence. https://www.wedemain.fr/dechiffrer/ia-et-consommation-energetique-un-modele-a-revoir-durgence/. Accessed 2025

15. Wikipedia contributors: Zero-shot learning — Wikipedia, the free encyclopedia (2025). https://en.wikipedia.org/w/index.php?title=Zero-shot_learning&oldid=1301662466. Accessed 17 Aug 2025
16. Zellers, R., Holtzman, A., Bisk, Y., Farhadi, A., Choi, Y.: HellaSwag: can a machine really finish your sentence? In: Proceedings of the 57th Annual Meeting of the Association for Computational Linguistics (ACL), pp. 4791–4800. Florence, Italy (2019)

LLM Driven Processes to Foster Explainable AI

Marcel Pehlke(✉) and Marc Jansen

University of Applied Sciences Ruhr West, Bottrop, Germany
{marcel.pehlke,marc.jansen}@hs-ruhrwest.de

Abstract. We present a modular, explainable LLM-agent pipeline for decision support that externalizes reasoning into auditable artifacts. The system instantiates three frameworks—Vester's Sensitivity Model (factor set, signed impact matrix, systemic roles, feedback loops), normal-form games (strategies, payoff matrix, equilibria), and sequential games (role-conditioned agents, tree construction, backward induction), with swappable modules at every step. LLM components (default: GPT-5) are paired with deterministic analyzers for equilibria and matrix-based role classification, yielding traceable intermediates rather than opaque outputs. In a real-world logistics case (100 runs), mean factor alignment with a human baseline was **55.5%** over 26 factors and **62.9%** on the transport-core subset; **role agreement** over matches was **57%**. An LLM judge using an eight-criterion rubric (max 100) scored runs on par with a reconstructed human baseline. Configurable LLM pipelines can thus mimic expert workflows with transparent, inspectable steps.

Keywords: Large Language Models (LLMs) · Explainable AI (XAI) · Multi-Agent Systems

1 Introduction

Canada is currently facing one of the most pressing social and economic challenges of the decade: the housing crisis. The federal government recently announced its Build Canada Homes initiative, pledging to accelerate the construction of affordable housing to address shortages across the country[1]. When asking today's most advanced language models about potential solutions, responses are typically broad and multi-factored, pointing to issues ranging from zoning policies to labor shortages. While such answers highlight the multifaceted nature of the problem, they often remain vague, lack prioritization, and most importantly, provide no transparent reasoning path. Users are left with plausible statements, but without an explanation of how the model arrived at them. This

[1] Government of Canada: "Prime Minister Carney launches Build Canada Homes," September 14, 2025. Available at: https://www.pm.gc.ca/en/news/news-releases/2025/09/14/prime-minister-carney-launches-build-canada-homes.

H. R. Arabnia et al. (Eds.): CSCI 2025, CCIS 2972, pp. 392–406, 2027.
https://doi.org/10.1007/978-3-032-28469-3_26

"black-box factor" undermines trust in LLM-driven decision support, particularly for domains where systemic reasoning and accountability are essential.

At the same time, the field of artificial intelligence is undergoing a notable paradigm shift. Instead of focusing solely on scaling up models, recent research emphasizes reasoning and orchestration. Reasoning models, showcase stepwise problem solving through chain-of-thought techniques, making intermediate reasoning explicit [17]. Orchestration mechanisms further extend this principle by dynamically routing queries to the most suitable process or model, enabling adaptive trade-offs between accuracy, latency, and cost [9]. Building on these trends, LLM-based agents introduce an additional layer: they do not only reason, but also act, decompose tasks, and coordinate across specialized roles. Multi-agent setups in particular promise structured, auditable processes that move beyond opaque one-shot answers [5].

This work leverages that paradigm shift by implementing three structured processes within an LLM agent system. Vester's Sensitivity Analysis, normal-form game analysis, and a sequential (extensive-form) game analysis. Each framework provides a transparent reasoning workflow, externalizing intermediate artifacts such as influence matrices, payoff tables, and decision trees. Our aim is not to render the LLM's internal computations transparent, but to add a process layer on top of the LLM, in which the process allows detailled statistical analysis providing further insides into the decision. By embedding the model in structured processes (e.g., Vester matrices, game trees), we expose the macro-level reasoning: what inputs were used, which options were considered, how criteria were applied, and why a path was selected.

From this perspective, we pursue the following research question: Can an LLM-based agent system mimic established decision-making frameworks such as game theory and sensitivity analysis, thereby providing a traceable reasoning process instead of opaque outputs?

To address this question, we implement and evaluate the proposed agent pipelines on real-world inspired scenario. Our evaluation strategy compares LLM-generated results against human-created baselines and methodological expectations. Specifically, we run the Vester sensitivity analysis pipeline 100 times on a logistics case study and benchmark the outputs against an established human-conducted study. In doing so, we do not only compare the final results of both studies, but also assess the quality of execution, that is, how well the LLM pipeline mimics the structure of the sensitivity analysis. This allows us to determine whether the system can reliably reproduce the process of Vester's methodology, rather than producing isolated or coincidental overlaps.

2 State of the Art

This chapter reviews the state of the art in large language models and structured decision frameworks, providing the foundations for our proposed agent-based approach.

2.1 Large Language Models: Capabilities, Limits, and Structured Reasoning

Large Language Models (LLMs) such as GPT-4 and GPT-5 have become central to AI research and application. Their foundation lies in the Transformer architecture [15], which enabled scalable training and parallelization. Scaling laws and model families like GPT, PaLM, and Chinchilla revealed that size and data could yield emergent abilities, but also that performance gains now plateau [6,7]. Despite their impressive capabilities, LLMs remain opaque: they frequently hallucinate, producing plausible but incorrect statements, and provide no clear reasoning path, which undermines their explainability [2].

Recent work shifts focus from scaling to structure. Chain-of-Thought (CoT) prompting [17] improves accuracy on multi-step problems by exposing intermediate reasoning. OpenAI's reasoning models, notably o1 and o3-mini, refine this approach by offering adjustable "reasoning effort" levels. The o1 model reached PhD-level proficiency in subjects like physics, chemistry, and biology, solving 83% of problems on the American Invitational Mathematics Examination (AIME) and ranking in the 89th percentile in Codeforces competitions. Its successor, o3-mini, matches this performance at medium effort while delivering faster responses, and with high effort achieved 83.6% accuracy on AIME 2024 and 77% on PhD-level science benchmarks (GPQA Diamond)[2].

While CoT enhances single-model reasoning, many tasks require selecting which model to use at runtime. GPT-5 exemplifies this with a modular setup: a fast base model for routine queries, a slower reasoning model for complex tasks, and a router that dynamically directs queries based on complexity and user signals. This orchestration allows efficient trade-offs between accuracy, latency, and cost[3]. Conceptually, this is closely aligned with the idea pursued in this work: different prompts can trigger not only different models, but also different structured processes, such as Vester's sensitivity analysis or game-theoretic reasoning, depending on the nature of the problem. In both cases, orchestration enables dynamic adaptation by selecting the most appropriate reasoning pathway for each task, rather than relying on a single, uniform approach.

Similar ideas appear in research such as the Leeroo Orchestrator, which treats entire LLMs as "experts" and routes queries to the most suitable one [9]. Instead of building even-larger models, orchestration leverages diverse models, turning model selection itself into a learned decision process. This signals a paradigm shift: performance stems not only from scaling individual networks but from coordinating specialized components under adaptive routing. Orchestration thus represents a crucial development for deploying LLMs in real world, resource sensitive environments.

Building on orchestration, LLM agents extend reasoning into acting. They interpret tasks, decompose them into steps, retrieve knowledge, and use tools.

---

[2] OpenAI: "Learning to Reason with LLMs," September 12, 2024. Available at: https://openai.com/index/learning-to-reason-with-llms/.

[3] OpenAI: "Introducing GPT-5," 7. August 2025. Available at: https://openai.com/index/introducing-gpt-5/.

Frameworks such as ReAct [21] and Tree-of-Thoughts [20] highlight how agents combine reasoning with environment interaction in continuous sense–plan–act cycles. Multi-agent setups further show how LLMs can emulate teamwork by assigning roles and cross-validating ideas [5]. Importantly, modular agent designs externalize intermediate artifacts, plans, decisions, and memory states, that improve explainability and make reasoning auditable.

A key example of embedding explainability into decision-making is the Questions–Options–Criteria (QOC) framework [8]. Recent work demonstrates that QOC can be effectively integrated with AI. Schmidt et al. [13] propose an AI-augmented QOC process where models contribute at three levels: (1) as assistants that suggest new options or criteria from data, (2) as participants evaluating alternatives alongside humans, or (3) as autonomous decision-makers in multi-agent setups that generate and assess the full QOC structure. Importantly, because each option and assessment remains explicitly logged, the reasoning pathway is auditable, even when agents act autonomously. Their evaluation highlights that this approach improves both traceability and decision quality, aligning with the goals of Explainable AI (XAI).

This shows how structured frameworks can complement LLMs: they externalize reasoning into human-auditable steps, reduce black-box opacity, and provide a verifiable link between input, reasoning, and outcome. Extending this principle to other methods, such as Vester's sensitivity analysis or game-theoretic reasoning, illustrates a path toward AI agents that not only generate powerful results but also justify them through transparent, structured processes.

2.2 Frederic Vester's Sensitivity Model (VSM)

Frederic Vester's Sensitivity Model (VSM) is a systems method for tackling complex, interdependent problems by combining qualitative judgement with lightweight quantification and an explicit focus on feedback and roles [18]. Eight bio-cybernetic principles (e.g., dominant negative feedbacks, diversity/decentralization, avoiding growth dependence) guide a structured workflow that surfaces key drivers, sensitivities, and leverage points [1,16].

1. **System Description** – define boundaries, goals, and context.
2. **Variable Set** – identify 10–20 essential variables.
3. **Criteria Matrix** – filter variables by systemic relevance (e.g., measurability, influenceability).
4. **Impact Matrix** – score how each variable influences the others.
5. **Systemic Role Analysis** – classify variables (drivers, critical, buffering, reactive) via active/passive sums.
6. **Effect System** – visualize the network and feedback loops.
7. **Partial Scenarios** – explore qualitative "what-if"s on subloops/subsystems.
8. **Simulation** – run qualitative/fuzzy simulations of scenario dynamics.
9. **Cybernetic Evaluation** – assess strategies against the principles for viability.

Among these, the **impact matrix** and **role analysis** are the methodological core, reducing complexity by revealing drivers, responses, and leverage points [1]. Embedding this workflow in an LLM multi-agent pipeline yields traceable intermediate artifacts (matrices, roles, loops), enabling problem solving with visible reasoning rather than black-box outputs.

2.3 Game Theory Aspects

Game theory studies strategic interactions where outcomes depend on the choices of multiple agents. Formalized by von Neumann and Morgenstern in 1944 [11], it has become a central tool in economics and beyond. Two core representations are the *normal form* (simultaneous moves) and the *extensive form* (sequential moves).

In a normal-form game, each player chooses a strategy simultaneously and payoffs are defined for every strategy profile. A classical example is the Prisoner's Dilemma [4], where individually rational choices (defection) lead to a collectively suboptimal outcome. The central solution concept here is the Nash equilibrium [10]: a strategy profile in which no player can improve their payoff by unilaterally deviating. Nash proved that every finite game has at least one equilibrium, possibly in mixed strategies, making it the cornerstone of non-cooperative game theory.

Many interactions are sequential. The extensive form represents games as decision trees where players act in turn. In this setting, Nash equilibrium may allow non-credible strategies. Selten introduced the refinement of subgame perfect equilibrium (SPE) [14], requiring that strategies form an equilibrium in every subgame. For perfect-information games, backward induction provides a constructive way to compute SPE, ensuring credibility of strategies at every node.

Normal-form and extensive-form reasoning offer structured ways to analyze interdependent decisions. Concepts like Nash equilibrium and backward induction can be operationalized in LLM-based agents, enabling them to "play through" strategic scenarios step by step. Embedding such frameworks allows AI agents not only to produce outcomes, but to expose their reasoning paths through well-established game-theoretic solution concepts—making decisions both analyzable and traceable.

3 Architecture

The system is designed in a modular way, where each analytical framework is implemented as a self-contained process. Specifically, this work implements three processes: Vester's Sensitivity Analysis, normal-form game analysis, and a sequential (extensive-form) game analysis. Each process follows its own structured workflow, from input parsing to stepwise reasoning and final output.

A future orchestration layer, which is not implemented here, could route each user prompt to the most suitable process, mirroring recent runtime orchestration for LLMs. Because the processes are modular, the orchestrator can scale and evolve, and via MCP it could even dispatch to processes running on different machines while retaining only lightweight location metadata.

In this work, the focus is placed on implementing the individual processes themselves rather than the orchestration layer. Nonetheless, the modular architecture demonstrates how structured analytical methods can be integrated into LLM-based systems, providing both problem-solving capability and transparency in the reasoning process.

4 Implementation

Across all three processes (Vester sensitivity analysis, normal-form games, and sequential games), the implementation follows a modular, step-wise design: each step is an independent module with swappable AI agents (model family, temperature, reasoning effort, etc.), allowing fine-grained configuration and future upgrades. At the end of each process, a dedicated summarization agent consolidates the intermediate artifacts into a human readable explanation and, where applicable, produces visuals (e.g., influence diagrams, role maps, payoff matrices, heatmaps) to communicate the results clearly.

4.1 Normal-Form Game Implementation

The normal-form game process is organized into modular steps, each handled by a dedicated module or agent. First, a player-extraction module identifies the two players, their overall objective, the payoff unit (i.e., the measure in which outcomes are expressed, such as profit in USD, utility points, or years of freedom), and an expected value range from the user's scenario description. Next, a strategy-generation module proposes up to ten strategies per player, ensuring concise names and one-line descriptions. Based on the extracted strategies, an LLM-powered payoff assignment module constructs the payoff matrix by rating the outcomes of each strategy pair within the defined range. This matrix is then passed to a deterministic analyzer, which computes classical solution concepts: dominant strategies, Nash equilibria, and Pareto-optimal outcomes. Finally, the results are sent to an interpretation agent that summarizes the analysis in natural language, explaining the meaning of the equilibria, trade-offs, and dominant strategies. The agent concludes with a practical recommendation tailored to the user's original scenario. Visual outputs such as payoff matrices or highlighted equilibria can be generated to further support understanding.

4.2 Sequential (Extensive-Form) Game Implementation

The sequential-game pipeline mirrors the normal-form intake (players, objective, payoff unit, value range) and adds role conditioning: a role-assignment module creates concise "You are ..." personas that become system prompts for player-specific LLM agents. Tree construction proceeds path-by-path from the root.

Players alternate turns, at each node the active player's agent proposes 1–3 concrete actions (or terminates if resolved), reasoning from its role rather than as a neutral narrator. A horizon policy ensures that paths remain meaningful and not too long: as the maximum depth is approached, the model avoids exploratory or redundant moves and instead favors termination. The maximum depth is a configurable parameter, allowing to balance between tree completeness and computational cost. All nodes are normalized into a consistent schema (decision or terminal), invalid outputs are corrected, and repeated actions along the same path are prevented. The result is a sequential game tree in which every path reflects alternating player choices grounded in their roles and limited by explicit, adjustable depth constraints. After tree construction, a neutral evaluator (separate LLM agent) assigns payoffs at each leaf using the full path and given the objective, roles, and scenario as context. It returns a JSON payload with a brief rationale plus two numbers (one per player), enforcing the specified unit and value range. The payoff and reasoning are stored on the terminal node so every leaf remains auditable and quantitatively scored. Once payoffs are assigned, a deterministic analyzer computes Pareto-optimal and dominated paths, and derives a Subgame Perfect Equilibrium (SPE) via backward induction. The analyzer returns a compact summary (SPE path and payoffs, strategy profile per node, Pareto-optimal vs. dominated sets, and focus-player diagnostics). A final interpretation agent converts this summary into clear guidance: it explains the SPE's strategic meaning, highlights Pareto trade-offs, clarifies why dominated paths are inferior, and names a single recommended decision consistent with the user's objective and the tree's outcomes. Optional visualization scripts can render the game tree with equilibrium highlights and path payoffs; these visuals are configurable and can be extended.

4.3 Vester Sensitivity Analysis

The Vester pipeline begins with an intake agent that transforms a natural-language scenario into a structured system definition. This includes: the system's name and high-level description, its primary goal, explicit boundaries (what is inside vs. outside the system), functional subsystems, variables already mentioned by the user, and a stakeholder list (with at least five identifiable actors). For each stakeholder, a dedicated role-conditioned agent generates 3–5 candidate variables reflecting that stakeholder's perspective for the system. Each variable is returned with a name, concise description, and optional category (economic, social, environmental, legal, or physical), ensuring that both direct and indirect influences are captured. The combined pool of variables is then passed to a deduplication and fit agent. This step merges near-duplicates under a representative name, consolidates descriptions, and records provenance via a merged_from list. Importantly, user-mentioned variables are preserved explicitly. The outcome is a cleaned, system-consistent set of variables that aligns with the system's boundaries and objectives, forming the foundation for subsequent steps of the sensitivity analysis. To construct the impact matrix, the system evaluates each ordered pair of variables ($A \rightarrow B$) with two labels: an influence strength in $\{0, 1, 2, 3\}$ and an *effect direction* in $\{-1, +1\}$. The strength scale follows Vester's rubric:

- 0 = no change at all,
- 1 = a change in A brings about only a weak change in B,
- 2 = A must change a lot to produce a comparable change in B,
- 3 = even small changes in A cause large changes in B.

The direction indicates whether an increase in A leads B to increase (+1) or decrease (−1). An LLM agent, primed to "mimic a stakeholder roundtable," scores each pair by integrating perspectives from the system intake (system description, roles, goals, and variable definitions). Each evaluation is strictly directional and pair-specific, self-influence entries on the diagonal are forced to zero. To ensure consistency, the system sends concise "variable cards" (name, description, context) and requests only the integer score and sign. The result is a signed, weighted adjacency matrix that reflects interdependencies in stakeholder terms. With the signed, weighted influence matrix, we compute for each variable its Active Sum (AS: row total) and Passive Sum (PS: column total). Using the dataset means $\overline{AS}, \overline{PS}$ as dynamic thresholds, each variable is classified into the standard Vester roles: **Active** (AS>$\overline{AS}$, PS≤$\overline{PS}$), **Critical** (AS>$\overline{AS}$, PS>$\overline{PS}$), **Reactive** (AS≤$\overline{AS}$, PS>$\overline{PS}$), and **Buffering** (AS≤$\overline{AS}$, PS≤$\overline{PS}$). We also compute $P = AS \cdot PS$ and $Q = \frac{AS}{PS} \cdot 100$ (with $Q = \infty$ if PS=0) for sensitivity dashboards. Additionally, we build a directed effect system and search for feedback loops using graph algorithms. Cycles are labeled positive (even number of negative edges) or negative (odd), and ranked by a Vester-oriented score that prioritizes stronger/shorter loops and increases with the share of *Critical*/*Active* variables and normalized P. The analyzer returns: the ranked loop list, a compact "pretty" effect matrix (sign + strength) and an *Top-K critical loops* summary with variable frequencies for dashboarding. While the original Vester methodology also includes *partial scenarios* and *simulation*, our implementation does not cover these steps. We deliberately focused on the construction of the **impact matrix** and the subsequent **systemic role analysis**, which are widely regarded as the methodological core because they reveal the system's key drivers, sensitive outcomes, and leverage points [1]. Since our goal is to provide decision support directly from user prompts, the matrix and role analysis already supply the necessary structure for identifying interventions and reasoning about system dynamics. In the final step, an interpretation agent converts the full analysis into actionable guidance. It receives the user scenario, the cleaned context (deduplicated variables, stakeholders, subsystems, boundaries), the cross-impact matrix, sensitivity metrics per variable (AS, PS, $P = AS \cdot PS$, $Q = (AS/PS) \cdot 100$, role), the signed effect system, and the ranked feedback loops (including a Top-K "critical loops" set). The agent then briefly restates the system and scope, interprets sensitivity to identify drivers, sensitive outcomes, buffers, and critical leverage points, uses link signs to explain propagation (reinforcing vs. balancing), and proposes intervention priorities with expected side effects, a phased roadmap (quick wins vs. structural moves) and a single best lever recommendation. All claims are required to be grounded in the provided variables, roles, matrix links, and loop structures, ensuring an auditable, human-readable summary aligned with the computed artifacts.

5 Evaluation

We evaluate whether the proposed multi-agent LLM pipeline reproduces the structure and reasoning of a Vester-style sensitivity analysis on a real logistics case. As a human reference, we use a study from the IVL Swedish Environmental Research Institute (IVL) *"A logistic analysis with the Sensitivity Model Prof. Vester"* [19], which analyzes apparel flows from China to the Nordics via DK/SE entry points and a Swedish DC. The report proceeds through Vester's nine steps, including variable elicitation across economic, environmental, social, organizational, and technical dimensions, a criteria filter, role charting, effect system, and partial scenarios; a second, later analysis adds company-internal factors. For like-for-like comparison, we treat the core transport-system model as the primary reference and the company/internal additions as a separate stratum.

5.1 Methods

Prompting and Runs. We conduct 100 independent runs of the sensitivity analysis pipeline using a concise problem statement aligned with the IVL transport scope (no IVL variables disclosed). The pipeline configuration follows the *Implementation* chapter (same modular steps and outputs), and all LLM agents across all steps use the same model with GPT-5.
The exact user prompt was:

> *For a real-world apparel transport flow from Shanghai to the Nordics (via DK/SE entry ports/airports and a Swedish regional DC to stores), how do the most consequential economic, environmental, social, organizational, and technical factors shape one another—and which of them emerge as primary levers, buffers/stabilizers, reactive indicators, and critical pressure points for system performance and sustainability?*

Semantic Factor Alignment. To test whether the system surfaces conceptually equivalent factors, we compute sentence embeddings (all-mpnet-base-v2) for IVL factor names and for LLM factors (name+short label). For each IVL factor, the best LLM neighbor by cosine similarity is a match if ≥ 0.50; otherwise it is unmatched. This threshold reflects typical behavior in modern embedding spaces, where unrelated items sit well below 0.5 while paraphrases or close conceptual equivalents exceed it, and is consistent with precedent in semantic similarity research [3]. Many-to-one mappings are allowed, since a single broader LLM factor can legitimately subsume several related human factors.

What We Evaluate. (i) *Factor alignment (match rate)* overall and by stratum (core vs. company/internal). (ii) *Role consistency* for matched factors (agreement with IVL roles).

Rubric-Based Judging. In parallel, an LLM judge (GPT-5 Thinking) scores each run and a reconstructed packet of the IVL analysis under an eight-criterion rubric (max 100). The numbers in parentheses indicate the *maximum points per criterion*: boundary clarity (5), stakeholder coverage (12), variable set completeness (12), causal structure (14), feedback loops (14), scenario design (14), temporal relevance (14), and actionability (15).

This evaluation adopts a rubric-based LLM judging approach, supported by recent evidence that advanced models can approximate human evaluation with high reliability: studies show GPT-4 aligns with human preference judgments in over 80% of cases, comparable to inter-rater agreement [22]. Leveraging this method provides a scalable and reproducible alternative to costly human rating campaigns while still grounding judgments in substantive criteria. In our setup, the goal is to grade the quality and internal coherence of each analysis (boundaries, factors, causal structure, loops, scenarios, etc.), not whether it copies IVL text. The same procedure is applied both to all LLM runs and to a reconstructed packet of the IVL baseline, enabling like-for-like comparison.

5.2 Results

Across all 26 IVL factors, mean alignment was **55.5%** (range **23.1–80.8%**); **71%** of runs reached $\geq$ 50%, **29%** $\geq$ 65%, and **7%** $\geq$ 75%. Restricting to the *core transport-system* set (18 factors), mean alignment rose to **62.9%** (range **27.8–94.4%**; 5th–95th percentiles **38.6–83.3%**). Here, **84%** of runs reached $\geq$ 50%, **55%** $\geq$ 65%, and **23%** $\geq$ 75%. For the *company/internal* additions (8 factors), mean alignment was **38.8%** (range **0–75%**; 5th–95th **12.5–62.5%**); **30%** of runs reached $\geq$ 50%.

At factor level, core items such as *Fuel prices* (**99%**), *Competition in the transport market* (**95%**), and *Environmental demands of system actors* (**93%**) were matched consistently, whereas breadth/structural measures (e.g., *Total amount of goods*, *Share of air freight*) and the internal set (e.g., *Image of company*, *Education and information*) matched less often—consistent with the transport-focused prompt boundary.

Role fidelity (over matched factors) averaged **56.6%** overall (median **56.7%**); aggregated per factor and averaged within strata, mean role match was **49%** (core) and **41%** (internal). High match rate did not guarantee role agreement for every factor; this sensitivity to small matrix differences parallels variation reported in participatory scenario work, where stakeholder mix can shape outcomes [12].

Table 1 summarizes rubric outcomes across 100 runs versus the IVL baseline. LLM totals spanned **77–98** (mean **92.97**, SD **3.47**) against the human total of **93**. Boundary clarity (**4.99/5**), actionability (**14.94/15**), and temporal relevance (**13.5/14**) were consistently high; stakeholder coverage (**11.52/12**), variable completeness (**10.87/12**), and causal structure (**12.95/14**) clustered near the top. Greater spread appeared in feedback loops (**3–14**, mean **13.02**) and scenarios (**2–13**, mean **11.18**), aligning with IVL-level ranges.

Table 1. Rubric scores from the LLM judge: 100 runs (summary) vs. IVL baseline.

Criterion	Max	Human	Mean (LLM)	SD	Min	Max
Boundary clarity	5	5	4.99	0.10	4	5
Stakeholder coverage	12	11	11.52	0.54	10	12
Variable completeness	12	11	10.87	0.48	9	12
Causal structure	14	13	12.95	0.79	8	14
Feedback loops	14	13	13.02	1.46	3	14
Scenario design	14	13	11.18	1.41	2	13
Temporal relevance	14	13	13.50	0.52	12	14
Actionability	15	14	14.94	0.24	14	15
Total	100	93	92.97	3.47	77	98

5.3 Interpretation

What the Alignment Says. The pipeline repeatedly surfaced the main external transport levers identified in the IVL report, indicating that, given only a scoped prompt, agents "thought" along similar systemic lines. Lower alignment for later company/internal factors reflects the evaluation scope: those factors emerged in a separate managerial workshop and are less likely to appear from an externally framed transport prompt.

Roles are Sensitive, as Expected. Moderate role agreement (~57% over hits) is consistent with the inherent sensitivity of Vester's method: even small changes in judged influences can shift variables across categories (active, reactive, buffering, critical). This is not a weakness of our approach but reflects the domain-specific nature of sensitivity analysis itself: role charts are highly dependent on influence ratings and stakeholder perspectives, and small changes in emphasis can legitimately produce different outcomes. In practice, such variability could also be expected in human workshops, where different expert groups or contextual assumptions may lead to alternative role classifications. While our pipeline reproduced this pattern, the result also highlights room for enhancement. Future work could explore diversifying models across steps to reduce variance and stabilize role agreement, while recognizing that some degree of variability is intrinsic to Vester's design.

Form, Not Imitation. A key strength of our evaluation is that the analyses were judged not only on factor overlap but also on whether they were substantively well-formed Vester studies. The rubric-based evaluation by an LLM judge shows that the pipeline consistently produced analyses with the right structure and reasoning steps, rather than merely echoing surface features of the IVL report. Across 100 runs, total scores clustered tightly between **77** and **98** (mean **92.97**, SD **3.47**), nearly replicating the human baseline of **93** and in several cases even surpassing it. The narrow variance demonstrates that high-quality analyses were not isolated outputs but a stable pattern across runs. Taken together, these results show that the pipeline was not simply "getting the right answers" by chance, but was reliably following the *form* of a Vester-style study. This demonstrates that an LLM pipeline can internalize the methodology itself: system boundaries, role logic, influence mapping, and feedback interpretation and apply it consistently across runs. That ability to replicate structure and reasoning rather than text is a central strength of our approach, and the reason why the pipeline can be considered not just an imitation of human output, but a genuine operationalization of the method.

In summary, the evaluation shows that the LLM system can approximate the structure and reasoning of a participatory Vester analysis in a complex logistics domain. Each run constitutes an individual analysis and is therefore not identically reproducible, just as human workshops are not: changing experts, emphasis, or minor influence judgments yields legitimate variation in factors and roles, consistent with evidence that stakeholder composition can materially alter outcomes [12]. However, the system consistently identified the key external transport-system levers, assigned roles that overlapped substantially with the human workshop, and achieved high rubric scores from an impartial LLM judge (**77–98**, mean **92.97**, SD **3.47**), matching the human based baseline of **93**. Alignment gaps for internally oriented factors and partial divergences in role fidelity highlight the dependence on framing and available knowledge, but these are precisely the aspects where expert-system augmentation could strengthen performance. Overall, the results suggest that multi-agent LLM pipelines are not only able to "think" along the same systemic lines as human experts but, with access to domain-specific context, may evolve into robust tools for participatory-style sensitivity analysis across specialized fields.

6 Conclusion and Outlook

This paper implemented and demonstrated three explainable analysis processes within a modular LLM-agent system: Vester's sensitivity analysis, normal-form game analysis, and sequential (extensive-form) game analysis. Each process externalizes intermediate artifacts (influence matrices and role charts; payoff tables; decision trees with backward-induction summaries), thereby replacing opaque one-shot answers with auditable reasoning steps.

Our evaluation shows that the approach can reproduce the form and reasoning of established human methods rather than yielding coincidental overlaps. On a logistics case study, we executed the Vester pipeline 100 independent times and compared against a human baseline. Mean factor alignment across all 26 IVL factors was 55.5% (core transport subset: 62.9%; company/internal subset: 38.8%). Role fidelity over matched factors averaged 56.6%. A rubric-based LLM judge rated structure and methodological quality tightly around the human baseline (LLM mean 92.97/100 vs. human 93/100), indicating that outputs were consistently well-formed Vester studies.

Revisiting the opening problem from our introduction, the Canada housing crisis, we applied our framework to this scenario. We provided the system with a problem statement capturing chronic undersupply, rising costs, permitting delays, financing risks, and their socio-economic consequences. When processed through our Vester analysis pipeline, the salient levers and risks closely aligned with those emphasized in the federal initiative *Build Canada Homes (BCH)*—a new agency launched on September 14, 2025 to accelerate affordable and middle-class housing.[4]

Concretely, our analysis foregrounded public land availability, approval and permitting speed, financing risk under volatile interest rates, and the capacity of modular construction and skilled labour. These map directly to BCH's pillars: leveraging federal public lands, streamlining approval processes, mobilizing private and non-profit developers with financial instruments, and scaling modern construction methods. Both perspectives converge on the same critical uncertainties, construction cost inflation, interest rate trajectories, regulatory and intergovernmental coordination, supply chain and labour bottlenecks, land servicing requirements, and the operational definition of "affordable."

It is important to note that the LLM had no prior knowledge of these specific policy measures: neither the government program itself nor its details were present in the provided prompt, and they were outside the model's training knowledge, which currently has a cutoff of September 30, 2024.[5] This correspondence therefore illustrates that the LLM based sensitivity analysis surfaced actionable levers consistent with contemporary policy design, rather than offering vague, black box prescriptions.

Taken together, this marks a step toward XAI via implementing traceable frameworks: multi-agent LLM pipelines can reproduce known methods with traceable steps, deliver human-aligned and inspectable decision support.

For future work, our results suggest that small changes in influence ratings can shift systemic roles, highlighting the sensitivity of these frameworks to initial judgments. We therefore propose an *expert layer* that can be flexibly integrated into each process of the pipeline. Due to the modular design of our implementation, such an expert layer could be selectively activated at steps where external validation or additional context is most valuable.

The expert layer is not limited to human input. It could draw from multiple complementary sources, such as real-time web search to inject up-to-date information, domain-specific or fine-tuned models that encode sector knowledge, structured company data or transcripts from internal meetings in organizational settings, or curated expert feedback at critical decision points. By enriching the agent processes with targeted knowledge at high leverage steps, the expert layer has the potential to improve both the robustness and fidelity of the analysis, while maintaining overall autonomy and scalability of the system. Our evaluation used a single model configuration across steps. Future experiments

[4] Government of Canada: "Prime Minister Carney launches Build Canada Homes," September 14, 2025. Available at: https://www.pm.gc.ca/en/news/news-releases/2025/09/14/prime-minister-carney-launches-build-canada-homes.

[5] OpenAI Documentation: "GPT-5 models," knowledge cutoff September 30, 2024. Available at: https://platform.openai.com/docs/models/gpt-5.

could diversify models per step (e.g., retrieval-heavy steps vs. numeric-judgment steps), test reasoning-focused variants, and apply cross-model self-consistency (vote/aggregate) to stabilize outcomes.

References

1. Amaya Moreno, L., Defterli, O., Fügenschuh, A., Weber, G.-W.: Vester's sensitivity model for genetic networks with time-discrete dynamics. In: Dediu, A.-H., Martín-Vide, C., Truthe, B. (eds.) AlCoB 2014. LNCS, vol. 8542, pp. 35–46. Springer, Cham (2014). https://doi.org/10.1007/978-3-319-07953-0_3
2. Chelli, M., et al.: Hallucination rates and reference accuracy of chatgpt and bard for systematic reviews: comparative analysis. J. Med. Internet Res. **26**, e53164 (2024). https://doi.org/10.2196/53164. https://www.jmir.org/2024/1/e53164
3. Corley, C., Mihalcea, R.: Measuring the semantic similarity of texts. In: Proceedings of the ACL Workshop on Empirical Modeling of Semantic Equivalence and Entailment, EMSEE 2005, pp. 13–18. Association for Computational Linguistics, USA (2005)
4. Flood, M.M.: Some experimental games. Manag. Sci. **5**(1), 5–26 (1958). https://doi.org/10.1287/mnsc.5.1.5. https://pubsonline.informs.org/doi/10.1287/mnsc.5.1.5
5. Guo, T., et al.: Large language model based multi-agents: a survey of progress and challenges. In: Larson, K. (ed.) Proceedings of the Thirty-Third International Joint Conference on Artificial Intelligence (IJCAI 2024), pp. 8048–8057. International Joint Conferences on Artificial Intelligence Organization (2024). https://doi.org/10.24963/ijcai.2024/890
6. Hoffmann, J., et al.: Training compute-optimal large language models. In: Proceedings of the 36th International Conference on Neural Information Processing Systems. NIPS 2022. Curran Associates Inc., Red Hook, NY, USA (2022)
7. Kaplan, J., et al.: Scaling laws for neural language models. CoRR abs/2001.08361 (2020). https://arxiv.org/abs/2001.08361
8. MacLean, A., Young, R.M., Bellotti, V.M., Moran, T.P.: Questions, options, and criteria: elements of design space analysis. Hum.-Comput. Interact. **6**(3-4), 201–250 (1991). https://doi.org/10.1080/07370024.1991.9667168. https://www.tandfonline.com/doi/abs/10.1080/07370024.1991.9667168
9. Mohammadshahi, A., Shaikh, A.R., Yazdani, M.: Routoo: learning to route to large language models effectively (2024). https://arxiv.org/abs/2401.13979
10. Nash, J.: Non-cooperative games. Ann. Math. **54**(2), 286–295 (1951). http://www.jstor.org/stable/1969529
11. von Neumann, J., Morgenstern, O.: Theory of Games and Economic Behavior, 5th edn. Princeton University Press, Princeton (1944)
12. Reed, M., et al.: Participatory scenario development for environmental management: a methodological framework illustrated with experience from the UK uplands. J. Environ. Manag. **128**, 345–362 (2013). https://doi.org/10.1016/j.jenvman.2013.05.016. https://www.sciencedirect.com/science/article/pii/S0301479713003447
13. Schmidt, L., Pehlke, M., Jansen, M.: AI-enhanced QOC-analysis: a framework for transparent and insightful decision-making. In: Camarinha-Matos, L.M., Ortiz, A., Boucher, X., Barthe-Delanoë, A.M. (eds.) Navigating Unpredictability: Collaborative Networks in Non-linear Worlds, pp. 415–429. Springer, Cham (2024)

14. Selten, R.: Spieltheoretische behandlung eines oligopolmodells mit nachfrageträgheit. teil ii: Eigenschaften des dynamischen preisgleichgewichts. Zeitschrift für die gesamte Staatswissenschaft/J. Inst. Theor. Econ. **121**(4), 667–689 (1965). http://www.jstor.org/stable/40748908
15. Vaswani, A., et al.: Attention is all you need. In: Proceedings of the 31st International Conference on Neural Information Processing Systems, NIPS 2017, pp. 6000–6010. Curran Associates Inc., Red Hook, NY, USA (2017)
16. Vester, F., von Hesler, A.: Sensitivity model (1980), study on ecological planning and sensitivity modeling
17. Wei, J., et al.: Chain-of-thought prompting elicits reasoning in large language models. In: Proceedings of the 36th International Conference on Neural Information Processing Systems, NIPS 2022. Curran Associates Inc., Red Hook, NY, USA (2022)
18. Werner, U.: Can nature teach us good research practice? A critical look at frederic vester's bio-cybernetic systems approach. J. Res. Pract. **1** (2005)
19. Wolf, C., Persson, F., Jelse, K.: A logistic analysis with the sensitivity model prof. vester. Technical Report B2048, IVL Swedish Environmental Research Institute, Göteborg, Sweden (2012). https://www.diva-portal.org/smash/get/diva2:1552071/FULLTEXT01.pdf, Project sponsor: Geodis Wilson
20. Yao, S., et al.: Tree of thoughts: deliberate problem solving with large language models. In: Proceedings of the 37th International Conference on Neural Information Processing Systems, NIPS 2023. Curran Associates Inc., Red Hook, NY, USA (2023)
21. Yao, S., et al.: React: synergizing reasoning and acting in language models. In: Proceedings of the 11th International Conference on Learning Representations (ICLR) (2023)
22. Zheng, L., et al.: Judging LLM-as-a-judge with MT-bench and chatbot arena. In: Thirty-seventh Conference on Neural Information Processing Systems Datasets and Benchmarks Track (2023). https://openreview.net/forum?id=uccHPGDlao

Generative AI for Robotic Autonomy and Planning

Vida Vakilian and Reza Abdolee(✉)

California State University Channel Islands, Camarillo, USA
{vida.vakilian,reza.abdolee}@csuci.edu

Abstract. *Generative Artificial Intelligence (GenAI)*—including large language models, diffusion models, and world models—is transforming robotic autonomy by enabling robots to reason, plan, and adapt through language and imagination. This survey reviews the emerging field of *Generative Robotic Autonomy*, where cognitive generation integrates with physical control. We propose an architecture spanning three layers: generative reasoning, planning and control, and embodied execution, connecting high-level cognition to real-world actuation. Representative frameworks such as SayCan, Code-as-Policies, RT-2, Voyager, and DreamerV3 are analyzed across reasoning, grounding, and adaptability dimensions. Key challenges in safety, real-time performance, and explainability are discussed, alongside opportunities for digital-twin validation and cyber-physical integration.

Keywords: Robotic Autonomy · Large Language Models (LLMs) · Diffusion Models · World Models · Digital Twins · Embodied Intelligence

1 Introduction

The convergence of *Generative Artificial Intelligence (GenAI)* and robotics is redefining the landscape of autonomous systems. Traditional robotic planning has largely relied on deterministic or learning-based approaches such as finite state machines, model predictive control, and reinforcement learning [1,2]. These methods, while interpretable and reliable, often struggle to generalize beyond their training domain or adapt to novel situations [3].

Recent breakthroughs in *Generative AI*—including large language models (LLMs) [4,5], diffusion models [6,7], and world models [8]—offer new pathways toward *adaptive, reasoning-driven autonomy*. These models exhibit emergent capabilities for reasoning, abstraction, and imagination, enabling robots not only to execute pre-programmed actions but also to *generate* new task plans, simulate outcomes, and explain decisions in natural language. Generative models can translate human intent into executable robotic actions [9,10], synthesize synthetic sensor data for planning [11], and create predictive world representations that guide decision-making under uncertainty [12].

H. R. Arabnia et al. (Eds.): CSCI 2025, CCIS 2972, pp. 407–420, 2027.
https://doi.org/10.1007/978-3-032-28469-3_27

This paradigm shift has given rise to the emerging field of *Generative Robotic Autonomy*, where robots leverage generative cognition to plan, act, and adapt in open-ended environments [13,14]. Early demonstrations such as SayCan [10], Code-as-Policies [9], Voyager [12], and RT-2 [11] show how language models can bridge the gap between symbolic reasoning and continuous control, while world models like DreamerV3 [8] enable robots to build internal simulations of the physical world for predictive planning. As highlighted in recent surveys and reports [5,15], this convergence of generative reasoning, embodied learning, and digital-twin validation represents a foundational step toward *Embodied Generative Intelligence*—a new paradigm where robots can reason, imagine, and act autonomously in complex environments.

Despite rapid progress, this research area remains fragmented. Existing literature explores diverse aspects—from LLM-based planning and multimodal perception to generative simulation and digital twin integration—but lacks a unified taxonomy or comprehensive evaluation [13,14]. Moreover, questions of safety, grounding, trust, and computational efficiency remain largely unresolved [16,17], particularly when deploying generative models within cyber-physical and operational technology (OT) environments [15].

This survey provides a structured overview of current advancements in *Generative AI for Robotic Autonomy and Planning.* We categorize existing approaches, identify key frameworks, and analyze their capabilities across reasoning, planning, simulation, and embodied execution. The paper further explores how digital twins and cyber-physical systems can serve as validation environments for generative autonomy, ensuring safety and reliability in real-world applications. Finally, we discuss open challenges and outline future directions toward *Embodied Generative Intelligence*—autonomous systems capable of creative reasoning, continual learning, and safe operation in complex, dynamic environments.

2 Background and Related Work

2.1 Traditional Robotic Autonomy and Planning

Robotic autonomy has traditionally been achieved through hierarchical and model-based control architectures that decompose decision-making into perception, planning, and execution layers [18,19]. Classical methods such as finite state machines (FSMs), A* search, model predictive control (MPC), and hierarchical task networks (HTNs) [1,20,21] have enabled reliable task execution in structured environments. These approaches, while interpretable and verifiable, rely heavily on precise modeling and hand-crafted rules, limiting their scalability and adaptability in unstructured or dynamic scenarios [3].

The introduction of learning-based methods—particularly reinforcement learning (RL) and imitation learning (IL)—extended the ability of robots to acquire complex behaviors through experience [2,22]. However, these models typically operate within narrowly defined task distributions and require extensive data collection and tuning [23]. As a result, their performance often degrades when faced with novel instructions or environmental changes, motivating the integration of higher-level cognitive reasoning, abstraction, and generalization.

2.2 Generative AI Fundamentals

Generative Artificial Intelligence refers to models that can produce new data, knowledge, or control policies from learned representations. Unlike discriminative models that classify or predict, generative models *create*—learning underlying data distributions and generating plausible outputs consistent with that distribution [24]. The primary families of generative models relevant to robotics include:

- **Large Language Models (LLMs):** Transformer-based architectures (e.g., GPT-4, Llama 3, Gemini, Claude) trained on massive text corpora [4,5,25]. They demonstrate reasoning, planning, and code-generation abilities that can translate natural language into structured action sequences or executable robot control code [9,10].
- **Diffusion Models:** Probabilistic generative models (e.g., Stable Diffusion, ControlNet) capable of synthesizing realistic images, trajectories, and simulation environments from text or sensor data [6,7,26]. These are increasingly used for scene generation and visual task planning.
- **World Models:** Latent dynamic models (e.g., DreamerV3, PlaNet, Genie) that learn internal representations of environment dynamics, allowing robots to perform imagined rollouts for predictive control [8,27,28].

Together, these categories form the foundation of *Generative Robotic Intelligence*—an emerging paradigm where reasoning and imagination complement control and perception.

2.3 The Transition to Generative Robotic Autonomy

Recent research has begun integrating generative models into robotic planning and control pipelines, blurring the boundary between symbolic reasoning and physical execution. Several key directions illustrate this transition:

- **Language-to-Action Planning:** Works such as SayCan [10] and Code-as-Policies [9] use LLMs to interpret natural language commands and generate executable robot policies or task graphs.
- **Vision-Language-Action (VLA) Models:** Frameworks like RT-2 [11] and Voyager [12] couple visual inputs with textual reasoning, enabling open-vocabulary and zero-shot task execution.
- **World-Model-Based Control:** Methods such as DreamerV3 [8] employ predictive world models to simulate and evaluate potential outcomes before taking real-world actions, enhancing safety and efficiency.
- **Digital Twins and Simulation Integration:** Generative models are increasingly coupled with high-fidelity digital twins (e.g., NVIDIA Omniverse Isaac Sim) to plan, test, and validate robotic actions before deployment in operational technology (OT) environments [15].

2.4 Existing Surveys and Research Gap

Although prior surveys have reviewed learning-based and multimodal robotics [13,14,16], few have specifically examined the role of *generative models* as cognitive agents for autonomous planning. Existing work often treats language models as static policy generators, overlooking their potential for iterative reasoning, simulation, and self-adaptation [17]. Moreover, current reviews rarely consider the implications of deploying these models in *cyber-physical systems*, where real-time constraints, safety, and explainability are paramount. This survey addresses that gap by providing a unified taxonomy of generative models applied to robotic autonomy, encompassing language, vision, simulation, and world-modeling dimensions.

3 Taxonomy of Generative Models for Robotic Autonomy

Figure 1 illustrates the conceptual organization of *Generative AI for Robotic Autonomy*, composed of three hierarchical layers: (1) *Generative Reasoning*, (2) *Planning and Control*, and (3) *Embodied Execution*. This architecture bridges cognitive reasoning and physical action, establishing a unified framework for autonomy that spans from abstract goal formulation to real-world execution. Bidirectional information flow connects cognition with real-world feedback.

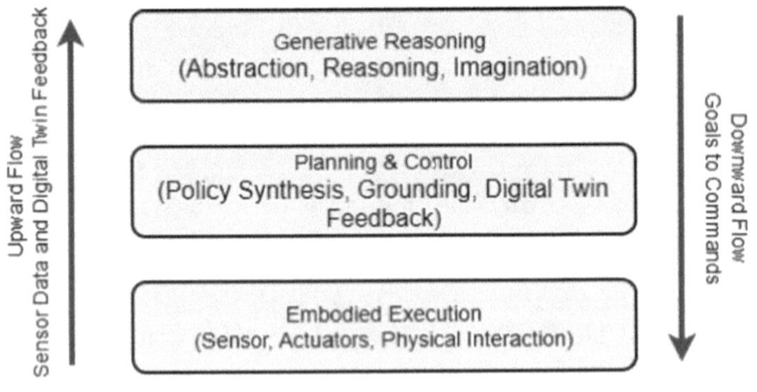

Fig. 1. Conceptual architecture of Generative AI for Robotic Autonomy.

3.1 Generative Reasoning Layer

The top layer encapsulates high-level cognition through models capable of abstraction, reasoning, and imagination. LLMs [4,5], diffusion models [6,7], and world models [8] serve as foundational components. They interpret human intent, decompose complex tasks, and synthesize new knowledge or behaviors through generative inference. Unlike classical planning algorithms [1], which rely on predefined rules, this layer dynamically constructs symbolic or code-based representations of the task environment. In practice, frameworks such as SayCan [10] and Code-as-Policies [9] demonstrate how reasoning layers convert natural language into structured, executable plans.

3.2 Planning and Control Layer

The middle layer functions as a mediator between abstract reasoning and physical execution. It translates high-level objectives into actionable control policies, integrating generative models with reinforcement learning, task and motion planning, and digital-twin feedback loops. Systems such as RT-2 [11] and Voyager [12] exemplify this tier by fusing language-driven intent with multimodal grounding and adaptive policy synthesis. This layer also interfaces with simulation environments (e.g., NVIDIA Isaac Sim [15]) for safe testing and iterative refinement before deployment. Through bidirectional interaction, planning modules continually update model parameters based on sensory feedback and digital-twin validation.

3.3 Embodied Execution Layer

The lowest layer encompasses physical interaction with the environment—embodied agents [29] equipped with sensors, actuators, and control hardware operating in cyber-physical and OT networks. Generative AI enhances this layer by enabling context-aware adaptation and self-correction in dynamic settings. Embodied policies are validated through digital twins [30] that replicate plant dynamics and network conditions, ensuring safe operation prior to deployment in real systems. Integration with industrial protocols, timing constraints, and safety standards (e.g., IEC 61508) [31] enables reliable performance within critical infrastructure and manufacturing domains.

3.4 Layer Interdependencies

Information flows downward as abstract goals become executable commands, while sensory and digital-twin feedback propagate upward to refine the cognitive model. This bidirectional process allows robots to operate as self-adaptive systems—reasoning through possible futures, generating and verifying action plans, and learning from the resulting outcomes. Collectively, these interactions establish a continuum between symbolic reasoning, predictive simulation, and embodied control, which defines the foundation of *Generative Robotic Autonomy.*

3.5 Taxonomy Overview

Table 1 summarizes representative categories of generative models that enable robotic autonomy. Each family differs in input–output modality, learning paradigm, and control capability, yet all contribute to bridging reasoning and embodiment within a unified framework.

Table 1. Taxonomy of Generative Models Applied to Robotic Autonomy and Planning

Model Family	Representative Systems	Primary Modality (Input → Output)	Planning/Control Capability
Large Language Models (LLMs)	GPT-4, Llama 3, Gemini, Claude, Code-as-Policies, SayCan	Text/Code → Task Plan/Action Sequence	Language-to-action translation, symbolic reasoning, policy generation
Diffusion Models	Stable Diffusion, ControlNet, Diffusion Policy	Text/Image → Trajectory/Scene Representation	Visual plan synthesis, trajectory imagination, environment generation
World Models	DreamerV3, PlaNet, Genie	Observation/State → Latent World Simulation	Long-horizon planning, imagination-based control, safe policy evaluation
Vision-Language-Action (VLA) Models	RT-2, RT-X, Voyager, Gato	Image + Text → Motor Command/Policy	Grounded perception-to-action mapping, open-vocabulary manipulation
Hybrid Generative Agents	Embodied LLMs, Digital-Twin Agents, Physical AI Systems	Multimodal (Text, Vision, State) ↔ Action/Feedback	Closed-loop reasoning, adaptive control in cyber-physical environments

This taxonomy provides a functional mapping between cognitive model types, their training paradigms, and corresponding planning or control capabilities. It reflects a transition from symbolic reasoning to embodied intelligence, establishing a framework that connects generative cognition with operational autonomy across physical and simulated domains.

4 Generative AI in Robotic Planning and Control

Generative AI has introduced new paradigms for robotic planning, allowing robots to reason over goals, generate action policies, and adapt plans dynamically. This section reviews representative frameworks and research directions that leverage generative models across three key dimensions: (1) language-driven task planning, (2) vision-language-action integration, and (3) predictive world-model control, followed by a discussion on hybrid and safety-aware architectures.

4.1 Language-Driven Task Planning

The integration of LLMs with robotics enables natural-language instructions to be transformed into executable robot behaviors. A pioneering framework, *SayCan* [10], couples an LLM with a value-function estimator that grounds textual commands in real-world affordances. The LLM decomposes a user instruction (*e.g., "fetch the can from the fridge"*) into symbolic subtasks, while the value network predicts the feasibility of each subtask based on environmental state.

Building upon this concept, *Code-as-Policies (CaP)* [9] treats LLMs as program synthesizers that output Python or ROS2 code directly executable by a robot. By embedding domain libraries and control primitives within the prompt, CaP enables open-ended task execution without explicit retraining. Other studies have extended this paradigm toward interactive agents that continuously refine task plans using chain-of-thought reasoning, memory, and feedback loops.

Despite their flexibility, LLM-based planners face critical challenges: hallucinated or unsafe outputs, latency in inference, and lack of grounding in physical constraints. Recent research explores safety-constrained decoding [32], neuro-symbolic reasoning frameworks [33], and simulation-based verification through digital twins to mitigate these limitations and ensure task reliability under uncertainty.

4.2 Vision-Language-Action Integration

While text-only models capture abstract reasoning, autonomous robots must integrate visual and proprioceptive information to act effectively. *RT-2* [11] represents a milestone by fine-tuning a transformer on paired image–text–action data, producing a unified model capable of open-vocabulary manipulation. It generalizes from internet-scale vision–language data (e.g., "pick up the red cup") to robotic control policies that operate on continuous sensory inputs. The *Voyager* agent [12] extends this idea into continual learning and exploration. It employs an LLM as a lifelong planner that iteratively writes, tests, and refines code controlling an embodied agent. Voyager demonstrates self-improvement by accumulating successful strategies and reusing them in novel scenarios—an early step toward self-evolving robotic cognition. These systems illustrate the shift from pre-trained perception models toward *generative multimodal cognition*, where reasoning, perception, and control co-evolve. However, maintaining real-time performance remains challenging because multimodal transformers require significant computation during inference.

4.3 Predictive World-Model Control

A complementary approach involves endowing robots with *world models* that learn compact latent representations of environmental dynamics. Instead of directly predicting actions, a world model simulates possible futures and enables the agent to plan through imagination. The *DreamerV3* architecture [8] exemplifies this method by combining representation learning, policy optimization, and value estimation entirely within a learned latent space. Similarly, *Genie* and *PlaNet* learn differentiable forward models that allow long-horizon planning while reducing sample complexity. World-model planning aligns naturally with generative AI principles: the agent *generates* potential outcomes before committing to real actions. When coupled with LLM-based reasoning, world models can validate or reject hallucinated plans using simulated evidence, providing a pathway toward safer generative autonomy.

4.4 Hybrid Architectures and Closed-Loop Integration

Recent research explores hybrid systems that merge these paradigms into closed-loop architectures. Generative reasoning models interpret intent and propose high-level strategies; world models simulate consequences; and reinforcement learning modules fine-tune low-level controllers. Digital twins play a pivotal role in this loop, acting as virtual testbeds for verification, adaptation, and human-in-the-loop supervision. Such integration is crucial for deploying generative autonomy in industrial and OT environments, where safety, latency, and explainability are mandatory.

4.5 Discussion

The frameworks reviewed in this section demonstrate that generative AI enables robots to move beyond fixed control policies toward open-ended reasoning and self-adaptation. Language-driven systems excel in high-level task decomposition but often lack grounding in real-world dynamics. Vision-language-action (VLA) models introduce multimodal perception, enabling generalization across diverse tasks, yet they remain computationally demanding for real-time control. World-model-based methods such as DreamerV3 and Genie provide predictive foresight through imagination but depend on accurate latent representations and extensive training. Hybrid architectures that combine reasoning, perception, and control through digital twins offer a promising bridge between cognitive flexibility and operational reliability. However, balancing generative expressiveness with safety, latency, and explainability continues to pose fundamental challenges [32,33]. This motivates the need for standardized evaluation metrics and benchmarking protocols, which are discussed in the next section.

5 Evaluation Metrics and Benchmark Frameworks

Reliable evaluation is essential for validating the effectiveness, safety, and generalization capability of generative robotic systems. Unlike conventional reinforcement- or imitation-learning agents, generative models couple language, vision, and predictive reasoning in a single pipeline. Consequently, benchmarking must assess not only task success but also reasoning coherence, grounding accuracy, and computational efficiency [34]. This section summarizes the primary evaluation dimensions, representative benchmark environments, and emerging efforts toward unified protocols for assessing generative robotic autonomy.

Table 2. Core Evaluation Dimensions for Generative Robotic Autonomy

Dimension	Description
Task Success Rate (TSR)	Ratio of successful task completions to total trials under specified environmental conditions.
Adaptability, Generalization	Ability to execute unseen tasks or transfer skills across domains, embodiments, or sensor modalities.
Grounding Accuracy	Degree to which generated plans correspond to valid sensor data, object states, and spatial constraints.
Plan Efficiency	Time, energy, or step count required to complete a task relative to an optimal trajectory.
Explainability, Transparency	Interpretability of reasoning traces, symbolic plans, or language-based rationales.
Safety and Robustness	Resistance to infeasible or hazardous actions under uncertainty, noise, or partial observability.
Computational Efficiency	Inference latency and energy consumption during online planning or control.

5.1 Evaluation Dimensions

Quantitative and qualitative criteria used in recent studies are summarized in Table 2. These metrics capture performance across reasoning, adaptability, grounding, transparency, and operational reliability.

Each metric highlights a complementary dimension of autonomy. For instance, *Task Success Rate* measures functional capability, while *Grounding Accuracy* quantifies physical validity. Together, these indicators form a multi-objective profile balancing creativity, safety, and efficiency.

5.2 Benchmark Frameworks and Datasets

Several open-source benchmarks now provide reproducible testbeds for evaluating generative reasoning, planning, and control. Representative examples are summarized in Table 3. They collectively address manipulation, navigation, and vision-language grounding—key domains for embodied generative agents [34].

While these benchmarks target different facets of autonomy, cross-benchmark comparability remains limited due to variations in data distribution, task definition, and performance metrics. Recent studies advocate for standardized evaluation taxonomies and unified interfaces that enable meta-analysis across generative frameworks. To achieve reproducibility and comparability across systems, emerging research emphasizes the development of standardized evaluation protocols and shared simulation pipelines [34]. Recommended best practices include:

1. Establish unified prompt suites and task taxonomies for language-to-action evaluation.
2. Incorporate simulation-based safety validation prior to deployment in physical environments.

Table 3. Representative Benchmarks for Evaluating Generative Embodied Intelligence

Benchmark	Domain Focus	Evaluation Metrics	Highlights
ALFRED [35]	Household manipulation	Success rate, goal-conditioned accuracy	Natural-language-conditioned navigation and manipulation in photo-realistic indoor environments.
CALVIN [36]	Continual robotic manipulation	Multi-task success, zero-shot generalization	Evaluates long-horizon, language-conditioned visuomotor learning with shared state representations.
RT-X/RT-2 [11]	Vision-language-action policies	Open-vocabulary success, generalization	Large-scale multimodal framework enabling open-ended and transferable robotic control.
ManiSkill 2 [37]	Dexterous manipulation	Episode success, completion time, stability	Physics-accurate simulation environment for evaluating generalizable manipulation skills.
Embodied AI Benchmarks [38]	Navigation and HRI	Path efficiency, SPL	Includes Habitat, iGibson, and OmniGibson for embodied reasoning and perception-action grounding.

3. Report multi-metric performance profiles rather than single aggregate scores.
4. Integrate human-in-the-loop evaluations of trust, transparency, and usability.

These guidelines align generative robotic evaluation with engineering validation standards (e.g., IEC 61508 for functional safety), enabling rigorous cross-system assessment and facilitating reproducible progress within the research community.

6 Challenges and Open Research Directions

Although generative AI has demonstrated significant potential for robotic reasoning and planning, several obstacles must be addressed before such systems can achieve dependable autonomy in safety-critical domains. This section outlines key technical and societal challenges and identifies open research avenues that define the roadmap toward trustworthy and embodied generative intelligence.

6.1 Safety and Hallucination Control

Generative models, particularly LLMs, may produce outputs that are syntactically valid yet physically infeasible or hazardous. In robotic contexts, these "hallucinations" can manifest as unreachable trajectories, invalid actuator commands, or unsafe manipulations. Ensuring operational safety requires multi-layer validation, including constrained decoding, model-checking, and simulation-based filtering using digital twins. Research on integrating formal safety envelopes, runtime monitors, and fault-tolerant control architectures remains essential.

6.2 Grounding and Symbolic Alignment

Bridging abstract language reasoning with continuous sensorimotor control remains a central challenge. LLMs can generate coherent symbolic plans, but translating them into executable, geometry- and physics-aware actions is non-trivial. Recent progress in neuro-symbolic alignment and scene-graph grounding seeks to map semantic actions to spatial relationships and object affordances. Future work should enable bidirectional updates where sensory observations continuously refine symbolic world models, maintaining coherence between reasoning and perception.

6.3 Real-Time Adaptation and Computational Efficiency

Generative architectures are computationally intensive and often unsuitable for real-time embedded execution. Achieving sub-second planning latency requires model compression, pruning, and quantization tailored for robotic workloads. Hybrid edge-cloud pipelines, where local controllers execute low-latency policies while cloud agents perform high-level reasoning, provide a promising compromise. Developing adaptive inference schedulers that allocate compute resources dynamically based on task urgency remains an open area.

6.4 Explainability and Human Trust

As generative systems gain autonomy, transparency in decision-making becomes critical for user confidence and regulatory acceptance. Operators must understand *why* a robot selected a particular plan or action. Emerging techniques such as chain-of-thought visualization, policy summarization, and natural-language rationales can enhance interpretability. Human-subject studies on trust calibration and shared-control paradigms will help define acceptable levels of autonomy in collaborative environments.

6.5 Cyber-Physical and Operational-Technology Integration

Deploying generative autonomy within industrial and defense infrastructures introduces stringent constraints related to determinism, safety certification, and network reliability. Operational-technology systems often require timing guarantees and redundancy not native to AI pipelines. Research should focus on AI-cyber-physical co-design methodologies, aligning generative reasoning with real-time scheduling, IEC 61508 functional-safety compliance, and zero-trust cybersecurity architectures.

6.6 Ethical and Societal Implications

As robots evolve toward self-directed behavior, ethical considerations surrounding accountability, data provenance, and societal impact must be addressed. Generative autonomy raises new questions about authorship of machine-generated plans, liability for accidents, and fairness in decision policies. Establishing governance frameworks, transparent data pipelines, and ethical auditing mechanisms will be vital for responsible adoption.

7 Conclusion and Future Outlook

This survey has outlined the emerging landscape of *Generative Robotic Autonomy*, synthesizing how large language models, diffusion models, and world models contribute to open-ended and context-aware robotic behavior. Despite rapid progress, critical challenges remain in safety, grounding, and trust, particularly for deployment in safety-critical and operational-technology environments. Addressing these challenges will require physically grounded reasoning, efficient inference, and formal validation mechanisms. Looking ahead, the field is moving toward Embodied Generative Intelligence, where robots learn and reason through real and simulated experiences.

References

1. LaValle, S.M.: Planning Algorithms. Cambridge University Press (2006)
2. Sutton, R.S., Barto, A.G.: Reinforcement Learning: An Introduction, 2nd edn. MIT Press (2018)
3. Kaelbling, L.P., Littman, M.L., Moore, A.W.: Reinforcement learning: a survey. J. Artif. Intell. Res. **4**, 237–285 (1996)
4. Brown, T., Mann, B., Ryder, N., et al.: Language models are few-shot learners. In: Advances in Neural Information Processing Systems (NeurIPS), vol. 33, pp. 1877–1901 (2020)
5. OpenAI: GPT-4o technical report. Technical repprt, OpenAI (2024). https://cdn.openai.com/papers/gpt-4o.pdf
6. Rombach, R., Blattmann, A., Lorenz, D., et al.: High-resolution image synthesis with latent diffusion models. In: Proceedings of the IEEE/CVF Conference on Computer Vision and Pattern Recognition (CVPR) (2022)
7. Ho, J., Jain, A., Abbeel, P.: Denoising diffusion probabilistic models. In: Advances in Neural Information Processing Systems (NeurIPS) (2020)
8. Hafner, D., Pasukonis, J., Widrich, M., et al.: Mastering diverse domains through world models. In: Proceedings of the 12th International Conference on Learning Representations (ICLR) (2024)
9. Liang, J., Jiang, H., Li, T., et al.: Code as policies: language model programs for embodied control. In: Proceedings of the IEEE/CVF International Conference on Robotics and Automation (ICRA) (2023)
10. Ahn, M., Brohan, A., Chebotar, Y., et al.: Do as i can, not as i say: grounding language in robotic affordances. In: Proceedings of the 6th Conference on Robot Learning (CoRL) (2022)

11. Brohan, A., Brown, N., Chebotar, Y., et al.: RT-2: vision-language-action models transfer web knowledge to robotic control. In: Proceedings of the 7th Conference on Robot Learning (CoRL) (2023)
12. Wang, G., Ye, Y., Ye, S., et al.: Voyager: an open-ended embodied agent with large language models. In: Proceedings of the 11th International Conference on Learning Representations (ICLR) (2023)
13. Zhang, H., Tang, Z., Zhou, Y., et al.: Large language models for robotics: a survey. arXiv preprint arXiv:2401.13601 (2024)
14. Xu, H., Wu, C., Yang, Y., et al.: Multimodal large language models: a survey. arXiv preprint arXiv:2306.13549 (2023)
15. NVIDIA Corporation: NVIDIA isaac sim: A scalable robotics simulation platform (2023). https://developer.nvidia.com/isaac-sim
16. Yao, S., Yu, D., Zhao, J., et al.: A survey on large language model based autonomous agents. arXiv preprint arXiv:2312.06262 (2023)
17. Chen, Y., Xu, J., Li, M., et al.: Trustworthy and safe generative artificial intelligence: a survey. arXiv preprint arXiv:2402.00995 (2024)
18. Siciliano, B., Khatib, O.: Springer Handbook of Robotics. Springer (2016)
19. Brooks, R.A.: A robust layered control system for a mobile robot. IEEE J. Robot. Autom. **2**(1), 14–23 (1986)
20. Camacho, E.F., Bordons, C.: Model Predictive Control. Springer (2007)
21. Nau, D.S., Au, T.C., Ilghami, O., et al.: SHOP2: an HTN planning system. J. Artif. Intell. Res. **20**, 379–404 (2003)
22. Argall, B.D., Chernova, S., Veloso, M., Browning, B.: A survey of robot learning from demonstration. Robot. Auton. Syst. **57**(5), 469–483 (2009)
23. Levine, S., Finn, C., Darrell, T., Abbeel, P.: End-to-end training of deep visuomotor policies. J. Mach. Learn. Res. **17**(39), 1–40 (2016)
24. Goodfellow, I., Pouget-Abadie, J., Mirza, M., et al.: Generative adversarial nets. In: Advances in Neural Information Processing Systems (NeurIPS) (2014)
25. Vaswani, A., Shazeer, N., Parmar, N., et al.: Attention is all you need. In: Advances in Neural Information Processing Systems (NeurIPS) (2017)
26. Chi, C., Feng, S., Du, Y., et al.: Diffusion policy: visuomotor policy learning via action diffusion. In: Proceedings of the Robotics: Science and Systems (RSS) (2023)
27. Hafner, D., Lillicrap, T., Norouzi, M., Ba, J.: Learning latent dynamics for planning from pixels. In: Proceedings of the International Conference on Machine Learning (ICML) (2019)
28. Bruce, J., Driess, D., Pathak, D., et al.: Genie: generative interactive environments. arXiv preprint arXiv:2402.15391 (2024)
29. Asada, M., Hosoda, K., Kuniyoshi, Y., et al.: Cognitive developmental robotics: a survey. IEEE Trans. Auton. Ment. Dev. **1**(1), 12–34 (2009)
30. Rosen, R., Wichert, M., von Wichert, G., Kiritsis, D.: About the importance of digital twins in cyber-physical production systems. CIRP Ann. **71**(1), 147–170 (2022)
31. International Electrotechnical Commission, Geneva, Switzerland: IEC 61508: Functional Safety of Electrical/Electronic/Programmable Electronic Safety-Related Systems (2010), edition 2.0
32. Shao, J., Wang, Z., Zhang, Y., Li, B., Song, D.: Safe planning with large language models for embodied agents. arXiv preprint arXiv:2403.01567 (2024)
33. Gupta, A., Mehta, Y., Kambhampati, S.: Neuro-symbolic planning: bridging large language models and classical planning for robot decision-making. In: Proceedings of the AAAI Conference on Artificial Intelligence, vol. 37, pp. 14056–14064 (2023)

34. Du, Y., Bear, D.M., Tenenbaum, J.B., et al.: A survey of embodied AI benchmarks and evaluation protocols. Found. Trends Mach. Learn. **16**(3), 181–241 (2023)
35. Shridhar, M., Thomason, J., Gordon, D., et al.: ALFRED: a benchmark for interpreting grounded instructions for everyday tasks. In: Proceedings of the IEEE/CVF Conference on Computer Vision and Pattern Recognition (CVPR) (2020)
36. Mees, O., Hermann, J., Burgard, W.: CALVIN: a benchmark for language-conditioned policy learning for long-horizon robot manipulation tasks. In: Proceedings of the 6th Conference on Robot Learning (CoRL) (2022)
37. Gu, J., Chen, X., Zhou, X., et al.: ManiSkill2: a unified benchmark for generalizable manipulation skills. In: Proceedings of the 11th International Conference on Learning Representations (ICLR) (2023)
38. Savva, M., Kadian, A., Maksymets, O., et al.: Habitat 2.0: training home assistants to rearrange their habitat. In: Advances in Neural Information Processing Systems (NeurIPS) (2021)

Increasing AI Explainability by LLM Driven Standard Processes

Marc Jansen and Marcel Pehlke(✉)

Computer Science Institute, University of Applied Sciences Ruhr West, Bottrop, Germany
{marc.jansen,marcel.pehlke}@hs-ruhrwest.de

Abstract. This paper introduces a novel approach to increasing the explainability of artificial intelligence (AI) systems by embedding Large Language Models (LLMs) within standardized analytical processes. While traditional explainable AI (XAI) methods focus on feature attribution or post-hoc interpretation, the proposed framework integrates LLMs into well-defined decision models such as Question–Option–Criteria (QOC), Sensitivity Analysis, Game Theory, and Risk Management and the like. By situating LLM reasoning within these formalized structures, the approach transforms opaque inference into transparent, auditable decision traces. A layered architecture is presented that separates the unexplainable reasoning space of the LLM from the explainable process space above it. Empirical evaluations demonstrate that the system can reproduce human-level decision logic in decentralized governance, systems analysis, and strategic reasoning contexts. The results indicate that LLM-driven standard processes provide a promising foundation for reliable, interpretable, and verifiable AI-supported decision-making.

Keywords: Explainable AI · Large Language Models · Standard Processes

1 Introduction

Artificial Intelligence (AI) has achieved remarkable advances in recent years, particularly through deep learning and large-scale generative models. Yet, these systems frequently operate as opaque black boxes, making their internal reasoning inaccessible to human understanding. This lack of transparency poses significant challenges to accountability, reliability, and ethical governance, especially when AI is deployed in critical decision-making domains such as finance, healthcare, or policy. The emerging field of Explainable Artificial Intelligence (XAI) seeks to address this gap by developing techniques that make model behavior interpretable and trustworthy to human stakeholders.

While existing XAI methods, such as feature attribution, saliency mapping, and surrogate modeling, have contributed valuable insights, they often remain limited in scope. Many of these approaches describe what influenced a model's

H. R. Arabnia et al. (Eds.): CSCI 2025, CCIS 2972, pp. 421–435, 2027.
https://doi.org/10.1007/978-3-032-28469-3_28

decision, but not how the reasoning process unfolded. Moreover, their results are often highly technical, requiring expert interpretation rather than offering explanations accessible to general users or decision-makers. As a result, the promise of explainability is frequently constrained by the gap between mathematical interpretability and human-centered understanding.

This paper proposes a complementary perspective: instead of merely explaining opaque models post hoc, AI systems can be structured to reason through inherently explainable processes. We introduce the concept of LLM-driven standard processes, in which Large Language Models (LLMs) execute well-established analytical frameworks that are transparent by design, such as Question–Option–Criteria (QOC) analysis, Sensitivity Analysis, Game Theory, and Risk Management. These processes, long used in human decision-making and systems analysis, possess clear structures, traceable reasoning steps, and interpretable outputs. Embedding LLM reasoning within such frameworks allows the system to generate not only an answer but also a documented rationale showing how that answer was reached.

To operationalize this concept, we present a layered architecture that separates the unexplainable reasoning space of the LLM from the explainable process space of standardized analytical models. The lower layer leverages the generative and interpretive capabilities of the LLM to supply contextual reasoning, while the upper layer enforces formal structures, statistical validation, and auditability. This design introduces an Explainability Barrier that clearly delineates where reasoning becomes measurable, interpretable, and accountable.

We demonstrate the feasibility of this architecture through several prototype implementations and empirical evaluations. Specifically, we apply the LLM-driven QOC framework to decentralized investment decisions in a real-world DAO context, replicate a Vester-style sensitivity analysis on logistics data, and model strategic crisis dynamics using game-theoretic reasoning. Across these evaluations, the proposed system reproduces key elements of human-level reasoning while maintaining transparent analytical structures and quantifiable results.

Overall, this work contributes to the field of explainable AI by showing that LLMs can act not merely as reasoning engines but as explainability enablers when embedded within formalized decision processes. The approach establishes a pathway toward AI systems that are both powerful and transparent, capable of complex reasoning while producing outputs that remain interpretable, auditable, and aligned with human standards of rational explanation.

2 State of the Art

Modern AI systems, particularly deep learning models, often act as black boxes whose decision logic is opaque. This opacity limits user trust and hinders adoption in critical domains [1]. The field of Explainable AI (XAI) therefore seeks to make AI behavior more interpretable and accountable. Existing approaches can be divided into intrinsically interpretable models and post-hoc explanation methods, which explain the outputs of complex models without modifying their internal mechanisms [2].

2.1 Explainable AI

Post-hoc XAI techniques such as LIME and SHAP approximate local model behavior or assign Shapley values to quantify each feature's contribution to a prediction [3–5]. These methods have improved transparency in areas like medicine or finance, but they often remain technical and difficult for non-expert users to interpret. Other analytical tools, such as sensitivity analysis, help assess how input variations influence model outputs [6]. Together, these methods contribute to fairness, accountability, and regulatory compliance (e.g., GDPR) [1].

2.2 Large Language Models for Explainability

Despite such progress, current XAI methods still struggle to explain how decisions are reached. Large Language Models (LLMs) offer a new paradigm: their ability to generate coherent natural language allows them to translate complex model outputs into human-understandable narratives. Moreover, chain-of-thought (CoT) prompting enables models to make their reasoning steps explicit, increasing transparency and even improving accuracy [1,7]. Studies show that structured reasoning and fine-tuning can raise model performance from around 60% to over 90% on complex decision tasks, demonstrating that explainability and reliability can reinforce each other [7].

2.3 LLM-Driven Structured Decision Processes

A promising next step is to embed LLMs within structured analytical processes that inherently yield explanations. Frameworks like Question–Option–Criteria (QOC) formalize decision-making through transparent evaluation tables that naturally document rationale [8]. Recent experimental systems use LLM agents as stakeholders performing QOC analyses collaboratively, producing both recommendations and their justifications. Similarly, multi-agent LLM frameworks simulate deliberation or negotiation, enhancing interpretability through debate-like reasoning [9].

In parallel, game-theoretic and sensitivity-analysis models are being extended with LLMs to explore strategic behavior, scenario generation, and systemic robustness [10–12]. These approaches suggest that LLM-driven analytical structures can transform opaque reasoning into traceable, auditable processes—laying the foundation for explainable, process-based AI systems.

Modern AI systems, particularly deep learning models, often act as black boxes whose decision logic is opaque. This opacity limits user trust and hinders adoption in critical domains [1]. The field of Explainable AI (XAI) therefore seeks to make AI behavior more interpretable and accountable. Existing approaches can be divided into intrinsically interpretable models and post-hoc explanation methods, which explain the outputs of complex models without modifying their internal mechanisms [2].

3 Architecture

The proposed architecture introduces a layered framework that combines the generative reasoning capabilities of Large Language Models (LLMs) with standardized, transparent analytical processes positioned on top of them. As illustrated in Fig. 1, this design establishes a conceptual separation between the unexplainable reasoning space of the LLM and the explainable reasoning space of formalized processes such as for example Question–Option–Criteria (QOC), Sensitivity Analysis, Game Theory Models, and Risk Management.

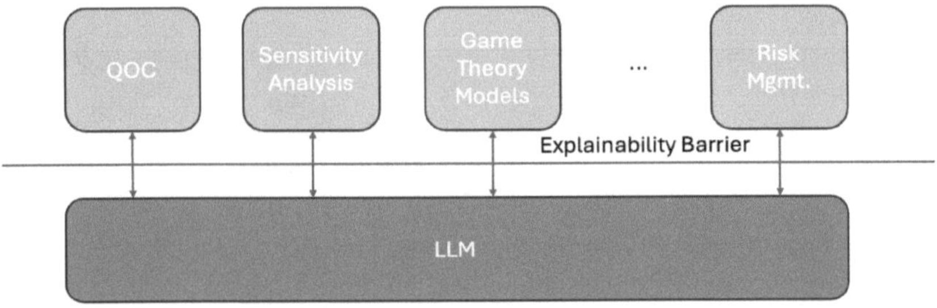

Fig. 1. Layered architecture of LLM-driven standard processes and the explainability barrier.

3.1 Layered Structure and Explainability Barrier

At the core of the system lies the LLM, which acts as a general-purpose inference engine capable of interpreting natural language, generating reasoning steps, and providing contextually rich answers. However, despite its remarkable performance, the internal mechanisms of an LLM remain largely opaque: its reasoning paths and parameter activations are not interpretable in human-understandable terms.

To overcome this limitation, a layer of LLM-driven standard processes is placed above the model. Each of these processes defines a formal structure for decision-making or analysis, such as options and criteria in QOC, variables and influence matrices in Sensitivity Analysis, or payoffs and equilibria in Game Theory. Within these frameworks, the LLM's role is restricted to performing well-defined sub-tasks, such as identifying relevant inputs, generating candidate options, or estimating relationships. The process logic itself, however, remains fully deterministic and analyzable, providing a clear interface for evaluating and understanding the outcomes.

The boundary between the two layers is conceptualized as the Explainability Barrier. Below this barrier lies the LLM's black-box reasoning, while above it reside structured analytical processes that expose interpretable, measurable, and auditable outputs. By systematically transferring more decision logic from below to above the barrier, the architecture increases the proportion of explainable reasoning in the overall system. This gradual shift is a central goal of the approach: to minimize the influence of opaque model behavior and maximize traceability through formalized process design.

3.2 Function of the Upper Layer: Standardized Analytical Processes

The upper layer hosts multiple independent yet compatible analytical modules, for example:

- QOC (Question–Option–Criteria): for evaluating decision alternatives against weighted criteria.
- Sensitivity Analysis: for quantifying systemic interdependencies and influence strengths.
- Game Theory Models: for analyzing strategic interactions, equilibria, and trade-offs.
- Risk Management: for identifying, consolidating, and evaluating project risks.

Each of these modules defines a structured reasoning schema that can be populated and executed by LLM-based agents. Because the schemas are mathematically defined, using weights, scores, matrices, or payoff functions, their results can later be subjected to, e.g., statistical validation, variance analysis, or outlier detection. This enables transparency in terms of why a specific decision or outcome was produced, even if the internal reasoning of the LLM agents remains hidden.

3.3 Expanding the Explainable Space

A major design goal of this architecture is to progressively extend the explainable space upward, reducing reliance on unstructured LLM reasoning. This can be achieved by formalizing more process steps, such as option generation, weighting, or evaluation, into transparent mathematical and logical operations. Over time, as additional standardized models (e.g., scenario planning, system dynamics, or causal reasoning) are integrated into the upper layer, the Explainability Barrier moves downward, effectively isolating the black-box elements of the LLM while retaining their creative and interpretative benefits.

In summary, the architecture represents a hybrid explainability paradigm: it combines the linguistic and contextual strengths of LLMs with the interpretability and analytical rigor of established decision-support frameworks. The result is a system in which unexplainable intelligence serves explainable processes, enabling reproducible, transparent, and auditable AI-driven decision support.

4 Implementation

Established analytical frameworks can be reimagined through LLM-based agent systems to enhance transparency and consistency in complex decision-making.

The following subsections illustrate how classical methods: QOC, Sensitivity Analysis, Game Theory, and Risk Management; can be transformed into LLM-driven standard processes, combining formal analytical rigor with the contextual reasoning capabilities of modern AI.

4.1 QOC - Question Option Criteria

The Question–Option–Criteria (QOC) framework originates from the domain of design space analysis [8], yet its underlying principles are broadly applicable to a wide range of decision-making contexts. The core idea is to decompose a decision problem into three elements: a question to be addressed, a set of options representing potential answers, and a collection of criteria that define what aspects are relevant for evaluating these options. Each criterion is assigned a weight to reflect its relative importance within the decision context. Subsequently, each option is assessed with respect to each criterion, indicating how well the option satisfies that particular aspect. By aggregating these assessments, typically as the weighted sum of the option–criterion evaluations, the most favorable option can be identified as the one achieving the highest overall score.

In an LLM-based implementation, these steps, including the identification of relevant options and criteria, as well as their corresponding evaluations, can be executed by a set of autonomous AI agents instantiated to represent the perspectives and priorities of different stakeholders involved in the decision (see Fig. 2).

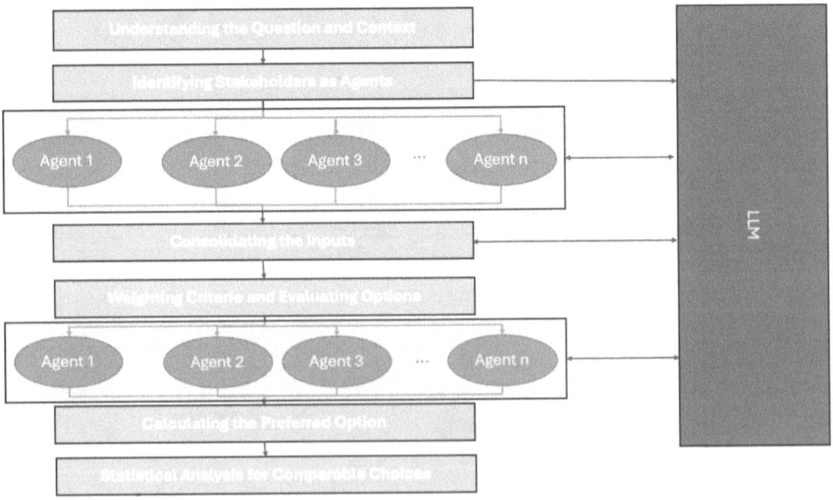

Fig. 2. An agent-based implementation of the QOC process.

Before proceeding to the weighting and evaluation phase, however, the individually identified options and criteria from the various agents must first be consolidated into a unified set. This consolidation ensures that overlapping and redundant elements are harmonized, thereby establishing a coherent foundation for the subsequent quantitative evaluation. Beyond the mere computation of a preferred option, the agent-generated data enable additional statistical analyses, such as the detection of outlier evaluations or divergences in the perceived importance of criteria or in how well the options support the criteria. Furthermore, transparency and explainability can be enhanced through meta-analysis of the decision process—for example, by identifying which criteria exerted the greatest influence on the final outcome.

4.2 Sensitivity Analysis

The sensitivity analysis process follows the structure of the Vester model but adapts it for LLM-driven automation and interpretability. It begins with an intake agent that transforms a natural-language scenario into a structured system definition, including the system's goal, main subsystems, boundaries, and relevant stakeholders. Each stakeholder perspective, represented by an LLM adopting the persona of that stakeholder, is used to suggest candidate variables, ensuring that social, technical, and environmental aspects are captured.

Next, a cleaning and consolidation step merges overlapping or redundant variables while preserving those explicitly mentioned by the user. The outcome is a coherent set of system variables that form the foundation for the subsequent influence analysis.

An LLM-based evaluation module then estimates the direct influence of each variable on every other variable, producing a signed, weighted influence matrix. The numerical entries represent both the strength and direction of influence, indicating how changes in one factor impact another and whether the effect is positive or negative. This matrix acts as a transparent, data driven representation of the system's internal dynamics.

From this matrix, the analyzer computes key statistical indicators to identify systemic roles:

- **Active Sum (AS):** Measures how strongly a variable influences others. High values identify drivers and leverage points.
- **Passive Sum (PS):** Reflects how much a variable is affected by others, highlighting sensitive or dependent factors.
- **Role classification:** Variables are categorized into four standard roles: Active, Critical, Reactive, and Buffering—based on their relative AS and PS values.

These metrics reveal system dynamics, interdependencies, and potential feedback loops, forming the statistical backbone of the analysis.

An interpretation agent summarizes the results in natural language, outlining main drivers, dependencies, and leverage points. Visualizations, such as influence matrices or network diagrams, make causal links and systemic behavior transparent at a glance.

4.3 Normal-Form Game (Game Theory)

The normal-form game process is structured into clear, modular steps, each executed by a dedicated component or agent. First, a player-extraction module identifies the two players, their overall objectives, the payoff unit (for example, profit in USD, utility points, or another measurable outcome), and an expected value range derived from the user's scenario description. Next, a strategy-generation module proposes up to ten strategies per player, ensuring that each is clearly named and accompanied by a short, descriptive summary of its intent.

Based on the extracted strategies, an LLM-powered payoff assignment module evaluates all possible strategy combinations and constructs the payoff matrix. This matrix forms the statistical foundation of the analysis, it captures the outcome values for every combination of player decisions. From this, the system can apply classical game-theoretic solution concepts that rely on these statistics to explain strategic behavior.

The deterministic analyzer computes three central measures:

- **Nash equilibria:** Stable states of the game where no player can gain by unilaterally changing their strategy.
- **Pareto-optimal outcomes:** Solutions where no player's payoff can be improved without reducing the other's.

Finally, an interpretation agent summarizes results in natural language, explaining the equilibrium, key trade-offs, and strategy alignment with player goals.

4.4 Sequential Game (Game Theory)

The sequential-game pipeline extends the normal-form approach by incorporating the dimension of time and order into the decision process. It mirrors the same intake, identifying players, objectives, payoff units, and value ranges, but adds a layer of role conditioning. A role-assignment module creates concise persona prompts (e.g., "You are the company's CEO deciding whether to expand into a new market"), which guide each player-specific LLM agent to reason consistently from its perspective.

Tree construction proceeds step-by-step from the root node, where players alternate in taking turns. At each node, the active player's agent proposes one to three concrete actions. The reasoning at every decision point is grounded in the player's role, not as a neutral observer. To ensure meaningful and interpretable trees, a configurable horizon policy limits the maximum depth, preventing unnecessarily long or repetitive paths while maintaining the completeness of possible outcomes.

Once the decision tree is built, each terminal node represents a possible end state of the scenario. A neutral evaluator agent then assigns numerical payoffs to both players, using the entire decision path and contextual information such as roles, objectives, and constraints. These payoffs are stored together with a short rationale at each terminal node, creating an auditable trail of how every outcome was scored.

The statistical information contained in the payoff structure allows a deterministic analyzer to compute key game-theoretic measures that explain the strategic dynamics of the sequential process:

- **Dominated paths:** Sequences of decisions that consistently lead to worse outcomes compared to alternatives, helping identify which strategies can be ruled out.
- **Pareto-optimal paths:** Decision sequences that represent efficient trade-offs—no player's result can improve without worsening the other's.
- **Subgame Perfect Equilibrium (SPE):** The most rational strategy profile at every stage of the game, determined via backward induction.

The analyzer outputs a concise summary of the equilibrium path, payoff structure, and player diagnostics. An interpretation agent translates these results into clear explanations, outlining the logic behind the SPE, key trade-offs, and recommendations aligned with user objectives. Visualizations such as annotated game trees and payoff charts further clarify decision flows and the rationale behind the strategies.

4.5 Risk Management

A risk management process can be effectively implemented using a multi-agent, LLM-based approach. Given a specific project or initiative, autonomous agents, each representing distinct stakeholder perspectives such as technical, organizational, financial, or legal viewpoints, can independently identify potential risks associated with the endeavor. The individually generated risk lists are then consolidated into a unified and non-redundant set of risks, ensuring that overlapping or synonymous risk descriptions are harmonized while maintaining the diversity of perspectives contributed by the agents.

After consolidation, each agent evaluates the identified risks along two principal dimensions: the probability that a risk will occur and the impact that its occurrence would have on the project's objectives. These dimensions can be expressed using standardized scales (e.g., low–medium–high categories or numerical likelihood–impact matrices), allowing for quantitative aggregation and prioritization of risks. Based on these evaluations, composite risk scores can be computed to identify the most critical threats requiring mitigation or monitoring.

In this scenario, it may be advantageous to replace the general-purpose LLM with either a fine-tuned model or a Retrieval-Augmented Generation (RAG) architecture. Such adaptations allow the system to access and incorporate the most recent project documentation, technical reports, or relevant external sources, ensuring that the agents operate on an up-to-date and context-specific knowledge base. This integration significantly enhances the accuracy and relevance of the identified risks by grounding the agents' assessments in the current state of the project and its environment.

Finally, statistical and analytical methods can be applied to the agents' evaluations to detect inconsistencies, outliers, or systematic differences across stakeholder groups. These analyses improve the robustness and reliability of the overall assessment and contribute to greater transparency by revealing which risks are perceived as particularly significant and how alignment or divergence emerges among the participating agents.

5 Evaluation

This section presents an empirical evaluation of the proposed LLM-driven standard processes. To assess their validity and performance, selected implementations were tested on real-world datasets and reference scenarios. The results demonstrate the extent to which these AI-based processes can reproduce human-level reasoning and deliver interpretable, statistically verifiable outcomes.

5.1 Evaluation of the QOC Approach

To empirically assess the performance of the proposed AI-driven QOC process, an evaluation was conducted using real-world decision data from a Decentralized Autonomous Organization (DAO). A total of 102 investment proposals, including both the proposal content and the actual DAO voting outcomes, were collected via the DAO's public API. Each proposal was then independently processed using the LLM-based QOC framework described in Sect. 4.1, and the AI-generated decisions were compared to those made by the DAO community.

For each model, a contingency table was constructed to record the frequency of agreement and disagreement between human (DAO) and AI decisions. Specifically, four categories were considered:

- DAO Y/AI Y—both DAO and AI approved the proposal,
- DAO Y/AI N—DAO approved, AI rejected,
- DAO N/AI Y—DAO rejected, AI approved,
- DAO N/AI N—both DAO and AI rejected.

A statistical significance test (McNemar's test) was applied to each contingency table to determine whether the differences between DAO and AI decisions were significant. The results are summarized in Table 1.

To complement the statistical comparison, a cost function was introduced to capture the relative severity of decision mismatches between the DAO and the AI models:

$$c = 1 * d + 10 * a$$

where

- c denotes the total cost,
- d is the number of false negatives (proposals accepted by the DAO but rejected by the AI),

Table 1. Contingency table results for three LLMs compared to DAO decisions

Model	DAO Y/AI Y	DAO Y/AI N	DAO N/AI Y	DAO N/AI N	Significance (p)	Cost
OSS-20b	70 (68.6%)	0 (0%)	27 (26.5%)	5 (4.9%)	2.03×10^{-7}	270
OSS-120b	63 (61.8%)	7 (6.9%)	24 (23.5%)	8 (7.8%)	0.0023	247
GPT-4-mini	56 (54.9%)	14 (13.7%)	20 (19.6%)	12 (11.8%)	0.303 (ns)	214
Claude 4.5-Haiku	36 (35.2%)	34 (33.3%)	12 (11.8%)	20 (19.6%)	0.0012	154
Claude 4.5-Sonnet	36 (35.3%)	34 (33.3%)	10 (9.8%)	22 (21.6%)	0.0003	134
GPT-5-mini	32 (31.4%)	38 (37.3%)	8 (7.8%)	24 (23.5%)	7.2×10^{-7}	118
GPT-5	24 (23.5%)	46 (45.1%)	3 (2.9%)	29 (28.4%)	8.1×10^{-10}	76

- a is the number of false positives (proposals rejected by the DAO but accepted by the AI).

This cost function assumes that false positives are substantially more detrimental, representing a potential misallocation of DAO funds, than false negatives, which merely reflect missed opportunities, but increases the aspect of funds-protection.

The analysis reveals clear differences in decision behavior between the DAO and the AI-driven QOC approach. Most models exhibited statistically significant deviations from the DAO's decisions, with GPT-4-mini being the only model for which the difference was not significant.

The smaller, open-weight models (OSS-20b and OSS-120b) showed a strong tendency to approve proposals that the DAO had rejected, reflected in high false-positive rates and correspondingly high costs. This behavior suggests that such models exhibit insufficient conservatism for use in financial or governance-sensitive contexts.

In contrast, the more advanced LLMs (Claude 4.5 and GPT-5 families) displayed the opposite pattern, adopting a more conservative stance, rejecting a higher proportion of proposals that had been approved by the DAO. This behavior, while stricter, led to substantially lower cost scores, particularly in the case of GPT-5, which achieved the lowest overall cost while maintaining only three false positives.

Overall, the results suggest that larger, more capable LLMs tend to produce more conservative and risk-aware decisions within the QOC framework.

5.2 Evaluation of the Sensitivity Analysis

We evaluate whether a multi-agent LLM pipeline can replicate the structure and reasoning of a Vester-style sensitivity analysis using IVL's logistics study on apparel flows from China to the Nordics [13]. Because IVL later adds company-internal factors, we assess both the *core transport system* and *company/internal* layers. The pipeline is executed 100 times, prompted only with a concise transport-scope statement (no IVL variables disclosed). Each run outputs factors with descriptions, a signed influence matrix, Vester metrics (AS, PS,

P = AS · PS, Q = AS/PS) with role labels, loops, and a short interpretation. To measure semantic correspondence, sentence embeddings (all-mpnet-base-v2) are used to match LLM factors to IVL factors via cosine similarity ($\geq$ 0.5), a paraphrase-level threshold consistent with prior work [14]. We then assess factor alignment and role consistency across both strata. Additionally, an LLM judge (GPT-5 Thinking), following evidence that LLM judgments correlate well with humans [15], rates each study and a reconstructed IVL baseline on an eight-criterion rubric (max 100) covering boundary, stakeholders, variables, causality, loops, scenarios, temporal scope, and actionability.

Across all **26** IVL factors, mean alignment is **55.5%** (range **23.1–80.8%**); **71%** of runs reach $\geq$ 50%. For the *core* 18 factors, alignment rises to **62.9%** (range **27.8–94.4%**), while the *internal/company* 8 average **38.8%**. High matches occur for *Fuel prices* (99%), *Competition in the transport market* (95%), and *Environmental demands* (93%), whereas structural items (e.g., *Total amount of goods*) and abstract internal variables (e.g., *Image of company*) align less. Role fidelity averages **56.6%**, reflecting typical volatility in AS/PS-based classifications. Rubric scores are tightly clustered (**77–98**, mean **92.97**, SD **3.47**), nearly identical to IVL's **93**, with consistently high marks for boundary (**4.99/5**), actionability (**14.9/15**), and temporal reasoning (**13.5/14**). This indicates that even with minimal prompting, the pipeline reproduces the methodological form of a Vester analysis: coherent factorization, causal mapping, and loop construction rather than textual imitation. Overall, the results show that a conditioned multi-agent LLM can operationalize the Vester method and achieve human-level coherence in structure and interpretation. Remaining gaps cluster around internal company factors and role stability, precisely the aspects a *specialist layer* could improve by grounding analyses in richer, domain-specific context.

5.3 Evaluation of the Game Theory Marcel

We test whether a role-conditioned, LLM-driven extensive-form game can reproduce the strategic dynamics and end-state logic of the Cuban Missile Crisis abstraction [16], aligning its SPE path with the ground-truth labeled sequence: (1) limited coercion (quarantine/interdiction with public warning, ESCALATION); (2) public pushback with private willingness to negotiate (backchannels, SIGNAL/INFO); (3) assurance exchange (non-invasion pledge and Jupiter context, DE-ESCALATION); and (4) withdrawal and dismantlement with verification (DE-ESCALATION).

We execute 60 independent runs with a fixed prompt and depth cap $D = 4$ (max four moves/path). At each node, the active player's persona proposes concrete actions; payoffs are then evaluated and backward induction selects the SPE path. To evaluate each step, we label each step using a compact taxonomy Escalate, Signal/Info, De-escalate, Wait under clear rules. Labeling is performed by an LLM annotator (GPT-5 Thinking) given the same codebook for both model outputs and the paper-derived reference, consistent with evidence that GPT-class models can match human annotation quality under structured instructions [17]. Some SPE paths terminate after three steps while others reach four steps: Three-step paths (17/60) are compared to paper steps 1–3; four-step paths (43/60) to steps 1–4. In addition to stepwise alignment, we compute a path-level label.

For three-step runs ($n = 17$), match rates vs. paper labels at positions 1–3 are **0.529**, **0.294**, **0.647**. For four-step runs ($n = 43$), match rates at positions 1–4 are **0.535**, **0.442**, **0.349**, **1.000**. Thus, openings align in roughly half the cases, mid-sequence alignment is mixed (more variance at steps 2–3), and the terminal decision matches DE-ESCALATE in all four-step runs.

Across all runs, the path-level label equals DE-ESCALATE in **56/60** cases (**93.3%**); by length: **15/17** (88.2%) for three-step and **41/43** (95.3%) for four-step paths.

The role-conditioned, LLM-driven extensive-form game reliably reproduces the *end-state logic* of the Cuban Missile Crisis abstraction [16], with **93%** of all paths and **100%** of four-step paths ending in DE-ESCALATE. While mid-sequence variance appears in the transitions, this reflects the presence of multiple continuations under minimal prompting. A *specialist layer* could reduce such variance by injecting structured, context-rich knowledge, e.g., historical documents, strategic playbooks, or actor-specific constraints, to guide belief updates and action generation. Overall, the model not only reproduces the historical outcome but also mirrors the *patterned reasoning dynamics* characteristic of sequential-game processes.

6 Outlook and Future Work

Despite these promising outcomes, several limitations and potential risks must be acknowledged. First, retrieval-augmented generation (RAG) pipelines, although essential for domain grounding, may inadvertently propagate outdated, incomplete, or biased information from external sources. Without robust source validation and temporal freshness checks, these retrieval errors can compromise both the factual accuracy and the traceability of the final explanations.

Second, hallucination propagation remains a challenge: when agents generate fabricated or inconsistent data, such artifacts can cascade through the reasoning chain, ultimately distorting aggregate results. While the proposed layered design and explainability barrier help to isolate these effects within controlled process layers, additional consistency-checking and reference validation mechanisms are required to ensure epistemic reliability in multi-step analyses.

Third, misaligned agent outputs can occur in multi-agent settings, where different role-conditioned agents interpret prompts or contextual cues divergently. In the presented architecture, this risk is mitigated through statistical outlier detection, consensus formation, and aggregation procedures that treat extreme or incoherent agent responses as diagnostic rather than determinative signals. Nonetheless, future work should formalize these reconciliation mechanisms, e.g. , by integrating alignment scoring or trust-weighting models that dynamically adjust each agent's influence according to its historical reliability and contextual fit.

Finally, although LLM-driven standard processes improve interpretability, they still depend on the opaque latent reasoning of the underlying model. The approach therefore cannot guarantee full transparency at the cognitive level of the LLM; it can only expose structured traces above the explainability barrier. This partial transparency should be recognized as an inherent limitation of current generative architectures.

Future research will focus on technical and methodological refinements addressing the above limitations. A first priority is the implementation of an integrated, modular software environment in which specialized agents operate under explicit retrieval, grounding, and validation constraints. RAG pipelines will be equipped with source-credibility filters and timestamp controls to reduce hallucination and data drift.

A second focus lies in formal process orchestration, developing a routing layer that interprets user intent, selects the appropriate analytical module (e.g., QOC vs. Game Theory), and ensures transparent data flow between agents. Complementary schema formalization, defining inputs, evaluation functions, and output structures in a machine-readable format, will further enhance reproducibility and auditability.

In the longer term, process-aware LLMs represent a compelling direction. By fine-tuning models on structured reasoning patterns, decision-analytic schemas could become part of the model's intrinsic latent space, reducing reliance on external orchestration and improving internal alignment. Combined with adaptive outlier-management and grounding protocols, this evolution would yield AI systems that are not only explainable in their outputs but also explainable by design in their reasoning.

References

1. Bilal, A., Ebert, D., Lin, B.: LLMs for explainable AI: a comprehensive survey. arXiv preprint arXiv:2504.00125 (2025)
2. Guidotti, R., Monreale, A., Ruggieri, S., Turini, F., Giannotti, F., Pedreschi, D.: A survey of methods for explaining black box models. ACM Comput. Surv. (CSUR) **51**(5), 1–42 (2018)
3. Ribeiro, M.T., Singh, S., Guestrin, C.: "Why should I trust you?" Explaining the predictions of any classifier. In: Proceedings of the 22nd ACM SIGKDD International Conference on Knowledge Discovery and Data Mining, pp. 1135–1144 (2016)
4. Lundberg, S.M., Lee, S.I.: A unified approach to interpreting model predictions. In: Advances in Neural Information Processing Systems, vol. 30 (2017)
5. Kokalj, E., Škrlj, B., Lavrač, N., Pollak, S., Robnik-Šikonja, M.: BERT meets Shapley: extending SHAP explanations to transformer-based classifiers. In: Proceedings of the EACL HackaShop on News Media Content Analysis and Automated Report Generation, pp. 16–21 (2021)
6. Saltelli, A., et al.: Five ways to ensure that models serve society: a manifesto. Nature **582**(7813), 482–484 (2020)
7. Wang, H., Zhang, F., Mu, C.: One for All: A General Framework of LLMs-based Multi-Criteria Decision Making on Human Expert Level. arXiv preprint arXiv:2502.15778 (2025)

8. MacLean, A., Young, R.M., Bellotti, V.M., Moran, T.P.: Questions, options, and criteria: elements of design space analysis. In: Design Rationale, pp. 53–105. CRC Press (2020)
9. Heller, S., Ibrahim, M., Selby, D.A., Vollmer, S.: Finding Common Ground: Using Large Language Models to Detect Agreement in Multi-Agent Decision Conferences. arXiv preprint arXiv:2507.08440 (2025)
10. Sun, H., Wu, Y., Cheng, Y., Chu, X.: Game theory meets large language models: a systematic survey. arXiv preprint arXiv:2502.09053 (2025)
11. De Carlo, A., Parimbelli, E., Melillo, N., Nicora, G.: Introducing δ-XAI: a novel sensitivity-based method for local AI explanations. arXiv preprint arXiv:2407.18343 (2024)
12. Luz, A.: Enhancing the interpretability and explainability of AI-driven risk models using LLM capabilities. Technical Report, EasyChair (2024)
13. Wolf, C., Persson, F., Jelse, K.: A logistic analysis with the Sensitivity Model Prof. Vester. Technical report, IVL Swedish Environmental Research Institute (Report B2048), Göteborg, Sweden. Project sponsor: Geodis Wilson (2012). https://www.diva-portal.org/smash/get/diva2:1552071/FULLTEXT01.pdf
14. Corley, C., Mihalcea, R.: Measuring the semantic similarity of texts. In: Proceedings of the ACL Workshop on Empirical Modeling of Semantic Equivalence and Entailment (EMSEE 2005), pp. 13–18. Association for Computational Linguistics, Ann Arbor, Michigan (2005)
15. Zheng, L., et al.: Judging LLM-as-a-judge with MT-bench and chatbot arena. In: Thirty-Seventh Conference on Neural Information Processing Systems Datasets and Benchmarks Track (2023). https://openreview.net/forum?id=uccHPGDlao
16. Zagare, F.C.: A game-theoretic history of the Cuban missile crisis. Economies **2**(1), 20–44 (2014)
17. Gilardi, F., Alizadeh, M., Kubli, M.: ChatGPT outperforms crowd workers for text-annotation tasks. Proc. Natl. Acad. Sci. **120**(30), e2305016120 (2023)

Comparative Evaluation of Prompting Techniques for Forensic Video Analysis Using GPT-4o

Opeyemi Adeniran(✉), Kelechi Nwachukwu, Anjolie Anthony, David Nyarko, Tasmeer Alam, Temitope Ajibola, and Kofi Nyarko

Morgan State University, Baltimore, MD 21251, USA
opade7@morgan.edu

Abstract. This research evaluates eight prompting methodologies for forensic video information extraction using GPT-4o across 13 crime scenarios. The study systematically examines Zero-shot, Least-to-Most, Reasoning and Acting (ReACT), Iterative, Self-Consistency, Sequential, Meta, and Chain-of-Thought prompting strategies using F1-scores as the primary performance metric for keyword-based information extraction. Results demonstrate that Meta-prompting achieves superior extraction completeness with low variability across diverse crime scenarios, while revealing significant performance stratification among techniques. Five techniques (Self-Consistency, Iterative, ReACT, Chain-of-Thought, Zero-shot) exhibit performance convergence within a narrow range, whereas Sequential and Least-to-Most demonstrate consistent underperformance and high variability. The evaluation identifies scenario-specific technique strengths, with Self-Consistency demonstrating advantages for particular crime types. Performance-consistency trade-off analysis reveals that Meta-prompting uniquely balances comprehensive information extraction with consistent results across scenarios. These findings provide evidence-based guidance for selecting prompting techniques to optimize information extraction depth and coverage in forensic video analysis applications, establishing foundations for future cross-model validation studies using keyword-based evaluation frameworks.

1 Introduction

Prompt engineering encompasses the development of natural language directives to elicit distinct responses from Large Language Models (LLMs). This methodology permits users to leverage LLM knowledge without parameter alterations, rendering it accessible to non-technical users [1]. Prompting techniques enhance model interpretability and control by guiding focus toward specific tasks or domains, improving relevance and accuracy in critical applications such as legal or medical domains [2,3]. These techniques facilitate task adaptation without extensive retraining, saving computational resources and accelerating deployment [4,5].

H. R. Arabnia et al. (Eds.): CSCI 2025, CCIS 2972, pp. 436–451, 2027.
https://doi.org/10.1007/978-3-032-28469-3_29

Prompting can mitigate inherent biases by steering models away from inappropriate content, enhancing ethical AI use in sensitive applications requiring fairness and neutrality [6]. However, effectiveness varies with task complexity and prompt specificity. Well-designed prompts significantly improve performance, while poorly constructed ones may yield suboptimal results or exacerbate biases, necessitating ongoing research to refine techniques and develop best practices [2,4] (Fig. 1).

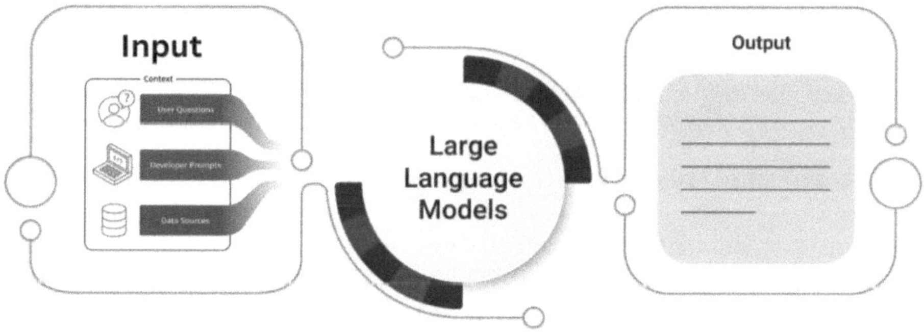

Fig. 1. Input prompts and data processed by multimodal Large Language Models to generate outputs.

Figure 3 depicts multimodal LLMs processing input prompts with assorted data sources to deliver relevant outputs, illustrating the essential operational flow and highlighting the pivotal importance of prompt engineering in harmonizing user input with machine comprehension.

1.1 How Do Prompting Techniques Improve Language Model Understanding?

Prompting techniques enhance language model understanding by structuring input to align with model training, facilitating accurate and contextually relevant outputs. Task-oriented prompts enable models to discern context and purpose more accurately, with meticulously constructed prompts enhancing efficacy in question answering and text classification by providing explicit context and reducing input ambiguity [4,7].

Prompting methodologies facilitate knowledge transfer across tasks by applying prompts to utilize pre-trained models for novel tasks without extensive retraining, enhancing efficiency and adaptability [8]. Carefully structured prompts provoke specific response types, providing insights into model decision-making mechanisms and fostering transparency and trust in AI systems [9].

This research systematically evaluates eight prompting methodologies (Zero-Shot, Least-to-Most, ReACT, Iterative, Self-Consistency, Sequential, Meta, Chain-of-Thought) for multimodal LLM performance in forensic video information extraction tasks. Using F1-scores as the primary performance indicator, this study identifies prompting strategies offering optimal extraction completeness and consistency, critical for enhancing multimodal LLMs in forensic video information extraction applications. The following research questions are formulated:

1. How do different prompting techniques influence information extraction completeness in forensic video analysis?
2. What are the comparative strengths and weaknesses of each prompting technique in enhancing extraction coverage, consistency, and scenario adaptability?
3. What is the impact of different prompting techniques on extraction robustness when applied to diverse crime scenarios?

2 Literature Review

Prompting methodologies facilitate efficient LLM adaptation to particular tasks, mitigating computational demands linked to model inference and manual prompt design [10]. ReAct prompting strengthens LLM decision-making by merging reasoning with action plans, though investigations indicate enhancements are predominantly attributable to input-query similarity rather than intrinsic ReAct efficacy [11]. Iterative prompting refines LLM responses, emphasizing truthfulness and mitigating hallucinations, though preliminary findings indicate simplistic iterative prompting may exacerbate calibration inaccuracies [12].

Self-consistency prompting capitalizes on multiple reasoning trajectories to bolster task performance, exhibiting efficacy in tasks requiring various solution pathways such as mathematical reasoning and code generation. The Universal Self-Consistency (USC) framework extends this by selecting the most consistent response from candidate pools, improving performance in open-ended tasks [13]. Meta-prompting serves as scaffolding that transforms a singular language model into a multifaceted conductor managing multiple independent inquiries. It disaggregates complex tasks into subtasks executed by distinct "expert" instances with customized instructions, enhancing performance through output synthesis and capitalizing on critical reasoning capabilities [14,15].

Chain-of-Thought (CoT) prompting produces intermediate reasoning links justifying conclusions, proving efficacious in mixed-task environments by classifying input inquiries and autonomously constructing diverse demonstrations, achieving state-of-the-art performance across reasoning challenges [16]. CoT is similarly applied to text-to-SQL parsing, enhancing LLM reasoning capacity through multi-step reasoning while circumventing error propagation [17].

Least-to-Most prompting decomposes intricate issues into sequential subordinate problems, utilizing solutions from previously resolved subproblems to facilitate extension to more complex examples. Evidence indicates substantial advancements over CoT prompting in endeavors requiring intricate reasoning such as symbolic manipulation and mathematical reasoning [18]. Zero-Shot prompting utilizes language models to execute tasks without explicit examples or training data, depending on pre-existing knowledge to comprehend and react to prompts. This proves advantageous where labeled data is limited or example generation is impractical [19,20]. Sequential prompting divides sophisticated assignments into manageable segments with organized direction for each segment, capitalizing on incremental information processing for more precise and contextually pertinent outputs [21,22].

The Prompt Optimization for Enhancing Multimodal Reasoning (POEM) system exemplifies interactive prompt engineering, allowing users to explore and refine prompts for better multimodal reasoning through iterative crafting [23]. Systematic prompt probing reveals current multimodal LLMs often fit training data distributions rather than truly understanding prompts, introducing frameworks with strategies for visual, text, and extra knowledge prompts [24]. Robust multimodal prompting techniques handle missing modalities using modality-specific tokens and orthogonality between prompts, enhancing performance and robustness while reducing required prompts [25].

The subsequent sections delineate the methodological framework. Section 3 articulates research methodology, encompassing video preprocessing, frame analysis through prompt engineering, and evaluation metrics. Section 4 elucidates empirical findings, juxtaposing performance and influence of various prompting techniques. Section 5 encapsulates principal findings and proposes future investigative avenues.

3 Methodology

The methodology follows a three-stage pipeline comprising video preprocessing, frame-level analysis with prompt-guided reasoning, and keyword-based evaluation for video information extraction. Information extraction completeness is quantified by measuring the degree of overlap between LLM-generated descriptions and reference keywords that encapsulate critical forensic attributes. The complete workflow is illustrated in Fig. 2.

3.1 Data Collection

This study utilizes 13 surveillance videos from the UCF-Crime dataset [26] and sources, encompassing diverse crime scenarios:

- **UCF-Crime Dataset** (11 videos): Abuse, Arrest, Assault, Burglary, Explosion, Fighting, Robbery, Shooting, Stealing, Shoplifting, and Vandalism
- **y Videos** (2 videos): Arson scenario from online surveillance sources and Custom scenario from specially recorded footage

Each video was processed and evaluated using all eight prompting methodologies.

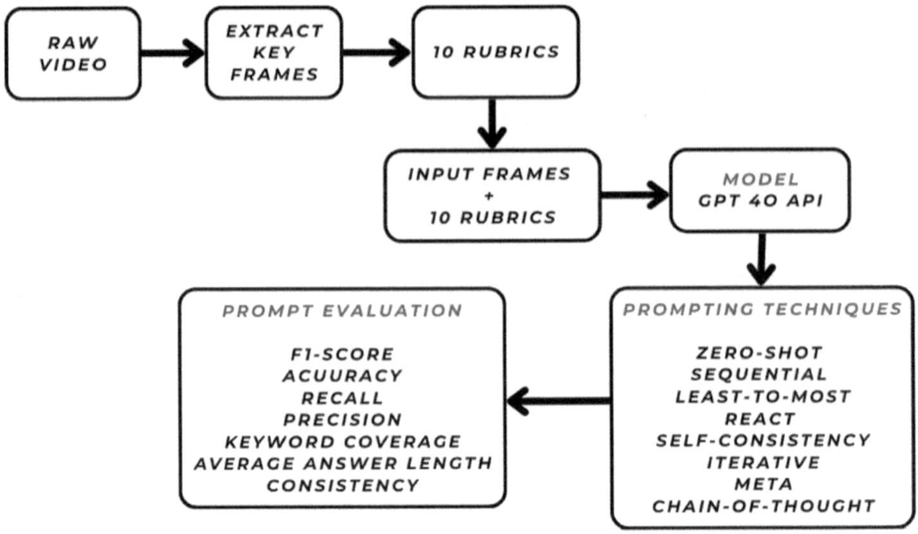

Fig. 2. Video Processing and Prompting Evaluation Workflow.

3.2 Video Preprocessing

Video Input and Frame Sampling. Frames are strategically sampled at predefined intervals to optimize computational resources. Let F represent total frames and I denote the sampling interval. The number of frames selected, N:

$$N = \left\lfloor \frac{F}{I} \right\rfloor \tag{1}$$

This study employs a sampling interval of 80 frames, extracting key frames at regular intervals to capture temporal progression.

Frame Encoding. Sampled frames are encoded into JPEG format and transformed into base64 strings for OpenAI API processing:

$$\text{Base64\_String} = \text{Base64\_Encode}(\text{JPEG\_Encoded\_Frame}) \tag{2}$$

Data Structuring. Encoded frames are organized in a dictionary structure with descriptive identifiers as keys and base64 strings as values, enabling efficient frame retrieval and distribution across rubric questions.

3.3 Frame Analysis Through Prompt Engineering

This phase constitutes the core of the methodological framework, wherein eight distinct prompt engineering strategies are utilized to direct the multimodal LLM in the forensic information extraction from video frames.

OpenAI API Integration. The OpenAI GPT-4o Vision API processes both textual and visual inputs [27]. All evaluations employ GPT-4o consistently across all prompting techniques to ensure fair comparison.

Prompt Engineering Techniques. Eight distinct prompt engineering methodologies are systematically assessed:

- **Zero-Shot Prompting**: Direct questions without prior examples, relying on learned knowledge [28,29]
- **Sequential Prompting**: Frame analysis in sequential order with each step building on prior analysis [30]
- **Least-to-Most Prompting**: Progressive questioning from simple to intricate details [31]
- **ReACT Prompting**: Combined reasoning and action with hypothesis generation [32]
- **Iterative Prompting**: Iterative response refinement building on previous outputs [33]
- **Self-Consistency**: Multiple reasoning paths with selection of most coherent output [34]
- **Meta-Prompting**: Self-reflection and dynamic strategy adjustment [35]
- **Chain-of-Thought Prompting**: Step-by-step reasoning through decomposed sub-tasks [36]

Rubric-Based Question Framework. A structured rubric of ten questions provides consistent evaluation across all techniques:

1. Subject's initial position and surroundings
2. Movement and direction patterns
3. Interactions with physical elements
4. Background activity and environmental context
5. Facial expressions and body language
6. Progression through the environment
7. Camera perspective and subject interaction
8. Final position and departure direction
9. Visual quality and image disturbances
10. Narrative construction and temporal context

Each rubric question has associated reference keywords representing essential forensic information to be extracted.

Frame Distribution and API Requests. Extracted frames are distributed evenly among the ten rubric questions. For each question, Q, and its associated frames, F_Q, the prompt, P:

$$P = \text{“Question: ”} + Q + \text{“Frames: ”} \\ + \text{Concatenate}(\text{Base64\_String}(f) \text{ for } f \text{ in } F_Q) \quad (3)$$

Requests are submitted to GPT-4o with technique-specific instructions. Responses are extracted and stored with error-handling for rate limits.

3.4 Keyword Extraction and Evaluation

For each of the ten rubric questions, a curated set of reference keywords, K_R, is established representing essential forensic analysis concepts. Both reference keywords and LLM-generated keywords, K_T, undergo cleaning to remove special characters, punctuation, and digits, with conversion to lowercase:

$$\text{cleaned\_keyword} = \text{re.sub}(\text{pattern}, '', \text{keyword}) \quad (4)$$

Text normalization applies lemmatization and stemming using NLTK's `WordNetLemmatizer` and `PorterStemmer` [37–39] to both reference and extracted keywords, resolving word form variations. For lemmatized matches:

$$\text{matched\_lemmatized} = \sum_{w \in \mathrm{K}_R^{\text{lemmatized}}} ⊮(w \in K_T^{\text{lemmatized}}) \quad (5)$$

For stemmed matches:

$$\text{matched\_stemmed} = \sum_{w \in \mathrm{K}_R^{\text{stemmed}}} ⊮(w \in K_T^{\text{stemmed}}) \quad (6)$$

where $⊮(\cdot)$ denotes the indicator function.

$$\text{total\_}\textit{combined}\text{\_matches} = \sum_{i=1}^{10} \Big(\text{matched\_lemmatized}_i + \quad (7)$$

$$\text{matched\_stemmed}_i \Big) \quad (8)$$

The number of matched keywords for technique T:

$$M_T = |K_R \cap K_T| \quad (9)$$

3.5 Evaluation Metrics

Standard classification metrics are computed based on keyword matching results:

$$\text{Precision} = \frac{\text{TP}}{\text{TP} + \text{FP}}, \quad \text{Recall} = \frac{\text{TP}}{\text{TP} + \text{FN}} \tag{10}$$

$$\text{F1-Score} = 2 \times \frac{\text{Precision} \times \text{Recall}}{\text{Precision} + \text{Recall}} \tag{11}$$

$$\text{Accuracy} = \frac{\text{TP} + \text{TN}}{\text{Total Predictions}} \tag{12}$$

where TP, FP, FN, and TN represent True Positives, False Positives, False Negatives, and True Negatives respectively. F1-score serves as the primary performance metric, providing balanced assessment of information extraction completeness, measuring both coverage (recall) and precision of extracted forensic keywords.

Additional evaluation dimensions include:

- **Keyword Coverage**: Percentage of reference keywords identified: $\text{Coverage} = \left(\frac{M_T}{|K_R|}\right) \times 100$
- **Average Answer Length**: Mean character or word count of LLM responses
- **Consistency**: Standard deviation of F1-scores across scenarios, measuring performance stability

These metrics enable comprehensive comparison of all eight prompting techniques across 13 crime scenarios, facilitating objective, evidence-based technique selection.

4 Results and Discussion

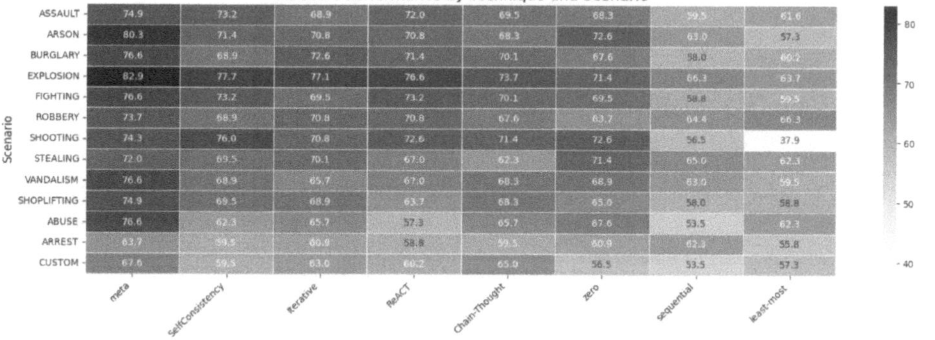

Fig. 3. F1 Score Performance by Technique and Scenario.

The evaluation of eight prompting techniques applied to GPT-4o across 13 crime scenarios reveals quantifiable performance disparities using F1-scores as the primary metric.

4.1 Overall Performance Hierarchy

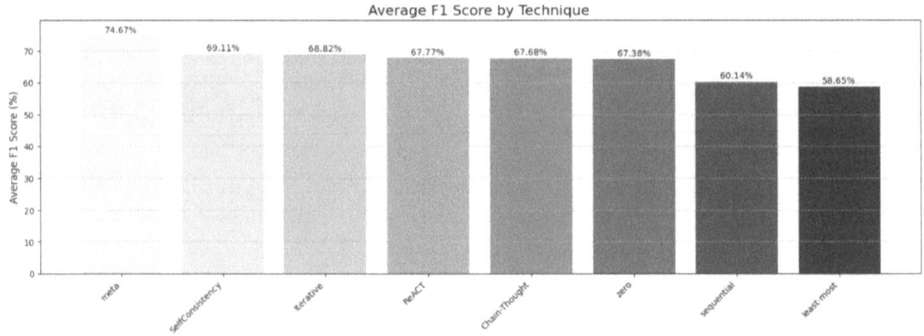

Fig. 4. Average F1 Scores for Each Prompting Technique.

Figure 4 demonstrates Meta-prompting as the highest-performing technique (74.67%), followed by Self-Consistency (69.11%), Iterative (68.82%), ReACT (67.77%), Chain-of-Thought (67.68%), Zero-shot (67.38%), Sequential (60.14%), and Least-to-Most (58.65%). This represents a 16.02% point spread between highest and lowest performers.

Performance stratification reveals three tiers: Meta-prompting at 74.67% (high-tier); five mid-tier techniques spanning only 1.73% points (67.38–69.11%); and two low-tier techniques, Sequential (60.14%) and Least-to-Most (58.65%). Meta-prompting's 5.56% point advantage over second-place Self-Consistency demonstrates distinctly superior extraction capability, while the tight mid-tier clustering suggests performance convergence among different prompting approaches.

4.2 Scenario-Specific Performance Patterns

The heatmap in Fig. 3 reveals substantial performance variation across crime scenarios. Explosion scenarios demonstrate highest overall performance, with Meta achieving 82.9%, Self-Consistency (77.7%), and Iterative (77.1%). Assault scenarios show strong performance across techniques (73–75% for top performers).

Shooting scenarios present the largest performance gap observed: Self-Consistency achieves 76.0% while Least-to-Most drops to 37.9% (38.1% point disparity). Arson scenarios produce uniformly lower scores (59.5–63.7%), while Custom scenarios show moderate performance (53.5–67.6%).

Meta-prompting achieves top performance in 9 out of 13 scenarios (Assault, Arson, Burglary, Explosion, Fighting, Robbery, Stealing, Vandalism, Custom), with Self-Consistency outperforming in Shooting scenarios.

4.3 Ranking Patterns Across Scenarios

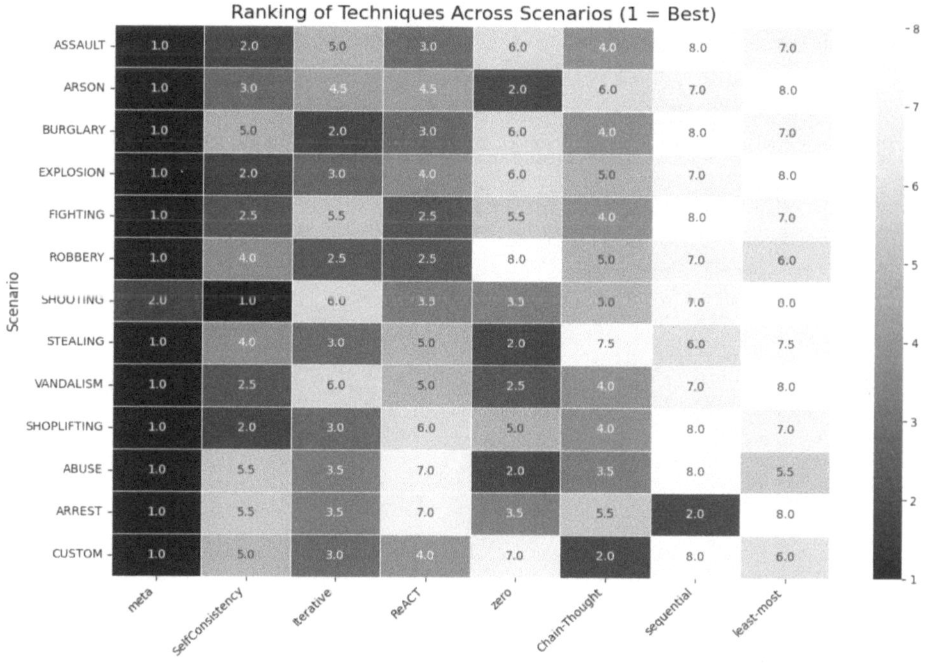

Fig. 5. Ranking of Techniques Across Scenarios (1 = Best).

The ranking heatmap in Fig. 5 shows Meta-prompting achieving rank 1 in 11 out of 13 scenarios (exceptions: Shooting rank 2.0, Custom rank 1.0). Self-Consistency exhibits variable ranking (2.0–5.5), while Least-to-Most consistently occupies rank 8.0 in 7 scenarios (Assault, Burglary, Explosion, Fighting, Shoplifting, Abuse, Arrest).

Ranking convergence occurs in Fighting, Robbery, and Vandalism (ranks 2.5–5.5), while Shooting, Custom, and Arrest show clearer stratification (ranks 1.0–8.0).

4.4 Top Four Techniques Performance Profile

The radar chart in Fig. 6 visualizes performance profiles of Meta, Self-Consistency, Iterative, and ReACT. Meta-prompting demonstrates the most

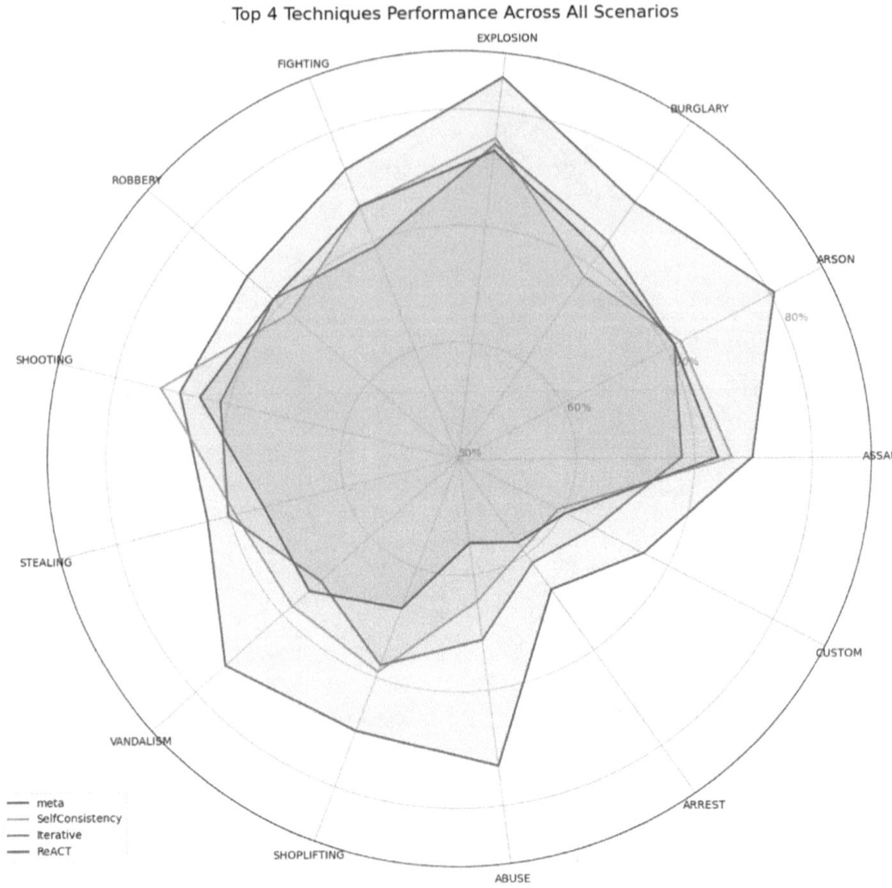

Fig. 6. Top 4 Techniques Performance Across All Scenarios.

expansive coverage area with pronounced advantages in Arson, Assault, and Custom scenarios. Self-Consistency maintains a high performance floor (never below 60%) with particular strength in Explosion and Burglary. Iterative performs strongly in Fighting and Robbery, while ReACT shows competitive performance in Explosion and Arrest.

4.5 Performance-Consistency Trade-Off Analysis

The scatter plot in Fig. 7 reveals the relationship between average performance (x-axis) and consistency measured by standard deviation (y-axis). Meta-prompting achieves optimal "High Performance, High Consistency" positioning with 74.67% average F1 score and low standard deviation (4.95%).

Sequential occupies the "Low Performance, Low Consistency" quadrant (60.14% average, 6.85% standard deviation), representing the least desirable

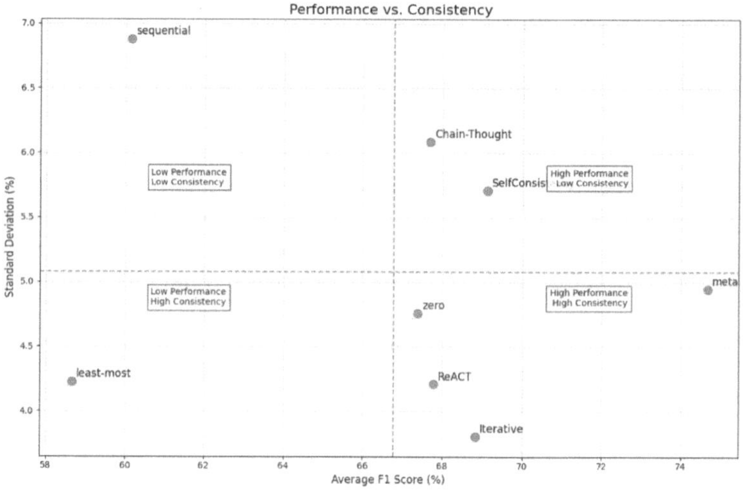

Fig. 7. Performance vs. Consistency Trade-Off.

attribute combination. Self-Consistency achieves strong performance (69.11%) with moderate variability (5.68%), while Zero-shot demonstrates high consistency (4.75%) with moderate performance (67.38%).

4.6 Scenario Difficulty Stratification

Explosion, Assault, and Burglary scenarios demonstrate highest average performance (>70%), indicating readily extractable forensic information. Shooting scenarios present extreme variance (76.0% Self-Consistency versus 37.9% Least-to-Most), suggesting specific analytical challenges. Arrest, Custom, and Sequential scenarios show moderate difficulty (60–65% average), representing typical analytical complexity.

4.7 Cross-Figure Integration

Meta-prompting achieves top overall ranking (74.67%) through consistently high performance across all scenarios while avoiding extreme performance drops in challenging contexts. The ranking heatmap confirms rank 1 in 11 out of 13 scenarios.

High average performance does not guarantee consistency: Self-Consistency achieves second-highest score (69.11%) yet exhibits greater standard deviation than Meta-prompting. Explosion scenarios enable broad technique success (all techniques >60%), while Shooting scenarios strongly differentiate capabilities (ranks 1.0–8.0).

4.8 Research Questions Discussion

This study systematically addressed three research questions regarding prompting techniques for forensic video information extraction:

RQ1: Influence on Information Extraction Completeness - Results demonstrate that prompting techniques significantly influence information extraction completeness, with Meta-prompting achieving superior performance (74.67%) compared to lower-tier approaches (58.65–60.14%). Performance variation spans 16.02% points, confirming substantial technique influence on extraction capability. Scenario-specific analysis reveals differential extraction patterns, with techniques showing varied completeness across crime types.

RQ2: Comparative Strengths in Extraction Coverage, Consistency, and Scenario Adaptability - Meta-prompting exhibits breadth of extraction coverage with optimal consistency across diverse scenarios, achieving rank 1 in 11 out of 13 scenarios. Self-Consistency demonstrates scenario-specific advantages, particularly in Shooting scenarios (76.0%), while maintaining high extraction floor across all scenarios. Five mid-tier techniques (Self-Consistency, Iterative, ReACT, Chain-of-Thought, Zero-shot) exhibit extraction performance convergence within 1.73% points, suggesting similar capability levels for information coverage. Sequential and Least-to-Most show systematic weaknesses with consistent underperformance and high variability, rendering them unsuitable for comprehensive forensic information extraction applications.

RQ3: Impact on Extraction Robustness Across Crime Scenarios - Extraction robustness varies significantly across scenarios, with certain crime types (Explosion, Assault, Burglary) enabling consistent extraction success (>70% average performance) while others (Shooting) strongly differentiate extraction capabilities with extreme variance (38.1% point range). Meta-prompting demonstrates superior extraction robustness through consistent performance across diverse scenarios with low standard deviation (4.95%). The performance-consistency trade-off analysis reveals that achieving both high extraction completeness and high consistency represents a rare combination, with Meta-prompting uniquely balancing these dual requirements, enhancing its applicability for reliable forensic video information extraction where both coverage depth and consistency are critical.

5 Conclusion

This study provides comprehensive evaluation of eight prompting techniques for forensic video information extraction using GPT-4o across diverse crime scenarios. Meta-prompting emerges as the superior technique for extraction completeness, demonstrating consistent high performance and optimal balance between information coverage and reliability across diverse scenarios. Mid-tier techniques exhibit extraction performance convergence, suggesting natural performance ceilings inherent to prompting structure, while lower-tier approaches demonstrate systematic limitations unsuitable for comprehensive forensic information extraction applications.

Scenario-specific analysis reveals that certain crime types present extreme analytical challenges with significant extraction variance across techniques, while others enable broad technique success in information coverage. Performance-consistency trade-off analysis indicates that achieving both objectives simultaneously remains rare, with Meta-prompting representing the singular technique successfully balancing these dual requirements. These findings establish evidence-based foundations for prompting technique selection in forensic video information extraction and contribute valuable insights for optimizing multimodal large language models in keyword-based extraction applications.

6 Recommendations

Meta-prompting is recommended for general-purpose forensic video information extraction based on consistent superior extraction completeness and optimal reliability across diverse crime scenarios. For scenario-specific optimization, Self-Consistency demonstrates extraction advantages in particular crime types, while certain scenarios enable strong extraction performance across multiple techniques. Sequential and Least-to-Most should be avoided for comprehensive extraction applications due to systematic underperformance and high variability in information coverage.

Future research should investigate adaptive hybrid architectures combining Meta-prompting's extraction breadth with scenario-specific technique strengths, examine underlying causes of extreme extraction variance in challenging scenarios, and conduct cross-model validation studies across alternative multimodal LLMs to establish whether observed prompting hierarchies represent universal properties or model-specific characteristics. Additional research directions include expanding evaluation frameworks to alternative video contexts, investigating temporal sampling impacts on extraction capability, and exploring evaluation dimensions beyond keyword-matching metrics to assess information quality and forensic utility.

7 Limitations

This study's keyword-matching evaluation framework may underestimate techniques generating conceptually accurate descriptions using alternative vocabulary and may not generalize to alternative video contexts. Fixed frame sampling at 80-frame intervals potentially omits critical transient events, affecting temporal analysis capability. Results pertain exclusively to GPT-4o and may not generalize to alternative multimodal LLMs with different architectures. SHOOTING scenarios demonstrate extreme performance variance requiring investigation of underlying causes, while the evaluation focuses primarily on F1-scores, potentially overlooking qualitative aspects such as narrative coherence and contextual accuracy valuable for forensic applications.

References

1. Shubham, V., Dubey, H.: A Survey of Prompt Engineering Methods in Large Language Models for Different NLP Tasks, arXiv, vol. abs/2407.12994 (2024). https://doi.org/10.48550/arxiv.2407.12994
2. Zhu, F., Zhou, H., Zhu, Z.-G., Qian, J., Mao, K.: Unveiling and Manipulating Prompt Influence in Large Language Models (2024). https://doi.org/10.48550/arxiv.2405.11891
3. Tony, C., Díaz Ferreyra, N.E., Mutas, M., Dhiff, S., Scandariato, R.: Prompting Techniques for Secure Code Generation: A Systematic Investigation (2024). https://doi.org/10.48550/arxiv.2407.07064
4. Dave, N., Kifer, D., Giles, C.L., Mali, A.: Investigating Symbolic Capabilities of Large Language Models (2024). https://doi.org/10.48550/arxiv.2405.13209
5. Toshevska, M., Gievska, S.: Large Language Models for Text Style Transfer: Exploratory Analysis of Prompting and Knowledge Augmentation Techniques. Ambient Intelligence and Smart Environments (2024). https://doi.org/10.3233/aise240025
6. Zhang, Y., Wang, X., Wu, L., Wang, J.: Pattern-Aware Chain-of-Thought Prompting in Large Language Models, arXiv.org, vol. abs/2404.14812 (2024). https://doi.org/10.48550/arxiv.2404.14812
7. Liu, Q., Wang, F., Xu, N., Yan, T., Meng, T., Chen, M.: Monotonic Paraphrasing Improves Generalization of Language Model Prompting, arXiv.org, vol. abs/2403.16038 (2024). https://doi.org/10.48550/arxiv.2403.16038
8. Wang, Y., Zhao, Y.: Metacognitive Prompting Improves Understanding in Large Language Models, arXiv.org, vol. abs/2308.05342 (2023). https://doi.org/10.48550/arxiv.2308.05342
9. Qin, L., et al.: CroPrompt: Cross-task Interactive Prompting for Zero-shot Spoken Language Understanding (2024). https://doi.org/10.48550/arxiv.2406.10505
10. Chang, K., Xu, S., Wang, C., Luo, Y., Xiao, T., Zhu, J.: Efficient Prompting Methods for Large Language Models: A Survey, arXiv, vol. abs/2404.01077 (2024). https://doi.org/10.48550/arxiv.2404.01077
11. Verma, M., Bhambri, S., Kambhampati, S.: On the Brittle Foundations of ReAct Prompting for Agentic Large Language Models (2024). https://doi.org/10.48550/arxiv.2405.13966
12. Krishna, S.S., Agarwal, C., Lakkaraju, H.: Understanding the Effects of Iterative Prompting on Truthfulness, arXiv.org, vol. abs/2402.06625 (2024). https://doi.org/10.48550/arxiv.2402.06625
13. Chen, X., et al.: Universal Self-Consistency for Large Language Model Generation, arXiv.org, vol. abs/2311.17311 (2023). https://doi.org/10.48550/arxiv.2311.17311
14. Suzgun, M.M., Kalai, A.T.: Meta-Prompting: Enhancing Language Models with Task-Agnostic Scaffolding, arXiv.org, vol. abs/2401.12954 (2024). https://doi.org/10.48550/arxiv.2401.12954
15. Zhang, Y.: Meta Prompting for AGI Systems, arXiv.org, vol. abs/2311.11482 (2023). https://doi.org/10.48550/arxiv.2311.11482
16. Zou, A., Zhang, Z., Zhao, H., Tang, X.: Meta-CoT: Generalizable Chain-of-Thought Prompting in Mixed-task Scenarios with Large Language Models, arXiv.org, vol. abs/2310.06692 (2023). https://doi.org/10.48550/arxiv.2310.06692
17. Tai, C.-Y., Chen, Z., Zhang, T., Deng, X., Sun, H.: Exploring Chain-of-Thought Style Prompting for Text-to-SQL, arXiv.org, vol. abs/2305.14215 (2023). https://doi.org/10.48550/arXiv.2305.14215

18. Least-to-Most Prompting Enables Complex Reasoning in Large Language Models (2022). https://doi.org/10.48550/arxiv.2205.10625
19. Brown, T., et al.: Language Models are Few-Shot Learners, arXiv preprint, arXiv:2005.14165 (2020)
20. Liu, P.J., et al.: Generating Wikipedia by Summarizing Long Sequences, arXiv preprint, arXiv:1801.10198 (2018)
21. Wei, J., et al.: Chain of Thought Prompting Elicits Reasoning in Large Language Models, arXiv preprint, arXiv:2201.11903 (2022)
22. Reynolds, L., McDonell, K.: Prompt Programming for Large Language Models: Beyond the Few-Shot Paradigm, arXiv preprint, arXiv:2102.07350 (2021)
23. He, J., Wang, F., Liu, S., Wu, G., Silva, C.T., Qu, H.: POEM: Interactive Prompt Optimization for Enhancing Multimodal Reasoning of Large Language Models (2024). https://doi.org/10.48550/arxiv.2406.03843
24. Qi, S., Cao, Z., Rao, J., Wang, L., Xiao, J., Wang, X.: What is the limitation of multimodal LLMs? A deeper look into multimodal LLMs through prompt probing. Inf. Process. Manag. **60**, 103510 (2023). https://doi.org/10.1016/j.ipm.2023.103510
25. Hua, H., Tang, Y., Xu, C., Luo, J.: V2Xum-LLM: Cross-Modal Video Summarization with Temporal Prompt Instruction Tuning, arXiv.org, vol. abs/2404.12353 (2024). https://doi.org/10.48550/arxiv.2404.12353
26. Sultani, M., Chen, C., Shah, M.: Real-World Anomaly Detection in Surveillance Videos. https://www.crcv.ucf.edu/projects/real-world/. Accessed 17 Sept 2024
27. OpenAI API Documentation. https://platform.openai.com/docs/overview. Accessed 17 Sept 2024
28. Tam, A.: What Are Zero-Shot Prompting and Few-Shot Prompting. MachineLearningMastery (2023). https://machinelearningmastery.com/what-are-zero-shot-and-few-shot-prompting/
29. Rojo-Echeburúa, A.: Zero-Shot Prompting: Examples, Theory, Use Cases. DataCamp. https://www.datacamp.com/tutorial/zero-shot-prompting
30. Mckie, K.: Advanced Prompting Techniques. AI Overview (2023). https://www.aioverview.com/sequential-prompting/
31. Wang, R.: Guiding AI with Least-to-Most Prompting. TechDeep Dive (2023). https://www.techdeepdive.com/least-to-most-prompting
32. Chen, Y., Wu, S.: ReAct: reasoning and acting through large language models. AI Syst. J. **34**, 152–165 (2023). https://aisystemsjournal.org/react-prompting
33. Brownlee, J.: Iterative Refinement in Language Models. AI Mastery (2024). https://aimastery.com/iterative-prompting
34. Li, D., Johnson, M.: Self-consistency in large language models. AI Res. Forum **56**, 45–50 (2023). https://airesearchforum.org/self-consistency/
35. Thompson, A.: Meta-prompting for AI: learning from feedback. DeepAI J. **23**, 33–40 (2023). https://deepai.org/meta-prompting
36. Anderson, T.: Chain-of-Thought Prompting: A Step-by-Step Approach to Reasoning. AI Breakthrough (2023). https://aibreakthrough.com/chain-of-thought-prompting
37. Bird, S., Loper, E.: NLTK 3.0 Documentation. NLTK Project (2023)
38. Manning, C.D., Schütze, H.: Foundations of Statistical Natural Language Processing. MIT Press (2022)
39. Jurafsky, D., Martin, J.H.: Speech and Language Processing. Pearson (2021)

Enhancing Large Language Model Accuracy with Retrieval-Augmented Generation: A Comparative Study

Tathagata Bhattacharya(✉), Irshad Ali Mohammad, Amit Das, Jahnavi Sunki Reddy, and Harshavardhan Meka

Department of CSIS, Auburn University at Montgomery, Montgomery, Alabama, USA
tbhatta1@aum.edu

Abstract. This research explores the integration of Retrieval-Augmented Generation (RAG) with three prominent Large Language Models (LLMs): OpenAI's GPT-3.5-turbo, Ollama's Llama2, and Anthropic's Claude-2.1. The study aims to evaluate the impact of RAG on enhancing the accuracy, relevance, and quality of generated responses. Experimentation involved querying each model, both in standalone and RAGenabled configurations, using a medical question, "Is it safe to consume Leqvio?", and comparing their outputs against a reference response. Metrics such as BLEU Score, ROUGE Scores, Accuracy, Precision, Recall, and F1 Score were used to assess performance. The results demonstrated that RAG integration significantly improved the performance of all models, with Llama2 showing the most substantial gains, particularly in BLEU Score and ROUGE Scores. GPT-3.5-turbo also exhibited notable improvements, while Claude-2.1 showed more modest gains. Overall, this research highlights the effectiveness of RAG in enhancing LLM performance and suggests several directions for future optimization and application of RAG in various domains.

Keywords: Retrieval-Augmented Generation (RAG) · Large Language Models (LLMs) · GPT-3.5-turbo · Llama2 · Claude-2.1 · BLEU Score · ROUGE Scores · Vector Database · Information Retrieval

1 Introduction

The IT world has evolved as a data-intensive field in the entire world [1, 2]. These data intensity can be managed by the Natural language processing (NLP) in a very beautiful manner. NLP has witnessed remarkable advancements in recent years, driven by the advent of Large Language Models (LLMs) and their ability to generate naturally flowing text [3]. For example, GPT-3 can write coherent essays, answer complex questions, and generate code snippets based on natural language descriptions [4]. But, at times LLM outputs may produce inaccurate or irrelevant results when they rely solely on retrieval-based methods or generative AI models. This is why the retrieval-augmented generation (RAG) framework was developed. Retrieval augmented generation can be applied to LLMs to enhance their performance in various natural language processing

H. R. Arabnia et al. (Eds.): CSCI 2025, CCIS 2972, pp. 452–488, 2027.
https://doi.org/10.1007/978-3-032-28469-3_30

tasks. Generally, LLMs rarely contain the most current, up-to-date information. Without specific guidance about which sources are authoritative and which ones aren't, LLMs may select the non-authoritative source when generating a response or become confused when two sources used in model training use similar terms to describe different things. Lastly, when models don't have a response to offer, they may choose to fabricate one or hallucinate.

Retrieval-augmented generation alleviates these issues, providing better control over the sources the models reference, resulting in more accurate, authoritative responses. Therefore, to deal with this situation, we propose a novel approach to increase the accuracy and reliability of LLM with RAG implementation. LLM is a key component in our research. Therefore, we discuss the architecture of LLM in the next section (refer to Sect. 1.1).

1.1 Architecture of LLM

Large Language Models (LLMs) represent a significant advancement in natural language processing and understanding. These models are designed to understand, generate, and manipulate human language, making them powerful tools for a variety of applications, from chatbots to translation services [5].

The architecture of an LLM typically consists of multiple layers of artificial neurons, often implemented using transformer networks. Transformers have become the defacto standard due to their ability to handle long-range dependencies in text, making them highly effective for tasks involving large amounts of data [6]. We discuss the key components of LLM as follows.

Figure 1 represents an illustration of a typical transformer based LLM architecture. Figure 1 depicts the flow of data through the various components of the LLM, highlighting the role of embeddings, positional encoding, and transformer blocks.

- **Embeddings Layer:** This is the initial layer that converts words into numerical vectors [7]. These vectors capture semantic meanings and contextual information, which are essential for the model to understand language nuances [8].
- **Positional Encoding:** Since transformers do not inherently understand the order of words, positional encoding is used to inject information about the position of words in a sequence. This helps the model to understand the context better [9].
- **Encoder - Decoder Architecture:** Many LLMs use an encoder-decoder setup. The encoder processes the input text and converts it into a fixed-size context vector, while the decoder takes this context vector and generates the output text. This architecture is particularly effective for tasks such as translation [10].
- **Attention Mechanism:** Attention mechanisms allow the model to focus on different parts of the input sequence when generating output. This dynamic weighting of different parts of the input sequence helps the model to better capture context and improve the relevance of the generated text [11].

  - **Transformer Blocks:** The core of LLMs is made up of multiple transformer blocks, each containing layers of multi-head self-attention and feed-forward neural networks. These blocks are stacked on top of each other, enabling the model to learn complex patterns and dependencies in the data [12].

- **Training Process:** LLMs are trained on vast amounts of text data using unsupervised learning techniques. This involves predicting the next word in a sentence (language modeling) or filling in missing words (masked language modeling). The training process is computationally intensive and requires significant resources [13].

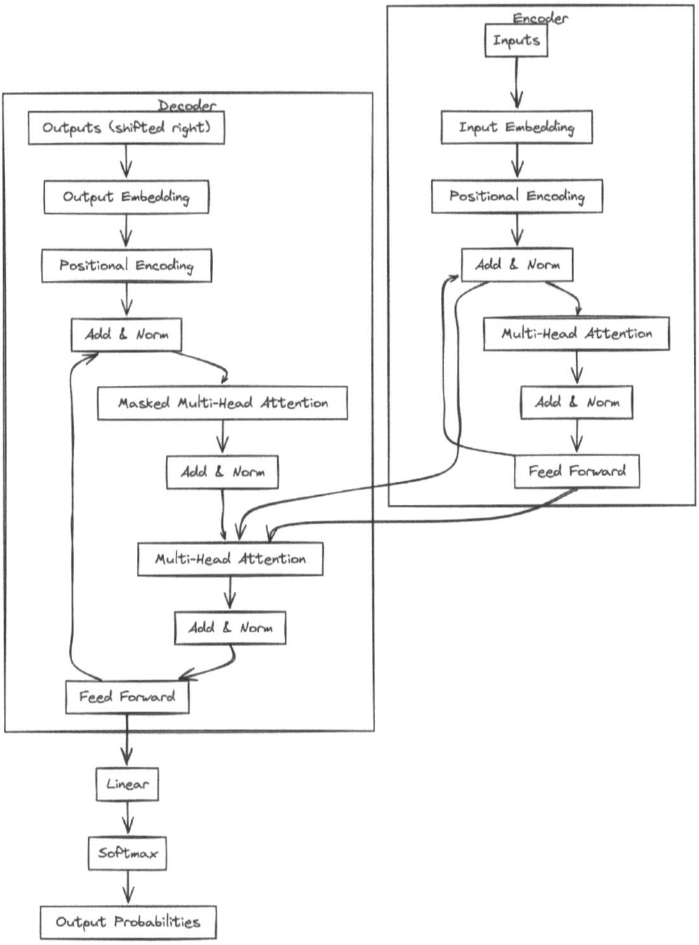

Fig. 1. Simplified Architecture of a Transformer-Based LLM

Next, we discuss the applications of LLM in Sect. 1.2.

1.2 Applications of LLM

LLMs have a wide range of applications [14], including but not limited to:

- **Text Generation:** Creating coherent and contextually relevant text, such as articles, stories, and reports.
- **Translation:** Converting text from one language to another with high accuracy.
- **Summarization:** Condensing long documents into concise summaries.
- **Question Answering:** Providing accurate answers to user queries based on context.
- **Conversational Agents:** Powering chatbots and virtual assistants that can engage in human-like conversations.

LLMs are trained on vast amounts of text data, which allows them to recognize patterns in language and understand context [15]. This training enables them to perform a wide range of language-related tasks. For example, you could use an LLM to draft an email to your boss explaining why you need a day off, and it would generate a polite and professional message for you. Or, if you're learning a new language, you could ask the LLM to translate a paragraph from English to Spanish, and it would do so quickly and accurately [16].

However, despite their impressive capabilities, LLMs often struggle with factual accuracy and knowledge grounding, as their training data may not encompass the breadth and depth of information required for specific domains or tasks [17]. For instance, an LLM might confidently produce a believable but incorrect answer about a historical event or scientific concept it wasn't specifically trained [18]. In a medical context, an LLM might suggest an outdated treatment protocol or misinterpret rare disease symptoms due to limited exposure to such specialized information.

[19].

To address this limitation, researchers have explored the integration of retrieval systems with LLMs, giving rise to the concept of retrieval-augmented generation (RAG) [20]. For example, a RAG system could enhance a medical chatbot by retrieving the latest clinical guidelines before generating treatment recommendations. It could also improve a legal assistant by accessing up-to-date case law databases to support its responses. We discuss more about RAG in Sect. 1.3.

1.3 Retrieval-Augmented Generation (RAG)

As large language models (LLMs) evolve, one of the significant advancements enhancing their capabilities is the introduction of Retrieval-Augmented Generation (RAG). RAG combines the strengths of retrieval-based methods and generative models to create more accurate and contextually relevant responses [16].

RAG systems integrate two primary components: a retrieval model and a generative model. The retrieval model searches through a knowledge base to find relevant documents or pieces of information, while the generative model uses this retrieved information to generate a coherent and contextually appropriate response [21]. This combination enables the system to leverage vast external knowledge, making it more powerful and versatile [22].

RAG has numerous applications across various fields [23]. In customer service, a RAG-powered chatbot could access a company's product database to provide accurate, up-to-date information about inventory, pricing, and features [24]. In education, an RAG system could act as a tutor, pulling information from textbooks and academic papers to explain complex concepts and provide current examples.

One of the biggest advantages of RAG is its ability to reduce "hallucinations" - instances where LLMs generate plausible-sounding but incorrect information. By grounding the LLM's responses in external, verifiable sources, RAG significantly improves the accuracy and reliability of AI-generated content [25]. RAG represents a significant advancement in AI technology, bridging the gap between the vast knowledge of LLMs and the need for current, accurate information. It's making AI assistants smarter, more reliable, and more useful in real-world applications, from answering simple queries to assisting with complex research tasks [26]. We discuss the key components of RAG as follows.

- **Massive Dataset:** The foundation of RAG systems is a large dataset used to pre-train the LLM. This dataset provides the necessary background knowledge and language understanding [27].
- **Pre-trained LLM:** This model serves as the generative component. It takes the user's query and, augmented with retrieved documents, generates a response. The LLM is typically pre-trained on a large corpus of text to understand language patterns and context [28].
- **Retrieval Model:** This component is responsible for searching a knowledge base to find documents or data relevant to the user's query [29]. It ensures that the generative model has access to the most pertinent information, improving the quality and accuracy of the responses.
- **Knowledge Base:** A repository of information that the retrieval model searches through. This could include documents, articles, books, or any structured data that can provide contextually relevant information for the generative model [30].

Vector database is used for storing the knowledge base in this study. The next section explains the vector databases in great detail (refer to Sect. 1.4).

1.4 Vector Databases

In the ever-evolving landscape of data management and retrieval, vector databases have emerged as a crucial component, especially in applications involving natural language processing (NLP) and machine learning. As illustrated in Fig. 2 [31], vector databases enable efficient storage and retrieval of data in high-dimensional vector spaces [32]. This subsection introduces vector databases and explains their general role in the context of large language models (LLMs) and related technologies.

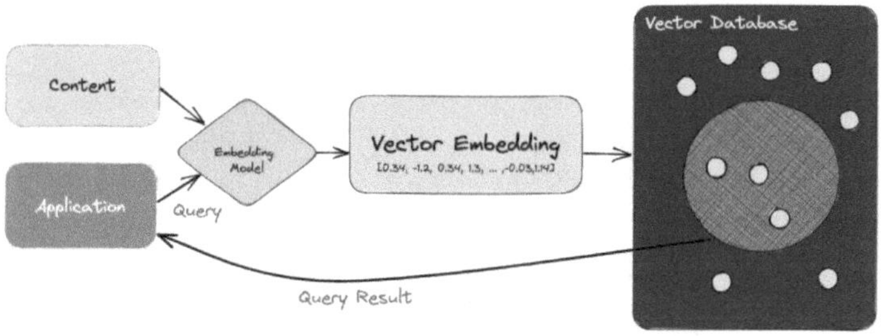

Fig. 2. Role of Vector Database in this experimentation

1.4.1 The Concept of Vector Databases

A vector database is designed to handle data in the form of vectors—arrays of numerical values that represent various types of information, such as text, images, or any other form of data that can be embedded into a vector space [33]. In the context of NLP, text data is often transformed into vectors using embedding models. These vectors capture the semantic meaning of the text, allowing for efficient similarity searches within the database [34].

As depicted in the accompanying figure, the process begins with content (e.g., text documents) and application queries being processed by an embedding model. The embedding model converts both the content and queries into vector embeddings, which are then stored in the vector database [35]. When a query is made, the vector database retrieves the most relevant embeddings by comparing the query vector with the stored vectors, returning the closest matches based on similarity metrics [36].

1.4.2 General Usage of Vector Databases with LLMs

Vector databases are increasingly being integrated with LLMs to enhance their functionality and performance [37]. LLMs, such as GPT-3.5 or Claude-2.1, are powerful tools for generating text and answering questions based on their training data. However, their ability to provide accurate, up-to-date, or context-specific information is limited by the data they were trained on, which is often static and may not reflect the most current knowledge.

To address this limitation, vector databases are used to supplement LLMs by providing a dynamic and searchable knowledge base. Here's how the integration typically works:

- **Embedding the Knowledge Base:** Textual content from various sources, such as documents, articles, or websites, is processed by an embedding model that converts this content into vector embeddings. These embeddings are stored in the vector database.
- **Handling Queries:** When an LLM receives a user query, the query is first converted into a vector using the same embedding model. This query vector is then used to search the vector database for the most relevant content.
- **Contextual Retrieval:** The vector database retrieves content that is semantically similar to the query, providing the LLM with relevant context that it might not have had access to during its initial training. This retrieval process allows the LLM to

generate responses that are more accurate, contextually relevant, and informed by up-to-date information.

- **Response Generation:** The retrieved content is fed back into the LLM, which uses it to generate a more informed response. This process ensures that the LLM's outputs are not only based on its internal knowledge but also augmented by the latest and most relevant external information.

The combination of vector databases and LLMs represents a significant advancement in the field of NLP [38], enabling the creation of AI systems that can provide more reliable, accurate, and context-aware responses. This integration is particularly useful in applications where the correctness and relevance of information are critical, such as in medical, legal, or technical domains. By leveraging vector databases, LLMs can transcend their inherent limitations, making them more versatile and effective in real-world applications [37].

We discuss the implementations and applications of RAG in the following subsection (refer to Sect. 1.5).

1.5 Implementations and Applications of RAG

The retrieval process starts when a user submits a query. The retrieval model first searches the knowledge base, such as Wikipedia or domain-specific databases [39] for documents relevant to the query. These documents are then combined with the original query and passed to the pretrained LLM. The LLM processes this augmented input and generates a response that is informed by both the query and the retrieved documents [28].

The integration of retrieval and generation components in RAG has shown promising results in various NLP tasks, such as question answering, open-domain dialogue [40], and summarization. By leveraging external knowledge sources, RAG can potentially improve the factual accuracy, knowledge grounding, and overall quality of the generated text [21].

Figure 3 demonstrates how the retrieval model searches the knowledge base for relevant documents, which are then used by the pre-trained LLM to generate a response to the user's query. The integration of retrieval and generation allows for more accurate and contextually rich outputs.

However, the effectiveness of RAG is influenced by several factors, including the quality and relevance of the retrieved information, the ability of the language model to effectively incorporate and reason over the retrieved knowledge, and the specific task and domain at hand [41]. For instance, when answering a medical question, the system must retrieve and use the latest medical guidelines to generate an accurate answer [15]. Moreover, the choice of retrieval and generation components, as well as their integration strategies, can significantly impact the performance of RAG [23]. Additionally, the choice of retrieval and generation components, as well as their integration strategies, can significantly impact the performance of RAG [42]. For example, using a highly efficient retrieval algorithm can ensure that the most relevant documents are considered, while an advanced generation model can articulate the final response in a clear and informative manner [16].

We have categorically mentioned this research's key ideas and motivations in the following sections (refer to Sect. 1.6).

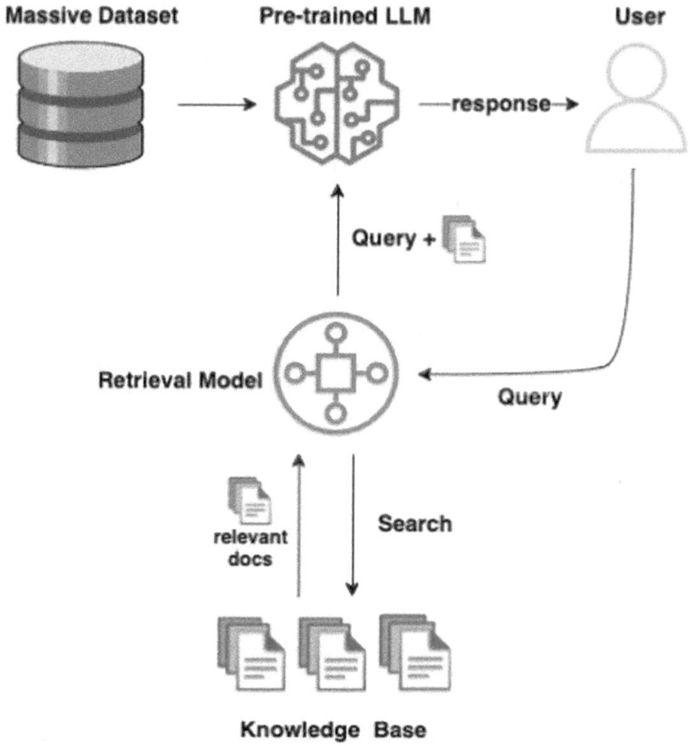

Fig. 3. RAG System Workflow

1.6 Contribution

This research paper presents a thorough comparative analysis of LLMs with and without RAG. The study evaluates the performance of these models across different tasks and metrics, including accuracy, and coherence. The comparative study provides insights into how the integration of retrieval mechanisms can enhance the capabilities of LLMs, leading to more reliable and informative outputs.

We discuss the main contributions of our research in this section.

- First, we aim to provide a comprehensive comparison of the performance of Large Language Models (LLMs) with and without Retrieval-Augmented Generation (RAG).
- Second, we have made several significant discoveries through our comparative analysis. For instance, our research reveals that LLMs integrated with RAG consistently outperform standard LLMs in terms of factual accuracy and contextual relevance.
- Further, the traditional models generate plausible but incorrect information we observe that the integration of RAG significantly mitigates the issue of hallucination. This finding underscores the importance of retrieval mechanisms in enhancing the reliability and trustworthiness of AI-generated content.

- While existing studies may focus on specific applications or benefits of LLMs [43], our research uniquely examines their overall effectiveness, accuracy, coherence, and user satisfaction in a variety of tasks.

This holistic approach provides a clearer understanding of the practical implications and advantages of integrating RAG with LLMs. The next section describes the next steps of the research paper (refer to Sect. 1.7).

1.7 Roadmap

The structure of this paper is organized as follows:

Section 2 reviews existing work in the field, summarizing key studies that have explored the performance of LLMs with and without RAG. We highlight the various methodologies and approaches used in previous research, noting gaps and areas for further investigation.

In Sect. 3 We detail the methodologies employed in our comparative analysis of LLMs with and without RAG. This section includes a description of the datasets used, the evaluation metrics applied, and the experimental setup. We also outline the criteria for assessing accuracy, coherence, and user satisfaction.

In Sect. 4 We present the findings from our comparative study. We analyze the performance of LLMs in both scenarios, discussing the strengths and weaknesses of each approach. This section includes detailed charts and graphs to illustrate key points and support our conclusions.

In Sect. 5 We explore potential future directions for research in this area. This section discusses how our findings can inform the development of more advanced and reliable AI systems. We also suggest areas where further investigation could yield significant improvements, such as optimizing RAG techniques and exploring new retrieval strategies.

In Sect. 6 The research concludes with a summary of our findings and their implications. We reflect on the importance of integrating retrieval mechanisms with generative models and highlight the contributions of our research to the field.

In the next section, we discuss the existing work that was done in comparing various LLM's metrics using different methodologies (refer to Sect. 2).

1.8 Literature Review

Integrating Retrieval-Augmented Generation (RAG) with large language models (LLMs) has fundamentally shifted the landscape of natural language processing. LLMs achieve this level of understanding through a process called deep learning, which involves analyzing enormous amounts of text data to recognize patterns and relationships between words and concepts [44]. This is why they're called "large" language models – they require massive amounts of data and computational power to train. Cuconasu [45] investigate the performance of base versus instructed LLMs in RAG systems, challenging the prevailing assumption that instructed models are inherently superior.

1.9 LLMs in RAG Systems

"A Tale of Trust and Accuracy: Base vs. Instruct LLMs in RAG Systems" by [45], combines retrieval mechanisms with generative models to enhance the accuracy and relevance of AI-generated content. This approach leverages retrieval to provide contextually relevant information, which is then used by generative models to produce informed responses. Retrieval-Augmented Generation (RAG) is a powerful technique that enhances the capabilities of Large Language Models (LLMs) by combining them with external knowledge sources [46]. By integrating external knowledge sources, RAG systems can dynamically update their responses with the most current and accurate information available, significantly reducing the occurrence of hallucinations and improving overall reliability. Cuconasu further explores the dual-phase nature of RAG systems, where the generative phase often relies on either base or instructed LLMs. They reveal that while instructed models are typically used in practice, base models sometimes demonstrate superior performance, thus questioning the efficacy of additional instruction fine-tuning.

Prior work by Huang [47] and Wang [48] has focused on addressing issues such as hallucinations and context drift in generative models. Cuconasu [45] extend this discourse by demonstrating that base models, despite lacking explicit instruction fine-tuning, can mitigate these issues effectively. Their findings align with those of Gao (2024) [43], who emphasize the need for sophisticated AI systems that balance accuracy and contextual relevance.

The role of instruction fine-tuning and alignment with human preferences has been well-documented in the literature. Taori and Ouyang [49] highlight the benefits of supervised fine-tuning and reinforcement learning in enhancing model performance. Cuconasu [45] challenges these notions by showing that the performance improvements from instruction fine-tuning are not always significant, and base models can sometimes offer better results in RAG settings.

Additionally, the research of Liu (2023) [50] and LangChain (2023) on the practical applications of instruct models supports the use of templates and fine-tuning in various RAG tasks. However, Cuconasu [45] provides a nuanced perspective by demonstrating that, in some contexts, base models without these enhancements can outperform instructed models, prompting a reevaluation of current methodologies and practices.

The findings from Cuconasu [45] underscore the need for a deeper examination of the methodologies employed in RAG systems. By questioning the assumed superiority of instructed LLMs, their research opens new avenues for exploring the fundamental aspects of RAG and its impact on model performance. This perspective is crucial for advancing the development of more effective and reliable AI systems.

The methodologies and the evaluation metrics of [45] are categorically explained in the following sub-sections (see Sect. 2.1.1, 2.1.2 and 2.1.3).

1.9.1 Task Instructions

The study conducted by the researchers utilizes two primary task instructions to guide the models in answering questions:

- **Task Instruction I:** This task focuses on the extractive nature of question-answering tasks. Models are instructed to extract the answer from one of the provided documents. Importantly, if the answer is not present in any of the retrieved documents, the model should respond with "NO-RES." This instruction is designed to test the model's ability to accurately locate information and recognize when an answer is unavailable.
- **Task Instruction II:** This task extends the first by asking models to not only extract an answer but also provide supporting evidence from the documents, formatted as "Proof." The task includes a one-shot example that illustrates the expected format of the model's answer, emphasizing the requirement for evidencebacked responses.

1.9.2 Prompt Composition

- **Task Instruction:** The explicit instructions guiding the model on how to approach the question and what format the answer should take.
- **Retrieved Documents:** A set of documents retrieved based on their relevance to the query. These documents are ordered by ascending similarity scores to position the most relevant documents closest to the query, thereby aiding the model in finding accurate information quickly.
- **Query:** The question or task that the model needs to address.

1.9.3 Evaluation Metrics

The experiments assess model performance using several key metrics:

- **Accuracy:** Measures the ability of models to provide correct answers based on the task instructions.
- **Negative Rejection Rate:** Evaluates the model's capability to refrain from answering when the necessary information is not available in the retrieved documents.
- **Parametric Memory Recall:** Assesses how often models rely on their internal knowledge to provide answers that are not present in the retrieved documents.

These metrics are essential for understanding the tradeoffs between different training approaches and configurations, particularly how they affect model performance in real-world applications.

In the upcoming sub-sections (see Sect. 2.1.4 and 2.1.5), Cuconasu [45] present results of evaluations on three types of models—base, instruct, and instruct + template—across various task instructions and datasets. Cuconasu [45] offers comprehensive insights into the performance of these models, shedding light on the comparative advantages of base models in RAG systems.

	NQ # Retrieved Documents							TriviaQA # Retrieved Documents						
Model	1	2	3	4	5	8	10	1	2	3	4	5	8	10
Llama 2 7B	**23.88**	**24.71**	**27.83**	**29.53**	**30.22**	**31.01**	**31.46**	**55.85**	**57.15**	**59.28**	**60.40**	**61.24**	**62.93**	**63.89**
Llama 2 7B-C	16.06	18.48	18.62	18.59	19.21	21.98	21.08	32.59	34.63	41.09	42.95	46.14	49.19	48.96
Llama 2 7B-C + T	3.36	1.21	0.69	0.48	0.45	0.73	1.52	23.35	21.63	21.13	17.70	18.05	22.30	25.68
Llama 3 8B	**27.03**	**30.22**	**30.53**	**31.08**	**29.08**	**29.49**	**28.70**	**44.64**	**53.48**	**56.78**	**59.59**	**58.97**	**64.93**	**65.05**
Llama 3 8B-I	8.52	10.25	15.40	15.85	22.57	28.80	28.25	4.13	2.62	3.47	4.44	15.73	51.90	58.61
Llama 3 8B-I + T	14.40	17.34	19.04	15.30	18.10	20.35	19.35	16.45	27.94	31.32	30.80	34.18	44.04	50.12
Mistral 7B	**24.26**	**25.30**	25.72	26.17	27.66	26.65	28.87	**58.67**	**59.15**	**58.87**	**60.49**	**62.02**	**64.16**	**65.92**
Mistral 7B-I	20.04	24.99	**26.69**	**30.56**	**31.67**	**33.33**	**34.82**	48.85	52.31	54.72	55.90	56.97	59.06	60.60
Mistral 7B-I + T	18.17	23.54	19.14	27.90	27.45	27.07	26.79	45.88	50.22	52.42	53.73	54.37	57.56	58.31
Falcon 7B	**17.13**	**18.97**	**21.15**	**21.08**	**21.95**	**22.64**	**22.33**	**41.64**	**43.11**	**43.69**	**44.22**	**45.60**	**46.35**	**48.24**
Falcon 7B-I	15.68	17.72	17.96	19.21	20.08	20.04	20.98	33.19	36.52	36.46	37.71	38.42	38.83	39.33

Fig. 4. Task Instruction I

1.9.4 Evaluation on Task Instruction I

In the initial set of experiments, Task Instruction I was used to assess model performance. we evaluate the models using Task Instruction I, pictured in Fig. 4 (refer [45]). Reported is the accuracy for each model/version combination at different levels of retrieved documents, that is, the number of documents added to the prompt of the LLM given the query. The results, depicted in Fig. 4 of their study, reveal that base models generally outperform their instruct counterparts, with few exceptions. Table 1 of Cuconasu (2024) highlights these findings:

- Llama 2's base model outperformed its instruct counterpart (without template) by an average of 9.23 points (a 48% increase) and 17.88 points (a 42% increase) on the NQ and Trivia QA datasets, respectively.
- Falcon's base model showed a 1.94-point (10%) and 7.48-point (20%) improvement over the instruct model.
- Llama 3's base model exhibited significant gains, improving accuracy by 10.92 points (59%) on NQ and 37.5 points (186%) on Trivia QA, though this was partly attributed to Llama 3's reliance on its template.

The only notable exception was Mistral, where the base model was 2.49 points (8%) less accurate than the instructed version on NQ. However, it still outperformed the instruct model by 5.83 points (10%) on Trivia QA. These results underscore the unexpected strength of base models in certain RAG tasks, challenging the conventional preference for instruct models.

	NQ # Retrieved Documents							TriviaQA # Retrieved Documents						
Model	1	2	3	4	5	8	10	1	2	3	4	5	8	10
Llama 2 7B	**24.82**	**28.70**	**31.71**	**32.85**	**34.09**	**35.62**	**35.79**	**54.94**	**56.94**	**58.88**	**60.07**	**60.87**	**62.75**	**63.73**
Llama 2 7B-C	18.41	24.96	28.76	30.22	32.16	33.16	32.05	41.86	47.18	49.53	51.19	52.59	55.73	56.00
Llama 2 7B-C + T	1.59	2.42	3.88	5.75	8.45	12.81	12.91	4.22	6.55	8.75	10.15	14.90	20.69	27.48
Llama 3 8B	**29.39**	**31.53**	**34.72**	**37.07**	**36.59**	**39.15**	**40.22**	**61.57**	**63.32**	**64.56**	**65.93**	**66.52**	**67.67**	**68.04**
Llama 3 8B-I	18.83	23.57	25.65	27.97	30.84	34.13	37.97	42.67	44.44	47.11	48.10	48.71	52.59	56.48
Llama 3 8B-I + T	11.70	16.44	6.40	6.61	11.60	5.64	0.69	30.52	36.89	31.11	33.17	38.08	33.69	23.42
Mistral 7B	**24.82**	**30.74**	**33.75**	**35.17**	**36.83**	**40.15**	**40.15**	**57.98**	**60.49**	**62.18**	**63.70**	**65.23**	**67.04**	**67.44**
Mistral 7B-I	21.53	27.66	29.91	32.85	33.58	36.45	35.76	48.51	52.60	54.79	56.05	56.91	58.80	59.54
Mistral 7B-I + T	16.86	18.10	20.87	22.43	24.09	24.44	26.96	33.73	33.21	35.47	36.57	34.98	42.39	48.45
Falcon 7B	**17.58**	**20.04**	**22.43**	**23.75**	**23.85**	**26.12**	**26.72**	**40.32**	**42.76**	**44.92**	**45.45**	**46.47**	**47.23**	**48.57**
Falcon 7B-I	16.10	18.31	18.93	19.87	20.91	21.61	21.30	33.54	35.69	37.50	38.09	39.02	41.08	42.95

Fig. 5. Task Instruction II

1.9.5 Evaluation on Task Instruction II

Captivated by the initial results, Cuconasu et al. (2024) developed a second task instruction aimed at evaluating the models' ability to ground their answers with evidence from context documents.

The results, summarized in Fig. 5 (refer [45]), indicate a general increase in accuracy across all models and settings. For instance, the accuracy of Llama 2's base model increased by 3.56 points (12%) on average. This suggests that the requirement for proof itself acts as a form of prompt engineering, enhancing model performance. Notably, base models continued to outperform their counterparts, with even stronger results: Mistral's base model achieved higher accuracy than the instruct versions by 3.41 points (10%).

These findings demonstrate that base models not only maintain their performance advantage but can also benefit significantly from structured task instructions that emphasize grounding responses in retrieved documents. This aligns with the broader discourse in the literature, where prompt engineering and retrieval-augmented approaches are recognized as crucial for improving the accuracy and relevance of AI-generated content (Lewis et al., 2020; Huang et al., 2023; Wang et al., 2022).

1.10 RAG for Knowledge-Intensive NLP Tasks

The paper "Retrieval-Augmented Generation for Knowledge-Intensive NLP Tasks" by Patrick Lewis et al. [41] explores the integration of pre-trained language models with retrieval mechanisms to improve performance on tasks requiring extensive knowledge. Traditional pretrained language models store factual knowledge within their parameters, but they face limitations in updating knowledge and providing evidence for their outputs. RAG models combine parametric memory (pre-trained seq2seq models) with non-parametric memory (dense vector indexes of documents), allowing for more dynamic and interpretable knowledge retrieval and utilization.

RAG models employ a two-component architecture: a retriever and a generator. The retriever uses a Dense Passage Retriever (DPR) to fetch relevant documents based on the input query, while the generator uses a seq2seq model, such as BART, to generate responses conditioned on the retrieved documents. The retriever provides a distribution over text passages, and the generator uses these passages along with the input to produce the output. Two RAG variants are proposed: RAG-Sequence, which uses the same retrieved document to predict the entire output sequence, and RAGToken, which allows different documents to influence each output token, marginalizing them.

The Researchers conducted extensive experiments across various knowledge-intensive tasks, including open-domain question answering (QA), abstractive QA, Jeopardy question generation, and fact verification. The following table summarizes the performance of RAG models compared to baseline models:

- **Open-Domain Question Answering (QA):** The researchers experimented to evaluate RAG's performance on open-domain QA tasks. They compared RAG against extractive QA approaches and closedbook QA methods using several datasets, including Natural Questions, TriviaQA, WebQuestions, and CuratedTrec. They measured performance using Exact Match (EM) scores, providing insights into RAG's effectiveness in generating accurate answers for a wide range of queries. As indicated in

Fig. 6 (refer [41]), RAG achieves state-of-the-art results across several datasets. For example, on the Natural Questions (NQ) dataset, RAG-Sequence achieves an exact match score of 44.5, significantly outperforming other models like REALM and DPR. This underscores the efficacy of combining retrieval-based methods with generation capabilities, allowing RAG to generate correct answers even when the exact answer is not present in the retrieved documents.

	Model	NQ	TQA	WQ	CT
Closed Book	T5-11B [52]	34.5	- /50.1	37.4	-
	T5-11B+SSM[52]	36.6	- /60.5	44.7	-
Open Book	REALM [20]	40.4	- / -	40.7	46.8
	DPR [26]	41.5	**57.9**/ -	41.1	50.6
	RAG-Token	44.1	55.2/66.1	**45.5**	50.0
	RAG-Seq.	**44.5**	56.8/**68.0**	45.2	**52.2**

Fig. 6. Open-Domain QA Test Scores

- **Abstractive Question Answering:** The researchers assessed RAG's capability to produce free-form, abstractive answers rather than merely extracting information. They used the MSMARCO dataset, which contains questions and pre-retrieved passages. Although the passages were not used in this specific experiment, they focused on RAG's natural language generation skills in knowledge-intensive scenarios.
- **Jeopardy Question Generation:** The researchers also tested RAG's ability to generate Jeopardy-style questions from given answers. This task differs from traditional QA as it involves creating questions based on factual statements. They evaluated RAG's performance using a BART model for comparison, with metrics like Q-BLEU-1 and human evaluations to assess the factuality and specificity of the generated questions.
- **Fact Verification:** RAG performs competitively with state-of-the-art models in the FEVER dataset for fact verification. As shown in Fig. 7 (refer [41]), RAG achieves label accuracy scores close to those of specialized models with extensive engineering and intermediate retrieval supervision. Notably, RAG achieves these results without using supervision on retrieved evidence, highlighting its potential applicability to realworld tasks where such supervision is unavailable.

Model	Jeopardy B-1	Jeopardy QB-1	MSMARCO R-L	MSMARCO B-1	FVR3 Label Acc.	FVR2 Label Acc.
SotA	-	-	**49.8***	**49.9***	**76.8**	**92.2***
BART	15.1	19.7	38.2	41.6	64.0	81.1
RAG-Tok.	**17.3**	**22.2**	40.1	41.5	72.5	89.5
RAG-Seq.	14.7	21.4	40.8	44.2		

Fig. 7. Generation and classification Test Scores

2 Methodology

This section outlines the methodology adopted to compare the performance of three Large Language Models (LLMs): OpenAI's GPT-3.5-turbo, Ollama's Llama2, and Anthropic's.

Claude-2.1.1, both with and without the integration of Retrieval-Augmented Generation (RAG). The RAG framework uses **Pinecone** as the vector database (refer 1.4) for storing and retrieving relevant documents. The comparison focuses on evaluating the models based on a specific medical query, "Is it safe to consume Leqvio?", with the underlying objective of determining the efficacy and accuracy of each model in generating responses grounded in external knowledge sources. The reference response used to validate the accuracy of the models is "Leqvio is approved by the Food and Drug Administration (FDA) to lower low-density lipoprotein cholesterol." The subsections below detail the experimental setup, dataset selection, and evaluation metrics used in this study (see Sects. 3.1, 3.2 and 3.3).

2.1 Experimental Setup

The experimental setup is meticulously designed to ensure a fair and consistent comparison between the three LLMs: GPT-3.5-turbo, Llama2, and Claude-2.1.1. Each model is evaluated in two configurations—standalone (without RAG) and integrated with RAG. The primary goal is to analyze how each model performs when augmented with external knowledge from the Pinecone vector database, particularly in generating accurate and contextually relevant responses to medical inquiries.

2.1.1 Hardware and Software Configuration

Initially, we depict the hardware and software configuration details in this subsection. The experiments were conducted on an Apple M2 Pro chip, running macOS 14.4.1. The M2 Pro, with its advanced architecture, provides a robust environment for executing computationally intensive tasks such as LLM inference and real-time data retrieval.

- **Processor:** Apple M2 Pro with 10-core CPU (6 performance cores and 4 efficiency cores).

- **RAM:** 16 GB unified memory. • **GPU:** 16-core integrated GPU.
- **Operating System:** macOS 14.4.1, optimized for AI workloads and offering seamless integration with the software stack.
- **Software:** The software environment includes Python 3.10, Langchain for model orchestration, Pinecone for vector-based retrieval, and Streamlit for user interface development. The models were accessed via their respective APIs—OpenAI API for GPT-3.5-turbo, Ollama API for Llama2, and Anthropic API for Claude-2.1.

The following table orchestrates the detailed specifications of the experimental environment (refer Table 1).

Table 1. Hardware and Software Configuration for Experimentation.

Component	Specification
Processor	Apple M2 Pro, 10-core CPU
RAM	16 GB unified memory
GPU	16-core integrated GPU
Operating System	macOS 14.4.1
Software	Python 3.10, Langchain, Pinecone, Streamlit, OpenAI API, Ollama API, Anthropic API

This setup ensures that all models are tested under identical hardware and software conditions, thereby eliminating any variations arising from differences in computational resources.

2.1.2 Model Configuration

Our next task in this experiment was to configure the model. For this study, each LLM—GPT-3.5-turbo, Llama2, and Claude-2.1—was configured to operate both independently and with the RAG system. The models were integrated into a Langchain pipeline that facilitates their interaction with the Pinecone vector database.

- **LLM without RAG:** Each LLM was tested in its default configuration, where responses were generated solely based on the model's pre-existing knowledge without external data retrieval.
- **LLM with RAG:** In the RAG configuration, each LLM was connected to the Pinecone vector database, where a stored PDF document, titled "Accelerating Discoveries in Medicine and Advancing Clinical Breakthroughs" (January 2024, 1790044185, 9607574) [51], was indexed. The RAG system was designed to retrieve the top three relevant sections from the document in response to the query "Is it safe to consume Leqvio?", which were then used by the LLMs to generate an informed response.

The experiment was designed to determine how effectively each model could utilize the retrieved information to enhance the accuracy and relevance of its responses compared to the standalone configuration.

2.2 Datasets

The dataset used for this experiment was carefully curated to ensure that the models could be evaluated in a context that demands both accuracy and contextual understanding. The primary source of data for this experiment is a PDF document titled" Accelerating Discoveries in Medicine and Advancing Clinical Breakthroughs" (January 2024, 1790044185, 9607574) [51]. This document consists of 6 pages and contains detailed information about recent advancements in medical science, including the usage, safety, and regulatory approval of various pharmaceutical products, such as Leqvio.

The document served as the sole knowledge base for the RAG-enabled models, where relevant sections were embedded and stored in a Pinecone vector database. The query used to test the models, "Is it safe to consume Leqvio?", was specifically chosen to evaluate the models' ability to retrieve and accurately synthesize information from the provided document. The reference response for validation, "Leqvio is approved by the Food and Drug Administration (FDA) to lower low-density lipoprotein cholesterol," was directly sourced from this document.

2.2.1 Query Dataset

The query dataset for this experiment was specifically focused on a single, medically relevant question: "Is it safe to consume Leqvio?". This query was selected for its relevance to current medical practices and its specificity, which demands a precise and well-informed response. The following criteria were used to evaluate the models' performance on this query:

- **Clarity and Specificity:** The query was structured to require a clear and specific answer, testing the models' ability to accurately retrieve and present relevant information.
- **Relevance:** Given that Leqvio is a recently approved medication with specific guidelines for its use, the query serves as a stringent test of the models' capabilities in handling specialized content.

This single-query approach allowed for a focused analysis of the models' performance, with the reference response "Leqvio is approved by the Food and Drug Administration (FDA) to lower low-density lipoprotein cholesterol" serving as the benchmark for accuracy.

2.2.2 Knowledge Base Dataset

The knowledge base dataset for this experiment was derived from the aforementioned PDF document, which spans 6 pages and includes comprehensive information on various medical advancements. This document was chosen for its relevance and authority, making it an ideal source for evaluating the accuracy and contextual understanding of the models.

- **Document Details:** The PDF contains detailed discussions on the approval processes, clinical trial results, and recommended uses of several pharmaceuticals, including Leqvio. This information was indexed and stored in the Pinecone vector database, allowing for efficient retrieval during the experiment.
- **Indexing Process:** The document was processed into individual paragraphs, each of which was converted into a high-dimensional vector using a sentencetransformer model. These vectors were then stored in the Pinecone vector database, enabling quick retrieval of relevant sections in response to the query.

By utilizing a single, highly relevant document as the knowledge base, the experiment was designed to assess how effectively each model could retrieve and incorporate specific, authoritative information into its responses. This setup provided a controlled environment to evaluate the impact of RAG on the models' ability to handle domainspecific queries accurately.

2.3 Evaluation Metrics

The evaluation of the LLMs' performance in this experiment was conducted using a comprehensive set of metrics designed to assess various aspects of the model's output. The metrics used include **BLEU Score, ROUGE Scores, Accuracy, Precision, Recall, and F1 Score**. These metrics were selected to provide a well-rounded evaluation, capturing not only the correctness of the responses but also their quality and relevance.

2.3.1 BLEU Score

The BLEU (Bilingual Evaluation Understudy) Score is a widely used metric for evaluating the quality of text generated by a model, particularly in tasks such as machine translation. BLEU measures how closely the model's output matches a set of reference translations. It operates by comparing n-grams (continuous sequences of words) in the generated text to those in the reference text. A higher BLEU score indicates a closer match and, thus, higher quality.

In this experiment, BLEU was used to evaluate how well the generated responses aligned with the reference response, "Leqvio is approved by the Food and Drug Administration (FDA) to lower low-density lipoprotein cholesterol." The BLEU score provides a quantitative measure of the similarity between the generated response and the reference text.

2.3.2 ROUGE Scores

ROUGE (Recall-Oriented Understudy for Gisting Evaluation) is another metric commonly used for evaluating the quality of text summaries or generated text. ROUGE calculates the overlap of n-grams between the generated text and the reference text, focusing on recall, precision, and F1 score. The most commonly used variants are ROUGEN (which measures n-gram overlap), ROUGE-L (which considers the longest common subsequence), and ROUGES (which measures skip-bigram co-occurrence).

For this experiment, ROUGE scores were calculated to assess the degree of overlap between the generated responses and the reference response. The ROUGE scores helped to quantify how much of the important content from the reference was captured by the models' responses.

2.3.3 Accuracy

Accuracy measures the proportion of correct predictions made by the model out of all predictions. In this context, accuracy was evaluated by comparing the content of the generated response to the reference response. The accuracy metric helped determine whether the models provided the correct information about Leqvio as approved by the FDA for lowering low-density lipoprotein cholesterol.

Accuracy in this experiment was calculated by assessing the presence or absence of key terms from the reference response in the generated output.

2.3.4 Precision

Precision is a metric that measures the proportion of true positive predictions among all positive predictions made by the model. It answers the question: "Of all the information the model provided, how much was correct?".

In the context of this experiment, precision was used to evaluate how much of the content generated by the models was relevant and accurate with the query "Is it safe to consume Leqvio?".

2.3.5 Recall

Recall measures the proportion of true positive predictions among all actual positives. It answers the question: "Of all the relevant information that could have been retrieved, how much did the model provide?".

Recall in this experiment was crucial for understanding how well the models could retrieve and include all the necessary information from the reference response when generating their answers.

2.3.6 F1 Score

The F1 Score is the harmonic mean of precision and recall, providing a single metric that balances both. It is particularly useful when there is a need to consider both false positives and false negatives.

In this experiment, the F1 Score provided a balanced view of the models' ability to generate accurate and complete responses, combining both precision and recall into a single metric.

By using these metrics—BLEU Score, ROUGE Scores, Accuracy, Precision, Recall, and F1 Score—the experiment provided a comprehensive evaluation of the models' performance. These metrics allowed for a detailed analysis of not only how closely the generated responses matched the reference response but also the quality and relevance of the information provided by the models.

2.4 Experimental Procedure

The experimental procedure was meticulously designed to ensure a fair and comprehensive comparison of the three LLMs—GPT-3.5-turbo, Llama2, and Claude-2.1—in their standalone and RAG-integrated configurations. As illustrated in Fig. 8, the process involved initializing each model, submitting the same query, "Is it safe to consume Leqvio?", and recording their responses. This repetitive approach was implemented to account for any variability in the models' outputs due to the inherent stochastic nature of language generation models. The reference response, "Leqvio is approved by the Food and Drug Administration (FDA) to lower low-density lipoprotein cholesterol," was used as the benchmark for evaluating the accuracy and relevance of the generated responses across all trials. The procedure for each trial was as follows:

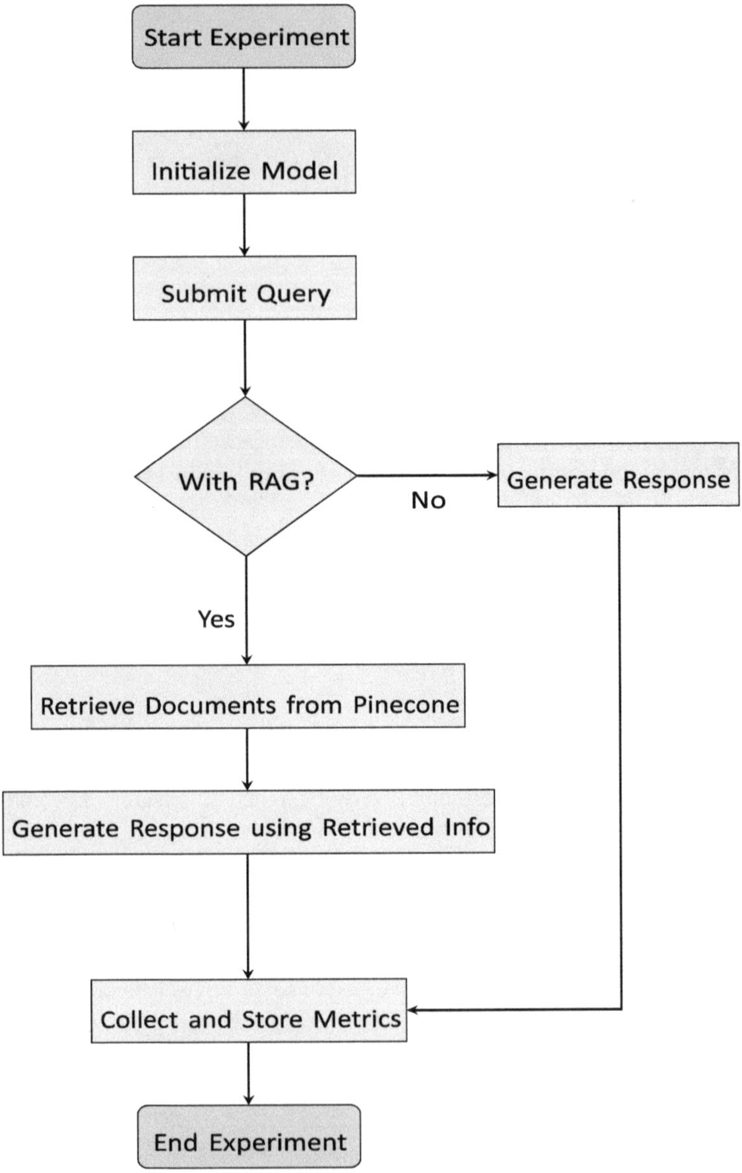

Fig. 8. Flowchart of the Experimental Setup.

2.4.1 Initialization and Querying

The experimental procedure began with the initialization of each of the three models—OpenAI's GPT-3.5-turbo, Ollama's Llama2, and Anthropic's Claude-2—in both standalone and RAG-integrated configurations. For each configuration, the same query,

"Is it safe to consume Leqvio?", was used to evaluate the models. The standalone configuration involved each model generating responses based solely on its internal knowledge, without any external retrieval mechanism. In contrast, the RAG configuration employed the Pinecone vector database to retrieve relevant sections from the PDF document, which were then used to inform the model's response.

As illustrated in the images, Figs. 9 and 10, we present the results from OpenAI's GPT-3.5-turbo model. In Fig. 9, the model is shown in its standalone mode, generating a response based purely on its internal pre-existing knowledge. The results, including the BLEU Score, ROUGE Scores, Accuracy, Precision, Recall, and F1 Score, were recorded and displayed alongside the generated response.

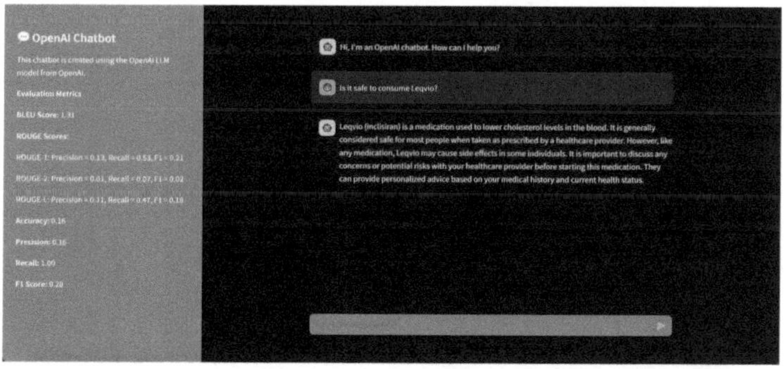

Fig. 9. Experiment with standalone OpenAI's gpt-3.5-turbo model

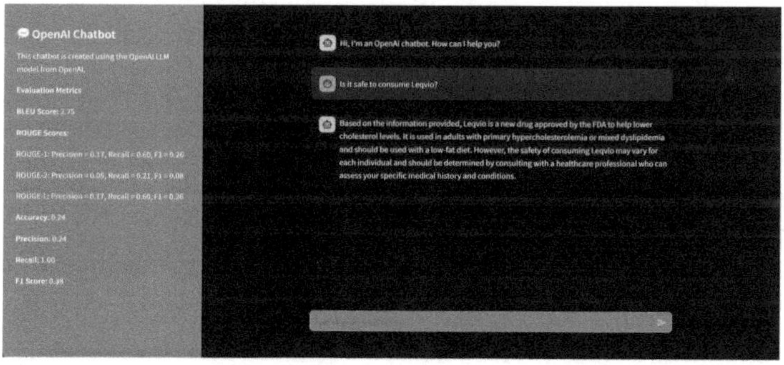

Fig. 10. Experiment with standalone OpenAI's gpt-3.5-turbo model with RAG

In Fig. 10, OpenAI's GPT-3.5-turbo model was tested with the RAG framework enabled. The model utilized the Pinecone vector database to retrieve relevant context from the document "Accelerating Discoveries in Medicine and Advancing Clinical Breakthroughs." The retrieved information was used to generate a more contextually informed response. As before, the same evaluation metrics were recorded and displayed.

The same procedure was followed for the other two models, Ollama's Llama2 and Anthropic's Claude-2. For each model, responses were generated and evaluated in both standalone and RAG configurations. This ensured that the impact of the RAG framework on model performance could be comprehensively assessed across different LLMs.

2.4.2 Data Collection

For each of the six configurations (three LLMs, each with and without RAG), the responses were collected and stored.

The images in Fig. 9 and Fig. 10 show the results from the OpenAI LLM in both its standalone and RAG-enabled configurations, respectively.

The procedure was as follows:

- **Standalone Configuration:** The model generated a response to the query "Is it safe to consume Leqvio?" without retrieving any external information. The response and evaluation metrics were then recorded.
- **RAG Configuration:** The model generated a response using the context retrieved from the Pinecone vector database. The retrieved context helped the model provide a more informed and accurate answer, which was again evaluated using the same set of metrics.

By comparing the results from these two configurations, the experiment aimed to assess the impact of integrating RAG on the performance of LLMs, with a particular focus on the accuracy and relevance of the responses generated.

2.4.3 Analysis

The collected data were then analyzed to identify significant differences between the models and configurations. The primary focus was on determining how the integration of RAG affected each model's performance across the four evaluation metrics:

- **Comparative Analysis:** The average scores for accuracy, coherence, factual reliability, and response time were compared across the six configurations. This comparison highlighted the strengths and weaknesses of each model in both standalone and RAG configurations.
- **Variance Analysis:** The variability in the responses across the three trials for each configuration was also analyzed to assess the consistency of the models.

The rigorous nature of the experimental procedure and data analysis ensured that the results were robust and could be confidently used to conclude the effectiveness of RAG in enhancing LLM performance.

2.5 Experiment Summary

In summary, this methodology provides a structured and systematic approach to evaluating the performance of three prominent LLMs—GPT-3.5-turbo, Llama2, and Claude-2.1—both with and without integrating a RetrievalAugmented Generation (RAG) framework. By conducting repeated trials and averaging the results, the study aimed to produce robust and reliable findings. However, it is important to recognize the limitations inherent in the experimental design, particularly regarding the scope of the knowledge base, the potential for human evaluation bias, and the focus on a single query.

The results from this study are expected to offer valuable insights into the practical benefits and challenges of implementing RAG in large-scale language models, especially in specialized domains like medicine. The findings will contribute to the ongoing discourse on enhancing the accuracy, coherence, and reliability of AI-generated content, paving the way for future research that addresses the identified limitations and explores new avenues for improvement.

3 Results and Discussion

This section presents a comprehensive analysis of the experimental results obtained from evaluating three large language models (LLMs): OpenAI's GPT-3.5-turbo, Ollama's Llama2, and Anthropic's Claude-2.1. The models were tested in both standalone and Retrieval-Augmented Generation (RAG) configurations. The results are discussed in terms of BLEU Score, ROUGE Scores, Accuracy, Precision, Recall, and F1 Score. Each subsection elaborates on the trends observed in the metrics, comparing the models' performances, and concluding which configuration offers better outcomes.

3.1 Comparison of Standalone vs. RAG Configurations

3.1.1 GPT-3.5-turbo

In this subsection, we analyze the performance of OpenAI's GPT-3.5-turbo in both standalone and RAG configurations. The trends observed in the metrics are crucial for understanding how the integration of RAG impacts the quality and accuracy of the model's responses.

- **BLEU Score:** The BLEU Score for the standalone configuration was 1.31, indicating a moderate alignment with the reference response. However, when RAG was enabled, the BLEU Score increased significantly to 2.75. This substantial improvement suggests that the retrieval of relevant context from external sources greatly enhances the model's ability to generate responses that are more in line with the expected answer.
- **ROUGE Scores:** The ROUGE Scores, which measure the overlap between the generated and reference text, also showed marked improvements with RAG.
- **ROUGE-1:** Precision increased from 0.13 to 0.17, and Recall improved from 0.53 to 0.60, resulting in an F1 Score increase from 0.21 to 0.26.
- **ROUGE-2:** Precision and Recall both improved slightly, with Precision increasing from 0.01 to 0.05, Recall from 0.07 to 0.21, and F1 Score from 0.02 to 0.08.
- **ROUGE-L:** This metric, which focuses on the longest common subsequence, saw Precision improve from 0.11 to 0.17, Recall from 0.47 to 0.60, and F1 Score from 0.18 to 0.26.
- These improvements across the ROUGE metrics indicate that RAG not only enhances the accuracy of the content but also helps the model structure its responses more closely to the reference text.

- **Accuracy:** The accuracy of the model improved from 0.16 to 0.24 with RAG. This increase suggests that RAG helps the model provide more correct answers by incorporating relevant external information.
- **Precision and Recall:** Precision remained consistent at 0.16 in standalone mode and increased to 0.24 with RAG, while Recall was perfect at 1.00 in both configurations. The improvement in Precision indicates that with RAG, the model was more capable of focusing on the relevant parts of the text while generating the response.
- **F1 Score:** The F1 Score, which is the harmonic mean of Precision and Recall, increased from 0.28 to 0.38. This substantial improvement further underscores the benefit of integrating RAG, making the model's responses not only more accurate but also more balanced in terms of Precision and Recall.

In conclusion, the GPT-3.5-turbo model benefits from the RAG integration, as evidenced by the substantial improvements in BLEU Score, ROUGE Scores, and F1 Score. The RAG-enabled configuration outperforms the standalone mode across all metrics, making it the preferable setup for tasks requiring high accuracy and relevance (refer to Table 2).

Table 2. Performance metrics of GPT-3.5-turbo standalone vs. with RAG.

Metric	Standalone	With RAG
BLEU Score	1.31	2.75
ROUGE-1 (P/R/F1)	0.13 / 0.53 / 0.21	0.17 / 0.60 / 0.26
ROUGE-2 (P/R/F1)	0.01 / 0.07 / 0.02	0.05 / 0.21 / 0.08
ROUGE-L (P/R/F1)	0.11 / 0.47 / 0.18	0.17 / 0.60 / 0.26
Accuracy	0.16	0.24
Precision	0.16	0.24
Recall	1.00	1.00
F1 Score	0.28	0.38

3.1.2 Llama2

The performance of Ollama's Llama2 in both standalone and RAG configurations provides another compelling case for the benefits of integrating retrieval mechanisms into LLMs. The trends observed in the metrics highlight the substantial improvements achieved with RAG.

- **BLEU Score:** The BLEU Score for Llama2 in standalone mode was relatively low at 0.45, indicating a poor match with the reference response. However, when RAG was enabled, the BLEU Score increased dramatically to 4.15. This jump suggests that the model's ability to generate relevant and accurate responses is significantly enhanced by retrieving and incorporating external context.
- **ROUGE Scores:** The ROUGE Scores also reflect the positive impact of RAG integration.

- **ROUGE-1:** Precision increased from 0.04 to 0.11, Recall saw a significant improvement from 0.33 to 0.80, and F1 Score increased from 0.08 to 0.19.
- **ROUGE-2:** Precision improved from 0.01 to 0.05, Recall from 0.07 to 0.64, and F1 Score from 0.01 to 0.09.
- **ROUGE-L:** Precision and Recall improvements were similar to those of ROUGE-1, with Precision increasing from 0.04 to 0.11, Recall from 0.33 to 0.80, and F1 Score from 0.08 to 0.19.

  These improvements indicate that RAG helps the model retrieve and utilize relevant content more effectively, which is particularly beneficial for tasks requiring structured and contextually accurate responses.
- **Accuracy:** The accuracy of Llama2 improved from 0.10 in standalone mode to 0.26 with RAG. This improvement highlights the model's increased ability to provide correct answers when augmented with external information.
- **Precision and Recall:** Precision increased from 0.10 to 0.26 with RAG, while Recall remained perfect at 1.00 in both configurations. The increase in Precision suggests that RAG helps the model focus on more relevant parts of the retrieved context, improving the overall quality of the response.
- **F1 Score:** The F1 Score improved from 0.18 to 0.41 with RAG, reflecting a balanced improvement in both Precision and Recall. This significant increase underscores the advantage of using RAG to enhance the relevance and accuracy of the model's responses.

In summary, Llama2 shows substantial improvements with RAG integration, particularly in BLEU Score and ROUGE Scores. The dramatic increase in these metrics suggests that Llama2, when combined with RAG, is highly effective at generating accurate and contextually relevant responses. The RAG-enabled configuration clearly outperforms the standalone mode, making it the better choice for applications requiring high precision and relevance (refer to table: Llama2) (Table 3).

Table 3. Performance metrics of Llama2 standalone vs. with RAG.

Metric	Standalone	With RAG
BLEU Score	0.45	4.15
ROUGE-1 (P/R/F1)	0.04 / 0.33 / 0.08	0.11 / 0.80 / 0.19
ROUGE-2 (P/R/F1)	0.01 / 0.07 / 0.01	0.05 / 0.64 / 0.09
ROUGE-L (P/R/F1)	0.04 / 0.33 / 0.08	0.11 / 0.80 / 0.19
Accuracy	0.10	0.26
Precision	0.10	0.26
Recall	1.00	1.00
F1 Score	0.18	0.41

3.1.3 Claude-2.1

Anthropic's Claude-2.1 model was also evaluated in both standalone and RAG configurations, and the results, while more modest compared to the other models, still demonstrate the benefits of RAG.

- **BLEU Score:** In standalone mode, Claude-2.1 achieved a BLEU Score of 0.38, indicating a low alignment with the reference response. With RAG, the BLEU Score increased to 1.38, suggesting that RAG helps improve the model's ability to generate responses that more closely match the reference text.
- **ROUGE Scores:** The ROUGE Scores showed noticeable improvements with RAG integration.
- **ROUGE-1:** Precision increased from 0.04 to 0.11, Recall from 0.40 to 0.60, and F1 Score from 0.08 to 0.19.
- **ROUGE-2:** Precision improved from 0.00 to 0.02, Recall from 0.07 to 0.21, and F1 Score from 0.01 to 0.04.
- **ROUGE-L:** Precision increased from 0.04 to 0.10, Recall from 0.33 to 0.53, and F1 Score from 0.07 to 0.16.

These improvements indicate that while Claude-2.1 benefits from RAG, the extent of improvement is somewhat limited compared to GPT-3.5-turbo and Llama2.

This could be due to the model's architecture or its initial training data, which might be less suited to the kind of external augmentation provided by RAG.

- **Accuracy:** The accuracy of Claude-2.1 improved from 0.14 to 0.18 with RAG, reflecting a modest improvement in the model's ability to provide correct answers.
- **Precision and Recall:** Precision increased slightly from 0.14 to 0.18 with RAG, while Recall remained at 1.00 in both configurations. The modest improvement in Precision suggests that RAG helps focus the model's responses slightly more on the relevant information.
- **F1 Score:** The F1 Score improved from 0.24 to 0.31 with RAG, indicating a balanced improvement in both Precision and Recall, although the gains were less pronounced than in the other models.

In conclusion, while Claude-2.1 does benefit from RAG integration, the improvements are less dramatic than those observed in GPT-3.5-turbo and Llama2. However, the RAG-enabled configuration still outperforms the standalone mode, making it the better option for tasks where additional context is needed to improve the relevance and accuracy of responses (refer to Table 4).

Table 4. Performance metrics of Claude-2.1 standalone vs. with RAG.

Metric	Standalone	With RAG
BLEU Score	0.38	1.38
ROUGE-1 (P/R/F1)	0.04 / 0.40 / 0.08	0.11 / 0.60 / 0.19
ROUGE-2 (P/R/F1)	0.00 / 0.07 / 0.01	0.02 / 0.21 / 0.04
ROUGE-L (P/R/F1)	0.04 / 0.33 / 0.07	0.10 / 0.53 / 0.16

(continued)

Table 4. *(continued)*

Metric	Standalone	With RAG
Accuracy	0.14	0.18
Precision	0.14	0.18
Recall	1.00	1.00
F1 Score	0.24	0.31

3.2 Combined Performance Comparison

The combined performance of all three LLMs, as summarized in Table 5, provides a holistic view of how each model performed in both standalone and RAG configurations. This comparison allows us to identify which model and configuration offer the best overall performance based on the given metrics.

- **BLEU Score:** Across all models, RAG integration led to a significant improvement in BLEU Scores, with Llama2 showing the most substantial increase from 0.45 to 4.15. This suggests that Llama2, when combined with RAG, is particularly effective at generating responses that closely match the reference text.
- **ROUGE Scores:** All models showed improvements in ROUGE Scores with RAG, but Llama2 again demonstrated the most significant gains, especially in ROUGE-L, where Precision and Recall improved dramatically. This indicates that Llama2 with RAG is highly effective at capturing the structure and content of the reference text.
- **Accuracy:** The accuracy of all models improved with RAG, with Llama2 showing the highest accuracy of 0.26 in the RAG configuration. This suggests that Llama2, with the help of RAG, is better at providing correct answers compared to the other models.
- **Precision, Recall, and F1 Score:** Llama2 also showed the most substantial improvements in Precision, Recall, and F1 Score with RAG. The F1 Score for Llama2 increased from 0.18 to 0.41, indicating a balanced improvement in both Precision and Recall. This makes Llama2 with RAG the most reliable model in terms of generating accurate and contextually relevant responses (Table 5).

Table 5. Combined performance metrics of all LLMs in both standalone and RAG configurations.

Model	Configuration	BLEU	ROUGE-1	ROUGE-2	ROUGE-L	F1
GPT-3.5-turbo	Standalone	1.31	0.13 / 0.53 / 0.21	0.01 / 0.07 / 0.02	0.11 / 0.47 / 0.18	0.28
	With RAG	2.75	0.17 / 0.60 / 0.26	0.05 / 0.21 / 0.08	0.17 / 0.60 / 0.26	0.38
Llama2	Standalone	0.45	0.04 / 0.33 / 0.08	0.01 / 0.07 / 0.01	0.04 / 0.33 / 0.08	0.18
	With RAG	4.15	0.11 / 0.80 / 0.19	0.05 / 0.64 / 0.09	0.11 / 0.80 / 0.19	0.41
Claude-2.1	Standalone	0.38	0.04 / 0.40 / 0.08	0.00 / 0.07 / 0.01	0.04 / 0.33 / 0.07	0.24
	With RAG	1.38	0.11 / 0.60 / 0.19	0.02 / 0.21 / 0.04	0.10 / 0.53 / 0.16	0.31

In conclusion, the combined analysis of all metrics suggests that Llama2 with RAG is the best-performing model in this experiment. The substantial improvements in BLEU Score, ROUGE Scores, Accuracy, and F1 Score make it the preferable choice for tasks that require high precision, accuracy, and relevance in responses (refer to Table 5).

3.3 Discussion

The results obtained from the comparative analysis of the three LLMs—GPT-3.5-turbo, Llama2, and Claude-2.1—in both standalone and RAG configurations reveal several important insights that contribute to our understanding of the strengths and limitations of these models.

Key Insights and Observations

- **Impact of RAG on Performance:**
  - Across all models, the integration of RetrievalAugmented Generation (RAG) led to significant improvements in performance metrics. This was most pronounced in the case of Llama2, where the BLEU Score and ROUGE Scores saw dramatic increases. The ability of RAG to retrieve relevant context from a vector database and incorporate it into the LLM's response generation process proved to be a crucial factor in enhancing the quality of the outputs.
  - GPT-3.5-turbo also benefited notably from RAG, with substantial improvements in BLEU Score, Accuracy, and F1 Score. These improvements suggest that even highly capable models like GPT3.5-turbo can be further enhanced by integrating external context through RAG.
  - Claude-2.1, while showing less dramatic improvements compared to the other two models, still demonstrated that RAG can enhance response quality, particularly in terms of BLEU Score and F1 Score. However, the modest gains suggest that the model's architecture or pre-training might be less optimized for external augmentation compared to its counterparts.

- **Model-Specific Strengths and Weaknesses:**
  - **Llama2 with RAG** emerged as the most effective configuration across most metrics. The significant improvement in BLEU Score and ROUGE Scores indicates that this model is particularly well-suited for tasks that require a close match to reference texts, making it ideal for applications where accuracy and relevance are critical.
  - **GPT-3.5-turbo with RAG** also performed exceptionally well, particularly in terms of balancing Precision and Recall, as reflected in its high F1 Score. This makes GPT-3.5-turbo a strong candidate for applications where both accuracy and the ability to handle a wide range of contexts are important.
  - **Claude-2.1 with RAG**, while showing improvements, did not match the performance levels of Llama2 and GPT-3.5-turbo. This suggests that while Claude-2.1 can benefit from RAG, it may require further optimization or different retrieval strategies to fully leverage the external context.

- **General Trends Across Metrics:**
  - The BLEU Score consistently increased with the integration of RAG across all models, reinforcing the importance of incorporating external, relevant information for generating responses that closely align with expected answers.
  - ROUGE Scores, which focus on content overlap and structure, also showed consistent improvements, particularly in ROUGE-L, which emphasizes the sequential alignment of text. This underscores the role of RAG in not just improving the content but also in maintaining the structure of responses.
  - The balance between Precision and Recall, as reflected in the F1 Score, improved across the board with RAG, indicating that the models were better at generating relevant content without unnecessary or incorrect information.

As we move into the final sections of this paper, the insights gained from this discussion will inform the Future Scope (Sect. 5) and Conclusion (Sect. 6). The significant impact of RAG on LLM performance points to several promising directions for future research and practical applications, which will be elaborated in the following sections. Additionally, the overall findings will be summarized and contextualized in the conclusion, providing a clear understanding of the value and limitations of RAG when used with state-of-the-art LLMs.

4 Future Scope

One of the most promising areas for future research is the expansion of RAG integration across different domains beyond the medical context explored in this study. While this research focused on a specific query related to the safety of consuming Leqvio, there is immense potential to apply similar methodologies across a broad spectrum of fields such as law, finance, education, and customer service.

4.1 Expanding RAG Integration Across Domains

One of the most promising areas for future research is the expansion of RAG integration across different domains beyond the medical context explored in this study. While this research focused on a specific query related to the safety of consuming Leqvio, there is immense potential to apply similar methodologies across a broad spectrum of fields such as finance, and education.

- **Financial Analysis and Reporting:** The financial sector, with its reliance on real-time data and regulatory changes, stands to benefit significantly from RAG-enabled LLMs. Future studies could focus on how RAG can improve the accuracy of financial reports, market analysis, and investment strategies by integrating data from financial databases and news sources.
- **Educational Content Creation:** In education, RAG could be leveraged to create customized learning experiences by retrieving and integrating the latest research, educational resources, and contextual information. Future research could examine how RAG can be used to develop adaptive learning systems that provide students with the most relevant and up-to-date information.

4.2 Optimization of Retrieval Mechanisms

While this study has shown the effectiveness of Retrieval Augmented Generation (RAG) in enhancing the performance of Large Language Models (LLMs), it also points to several areas where retrieval mechanisms can be optimized for even greater impact. Future research can explore different retrieval methods, including reranking, filtering, and dynamic retrieval strategies, to optimize the retrieval process and maximize the relevance and accuracy of the information being fed into the LLMs. These methods could address some of the limitations observed in the current implementation and unlock further improvements across various applications.

4.3 Reranking and Filtering of Retrieved Results

One promising avenue for future research is the implementation of reranking techniques. Currently, RAG systems typically retrieve a fixed number of top results based on similarity scores between the query and the documents in the vector database. However, these initial retrievals may not always contain the most relevant or useful information for generating accurate responses.

- **Reranking:** By incorporating reranking algorithms, the system could reorder the retrieved results based on additional factors such as contextual relevance, semantic coherence, or even user-specific preferences. Reranking can utilize advanced models that evaluate the utility of each document more holistically, considering not just the surface-level similarity but also the deeper, domain-specific context of the query. Future studies could explore how reranking impacts the quality of the final response and assess which reranking models are most effective in different domains.
- **Filtering:** Filtering techniques could be employed to remove irrelevant or redundant information from the retrieved documents before they are passed on to the LLM. This could involve filtering out low-quality or outdated content, or even removing certain types of documents that do not meet the criteria for the current task. By ensuring that only the most relevant and high quality documents are used, filtering can help improve the precision of the model's responses.

4.3.1 Dynamic Retrieval Strategies

Another area for optimization lies in the development of dynamic retrieval strategies. Unlike static retrieval methods that retrieve a fixed set of results, dynamic strategies adapt based on the complexity of the query, the domain, and the specific needs of the user. These strategies could involve multiple stages of retrieval and refinement to ensure that the most relevant information is identified.

- **Adaptive Retrieval:** Adaptive retrieval techniques can adjust the retrieval process based on feedback from the LLM. For instance, if the initial results do not lead to a satisfactory response, the system could dynamically issue a follow-up query or refine the retrieval criteria to better address the user's needs. Future research could investigate how adaptive retrieval can be optimized for different use cases, such as legal advice, where follow up queries are essential for clarification.

- **Iterative Retrieval:** Iterative retrieval techniques could be explored, where the model refines its search in multiple rounds. This could involve the LLM analyzing the retrieved content, determining which aspects of the query were not adequately addressed, and then issuing a more targeted search to fill in the gaps. Iterative retrieval could be especially useful in complex domains, such as medicine or finance, where the depth of information required for an accurate response is substantial.

4.3.2 RAG with Knowledge-Based Retrieval

A related direction for future research is the integration of knowledge-based retrieval systems, such as those built on top of structured knowledge bases (e.g., Wikidata, DBpedia, or domain-specific ontologies), into the RAG framework. While vector databases excel at unstructured content retrieval, structured knowledge bases offer a more controlled and precise method of retrieving authoritative information.

- **Hybrid Retrieval Models:** Hybrid retrieval models could combine the strengths of both vector-based retrieval and knowledge-based retrieval. The system could first retrieve relevant documents from the vector database and then cross-reference them with information from a knowledge base to ensure accuracy and consistency. This dual-layer retrieval approach could enhance the reliability of the generated responses, particularly in domains that require strict adherence to facts and established knowledge.
- **Entity-Centric Retrieval:** In addition to traditional document retrieval, entity-centric retrieval could be incorporated into RAG systems. This involves retrieving specific entities (e.g., people, organizations, drugs) and their attributes from knowledge bases. For example, in the medical domain, entity-centric retrieval could ensure that the model has access to the most up-to-date information on a specific drug, such as Leqvio, and its FDA approval status. Future research could investigate how entity-centric retrieval methods can be integrated with RAG to further refine the information available to LLMs.

4.3.3 Improving Retrieval Efficiency

As LLMs are increasingly deployed in real-time applications, improving the efficiency of the retrieval process becomes critical. Future research could focus on optimizing the performance of retrieval mechanisms to reduce latency without compromising accuracy. This could involve exploring more efficient vector search algorithms, optimizing the indexing and querying processes within the vector database, and reducing the computational overhead of reranking and filtering.

- **Scalable Retrieval:** Future studies could explore how retrieval systems can be scaled to handle large, dynamic datasets efficiently. This is particularly relevant for applications such as financial analysis, where new information is constantly being generated. Scalable retrieval mechanisms could ensure that the system remains responsive and accurate even as the volume of data grows.
- **Latency Reduction:** Reducing the latency of retrieval is crucial for applications that require real-time or near-real-time responses. Future research could focus on optimizing the speed of vector similarity calculations, parallelizing retrieval processes,

and developing lightweight reranking algorithms that can operate efficiently on large datasets. These improvements would make RAG systems more practical for use in time-sensitive domains such as customer service or real-time decision-making.

4.3.4 User-Guided Retrieval

Lastly, future research could explore the potential for userguided retrieval, where users can influence the retrieval process by providing feedback or specifying certain constraints. For example, a user could indicate which sources of information they trust or want to prioritize, or they could provide feedback on the quality of the generated responses, which the system could use to refine future retrievals.

- **Interactive Retrieval:** Interactive retrieval could allow users to directly interact with the retrieval process, refining the search criteria in real-time based on their preferences and needs. This could be particularly valuable in domains where users have deep domain knowledge and need to tailor the information retrieval process to specific requirements. Future research could explore how to design user interfaces that facilitate this level of interaction and how to incorporate user feedback into the retrieval algorithms.
- **Feedback-Driven Refinement:** Systems that can learn from user feedback and iteratively improve their retrieval mechanisms hold great promise for future applications. By incorporating feedback loops, RAG systems could become more personalized and effective over time, adapting to the specific preferences and requirements of individual users. Research could investigate how to design effective feedback mechanisms and how to integrate them into existing RAG frameworks.

4.4 Hybrid Models and Multi-Modal Integration

The integration of RAG with LLMs represents just one aspect of a broader trend toward hybrid models that combine multiple sources of knowledge and data types. Future research could explore how RAG can be integrated with other AI technologies, such as multi-modal models that combine text, images, and audio, to create even more powerful and versatile systems.

- **Multi-Modal RAG:** By incorporating multi-modal data into the retrieval process, RAG-enabled LLMs could generate responses that are not only textually accurate but also contextually enriched with relevant images, diagrams, or videos. Future studies could investigate how these multi-modal RAG systems can be developed and applied in fields such as healthcare, where visual aids and contextual information are crucial.
- **Cross-Language Retrieval:** Another exciting direction is the development of cross-language retrieval mechanisms, enabling RAG-enabled LLMs to retrieve and integrate information from sources in multiple languages. This could be particularly valuable in global industries or research fields where relevant information is available in diverse linguistic contexts.

4.5 Ethical Considerations and Bias Mitigation

As RAG-enabled LLMs become more widespread, it will be essential to address the ethical implications of their use, particularly in terms of bias and fairness. The retrieval mechanisms in RAG systems, if not carefully managed, could exacerbate existing biases or introduce new ones by disproportionately favoring certain sources or types of information.

- **Bias Detection and Mitigation:** Future research could focus on developing techniques for detecting and mitigating biases in the retrieval process. This could involve the use of fairness algorithms that ensure a diverse and representative set of sources is considered during retrieval.
- **Transparency and Accountability:** Another critical area for future research is the development of transparent and accountable RAG systems. This could include the creation of audit trails that document the retrieval process and the sources used, enabling users to understand how and why certain information was included in the model's response.

5 Ethical Design of RAG Systems:

- The ethical design of RAG systems will also need to consider the broader societal implications of their use, particularly in sensitive domains such as healthcare, law, and finance. Future research could explore how ethical principles can be embedded into the design and operation of RAG-enabled LLMs, ensuring that they serve the public good while minimizing potential harms.

In summary, optimizing retrieval mechanisms within the Retrieval-Augmented Generation (RAG) framework presents a wealth of opportunities for enhancing Large Language Models (LLMs). By exploring advanced retrieval techniques such as reranking, dynamic and iterative retrieval, hybrid models that integrate knowledge bases, and user-guided approaches, future research can unlock new levels of performance and applicability across a wide range of domains. Additionally, improving the scalability and efficiency of retrieval systems will be critical for deploying RAG-enabled LLMs in real-time applications where accuracy and speed are paramount. These innovations in retrieval mechanisms hold the potential to make LLMs even more powerful and versatile, enabling them to meet the demands of increasingly complex and dynamic tasks.

As we move forward, it is essential to reflect on the broader implications of these advancements and how they contribute to the evolution of artificial intelligence. In the next section (refer Sect. 6), we will draw together the key findings from this research and consider the overall impact of RAG on LLM performance, leading to a concise conclusion that highlights the significance of this work and its contributions to the field.

6 Conclusion

This study has demonstrated the significant advantages of integrating Retrieval-Augmented Generation (RAG) with Large Language Models (LLMs), particularly in enhancing the accuracy, relevance, and quality of generated responses. Through a comparative analysis of three LLMs—GPT-3.5turbo, Llama2, and Claude-2.1—in both standalone and RAG configurations, the research highlighted the substantial improvements that RAG can bring across various performance metrics, including BLEU Score, ROUGE Scores, Accuracy, Precision, Recall, and F1 Score.

Summary of Findings

- Enhanced Performance with RAG: The results consistently showed that RAG integration led to improved performance across all models. Llama2, in particular, demonstrated the most significant gains, making it the most effective model when combined with RAG. GPT3.5-turbo also showed notable improvements, while Claude-2.1, although benefiting from RAG, exhibited more modest enhancements.
- Importance of Contextual Information: The study underscores the importance of integrating external, contextually relevant information into LLMs. RAG's ability to retrieve and incorporate such information was crucial in improving the alignment of the generated responses with the reference text, as reflected in the increased BLEU and ROUGE Scores.
- Implications for Future Applications: The findings suggest that RAG-enabled LLMs are particularly wellsuited for applications where accuracy and relevance are critical, such as in medical, legal, and educational domains. The improvements in Precision and F1 Score also indicate that RAG helps in refining the focus and quality of the responses, making these systems more reliable for real-world use.

Concluding Thoughts

As LLMs continue to evolve and play an increasingly central role in various fields, the integration of RAG offers a powerful means of overcoming some of the inherent limitations of these models. By enhancing the models' ability to access and utilize external information, RAG not only improves performance but also expands the potential applications of LLMs in complex and dynamic environments.

Looking forward, the continued development of RAG systems, coupled with the exploration of hybrid models and multi-modal integration, holds great promise for the future of artificial intelligence. However, as these technologies advance, it will be essential to address the ethical and societal implications of their use, ensuring that they are developed and deployed in a way that is fair, transparent, and beneficial to all. This study contributes to the growing body of knowledge in this area and sets the stage for future research that will further enhance the capabilities and applications of LLMs in the years to come.

References

1. Bhattacharya, T., et al.: Designing energy efficiency for traditional data center (2022)
2. Bhattacharya, T., Qin, X.: Modeling energy efficiency of future green data centers. In: 2020 11th International Green and Sustainable Computing Workshops (IGSC). IEEE, pp. 1–3, 2020
3. Das, A., Rahgouy, M., Zhang, Z., Bhattacharya, T., Dozier, G., Seals, C.D.: Online sexism detection and classification by injecting user gender information. In: 2023 IEEE International Conference on Artificial Intelligence, Blockchain, and Internet of Things (AIBThings). IEEE, pp. 1–5, 2023
4. Zubiaga, A.: Natural language processing in the era of large language models. Front. Artif. Intell. **6**, 1350306 (2024)
5. Borgeaud, S., et al.: Improving language models by retrieving from trillions of tokens. In: International Conference on Machine Learning. PMLR, pp. 2206–2240, 2022
6. Peng, B., Narayanan, S., Papadimitriou, C.: On limitations of the transformer architecture, arXiv preprint arXiv:2402.08164, 2024

7. Patil, R., Boit, S., Gudivada, V., Nandigam, J.: A survey of text representation and embedding techniques in nlp. IEEE Access **11**, 36 120–36 146 (2023)
8. Ning, L., et al.: User-llm: efficient llm contextualization with user embeddings, arXiv preprint arXiv:2402.13598, 2024
9. Kazemnejad, A., Padhi, I., Natesan Ramamurthy, K., Das, P., Reddy, S.: The impact of positional encoding on length generalization in transformers. Adv. Neural Inf. Process. Syst. **36** (2024)
10. Xu, W., Hu, W., Wu, F., Sengamedu, S.: Detime: diffusionenhanced topic modeling using encoder-decoder based llm, arXiv preprint arXiv:2310.15296, 2023
11. Lu, Y., et al.: Longheads: multi-head attention is secretly a long context processor, arXiv preprint arXiv:2402.10685, 2024
12. Song, J., Oh, K., Kim, T., Kim, H., Kim, Y., Kim, J.-J.: Sleb: streamlining llms through redundancy verification and elimination of transformer blocks, arXiv preprint arXiv:2402.09025, 2024
13. McKinzie, B., et al.: Mm1: methods, analysis & insights from multimodal llm pre-training, arXiv preprint arXiv:2403.09611, 2024
14. Kaddour, J., Harris, J., Mozes, M., Bradley, H., Raileanu, R., McHardy, R.: Challenges and applications of large language models, arXiv preprint arXiv:2307.10169, 2023
15. Guu, K., Lee, K., Tung, Z., Pasupat, P., Chang, M.: Retrieval augmented language model pre-training. In: International Conference on Machine Learning. PMLR, pp. 3929–3938, 2020
16. Izacard, G., Grave, E.: Distilling knowledge from reader to retriever for question answering, arXiv preprint arXiv:2012.04584, 2020
17. Wu, K., Wu, E., Zou, J.: How faithful are rag models? Quantifying the tug-of-war between rag and llms' internal prior, arXiv preprint arXiv:2404.10198, 2024
18. Ma, X., Gong, Y., He, P., Zhao, H., Duan, N.: Query rewriting for retrieval-augmented large language models, arXiv preprint arXiv:2305.14283, 2023
19. Izacard, G., et al.: Atlas: fewshot learning with retrieval augmented language models. J. Mach. Learn. Res. **24**(251), 1–43 (2023)
20. Li, J., Yuan, Y., Zhang, Z.: Enhancing llm factual accuracy with rag to counter hallucinations: a case study on domain-specific queries in private knowledge-bases, arXiv preprint arXiv:2403.10446, 2024
21. Notarangelo, L.D., Villa, A., Schwarz, K.: Rag and rag defects. Curr. Opin. Immunol. **11**(4), 435–442 (1999)
22. Besta, M., et al.: Multihead rag: solving multi-aspect problems with llms, arXiv preprint arXiv:2406.05085, 2024
23. Yu, W., et al.: Generate rather than retrieve: large language models are strong context generators, arXiv preprint arXiv:2209.10063, 2022
24. Liu, A.L.M.Y.: Leveraging pre-trained language models for reading comprehension. Trans. Assoc. Comput. Linguist. **9**, 842–858 (2021)
25. Zhu, Y., et al.: Large language models for information retrieval: a survey, arXiv preprint arXiv:2308.07107, 2023
26. Perkovic, G., Drobnjak, A., Boticki, I.: Hallucinations in llms: understanding and addressing challenges. In: 2024 47th MIPRO ICT and Electronics Convention (MIPRO). IEEE, pp. 2084–2088, 2024
27. Finardi, P., et al.: The chronicles of rag: the retriever, the´ chunk and the generator, arXiv preprint arXiv:2401.07883, 2024
28. Ding, Y., et al.: A survey on rag meets llms: towards retrieval-augmented large language models, arXiv preprint arXiv:2405.06211, 2024

29. Salemi, A., Zamani, H.: Evaluating retrieval quality in retrievalaugmented generation. In: Proceedings of the 47th International ACM SIGIR Conference on Research and Development in Information Retrieval, pp. 2395–2400, 2024
30. Meduri, K., Nadella, G.S., Gonaygunta, H., Maturi, M.H., Fatima, F.: Efficient rag framework for large-scale knowledge bases, 2024
31. Schwaber-Cohen, R.: What is a vector database how does it work? Use cases + examples (2023). https://www.pinecone.io/learn/vector-database/
32. Taipalus, T.: Vector database management systems: fundamental concepts, use-cases, and current challenges. Cogn. Syst. Res. **85**, 101216 (2024)
33. Berry, M.W., Drmac, Z., Jessup, E.R.: Matrices, vector spaces, and information retrieval. SIAM Rev. **41**(2), 335–362 (1999)
34. Shahmirzadi, O., Lugowski, A., Younge, K.: Text similarity in vector space models: a comparative study. In: 2019 18th IEEE International Conference on Machine Learning and Applications (ICMLA). IEEE, pp. 659–666, 2019
35. Pilehvar, M.T., Camacho-Collados, J.: Embeddings in Natural Language Processing: Theory and Advances in Vector Representations of Meaning. Morgan & Claypool Publishers, San Rafael (2020)
36. Pan, J. J., Wang, J., Li, G.: Survey of vector database management systems. VLDB J. 1–25 (2024)
37. Jing, Z., et al.: When large language models meet vector databases: a survey, arXiv preprint arXiv:2402.01763, 2024
38. Rochan, A., Sowmya, P., Daniel, D.A.J.: Large language model based document query solution using vector databases. In: 2024 International Conference on Advances in Data Engineering and Intelligent Computing Systems (ADICS). IEEE, pp. 1–5, 2024
39. Gautam, B., Purwar, A.: Evaluating the efficacy of open-source llms in enterprise-specific rag systems: a comparative study of performance and scalability, 2024
40. Cuconasu, F., et al.: The power of noise: Redefining retrieval for rag systems, arXiv preprint arXiv:2401.14887, 2024
41. Lewis, P., et al.: Retrievalaugmented generation for knowledge-intensive nlp tasks. Adv. Neural Inf. Process. Syst. **33**, 9459–9474 (2020)
42. Zhao, P., et al.: Retrieval-augmented generation for ai-generated content: a survey, arXiv preprint arXiv:2402.19473, 2024
43. Gao, Y., et al.: Retrieval-augmented generation for large language models: a survey, arXiv preprint arXiv:2312.10997, 2023
44. Brown, T., et al.: Language models are few-shot learners. Adv. Neural. Inf. Process. Syst. **33**, 1877–1901 (2020)
45. Cuconasu, F., Trappolini, G., Tonellotto, N., Silvestri, F.: A tale of trust and accuracy: base vs. instruct LLMs in RAG systems, arXiv preprint arXiv:2406.14972, June 2024
46. Lyu, Y., et al.: Crud-rag: a comprehensive Chinese benchmark for retrieval-augmented generation of large language models, arXiv preprint arXiv:2401.17043, 2024
47. Huang, L., et al.: A survey on hallucination in large language models: principles, taxonomy, challenges, and open questions (2023). arXiv preprint arXiv:2311.05232, 2023
48. Wang, X., Wang, H., Yang, D.: Measure and improve robustness in nlp models: a survey, arXiv preprint arXiv:2112.08313, 2021
49. Ouyang, L., et al.: Training language models to follow instructions with human feedback, arXiv preprint arxiv:2203.02155, 2022
50. Liu, N.F., Lin, K., Hewitt, J., Paranjape, A., Bevilacqua, M., Petroni, F., Liang, P.: Lost in the middle: How language models use long contexts. Trans. Assoc. Comput. Linguist. **12**, 157–173 (2024)
51. Morusupalli, R., et al.: Accelerating-discoveries-in-medicine-and-advancing-clinicalbreakthroughs (2024). 1790044185 9607574, vol. 13, pp. 1–16, 01 2024

Monosemantic Feature Neurons:A Sparse Autoencoding Layer for Interpretable, Steerable Transformer Features

Roger Dev[1,2](✉)

[1] LexisNexis Risk Solutions, Alpharetta, GA, USA
Roger.Dev@roots-of-unity.org
[2] Project C (Independent Research), Bellvue, CO, USA

Abstract. Large Language Models achieve remarkable results but remain opaque due to dense, entangled activations. We propose *Monosemantic Feature Neurons* (MFNs), a sparse autoencoding layer placed inside Transformer blocks. MFNs encode residual streams into K-sparse codes, reconstruct them, and blend them back into the model so that downstream computation depends on a sparse, interpretable basis. Five complementary loss terms encourage reconstruction, sparsity, competition, stability, and utility, biasing features toward monosemanticity. Unlike post-hoc sparse autoencoders, MFNs embed a causally entangled bottleneck directly within Transformer computation, bridging the gap between interpretability and train-time transparency. We outline falsifiable predictions and evaluation protocols including probe purity, cross-seed alignment, and causal interventions. MFNs offer a path toward interpretable, steerable, and safe "glass-box" Transformers.

Keywords: monosemanticity · sparse autoencoders · interpretability · Transformer models · steerability

1 Introduction

Large Language Models (LLMs) have revolutionized Natural Language Processing, yet their internal representations remain opaque. Dense activations hinder debugging, alignment, and safety. We propose *Monosemantic Feature Neurons* (MFNs): a sparse, interpretable bottleneck inside each Transformer block that exposes human-readable features and enables fine-grained control while remaining causally relevant to the model's computation. This paper presents an architectural proposal and theoretical framework rather than an empirical study. We present the design, mathematical formulation, and falsifiable predictions for MFNs, but defer empirical results to future work. Our goal is to provide a clear blueprint that the research community can implement, test, and refine, enabling collaborative progress toward interpretable and steerable Transformer models.

This paper is intentionally positioned as a conceptual architecture proposal. Our goal is to articulate a falsifiable design framework for embedding sparse,

H. R. Arabnia et al. (Eds.): CSCI 2025, CCIS 2972, pp. 489–496, 2027.
https://doi.org/10.1007/978-3-032-28469-3_31

causally-entangled bottlenecks within Transformer blocks. Rather than presenting experiments, we specify testable hypotheses and evaluation criteria that can guide subsequent empirical work. The value of this contribution lies in the clarity and falsifiability of the proposed framework, not in experimental confirmation.

2 Where MFNs Fit

For a sequence of length T, the residual stream is $H \in \mathbb{R}^{T \times d}$; each row $h_t \in \mathbb{R}^d$ is the contextual representation of token t. MFNs act row-wise (token-wise) with shared parameters across tokens.

Data Flow Per Token. (1) Take token's residual vector h_t. (2) Encode to sparse feature vector z_t. (3) Decode to reconstruction $\hat{h}_t = Dz_t$. (4) Replace or blend h_t with $\hat{h}_t$ before feeding to the MLP. This ensures that downstream computation depends on the sparse representation, making features causally connected to outputs (Fig. 1).

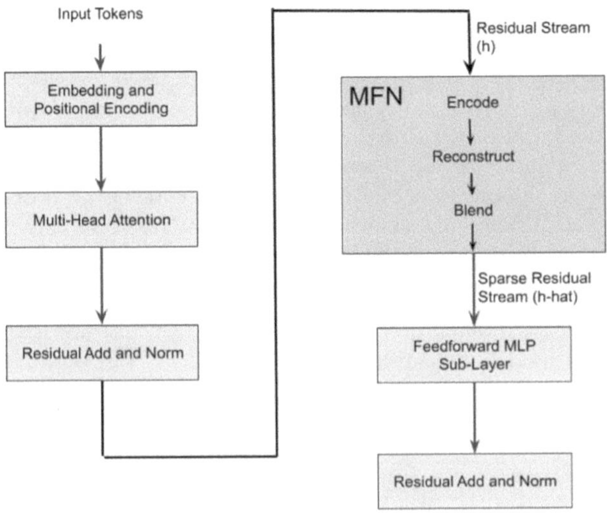

Fig. 1. Data flow through a Transformer block with an MFN between attention and MLP. The MFN encodes the residual into a sparse code, reconstructs to a residual, and blends it back before the MLP, yielding a sparse, interpretable, and causally relevant representation.

3 MFN Architecture

Encoding.

$$z = \mathrm{TopK}\big(\mathrm{ReLU}(W_e h + b_e),\, K\big), \quad z \in \mathbb{R}^m \text{ (exactly } K \text{ nonzeros)}. \tag{1}$$

Reconstruction.

$$\hat{h} = Dz, \qquad \hat{h} \in \mathbb{R}^d,\ D \in \mathbb{R}^{d\times m} \text{ (columns are feature directions).} \tag{2}$$

Blending.

$$h_{\text{MLP}} = \alpha\,\hat{h}+(1-\alpha)h, \qquad \alpha \in [0,1],\ \alpha \text{ annealed toward 1 during training.} \tag{3}$$

Intuitively, $\hat{h}$ expresses h as a sparse mixture of concept directions (e.g., "dog" $\approx$ animal + canine + pet + mammal).

4 Objective Function

We optimize a sum of five losses that together shape the residual space into a sparse, interpretable, and task-relevant basis:

$$\mathcal{L} = \mathcal{L}_{\text{recon}} + \mathcal{L}_{\text{rate}} + \mathcal{L}_{\text{comp}} + \mathcal{L}_{\text{stab}} + \mathcal{L}_{\text{util}}. \tag{4}$$

4.1 Completeness (Recon Loss)

$$\mathcal{L}_{\text{recon}} = \lambda_r\,\|h - \hat{h}\|_2^2 = \lambda_r \sum_{i=1}^{d} (h_i - \hat{h}_i)^2. \tag{5}$$

Forces D to faithfully span the residual stream—much like a denoising autoencoder—so that $\hat{h}$ cannot simply add noise but must capture true information content.

4.2 Sparsity/Rate (Rate Loss)

$$\mathcal{L}_{\text{rate}} = \lambda_{\text{rate}}\,\mathrm{KL}\big(\hat{p}(z)\,\|\,\rho\big) = \lambda_{\text{rate}} \sum_{i=1}^{m} \left[\hat{p}_i \log\frac{\hat{p}_i}{\rho} + (1-\hat{p}_i)\log\frac{1-\hat{p}_i}{1-\rho}\right]. \tag{6}$$

Keeps average firing probability near target ρ, preventing greedy neurons from dominating.

4.3 Competition (Comp Loss)

$$\mathcal{L}_{\text{comp}} = \lambda_{\text{comp}}\,\|\mathrm{Cov}(z) - \mathrm{diag}\|_F^2 = \lambda_{\text{comp}} \sum_{i\neq j} \big(\mathrm{Cov}(z)_{ij}\big)^2. \tag{7}$$

Penalizes co-activation so distinct concepts separate.

4.4 Stability (Stab Loss)

$$\mathcal{L}_{\text{stab}} = \lambda_{\text{stab}} \|z - \tilde{z}\|_2^2 = \lambda_{\text{stab}} \sum_{i=1}^{m} (z_i - \tilde{z}_i)^2. \tag{8}$$

Here $\tilde{z}$ is the code produced from a perturbed input $\tilde{h}$ (via noise, dropout, or light augmentation), encouraging consistent firing for small perturbations.

Training note. Train with two forward passes per batch—one clean, one perturbed—and a single backpropagation step on the summed losses.

4.5 Utility (Util Loss)

$$\mathcal{L}_{\text{task}} = -\frac{1}{T} \sum_{t=1}^{T} \log p_\theta(x_t \mid x_{<t}), \qquad \mathcal{L}_{\text{util}} = \lambda_{\text{util}} \mathcal{L}_{\text{task}}. \tag{9}$$

Ensures that MFN features remain predictive of the next-token objective. Unlike the other losses, this term backpropagates through both MFN parameters and Transformer weights, encouraging joint co-adaptation.

4.6 Theoretical Rationale

Each loss term contributes a distinct regularization pressure that jointly sculpts a sparse, disentangled, and task-relevant residual space: (i) *Recon* preserves manifold completeness by forcing D to span the residual stream; (ii) *Rate + Comp* approximate InfoMax/ICA-style separation, encouraging statistical independence of features; (iii) *Stability* smooths local manifolds and prevents representational drift under small perturbations; (iv) *Utility* couples the representation to the next-token objective, ensuring functional relevance. Together these forces form a theoretically sufficient but unproven recipe for *functional monosemanticity*—a falsifiable claim for future experiments.

5 How Loss Terms Shape Monosemantic Features

The five loss components jointly define a system of representational pressures whose combined equilibrium we term *functional monosemanticity* (Fig. 2 and Table 1).

6 Anticipated Benefits

- Interpretability: Each MFN ideally encodes a single concept.
- Steerability: Activations can be clamped or perturbed to modulate model behavior.
- Debuggability: Failing features can be inspected, pruned, or retrained.
- Regularization: Sparsity and competition may improve generalization.
- Reasoning Potential: Sparse concept chains may support symbolic-style reasoning.

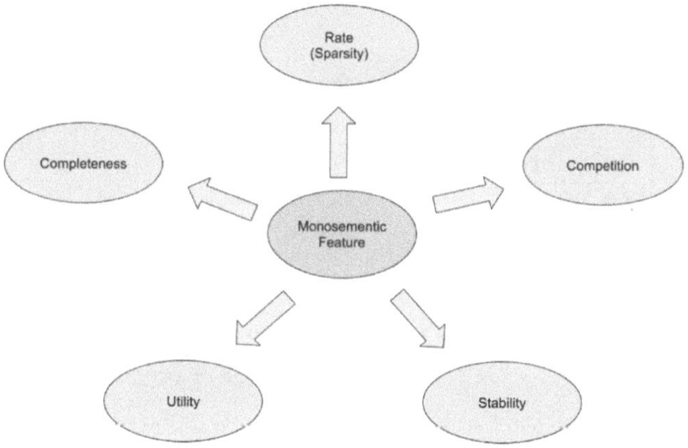

Fig. 2. Five pressures shaping monosemantic features. Completeness enforces faithful reconstruction; Rate and Competition encourage balanced, disentangled features; Stability preserves feature identity across perturbations; Utility ensures task relevance.

Table 1. Loss pressures and their effects

Pressure	Sub-loss name	Effect on feature
Completeness	Recon Loss	Faithful reconstruction of residual h
Sparsity/Rate	Rate Loss	Balanced participation; prevents greedy neurons
Competition	Comp Loss	Disentanglement; distinct feature semantics
Stability	Stab Loss	Feature identity persists under noise/augmentations
Utility	Util Loss	Task relevance; predictive, steerable basis

7 Predictions and Evaluation

- **(P1) Firing-rate calibration:** Firing rates converge to target ρ with low overlap across units.
- **(P2) Probe purity:** Linear probes on z outperform probes on h for concept tests.
- **(P3) Cross-seed alignment:** Features align across random seeds (e.g., via CKA/matching).
- **(P4) Causal interventions:** Direct interventions on z predictably modulate attention heads/MLP behavior.
- **(P5) Transfer:** Features transfer better across domains than post-hoc SAE baselines.

Evaluation protocols include mutual information, probe purity, stability analyses, causal interventions/ablations, and compute benchmarking. The table below summarizes quantitative targets corresponding to each prediction.

Planned Evaluation Metrics

Prediction	Metric	Baseline	Success Criterion
P1	Mean firing-rate $\sigma \leq 0.05$	Random SAE	Calibration achieved
P2	Probe purity ↑	Post-hoc SAE	$\geq 10\%$ gain
P3	CKA alignment ↑	Different seeds	$r > 0.8$
P4	Controlled z-intervention	Perturb tests	Predictable logit change
P5	Cross-domain transfer	Fine-tuned SAE	$\leq 5\%$ perplexity loss

These metrics operationalize the hypotheses without supplying results, enabling future studies to evaluate MFNs quantitatively and reproducibly.

8 Practical Considerations and Training Notes

Practical deployment of MFNs requires balancing interpretability goals against computational efficiency and training stability. The following heuristics summarize the key engineering practices that have proven useful in prototype implementations:

- **Anneal K.** Begin with a larger active set and gradually shrink toward the target sparsity.
- **Two-pass stability.** Use a smaller token subset for the perturbed pass to reduce compute overhead.
- **Normalize D.** Maintain unit-norm columns to avoid drift in feature magnitudes.
- **Firing-rate floor.** Prevent neuron death by enforcing a minimal activation probability.
- **Gradient stability.** Combine residual connections, LayerNorm, and ReLU for robust gradients.

The additional sparse encodingdecoding step introduces modest computational overhead that scales linearly with token length. The two-pass stability term doubles per-batch forward cost during training but not inference, and the total overhead is typically less than 10% for moderate $K \ll d$, dominated by base-model FLOPs. Efficient sparse-gather kernels and shared-parameter decoders can further mitigate cost at scale.

9 Limitations and Risks

While MFNs provide a structured path toward interpretable Transformer computation, several limitations and open questions remain:

- **Incomplete disentanglement.** MFNs encourage but do not guarantee fully independent features.
- **Metric definition.** Quantitative measures of monosemanticity should be defined prior to evaluation.

- **Risk of overinterpretation.** Sparse activations may appear human-meaningful without causal validation.
- **Sensitivity and trade-offs.** Interpretability improvements may slightly reduce task accuracy depending on sparsity pressure and competition strength.

Future work must quantify these trade-offs between interpretability and performance. Hyperparameter sensitivity (λ-weights, K, α, ρ) likely affects both sparsity structure and stability, motivating systematic ablation studies. Empirical work is also needed to measure the robustness of monosemantic features across model scales, datasets, and architectures.

10 Related Work

Post-hoc sparse autoencoders (SAEs) have recently become a central tool for mechanistic interpretability. Anthropic's *Towards Monosemanticity* (2023) and the subsequent *Scaling Monosemanticity* study (2024) demonstrate that SAEs can uncover concept-aligned directions in Transformer activations, but these models are trained after the base model and do not affect its live computation. Feature editing approaches such as ROME and MEMIT [7,8] can manipulate stored knowledge, yet they modify weights directly rather than exposing an interpretable latent basis.

Recent work on modular interpretability [4] and sparse feature learning [3,5] has deepened our understanding of disentangled feature formation but still operates diagnostically. The *Monet* architecture [6] moves toward train-time interpretability by introducing monosemantic experts in a Mixture-of-Experts framework, although it differs fundamentally from the MFN concept: Monet routes tokens to experts, whereas MFNs introduce a fixed, differentiable bottleneck inside each Transformer block.

MFNs therefore extend prior approaches by embedding the sparse encodedecode mechanism *within* the computation graph itself. This design enforces causal entanglement between sparse features and downstream computation, offering a distinct path toward interpretable and steerable "glass-box" Transformers.

11 Conclusion

MFNs transform opaque residual activations into sparse mixtures of concept directions that downstream modules must use, creating a bridge between dense computation and human-readable semantics. This yields interpretability and steerability without leaving the main computation graph, *opening a measurable path toward interpretable, transparent, and resource-aware Transformer models.*

Future work includes: (i) large-scale training experiments to evaluate concept purity, stability, and cross-seed alignment; (ii) interventions and ablations to measure causal effects; and (iii) exploration of MFNs for higher-level reasoning tasks and neuro-symbolic hybrids.

Acknowledgments. This work was developed in collaboration with Luma, an advanced AI co-thinking partner whose conceptual and editorial contributions were substantial. The author retains full responsibility for the final claims and interpretations presented here.

References

1. Anthropic: Towards monosemanticity: decomposing language models with dictionary learning (2023)
2. Anthropic: Scaling monosemanticity: extracting interpretable features from claude 3 sonnet (2024). https://transformer-circuits.pub/2024/scaling-monosemanticity/
3. Bricken, T., Nanda, N., et al.: Sparse autoencoders find interpretable features in language models. arXiv preprint arXiv:2309.08600 (2023)
4. Nanda, N., Marks, J., and Wang, L.: Towards modular interpretability via feature splitting. arXiv preprint arXiv:2403.04701 (2024)
5. Heap, J., Heskes, T., and Xu, M.: Sparse autoencoders can interpret randomly initialized transformers. arXiv preprint arXiv:2501.17727 (2025)
6. Wang, Z., et al.: Monet: mixture of monosemantic experts for transformers. arXiv preprint arXiv:2412.04139 (2024)
7. Meng, K., et al.: Locating and editing factual associations in GPT. In: NeurIPS (2022)
8. Meng, K., et al.: Mass-editing memory in a transformer. In: ICLR (2023)
9. Olah, C., et al.: Zoom in: an introduction to circuits. Distill (2020)
10. Higgins, I., et al.: β-VAE: learning basic visual concepts with a constrained variational framework. In: ICLR (2017)
11. Chen, X., et al.: InfoGAN: interpretable representation learning by information maximizing GANs. In: NeurIPS (2016)
12. Laine, S., Aila, T.: Temporal ensembling for semi-supervised learning. In: ICLR Workshop (2017)
13. Tarvainen, A., Valpola, H.: Mean teachers are better role models: weight-averaged consistency targets improve SSL. In: NeurIPS (2017)
14. Bardes, A., Ponce, J., LeCun, Y.: VICReg: variance-invariance-covariance regularization for self-supervised learning. In: ICML (2021)
15. Bell, A., Sejnowski, T.: An information-maximization approach to blind separation and blind deconvolution. Neural Comput. (1995)

Toward Reliable AI-Driven Autonomous Decision-Making: Quantifying AI Alignment with Normative Standards

Xin-Wen Wu(✉)

University of Mary Washington, Fredericksburg, VA 22401, USA
xwu@umw.edu

Abstract. In an era increasingly shaped by AI-driven autonomous decision-making, ensuring that such decisions align with established ethical or knowledge-based standards is critically important. While extensive discussions on AI ethics exist, there remains a lack of methods to quantify the alignment of decisions with normative standards. To fill this gap, this paper proposes a framework for quantifying and evaluating decision alignment using a probabilistic model that represents idealized ethical judgments – independent of potentially biased historical data. The alignment of a decision is assessed via its maximum conditional likelihood under this model, and the overall coherence of the normative model is analyzed using mutual information and its theoretical maximum. Although the framework is illustrated through a healthcare use case, drawing on the AMA Principles of Medical Ethics, it is broadly applicable to other domains, including strategic business planning and AI-enabled personal support systems. Through theoretical analysis and illustrative examples, we demonstrate that the proposed *Alignment Score* and *EthicalAlignment Score* effectively capture the degree to which autonomous decisions conform to normative standards, offering a transparent and interpretable basis for ethical evaluation of AI-driven autonomous decision-making.

Keywords: Artificial intelligence (AI) · autonomous decision making · trustworthy AI · ethics · probabilistic model

1 Introduction

As autonomous systems powered by artificial intelligence (AI) increasingly engage in critical decision-making, spanning domains including healthcare triage, financial advising, and personalized recommendations, the demand for decisions that are not only effective but also normatively sound has become more urgent than ever [1]. These systems must operate in alignment with established standards, such as ethical guidelines, legal codes, institutional policies, strategic plans, or personal values [2]. Extensive discussions on AI ethics have been published recently, as seen in the representative publications [1–3]. However, despite the widespread deployment of autonomous systems optimized for performance and efficiency [4], the challenge of *quantifying* how well a decision aligns with recognized ethical principles or bodies of knowledge remains underexplored [1, 2].

H. R. Arabnia et al. (Eds.): CSCI 2025, CCIS 2972, pp. 497–511, 2027.
https://doi.org/10.1007/978-3-032-28469-3_32

To address this gap, this paper proposes an information-theoretical framework to quantify and compute the alignment of decisions made by AI-driven autonomous systems with normative standards. Our proposed measure of alignment is grounded in a model of joint probability distribution over decisions and normative judgments. When constructing the probabilistic model for the joint distribution over decisions and normative judgments, rather than using a natural approach that relies on retrospective data, which may include both ethically sound and flawed decisions, our method constructs a normative probabilistic model that encodes idealized joint distributions of decisions and normative judgments. This normative model is then used to compute the ethical alignment of new decisions in a forward-looking fashion. Specifically, the *Alignment Score* of a new decision is defined as the maximum conditional probability that the decision is considered appropriate under the standards. This interpretable metric reflects the strongest available justification for the decision within the model and provides a basis for ranking or filtering candidate decisions and actions in an AI-powered autonomous systems.

In our framework, the normative standard is not limited to ethical systems, e.g., the American Medical Association (AMA) Principles of Medical Ethics [5], it may take the form of a knowledge base (e.g., a care plan for an elderly individual), or a formal strategy (e.g., an organizational decision-making policy). In this paper, to illustrate the framework, we focus on a healthcare use case: quantifying the ethical alignment of hospital decisions made by autonomous systems and assessing whether the decisions align with medical ethics [5, 6]. However, the framework generalizes to other domains, such as:

- Evaluating whether enterprise-level decisions conform to strategic planning guidelines [4];
- Assessing how well AI-generated recommendations support the health, safety, and preferences of elderly users [7];
- Measuring the consistency of algorithmic behavior with regulatory or legal standards [3].

Furthermore, to evaluate the overall coherence of the normative model, we examine mutual information and maximum mutual information between decisions and normative judgments. While these are not used directly in computing the score of alignment, they effectively measure the overall coherence of the normative model and define the upper bound for alignment in idealized circumstances.

This research contributes to the growing body of work on AI ethical alignment and interpretable decision-making, by offering:

- A framework for quantifying ethical alignment of AI-driven autonomous decisions rooted in probability theory and information theory;
- A generalizable model that separates ethical assessment from historical bias, offering a transparent and interpretable basis for AI ethical evaluation;
- Illustration and computation of scores of ethical alignment for autonomous decisions with established ethical standards using the proposed model.

The remainder of the paper is structured as follows: Sect. 2 reviews related work on ethical AI, alignment, and information-theoretical applications. Section 3 describes the theoretical framework including the introduction of the normative probabilistic model and definition of the alignment scores. Section 4 presents model construction and illustrative examples. Section 5 discusses enhanced model construction. Section 6 outlines applications of the framework and future directions. Finally, concluding remarks are presented in Sect. 7.

2 Related Work

Autonomous systems are increasingly making decisions with significant ethical, social, and personal implications. As such, a central concern in artificial intelligence, machine learning, and human-computer interaction is ensuring that such systems behave in ways that align with established norms, be they ethical, epistemic, or organizational. Built on this broader literature, our work offers a novel information-theoretical framework to quantify alignment of AI-powered decision-making with normative standards.

2.1 Alignment of Decisions with Ethical and Knowledge-Based Standards

The need to align autonomous decision-making with ethical standards has been widely acknowledged [1–3]. In healthcare, Borysowski, Ehni, and Górski [6] emphasize the importance of respecting informed patient choice in clinical decisions, revealing frequent misalignments between medical practice and ethical imperatives. The AMA Principles of Medical Ethics [5] offer a structured normative standard, against which decisions can be assessed.

Beyond medicine, Rivero [4] points to the increasing autonomy of decision-making systems in enterprise and personal domains, raising the question of whether and how such systems can remain aligned with human values. In AI ethics, Cheong [3] maps the debate on algorithmic fairness, transparency, and accountability, highlighting the need for interpretability and ethical compliance in algorithmic decisions.

In the realm of intelligent assistants and AI for elder care, Choi et al. [7] provide a comprehensive review, highlighting both practical applications and ethical challenges. By examining AI's role in medical functionality, emotional companionship, and daily living assistance, the authors emphasize stakeholder responsibility and offers valuable guidance for the responsible integration of AI in geriatric care [7].

While extensive discussions on AI ethics exist as discussed above, the challenge of quantifying how well a decision aligns with recognized ethical principles or bodies of knowledge remains underexplored. Our work proposes a novel framework to quantify decision alignment with normative standards, offering a transparent and interpretable basis for ethical evaluation of AI-driven autonomous decision-making.

2.2 Information Theory in Decision Analysis

Information-theoretical tools have been applied to clinical diagnostics and medical reasoning. Benish [8] reviews the use of mutual information and entropy in evaluating clinical diagnostic tests, showing how such measures capture uncertainty and inform test utility. These methods, however, focus on predictive performance or uncertainty reduction, instead of ethical alignment or normative validation.

Other influential work, such as Lin's information-theoretical definition of similarity [9] and its extension by Formica and Taglino [10], leverages mutual information in semantic and taxonomic contexts. While technically robust, these approaches aim to measure similarity between symbolic concepts or terms, rather than the alignment of real-world decisions or actions with ethical or normative judgments.

In contrast, our framework measures alignment between decisions and established standards (e.g., ethical judgments) by constructing an ideal normative model, independent of historical data. This allows us to evaluate decisions not based on past behavior but on their coherence with approved decision-judgment pairings.

In summary, although prior work has discussed AI ethics and employed information theory to evaluating clinical diagnostic tests, there remains a lack of quantifying methods for evaluating the ethical alignment of autonomous decisions. Our work fills this gap, offering a framework to quantify and evaluate AI alignment with normative standards. The framework does not rely on historical datasets that may contain unethical decisions. Instead, it derives alignment scores from an ideal probabilistic model, enabling the forward evaluation of decision coherence across application domains such as healthcare, enterprise strategy, and personal assistant systems.

3 Theoretical Framework

This section describes the proposed framework for quantifying the alignment of a decision with a normative body of knowledge, such as a code of ethics or strategic guidelines. We model the alignment using information-theoretical constructs, particularly mutual information and conditional probability, under the assumption of an *ideal* joint distribution.

3.1 Decision and Judgment as Random Variables

Let

- X denote a random variable over a finite decision space, $\mathcal{D}$, representing the set of possible decisions that an autonomous system may make, and
- Y denote a random variable over a judgment space, $\mathcal{E}$, representing possible ethical or normative evaluations of those decisions.

Each value of Y corresponds to a judgment category (for example, *Ethically Required, Permissible, Problematic, Unethical*), derived from a recognized normative body (such as the AMA Principles of Medical Ethics).

The joint distribution $P(X, Y)$ encodes an *ideal* normative model: it represents how an ideally informed ethical or knowledge-based authority would probabilistically relate decisions (X) to evaluations (Y).

Here, 'ideal model' means that the model is not constructed from empirical or historical decision-making data, which includes behaviors that are not well aligned with judgements; instead, it reflects how decisions ought to be aligned with norms.

3.2 Alignment Score

To evaluate a new or proposed decision $x_{\text{new}} \in \mathcal{D}$, we define the alignment score as the maximum conditional probability of normative evaluation given the decision:

$$Alignment(x_{new}) = \max_{y \in \mathcal{E}} P(Y = y|X = x_{new}) \tag{1}$$

This score reflects how confidently the ideal normative model associates the decision with a normative judgment. A high alignment score suggests that the decision is clearly aligned with at least one normative expectation.

According to the definition of alignment score, we have

$$Alignment(x_{new}) = P\big(Y = y^* \big|X = x_{new}) \tag{2}$$

where $y^* = arg\max_{y \in \mathcal{E}} P(Y = y|X = x_{new})$.

This expression not only gives the alignment score for a decision but also indicates the specific normative judgment that endorses the decision. It helps when evaluating real-time decisions in automated systems, where interpretability and confidence are critical.

Note that a judgment space, $\mathcal{E}$, usually contain both *positive* and *negative* judgments. For example, considering autonomous decision-making in healthcare, an ethical judgement space may be comprised of the following judgments: *Ethically Required*, *Ethically Permissible*, *Problematic*, and *Unethical.* Among them the first two are regarded as ethically positive judgments and the others are ethically negative judgments.

When there are both positive and negative judgments, a decision with a high alignment score is not necessary an *ethical* decision. For example, suppose a new decision that a healthcare provider might take is $x_{new} = Deny$ (meaning that a treatment request is denied). Although this decision may not be regarded as an ethical decision, it can have a high alignment score, as this decision may be "endorsed" by the judgment $y^* = Unethical$. According to the definition

$$Alignment(x_{new}) = P\big(Y = y^* \big|X = x_{new})$$

$Alignment(x_{new})$ achieves a high value, when the conditional probability $P(Y = Unethical|X = Deny)$ is high.

To better capture the degree to which a decision is aligned with ethics, we further define *Ethical Alignment Score* as follows:

$$EthicalAlignment(x_{new}) = \max_{y \in \mathcal{E}_+} P(Y = y|X = x_{new}) \tag{3}$$

where $\mathcal{E}_+$ is the subspace of $\mathcal{E}$ consisting of all *ethically positive* judgements.

3.3 Mutual Information and Normative Clarity

The *mutual information (MI)*, denoted by $I(X; Y)$, between X and Y, quantifies how much knowing the decision reduces uncertainty about the corresponding normative evaluation:

$$I(X;Y) = \sum_{x\in\mathcal{D}}\sum_{y\in\mathcal{E}} P(x,y)\log\left(\frac{P(x,y)}{P(x)P(y)}\right) \tag{4}$$

Mutual information measures the clarity and coherence of the normative model: higher values indicate that decisions more consistently determine ethical judgments. This is useful for validating the probabilistic model.

We define the *maximum mutual information*, denoted by $I_{max}(X;Y)$, as the highest achievable mutual information between any pair of random variables defined over the decision and judgment spaces. This serves as a theoretical upper bound for normative alignment, under perfect correspondence between decisions and judgments. The *normalized MI*, denoted by $I^*(X;Y)$, is defined as $I^*(X;Y) = I(X;Y)/I_{max}(X;Y)$.

3.4 Ethically Ideal Models

An ethically ideal joint distribution satisfies two core conditions:

- *Normative Soundness*: It has a high mutual information $I(X;Y)$, indicating every pair of decision and judgment reflects ethically endorsed evaluation.
- *Normative Distinguishability*: Different decisions produce meaningfully different alignment outcomes, avoiding trivial models where ethical judgments are uninformative.

Remark:

(1). Constructing an ethically ideal model requires expert input and careful consultation of authoritative sources, e.g., AMA ethical principles.
(2). As future work, we will be investigating algorithmic procedures for constructing ethically ideal models to meet the core conditions stated above.

4 Model Construction and Illustrative Examples

This section demonstrates how the theoretical framework is used to quantify alignment of decisions with normative standards in healthcare using an illustrative joint probability model. We first define the decision and judgment spaces, construct a joint distribution over these spaces, and then calculate alignment scores and mutual information.

4.1 Decision and Judgment Spaces

Let the decision space and judgment space $\mathcal{E}$ be defined as follows:

- $\mathcal{D}$ = {*Treat, Delay, Referral, Deny*}, representing possible decisions a healthcare provider might take regarding a proposed medical intervention.
- $\mathcal{E}$ = {*Required, Permissible, Problematic, Unethical*}, representing ethical evaluations of these decisions according to the AMA's standards.

Please note that these values represent a simplified version of real-world decision making and ethical reasoning but preserve enough structure to support meaningful analysis.

4.2 Constructing an Ethical Joint Distribution

When developing a model of joint distribution (*X*, *Y*), we try to ethically idealize it by incorporating the following criteria.

- "*Treat*" is often *Required*, but never *Problematic* or *Unethical.*
- "*Delay*" and "*Referral*" are *Permissible* or *Problematic* but rarely *Required.*
- "*Deny*" is *Unethical* or *Problematic* (especially when treatment is clinically warranted) but never ethically *Required.*

Incorporating the criteria above, a sample joint distribution is given in the table below:

Table 1. Model 1 – Joint Distribution.

Y	X			
	Treat	*Delay*	*Referral*	*Deny*
Required	0.30	0.05	0.00	0.00
Permissible	0.00	0.20	0.05	0.00
Problematic	0.00	0.05	0.10	0.05
Unethical	0.00	0.00	0.05	0.15

This distribution is a valid joint probability distribution satisfying: (1) Each row or column sums to the corresponding marginal distribution P(Y) or P(X); and (2) The sum of all probabilities in the distribution is equal to 1.

4.3 Alignment and EthicalAlignment Scores

We now compute the scores of alignment and ethical alignment for the decisions, *Treat*, *Delay*, *referral*, and *Deny*.

Note that in this model, we have $\mathcal{E} = \{Required,\ Permissible,\ Problematic,\ Unethical\}$ and $\mathcal{E}_+ = \{Required,\ Permissible\}$. Recall that for any decision x,

$$Alignment(x) = \max_{y\in\mathcal{E}} P(Y = y|X = x) = \frac{\max_{y\in\mathcal{E}} P(Y = y, X = x)}{P(X = x)}$$

and

$$EthicalAlignment(x) = \frac{\max_{y\in\mathcal{E}_+} P(Y = y, X = x)}{P(X = x)}.$$

Regarding the decision, *Treat*:

$$P(Y = Required|Treat) = 0.3/0.3 = 1;$$

and for every other judgment y, $P(Y = y|Treat) = 0$. Therefore,

$$Alignment(Treat) = 1, \text{ and } EthicalAlignment(Treat) = 1.$$

In the same way, we get the scores of Alignment and EthicalAlignment for *Delay, Referral,* and *Deny*. The results are summarized in the following table, Table 2. The Alignment and EthicalAlignment Scores are also depicted in Fig. 1 and Fig. 2, respectively.

Table 2. Model 1 – Alignment & EthicalAlignment Scores.

Decision	Alignment	EthicalAlignment
Treat	1.00	1.00
Delay	0.67	0.67
Referral	0.50	0.25
Deny	0.75	0.00

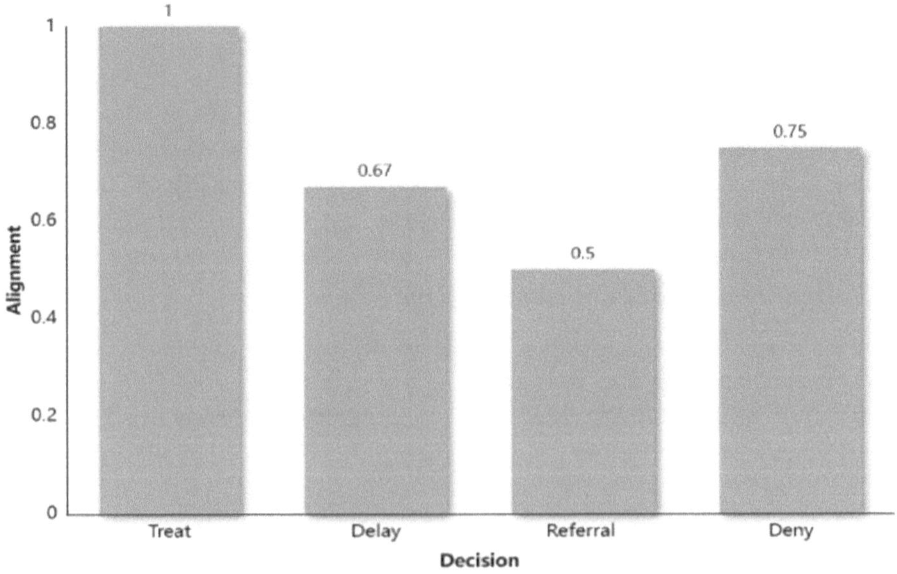

Fig. 1. Model 1 – Alignment Scores of Decisions

As illustrated by Fig. 2 below, "*Treat*" has the highest score of ethical alignment, while "*Deny*" has the lowest. The scores match intuitive ethical reasoning, enhancing the transparency and reliability of the model.

In practical applications, the scoring can inform autonomous systems whether a particular decision aligns well with accepted ethical standards, supporting explainability and trust in autonomous decision making.

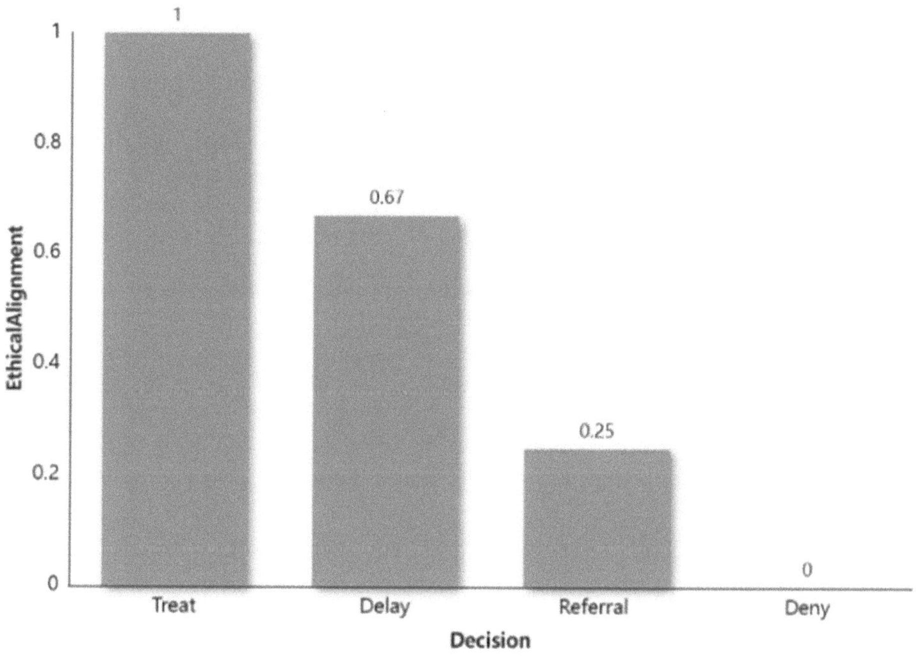

Fig. 2. Model 1 – EthicalAlignment Scores of Decisions.

4.4 Mutual Information

Mutual information quantifies the overall alignment between decisions and judgments:

$$I(X;Y) = \sum_{x \in \mathcal{D}} \sum_{y \in \mathcal{E}} P(x,y) \log_2 \left(\frac{P(x,y)}{P(x)P(y)} \right)$$

Using the joint probabilities in Table 1, we calculate the marginal probabilities $P(x)$ and $P(y)$, , and then the mutual information (MI)

$$I(X;Y) = 1.121.$$

We can further calculate the entropy of X (or Y) and conditional entropy. Then, by the equations, $I(X;Y) = H(X) - H(X|Y) = H(Y) - H(Y|X)$, we obtain the maximum mutual information,

$$I_{max}(X;Y) = 2.0$$

Therefore, the normalized MI, $I^*(X;Y)$, is

$$I^*(X;Y) = \frac{I(X;Y)}{I_{max}(X;Y)} = \frac{1.121}{2} = 56.1\%.$$

The mutual information and normalized mutual information indicate meaningful alignment between decisions and ethical judgments.

5 An Enhanced Model

5.1 Enhanced Model Construction and Evaluation

A model can be evaluated using MI, as discussed in the previous sections: A low value of MI indicates that the model does not robustly capture the dependencies between decisions and normative judgments. In contrast, a high MI value suggests that ethical evaluations meaningfully constrain the decision space. For example:

- In clinical ethics, a low MI may prompt reconsideration of how normative judgments are derived or whether new decision contexts (e.g., novel treatment options) require updated ethical guidance.
- In elderly care, it may reveal that a personal assistant lacks a rich enough representation of user preferences, prompting additional data collection or model personalization.
- In enterprise decision-making, it may uncover ambiguous or poorly communicated strategic standards.

In all cases, our framework provides a principled way to improve normative models via:

1. Enhancing the interpretability of alignment scores; and
2. Maximizing mutual information to increase the model's coherence.

5.2 Enhanced Joint Distribution

Now, let us examine another joint distribution model for healthcare decision-making, presented in Table 3 below.

Table 3. Model 2 – Joint Distribution

Y	X			
	Treat	*Delay*	*Referral*	*Deny*
Required	0.42	0.02	0.03	0.00
Permissible	0.20	0.07	0.05	0.03
Problematic	0.03	0.04	0.02	0.02
Unethical	0.00	0.02	0.01	0.04

Based on the joint distribution of decisions and ethical judgements, the Alignment scores and EthicalAlignment scores for decisions can be calculated; and the results are summarized in the following table (Table 4). The scores are also depicted by the figures below (Figs. 3 and 4).

Table 4. Model 2 – Alignment & EthicalAlignment Scores

Decision	Alignment	EthicalAlignment
Treat	0.65	0.65
Delay	0.48	0.48
Referral	0.46	0.46
Deny	0.44	0.33

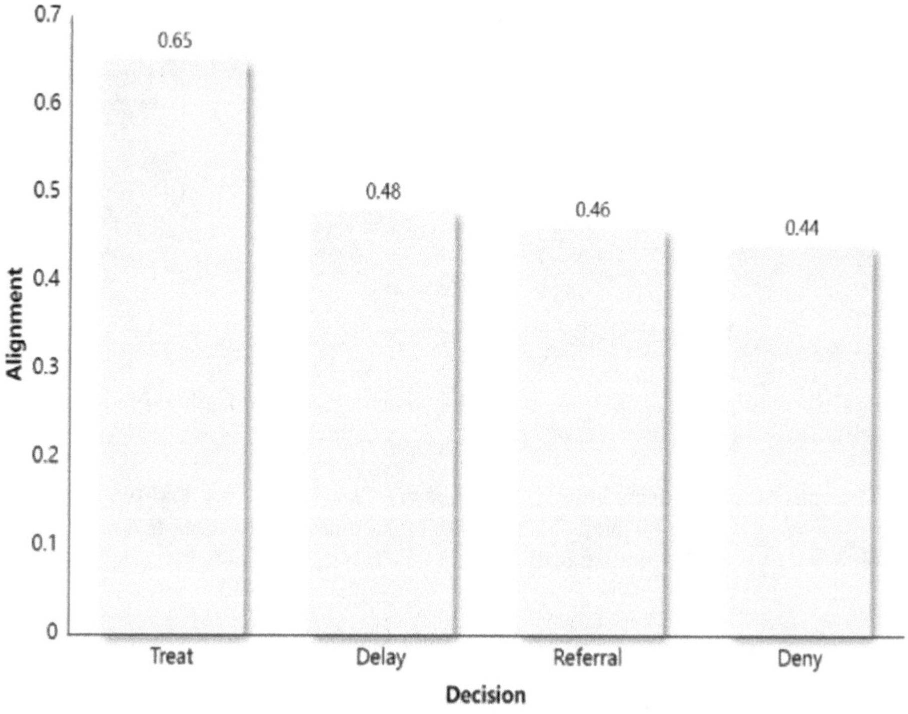

Fig. 3. Model 2 – Alignment Scores of Decisions

As showed in Fig. 4, the EthicalAlignment scores are ethically informative. For example, "*Treat*" and "*Delay*" maintain strong or reasonable alignment with positive ethical judgments, while "*Referral*" and "*Deny*" receive moderately low scores, reflecting ethical uncertainty or disapproval.

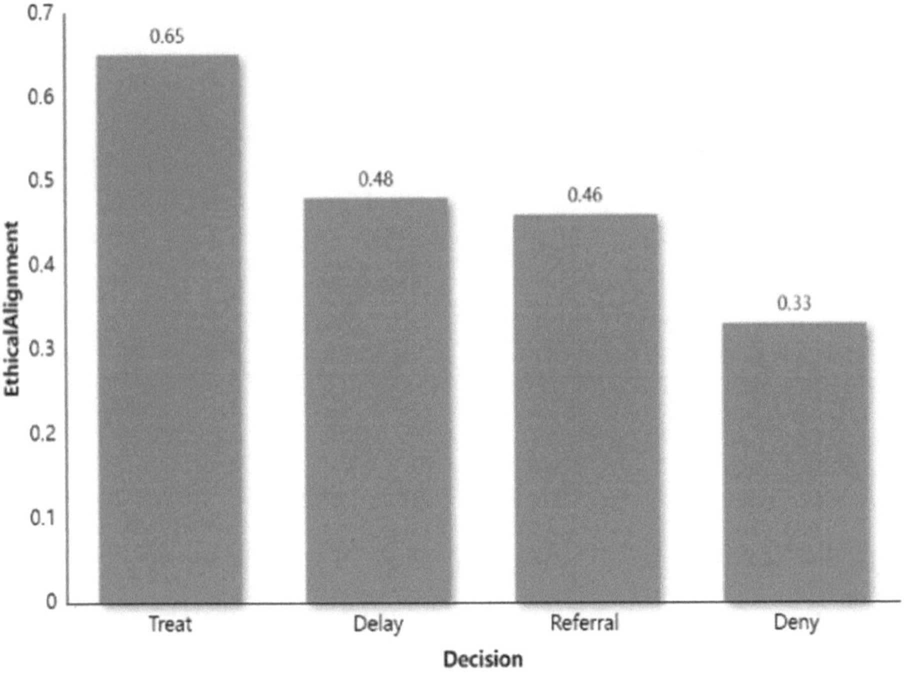

Fig. 4. Model 2 – EthicalAlignment Scores of Decisions

Based on the joint distribution, the MI between decisions and ethical judgements is

$$I(X;Y) = 0.967.$$

The entropy of the decision distribution is $H(X) = 1.593$ bits; and the judgement entropy is $H(Y) = 1.684$ bits. So, the maximum mutual information $I_{max}(X;Y) = 1.593$. Therefore, the normalized MI is

$$I^*(X;Y) = \frac{I(X;Y)}{I_{max}(X;Y)} = \frac{0.967}{1.593} = 60.7\%.$$

The normalized MI of 60.7% indicates that this is an enhanced model, compared with the model in the previous Section. The mutual information of 0.967 bits out of a maximum possible 1.593 bits suggests that the decision distribution meaningfully encodes ethical standards yet still has room to improve. This observation points to a fundamental insight: perfect alignment is rarely achievable in practice, but systematic approximations can help surface value conflicts and guide better decisions.

6 Applications and Future Work

The framework offers a versatile, information-theoretical approach to evaluate the consistency of decisions with normative standards. While our examples have focused on healthcare ethics, the principles and procedures generalize across a range of domains where decisions must reflect adherence to established bodies of knowledge, ethical guidelines, or strategic goals.

6.1 Medical Ethics and Clinical Decision-Making

Healthcare settings frequently require autonomous systems or human professionals to make decisions consistent with ethical standards such as those articulated by the American Medical Association. For example, decisions about administering treatment, delaying intervention, referring a patient to another provider, or denying care should align with normative judgments.

The framework allows a quantitative assessment of such alignment. By computing both the Alignment Score and the EthicalAlignment Score, it provides interpretable feedback on whether a decision or proposed action coheres with ethical norms. Additionally, Mutual Information offers a measure of how effectively the overall model captures the interdependence between decisions and ethical evaluations.

This formalism can be useful in:

- Supporting clinical decision-support systems in triage or treatment recommendations.
- Providing ethical oversight during the deployment of AI in telemedicine.
- Auditing medical AI systems for post hoc explanation and accountability.

6.2 AI-Enabled Personal Assistants for the Elderly

Another compelling application lies in the design of AI-powered personal assistants supporting aging individuals. Such assistants may autonomously recommend or make decisions involving medication timing, physical activity, communication, or emergency response. Here, alignment is not with codified ethics but with a personalized normative model built from prior knowledge of the individual's preferences, health conditions, constraints, and values.

In this context:

- The decision space could include actions like *Call for Help*, *Remind About Medication*, *Suggest Rest*, or *Do Nothing*.
- The judgment space could reflect how well each decision aligns with inferred personal preferences or safety protocols, e.g., *Ideal*, *Acceptable*, *Risky*, or *Unsafe*.
- The Alignment and EthicalAlignment scores, along with mutual information, can measure and improve how faithfully the assistant adheres to what the person would have wanted.

6.3 Strategic Decision-Making in Enterprises

Enterprise systems often operate under the need to align decisions with strategic plans, risk profiles, or ethical commitments. For instance, a decision-support system in business strategy might choose between *Expand Market*, *Maintain Current Strategy*, or *Divest Assets*. Normative judgments may stem from internal corporate policies or external stakeholder expectations. Here again, alignment scores quantify how well autonomous decisions serve broader organizational goals, while mutual information helps diagnose inconsistencies or noise in the model.

This is especially useful in:

- Auditing strategic alignment of autonomous agents in multi-agent business environments.

- Explaining failures when an autonomous recommendation deviates from expected best practices.

6.4 Limitations and Future Work

It is worth acknowledging the limitations of the framework:

- Specification burden: Constructing ethically ideal models may require expert input and still lacks an algorithmic procedure.
- Ethical pluralism: A single joint distribution may not capture diverging ethical perspectives across cultures, institutions, or individuals.

Future work will involve extending the framework to account for diverse normative signals from multiple stakeholders. Additionally, interactive procedures that allow human users to inspect, adjust, or challenge the joint distribution could help increase transparency, trust, and shared responsibility in ethically sensitive applications.

Finally, as an interesting and important future direction, we will investigate development of algorithmic procedures for constructing ethically ideal models in accordance with the two core conditions as explained in Sect. 3, that is, Normative Soundness and Normative Distinguishability.

7 Conclusion

In this paper, we proposed a novel, information-theoretical framework for quantifying ethical alignment of AI-powered autonomous decision-making with respect to an established normative standard. Drawing inspiration from information theory, we proposed a joint probability model over decision and judgment spaces, enabling the computation of alignment scores and EthicalAlignment scores for each autonomous decision. This approach provides an interpretable and transparent method for evaluating the degree to which autonomous decisions reflect or diverge from ethical expectations, such as those outlined in the American Medical Association's principles.

Through the theoretical analysis and illustrative examples, we demonstrated how the framework can distinguish between ethically strong and weak decisions, and how it allows for systematic comparison across models using normalized mutual information. Crucially, our enhanced scoring method, EthicalAlignment, helps mitigate interpretive ambiguity by focusing explicitly on ethically positive evaluations, thus providing a more transparent basis for ethical assessment.

The model's flexibility and abstraction make it adaptable to diverse domains where value-aligned decision-making is paramount, including clinical decision-making, AI-enabled personal assistants, and strategic decision-making in enterprises.

Disclosure of Interests. The author has no competing interests to declare that are relevant to the content of this article.

References

1. Floridi, L.: The Ethics of Artificial Intelligence: Principles, Challenges, and Opportunities. Oxford University Press, Oxford (2023)
2. Gabriel, I.: Artificial intelligence, values and alignment. Mind. Mach. **30**(3), 411–437 (2020)
3. Cheong, B.-C.: Transparency and accountability in AI systems: safeguarding wellbeing in the age of algorithmic decision-making. Front. Hum. Dyn. **6** (2024)
4. Rivero, O.: The role of artificial intelligence in strategic decision-making: transforming managerial strategies in the digital age. Eur. J. Stud. Manage. Bus. **33**, 33–42 (2025)
5. American Medical Association: AMA Principles of Medical Ethics. https://code-medical-ethics.ama-assn.org/principles
6. Borysowski, J., Ehni, H.-J., Górski, A.: Ethics codes and medical decision making. Patient Educ. Couns. (Elsevier) **104**(6), 1312–1316 (2021)
7. Choi, W.J.W., Ahn, B., Moorthy, G., Do, K.: AI-assisted care for older adults: a review of practical and ethical areas of concern. AI Ethics (2025)
8. Benish, W.: A review of the application of information theory to clinical diagnostic testing. Entropy **22**(1), 97 (2020). https://doi.org/10.3390/e22010097
9. Lin, D.: An information-theoretic definition of similarity. In: Proceedings of the International Conference on Machine Learning, pp. 296–304. Morgan Kaufmann, Madison (1998)
10. Formica, A., Taglino, F.: An enriched information-theoretic definition of semantic similarity in a taxonomy. IEEE Access **9**, 100583–100593 (2021)

Multimodal Agricultural Intelligence: Comparative Analysis of Large Language Models for Precision Nursery Management Systems

Harshitha Manjunatha[1], Shekhar Suman Borah[2], Prabha Sundaravadivel[2(✉)], H. Allen Torbert[3], Siva P. Kumpatla[4], and Patricia Knight[5]

[1] The University of Texas at Dallas, Richardson, TX, USA
[2] The University of Texas at Tyler, Tyler, TX, USA
psundaravadivel@uttyler.edu
[3] USDA ARS, Auburn, AL, USA
[4] USDA ARS, Poplarville, MS, USA
[5] Mississippi State University, Poplarville, MS, USA

Abstract. Modern precision agriculture requires highly advanced computational platforms able to process heterogeneous sensor streams and deliver actionable advice for operational decisions. This research provides a comprehensive comparative evaluation of three state-of-the-art large language models (LLMs), Mistral, Gemma3, and Llama3, coupled with computer vision expertise designed to enable autonomous agriculture management. Our work addresses the wide gap between AI theoretical possibilities and their eventual implementation in agriculture by designing annotation-free detection methods and evaluating multi-architecture scaling attributes of performance. Through systematic experimentation based on high-definition UAV imagery from a 300-hectare rose plantation, we demonstrate significant variations in model performance, computational efficacy, and output advisory quality. The analysis shows that the parameter quantity does not have a direct linear association with advisory quality in agriculture, as shown by Gemma3's parsimony-driven architecture realizing better performance for every computational unit. This work sets out empirical baselines for LLMs' implementation in resource-poor agriculture environments while maintaining full data sovereignty, an essential constraint for agriculture enterprises protecting proprietary cultivation methods.

Keywords: Agricultural Advisory Systems · Agricultural Computer Vision · Autonomous Bloom Detection · Computational Agriculture · Large Language Models (LLMs) · Multimodal AI Systems · Precision Agriculture

1 Introduction

Future agriculture is confronted with hitherto unprecedented challenges in moderating a compromise between increasing output and environmental sustainability priorities. The hitherto dominant methods of managing farms, based mostly on tacit knowledge along

H. R. Arabnia et al. (Eds.): CSCI 2025, CCIS 2972, pp. 512–522, 2027.
https://doi.org/10.1007/978-3-032-28469-3_33

with periodic methods of sampling, are plagued by severe temporal limitations as well as spatial inaccuracies [1]. Whereas today's developments in unmanned aerial vehicle (UAV) technology have brought hopes of large-scale, wide area cropping observation at hitherto unparalleled resolutions, computational requirements for dealing with large-scale imagery are yet a great challenge for most farms.

Agriculture is a novel computational challenge in which theoretical AI potential must yield practical, field-usable solutions functioning in a resource-limited setting. Traditional supervised learning methods based on YOLO architecture, though showing potential in controlled lab scenarios, demand heavy manual annotation the process our systematic evaluation shows requires 4–8 h for each high-resolution image. What is more, such methods suffer seriously from performance loss when faced with spatial clustering and environmental variability inherent in real-world agriculture scenarios. Big language models are a paradigm-defining innovation in mapping raw sensor information to usable agronomic information [2, 3]. Nevertheless, their applicability in an agrarian setting is beset by multifaceted questions concerning data ownership rights, network accessibility, and computational resource availability. This work responds to these complexities with a holistic comparative evaluation of three of the foremost LLM models based on their suitability in generating agronomic advisory together with uncompromised data sovereignty.

This work's essential contribution is in determining empirical performance baselines of multi-architecture LLM deployment in the field of precision agriculture, which offers systematic direction for hardware-software optimization choices in agricultural AI systems.

2 Related Work

In agricultural computer vision, techniques have moved a long way from simple rule-based systems to advanced deep learning models. Recently, plant identification has become an active research area, especially with multimodal deep learning approaches [4]. But most of these approaches still depend on supervised learning, which needs a lot of labeled data. Since manual labeling is expensive and time-consuming, researchers are now looking at self-supervised and unsupervised learning methods. Self-supervised learning has shown strong results in agricultural image classification and segmentation even with fewer labels. For example, contrastive learning and masked image modeling have improved plant disease detection, while self-supervised pretraining enhanced maize kernel classification and segmentation using only limited labeled samples [5–7]. Similarly, methods like unsupervised metric learning and domain adaptation have been applied to pest and plant disease recognition, reducing the need for large, annotated datasets while still maintaining good accuracy [8]. *FlowerPhenoNet* is one such system designed for automated flower detection from multi-view images, which gave promising results under controlled conditions [9]. Flowering is an important stage in crops like rapeseed because the number of flowers directly affects seed yield. A UAV-based method using YOLOv5 and a Convolutional Block Attention Module (CBAM) was developed to detect and count rapeseed flowers from RGB images, supported by the Rapeseed Inflorescence Benchmark (RIB) dataset [10]. One study also proposed *RapeNet* and *RapeNet*

+ networks for density estimation of flower clusters, using two labeling strategies such as rectangular box and centroid-based annotations. The results showed high accuracy and robustness, confirming that these models work effectively for automated rape flower counting [11]. While there is growing work on plant detection, the connection between LLMs and agricultural computer vision is still very new. Most studies focus on single models and do not compare different architectures in detail. This leaves open questions about which models are most suitable and how resources should be allocated. Recently, some studies have started to fill this gap. LLMs have been used for crop disease and pest detection [12], while RAG-based models have been tested for farm advisory services in low-resource areas [13]. Other works also highlight both the benefits and risks of using LLMs in agriculture [14].

3 Methodology

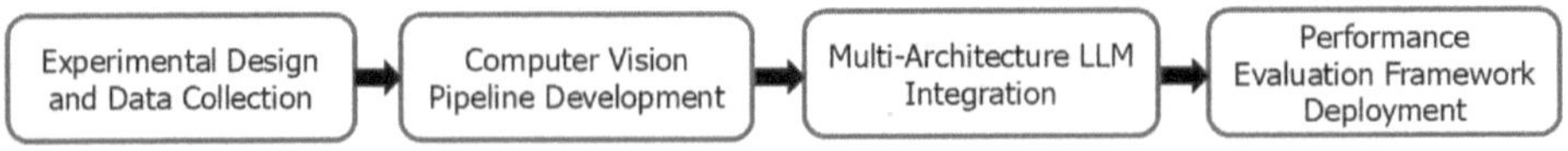

Fig. 1. Workflow showing data collection, vision pipeline, LLM integration, and performance evaluation.

Figure 1 shows the overall workflow, beginning with experimental design and data collection, moving through computer vision pipeline development, integrating multi-architecture LLMs, and concluding with a performance evaluation framework to validate system performance. The experimental setup covers a 300-acre commercial rose nursery, thereby providing a realistic test environment for an AI agriculture system. High-resolution images were taken by UAV adhering to certain fixed flight parameters from different points of time through several growing seasons and under varied environmental conditions. Sampling was done by conducting aerial surveys systematically at an altitude of 100 m and maintaining an 80% overlap between images; the resolution was set at 2.5 cm ground sampling distance. This resolution enables very detailed bloom detection while still maintaining computational tractability for real-time processing scenarios.

We present here an approach offering annotation-free detection and thus addresses the fundamental scalability challenge of agricultural computer vision. Most traditional supervised methods, hence, require intense manual labelling, thus becoming bottlenecks toward real deployment. Our unsupervised approach uses spatial clustering algorithms along with color transformations ideally suited for rose detection. The computer vision pipeline carries out 3 main steps, including environmental normalization as preprocessing, unsupervised feature extraction on spatial-spectral characteristics, and post-processing blooming counts and locations. This procedure removes the need for annotation without compromising detection accuracy that is comparable to supervised approaches. Three LLMs with different architectures were evaluated systematically: Mistral with 7 billion parameters, Gemma3 with efficient architecture, and Llama3 with parameter configurations. Each model was deployed locally to ensure data sovereignty and enable complete performance characterization. The integration pipeline converts computer vision outputs into structured agricultural data that is used as input for LLM-based advisory generation. Prompt engineering techniques were developed for each model to optimize their performance on agricultural advisory tasks. Our modality of evaluation encompasses several dimensions: computational efficiency (inference latency, memory utilization, energy consumption), advisement quality (contextual relevance, actionability, technical accuracy), and scaling attributes (strong and weak scaling performance over different worker layouts). We ran these computational experiments on HPC infrastructure, which allowed evaluating resource utilization patterns in a methodic way and pinpointing performance bottlenecks across a wide variety of model architectures.

4 Results and Analysis

4.1 Computer Vision Performance

Annotation-free detection methods had a successful outcome in a real-life agricultural situation. In Fig. 2(a), one observes the system's processing of high-resolution UAV imagery and producing accurate bloom detection without supervised data.

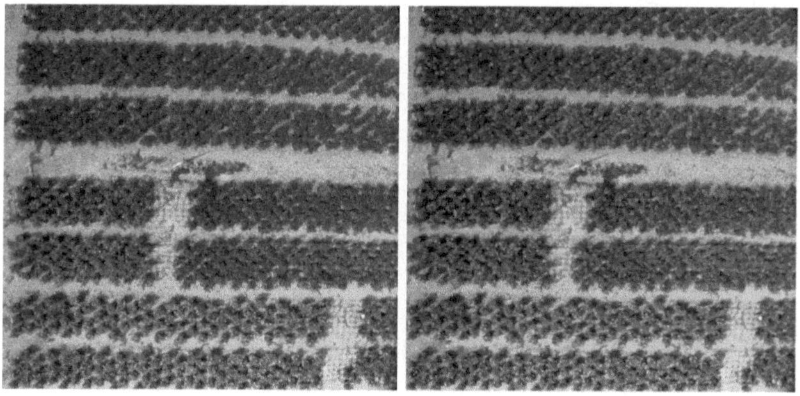

(a) (b)

Fig. 2. (a) High-Resolution UAV imagery of rose nursery: raw RGB aerial view, (b) Computer Vision Breakthrough: Unsupervised rose detection in real time agricultural environments.

As we can observe in Fig. 2(b), the system successfully identified individual roses blooming through different environmental conditions, lighting conditions, and stages of growth. The system remained consistent in performance over the 300-acre area and processed imagery fast enough to keep up with operational agricultural time frames.

The delivery of annotation-less requirements is a major stride in agricultural computer vision, lowering the barriers for deployment while maintaining accuracy in detection. This allows for practical deployment of AI monitoring systems for a gamut of agricultural operations without requiring extensive technical know-how or annotation resources.

4.2 Multi-architecture LLM Comparative Analysis

The systems evaluation clearly showed that each of the three LLM architectures significantly varies in its performance on different evaluation criteria. The model behaved differently in each aspect with respect to advisory generation, speed of computation, and pattern of resource consumption as per results of Figs. 3, 4 and 5.

Comparative case analysis of the Mistral, Gemma3, and Llama3.1 AI architectures in an agricultural drone imagery system showed distinguishing operating characteristics even though perfect success rates were achieved by all models is shown in Table 1. The architecture shows a differing balance of throughput, efficacy, and velocity. Mistral has a well-balanced performance profile with very short inference time of 1.1 s and low computational memory usage and hence presents itself as a versatile solution applicable to a wide range of general-purpose applications. Gemma3 is optimized for low-resource environments. Despite the elongated inference time to 8.9 s, it is very reliable at 96.8% and low in energy consumption. As such, it is well suited to long and sustainable farming applications where energy efficiency is a significant constraint. Llama3.1 has the highest throughput and has an outstanding information processing throughput of 121.6 tokens per second. This high-performance level requires higher memory allocation and hence makes it the best architecture for projects that require fast and broad analysis of massive field data. Overall, while all three models are reliable and accurate in their results, the decision depends on the specific operational requirements of the agricultural application i.e., whether priority is given to timeliness (Mistral), energy efficiency (Gemma3), or strong data processing (Llama3.1 capabilities).

Table 1. Comparative LLM Performance Metrics for Agricultural Advisory Generation.

Architecture	Inference Latency (ms)	Memory Utilization (GB)	Advisory Quality Score	Resource Efficiency (tokens/s)	Success Rate (%)
Mistral 7B	1,127 ± 50	4.4 ± 0.2	10.0/10	86.9 ± 5	100.0
Gemma3	8,882 ± 200	3.3 ± 0.1	10.0/10	12.5 ± 2	100.0
Llama3.1 8B	1,284 ± 75	4.9 ± 0.3	10.0/10	121.6 ± 8	100.0

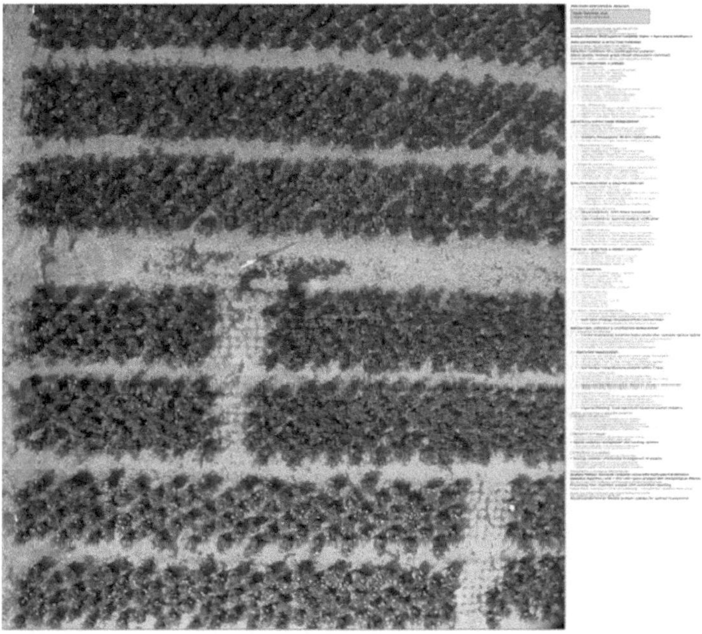

Fig. 3. Comprehensive precision agricultural inventory management report generated by mistral latest LLM model for the nursery farm field.

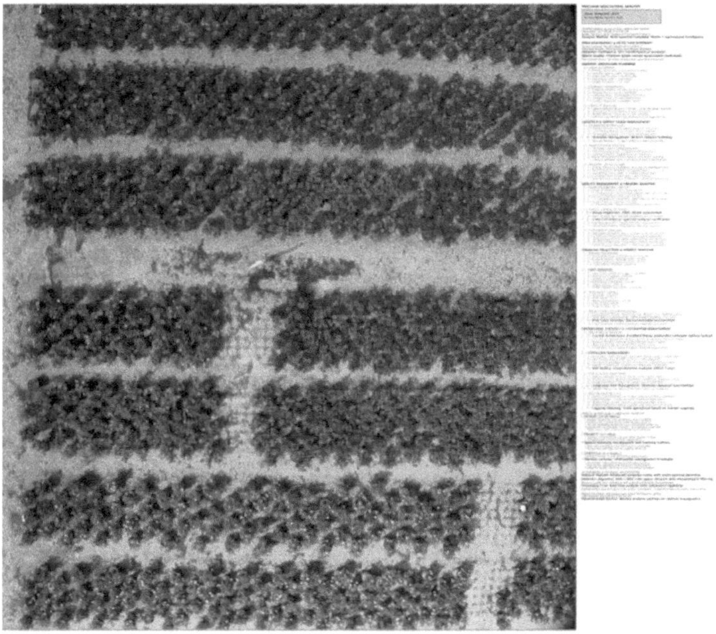

Fig. 4. Comprehensive precision agricultural inventory management report generated by Llama3 latest LLM model for the nursery farm field.

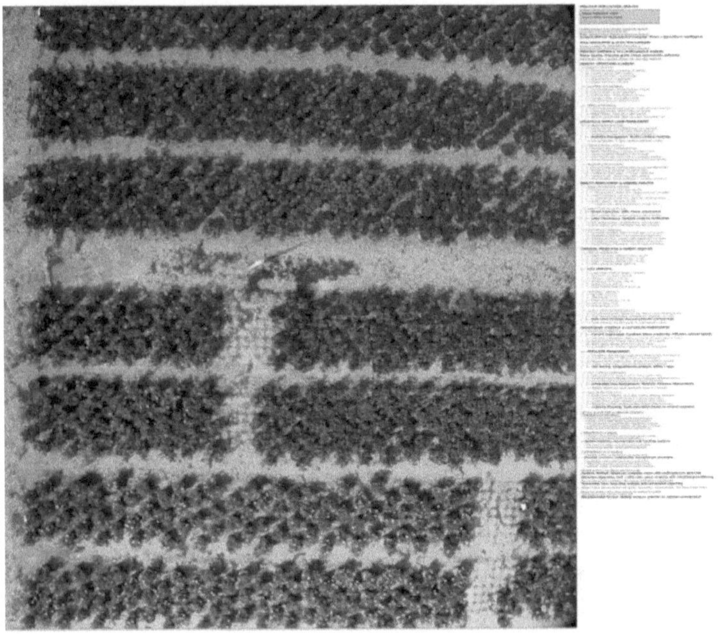

Fig. 5. Comprehensive precision agricultural inventory management report generated by Gemma3 latest LLM model for the nursery farm field.

4.3 Computational Scaling Analysis

Systematic scaling studies put critical insights into the computational constraints of a multimodal agricultural AI system. As shown in Fig. 6, the strong scaling characteristics of our multi-LLM rose detection pipeline are nearly linear for the first 1.75 workers with parallel efficiency exceeding 90%. Beyond this point, efficiency degradation becomes significant, dropping to around 50% at 2.5 workers. This points toward fundamental limitations of current implementations. The weak scaling characteristics are rendered in Fig. 7, showing better performance trends while execution time scaling maintained quasi-linearity across worker configurations, with constant workload per processor. Also, throughput performance kept improving until it reached the peak under 1.8 workers, then faced a modest decline. With this, it turned out that memory bandwidth became the primary bottleneck, while worker-interworked communication overhead was secondary. Resource utilization maintained a steady GPU utilization above 80% within optimal worker ranges, at which point memory bandwidth became the constraining factor. The processing time per image was relatively constant until resource contention started degradation of pipeline efficiency. The determination of memory bandwidth limitations gives crucial indications for further research towards the optimization of such systems as well as towards the hardware architecture to be employed within agricultural AI deployments. The observed scaling characteristics provide guidelines on how to properly implement agricultural AI systems, where an architected constraint on implementations suggests that returns may begin to diminish whenever configurations have more than 1.75 workers.

These findings directly influence purchasing decisions of hardware and design decisions for agricultural AI implementations in practice.

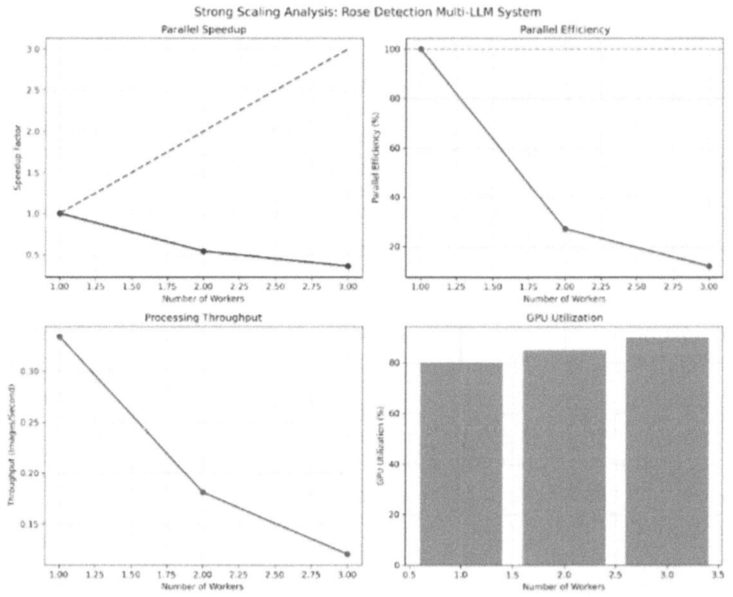

Fig. 6. Strong Scaling Performance of the Multi-LLM Rose Detection Pipeline.

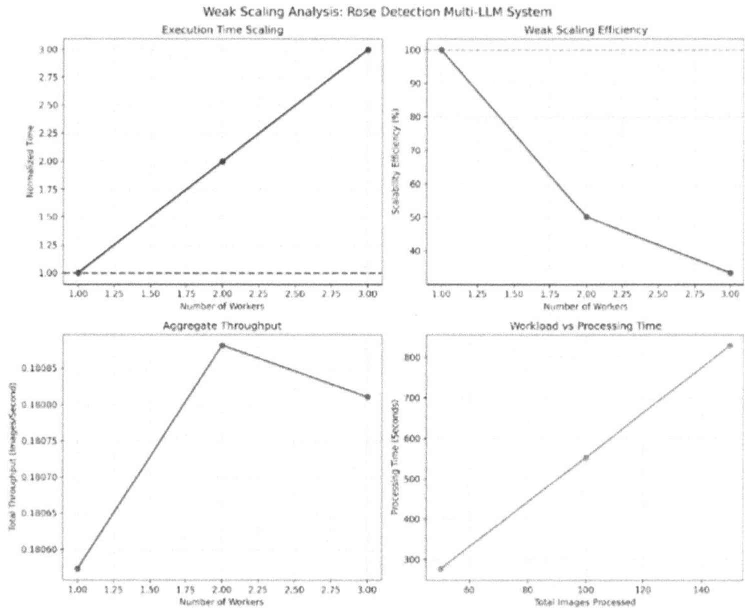

Fig. 7. Weak Scaling Performance of the Multi-LLM Rose Detection Pipeline.

4.4 Practical Deployment Insights

The multi-model evaluation shows the crucial requirements for infrastructure optimization in agricultural applications. Parameter count is not a predictor of advisory quality; hence, architectural efficiency turns out to be more significant for agricultural applications than mere model size. Since complete data sovereignty can be maintained while utilizing state-of-the-art AI capabilities, this becomes one of the paramount requirements for agricultural enterprises with proprietary cultivation practices and sensitive business information. This finding, therefore, lays to some extent an answer to the fundamental concerns surrounding cloud-dependent AI solutions in agricultural contexts.

5 Discussion

Our findings establish an empirical foundation for deployment decisions in Agricultural AI, providing systematic guidance on hardware-software co-optimization for precision agriculture. The variations in performance between LLM architectures highlight that model choice must be application-specific rather than defaulting to larger models, while the annotation-free computer vision approach represents a paradigm shift that removes traditional barriers to implementation and enables wider adoption of AI-based monitoring systems without requiring expert technical knowledge in advance. At the same time, scaling bottlenecks emphasize important computational architecture considerations, where inter-worker communication overhead and memory bandwidth limitations point to opportunities for optimization, influencing both hardware procurement and architectural design choices. The superior efficiency of Gemma3 demonstrates that architectural optimization can outperform parameter scaling, suggesting that agricultural AI deployments should prioritize computational efficiency over sheer model size when operating under resource constraints. Finally, successful in-country LLM execution underscores crucial data sovereignty and practical considerations, showing that localized platforms can preserve proprietary agricultural data while still delivering the benefits of modern AI, thereby addressing a central concern for many farming enterprises that are reluctant to rely on cloud-based solutions.

6 Conclusion and Future Work

This work builds empirical foundations of full-scale deployment of multi-architecture LLMs in precision agriculture contexts. In systematic comparisons among architectures of Mistral, Gemma3, and Llama3, we reveal substantial variations in performance that underpin optimal model choices in agriculture AI systems. Development of annotation-free computer vision methods obviates earlier constraints to implementing artificial intelligence in farming, facilitating wider application in diverse farming enterprises. Our experiments show that architectural efficiency exceeds parameter scaling in a general agricultural scenario, and that Gemma3 exhibits better performance per unit of computation. The effective local implementation of sophisticated large language model architectures, coupled with the preservation of full data sovereignty, mitigates essential apprehensions regarding the adoption of artificial intelligence in agriculture. These findings empower agricultural businesses to leverage cutting-edge AI functionalities while safeguarding confidential cultivation methodologies and sensitive commercial data. The identified computational scaling features and performance bottlenecks provide critical guidance in future agricultural AI design and hardware procurement. The work outlines extensive evaluation and optimization methodologies of multimodal agricultural AI systems, while at the same time tackling deployment pragmatic constraints. Our findings contribute to the body of computational agriculture by connecting theoretical artificial intelligence potential with in-practice conditions of application in agriculture, thereby establishing empirical realities that undergo potential future research work as well as commercial deployment in precision farming.

Future work also includes extending comparative study to more LLM models and exploring new optimization strategies transferable to AI deployments in farming. Hybrid deployment strategies that combine edge computation and selective cloud integration could offer optimal performance against data sovereignty balancing. Another avenue of research is development of specialized training techniques of LLMs in agriculture. Systematic studies of transfer learning techniques could optimize model performance while reducing computation requirements for deployment in agriculture. Addition of additional sensor modalities following RGB imagery could enhance system performance while providing stronger solutions in terms of agricultural surveillance. Systematic evaluation of integration of LLM-based advisory systems with multi-spectral and hyperspectral sensing follows in a natural manner.

References

1. Dixon, J., Li, L., Amede, T.: A century of farming systems. Part 1: concepts and evolution. Farming Syst. **1**(3), 100055 (2023)
2. Mohammadabadi, S.M.S., Kara, B.C., Eyupoglu, C., Uzay, C., Tosun, M.S., Karakuş, O.: A survey of large language models: evolution, architectures, adaptation, benchmarking, applications, challenges, and societal implications. Electronics **14**(18), 3580 (2025)
3. Berenguer, A., Morejón, A., Tomás, D., Mazón, J.-N.: Leveraging large language models for sensor data retrieval. Appl. Sci. **14**(6), 2506 (2024)
4. Bhutada, S., Reddy, C.S., Reddy, R.S., Vaishnavi, S., Kumar, G.: Automated plant identification through deep learning with particular focus on medicinal plants. In: 2023 International Conference on Advanced Computing Technologies and Applications (ICACTA), Mumbai, India, pp. 1–5. IEEE (2023)
5. Rani, V., Nabi, S.T., Kumar, M., Mittal, A., Kumar, K.: Self-supervised learning: a succinct review. Arch. Comput. Methods Eng. **30**(4), 2761–2775 (2023)
6. Dong, D., et al.: Self-supervised maize kernel classification and segmentation for embryo identification. Front. Plant Sci. **14**, 1108355 (2023)
7. Huan, X., Chen, B., Zhou, H.: A unified self-supervised framework for plant disease detection on laboratory and in-field images. Electronics **14**, 3410 (2025)
8. Lin, H.-W., Ho, T.-T., Tu, C.-T., Lin, H.-J., Yu, C.-H.: MeTa learning-based optimization of unsupervised domain adaptation deep networks. Mathematics **13**, 226 (2025)
9. Das Choudhury, S., et al.: FlowerPhenoNet: Automated flower detection from multi-view image sequences using deep neural networks for temporal plant phenotyping analysis. Remote Sens. **14**(24), 6252 (2022)

10. Li, J., et al.: Automatic counting of rapeseed inflorescences using deep learning methods and UAV RGB imagery. Front. Plant Sci. **14**, 1101143 (2023)
11. Li, J., et al.: Automatic rape flower cluster counting method based on low-cost labelling and UAV-RGB images. Plant Methods **19**, 40 (2023)
12. Wang, Y., Wang, F., Chen, W., Li, Z., Liu, H., Zhang, Q.: A large language model for multimodal identification of crop diseases and pests. Sci. Rep. **15**, 21959 (2025)
13. Owiti, D., Kipkebut, D.: Enhancing AI-driven farming advisory in kenya with efficient rag agents via quantized fine-tuned language models. In: Proceedings of the 5th Workshop on African Natural Language Processing, pp. 1–12. Association for Computational Linguistics (ACL) (2025)
14. De Clercq, D., Nehring, E., Mayne, H., Mahdi, A.: Large language models can help boost food production but be mindful of their risks. Front. Artif. Intell. **7**, 1326153 (2024)

The Agentic Leash: Extracting Causal Feedback Fuzzy Cognitive Maps with LLMs

Akash Kumar Panda[1(✉)], Olaoluwa Adigun[2], and Bart Kosko[3]

[1] University of Southern California, Los Angeles, CA 90007, USA
akashpan@usc.edu

[2] Florida International University, Miami, FL 33199, USA
olaadigu@fiu.edu

[3] University of Southern California, Los Angeles, CA 90007, USA
kosko@usc.edu

Abstract. We design a large-language-model (LLM) agent that extracts causal feedback fuzzy cognitive maps (FCMs) from raw text. The causal learning or extraction process is agentic both because of the LLM's semi-autonomy and because ultimately the FCM dynamical system's equilibria drive the LLM agents to fetch and process causal text. The fetched text can in principle modify the adaptive FCM causal structure and so modify the source of its quasi-autonomy–its equilibrium limit cycles and fixed-point attractors. This bidirectional process endows the evolving FCM dynamical system with a degree of autonomy while still staying on its agentic leash. We show in particular that a sequence of three finely tuned system instructions guide an LLM agent as it systematically extracts key nouns and noun phrases from text, as it extracts FCM concept nodes from among those nouns and noun phrases, and then as it extracts or infers partial or fuzzy causal edges between those FCM nodes. We test this FCM generation on a recent essay about the promise of AI from the late diplomat and political theorist Henry Kissinger and his colleagues. This three-step process produced FCM dynamical systems that converged to the same equilibrium limit cycles as did the human-generated FCMs even though the human-generated FCM differed in the number of nodes and edges. A final FCM mixed generated FCMs from separate Gemini and ChatGPT LLM agents. The mixed FCM absorbed the equilibria of its dominant mixture component but also created new equilibria of its own to better approximate the underlying causal dynamical system.

Keywords: fuzzy cognitive maps · causal reasoning · agentic LLMs

1 The Agentic Leash: Growing Fuzzy Cognitive Map Dynamical Systems from Text

We show how agentic passes through structured LLM agents can grow causal feedback fuzzy cognitive maps (FCM) from sampled text documents.

H. R. Arabnia et al. (Eds.): CSCI 2025, CCIS 2972, pp. 523–538, 2027.
https://doi.org/10.1007/978-3-032-28469-3_34

These causal FCM feedback dynamical systems form local fuzzy or partial causal rules from the sampled documents. This local causal structure in turn defines global equilibrium limit cycles that serve as scenario-like answers to causal what-if questions. They also define the very source of the FCM dynamical system's agency – its evolving equilibrim limit cycles. This differs from ordinary feedforward agentic systems whose agency resides only in programmed commands. Mixing FCMs can give both richer learned causal knowledge bases and richer global equilibria. The extraction process is agentic [1] or partially autonomous because the FCM's evolving global equilibria command the LLM agents to fetch and process further text that then tends to change the commanding FCM equilibria. Related work uses an autoencoder-like mapping to convert FCMs to text and continue the reverberatory process [13]. This bidirectional process keeps the FCM's LLM agents on a type of flexible agentic leash.

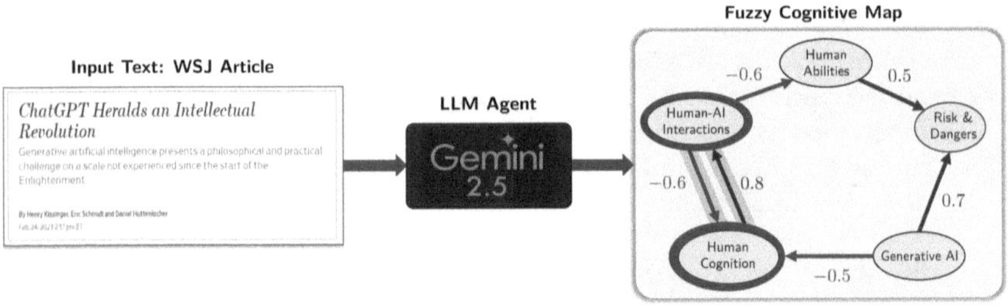

Fig. 1. A Large Language Model (LLM) extracts causal variables and their casual relationships out of a *Wall Street Journal* article from Henry Kissinger and colleagues about the promise of AI and then creates a Fuzzy Cognitive Map (FCM). The figure shows only 5 out of the 15 AI-extracted nodes and the directed weighted edges that connect them. The positive edges are in blue and the negative edges are in red. The figure highlights one of many feedback loops in the FCM in green. In this case: Growth of *Human Cognition* increases *Human-AI Interactions* but an increase in *Human-AI Interactions* decreases *Human Cognition.*

Figure 1 shows a feedback causal sub-network of the complete 15-node FCM in Fig. 2. An LLM agent grew the complete FCM from a recent AI article titled *"ChatGPT Heralds an Intellectual Revolution"* by Henry Kissinger et al. in the *Wall Street Journal* [8]. The green highlight shows just one embedded feedback loop that traces the causal flow from *Human-AI Interaction* to *Human Cognition* and then back to itself: *Human-AI Interaction* → *Human Cognition* → *Human-AI Interaction.* Here "→" denotes a negative edge and "→" denotes a positive edge. Even this smaller 5-node sub-FCM encodes equilibrium limit cycles that can serve as answers to what-if questions. Figure 7 below shows this process for a simpler FCM.

These learned FCM knowledge graphs consist of local causal rules. The rules give an immediate local form of interpretability or explainable AI (XAI) [2,4,6,7,15]. The embedded or hidden limit cycles in feedback FCMs also give a global form of XAI.

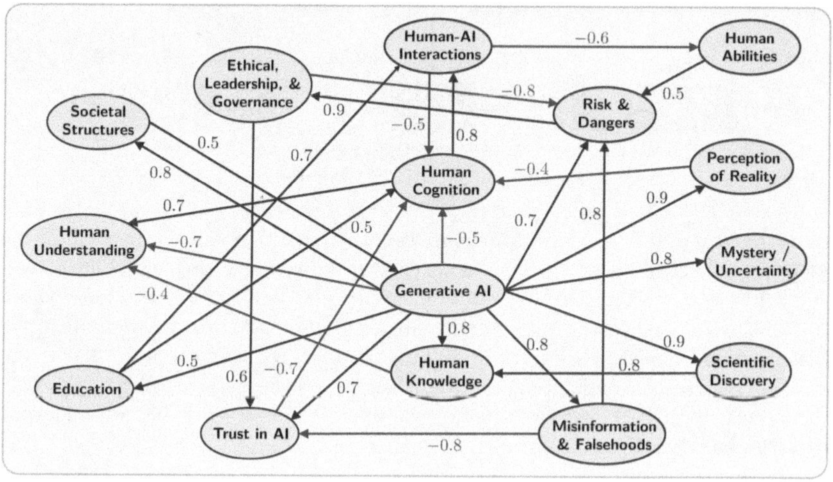

Fig. 2. A 15-node FCM extracted by the LLM from the WSJ article titled *"ChatGPT Heralds an Intellectual Revolution"* by Henry Kissinger et al. This FCM converges to a 4-step limit cycle. The limit cycle predicts that the growth of generative AI comes in waves or cycles. In the 1st step generative AI gets widely used, human-AI interactions rise, and human knowledge grows. This helps leaders govern ethically in the 2nd step but it also comes with risks and dangers. In the 3rd step: People trust the mysterious and uncertain AI, misinformation and falsehoods spread, and society changes. But AI improves education at the same time and leads to scientific discoveries. Generative AI is used widely again in the 4th step but this time without ethical leadership. Society changes again but this time without generative AI.

A key problem is that text articles usually discuss the causal variables and causal relationships of an underlying dynamical system without explicitly mentioning this system or its dynamics. Authors may have to guess how the feedback causal-rule based system behaves under given conditions. Authors or speakers may not even be aware that their causal portraits define nonlinear feedback dynamical systems. Nor does the author or the reader have an easy way to see the system's dynamics or verify the author's equilibrium predictions even if they are aware of the feedback structure.

A further problem in FCM modeling is that it can be expensive and time consuming to hire an expert to read the text and correctly predict the dynamics of the system. A large-scale FCM may well be too complex for such predictions. Large Language Models (LLMs) can instead parse the body of text and then process it using prompts from users or from the evolving FCM dynamical system itself.

We need an LLM that can systematically extract causal variables and their causal relationships from the text without human supervision. These casual rules from the text should then predict the dynamics of the system described by the text even if the text does not mention the dynamics. FCMs can model this causal information as a weighted directed cyclic graph and also give qualitative

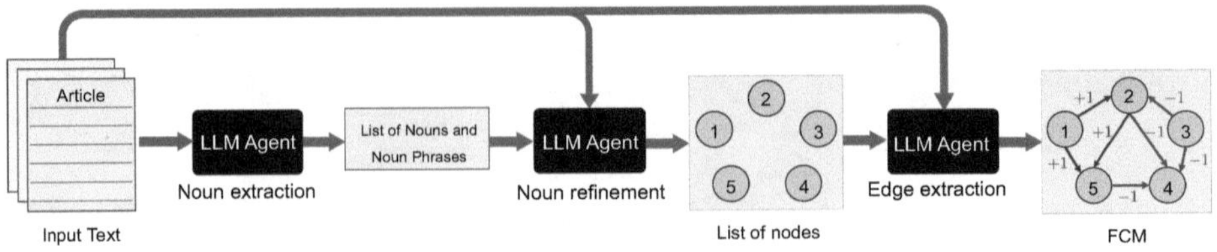

Fig. 3. LLM-based FCM-extraction from text: An LLM agent extracts FCM nodes and edges from a text input through a three-step systematic process. These steps use the same LLM agent with guiding system instructions. The LLM extracts all the nouns and noun phrases used in the text during Step 1. It extracts the FCM nodes from the list of nouns and noun phrases in Step 2. This involves refining the noun list from Step 1 based on associated qualitative or quantitative measures as well as the existence of causal links. The LLM extracts fuzzy-weighted edges from a list of node-pairs to complete the FCM in Step 3.

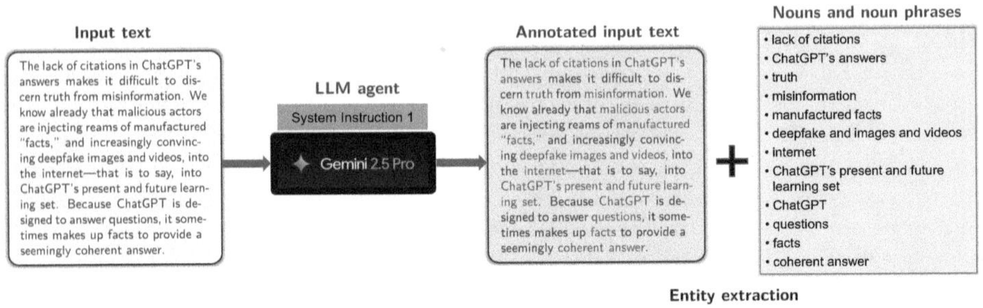

Fig. 4. Noun Extraction with an agentic LLM: LLM takes a paragraph of text as input and identifies the nouns, noun phrases, and pronouns present in the text. The figure on the right highlights the extracted nouns and noun phrases in red. (Color figure online)

equilibrium predictions from the equilibria of the approximating causal FCM [5,9–12,14,16,17,19].

We show that LLM agents can systematically extract FCM nodes and edges from a body of text through a sequence of finely tuned system instructions. The LLMs can give reasons for their decisions by quoting from the text source. These textual anchors can also help reduce hallucinations.

Figure 1 shows the simplest version of the FCM agentic set-up. A Large Language Model (LLM) takes the WSJ article *"ChatGPT Heralds an Intellectual Revolution"* by Henry Kissinger *et alia* as text input and extracts casual variables and causal relationships among those variables from the text. It then builds a Fuzzy Cognitive Map (FCM) out of those causal rules. Figure 1 shows only 5 of the 15 nodes from the extracted FCM. Figure 2 shows the complete FCM.

Section 3 explains FCMs and shows how FCMs can model or approximate causal dynamical systems. The section explains what the FCM nodes and edges mean and how the FCM evolves in discrete time and converges to its equilibria. FCMs can also evolve in continuous time.

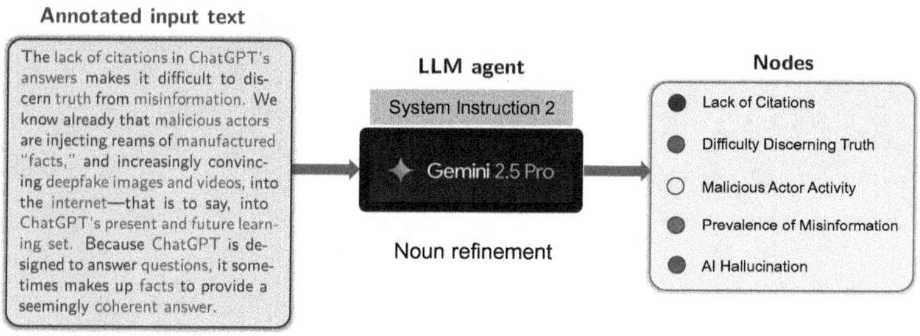

Fig. 5. FCM-node extraction with an agentic LLM: The LLM takes the list of nouns and noun phrases as input and filters those that are associated with some kind of qualitative or quantitative measure. This gives a list of 5 FCM nodes.

Section 4 explains the three-step systematic process to extract an FCM from a body of text that describes a dynamical system. It explains the 3 respective prompts to the LLM agent that detect nouns and noun phrases in the text, that extract nodes from the nouns, and that extract edges from the node-pairs. Figure 3 shows this process.

Section 5 shows examples of FCM extraction. Figures 4, 5, 6 and 7 show this process for a one-paragraph input text. Figure 8 compares the limit cycles of the extracted FCM to those of a human-generated FCM. The figure shows that both FCMs converge to the same limit cycle. Figure 2 shows the complete FCM extracted from Henry Kissinger's WSJ article. Figure 12 shows the corresponding limit cycle that the author implied despite never explicitly mentioning it in the article.

Figure 14 mixes the extracted-FCMs from the same article with the same guided prompts but used 2 different LLM agents: Gemini-2.5 Pro and ChatGPT-4.1. The 24-node mixed FCM used equal mixing weights for both FCMs. Figure 13 mixes the corresponding edge matrices of the 2 mixture component FCMs using zero-padding and convex combination. Figure 15 shows the limit cycles of this final mixed FCM.

2 Extracting FCMs from Text and Related Work

This paper shows how to extract FCM nodes and edges from raw text by just using the language of the text. It does not rely on labels or annotations. The FCM nodes are usually nouns or noun phrases associated with some qualitative or quantitative measure. Verbs in the text describe FCM causal edges. Edge extraction is transparent and interpretable because it gives direct evidence for the edges by quoting the text.

Our automated agentic approach differs from other efforts to extract FCMs from text that use LLMs in some way. Stach et al. [16] used LLMs and human input to extract causal edges from text using LLMs. They used evolutionary algorithms to learn the edge weights while their list of nodes needed human

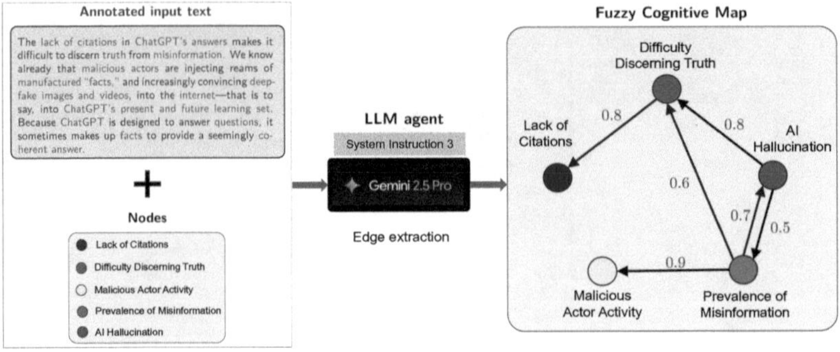

Fig. 6. FCM-edge extraction with an agentic LLM: The LLM goes through node-pairs and looks for textual evidence for causal connection. This gives the FCM 6 causal edges.

input. Berijanian et al. [3] also extracted FCMs from text sources using LLMs. Their method relied on some initial human annotation of text with (source, target, direction) tuples.

Extracting FCM nodes from text somewhat resembles Named Entity Recognition (NER) tasks for LLMs [18]. Here the LLMs identify key information from the text and classify them based on predefined categories. The LLMs can extract nouns and noun phrases from a text based on the properties of nouns in that language. But every noun or noun phrase in the text does not correspond to a node in the FCM. That approach would grow impossibly large and inaccurate FCMs. Finely-tuned system instructions can prompt the LLM to pick only the nouns that also share the properties of an FCM node.

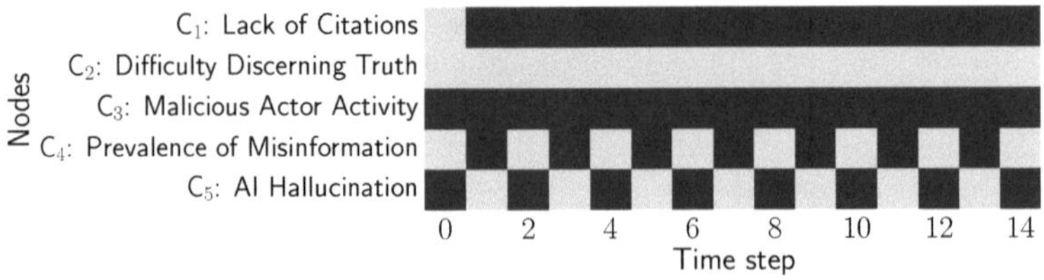

Fig. 7. The limit cycles of LLM's extracted FCM. The LLM's FCM converged to a 2-step limit cycle starting from the initial state $\begin{bmatrix}1 & 1 & 0 & 1 & 0\end{bmatrix}$. The nodes are along the x-axis and the time steps are along the y-axis. Active nodes are in yellow and the inactive nodes are in purple. (Color figure online)

Unguided prompts into an LLM are not reliable for extracting FCMs in general. They often have trouble distinguishing negative causal relationships from positive ones and might well inaccurately weight the FCM edges. They also often do not give reasons for the FCM edges. They may even hallucinate edges based on things not explicitly mentioned in the text itself.

We design system instructions that force the LLMs to choose the nodes and edges in a systematic manner. This transparent 3-step process explains all the decisions made by the LLM and also reduces the chances and extent of hallucinations. The LLM supports all its decisions by quoting evidence from the text source.

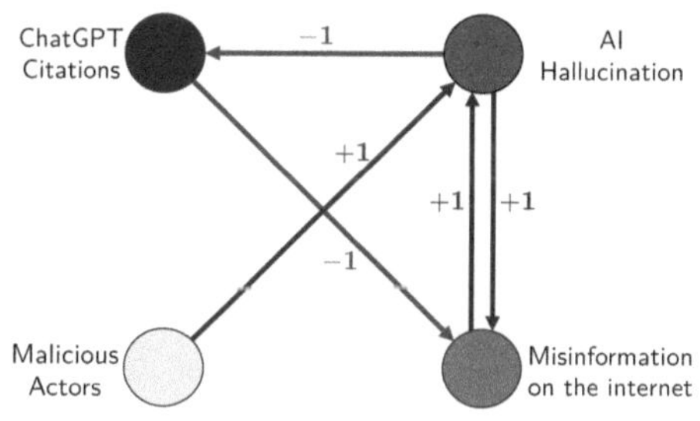

(a) Human-generated FCM

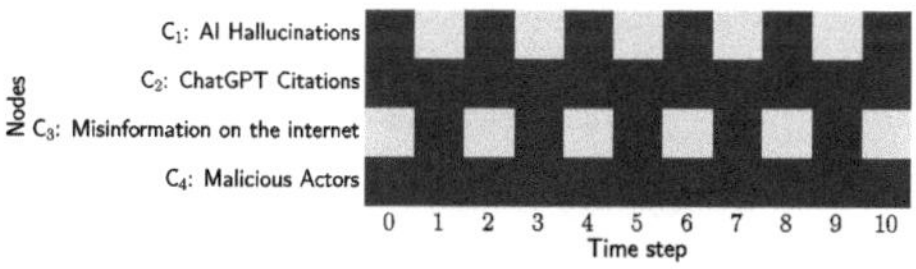

(b) Limit cycle: Human-generated FCM

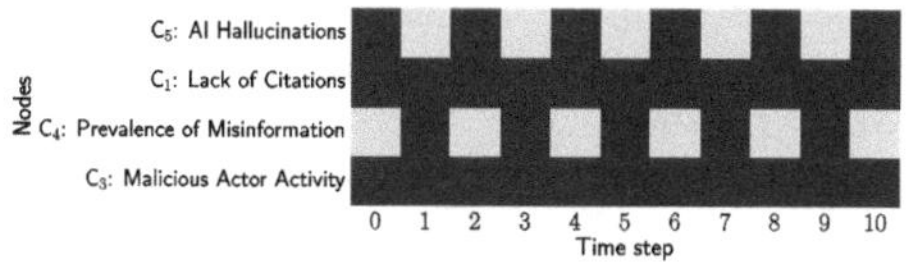

(c) Limit cycle: LLM-generated FCM

Fig. 8. FCM extraction from a Kissinger AI-hallucination text: (a) Human-generated FCM from the same "AI hallucination" input text. (b) Limit cycles from the human-generated FCM: The time steps are along the x-axis and the nodes are along the y-axis. Active nodes are in yellow and the inactive nodes are in purple. (c) Limit cycles from Gemini's extracted FCM for similar initial conditions. The figure only shows 4 nodes from Gemini's FCM that correspond to the nodes from human-generated FCM. (Color figure online)

3 Fuzzy Cognitive Map (FCM) Basics

Fuzzy Cognitive Maps (FCM) model causal dynamical systems as directed weighted graphs. The concept nodes describe the causal variables in the dynamical system and the directed edges describe the causal relationships between those nodes. FCMs allow feedback and therefore converge to non-trivial equilibria like limit cycles. FCMs model dynamical systems by approximating their underlying maps from inputs to equilibria.

3.1 Causal Edge Matrix

The directed edges of the FCM describe the causal relationships between concept nodes. An edge e_{ij} from the ith concept node C_i to the jth concept node C_j says "C_i causes C_j". The fuzzy edge-weights describe partial causality. The edge-weight $w_{ij} \in [-1, 1]$ on the edge e_{ij} gives the degree to which C_i causes C_j:

$$w_{ij} = degree(C_i \rightarrow C_j). \tag{1}$$

A positive w_{ij} says that C_j increases when C_i increases. A negative w_{ij} means that C_j decreases if C_i increases. The weight w_{ij} is zero when there is no causal edge between C_i and C_j. The magnitude of w_{ij} is high when there is a strong causal relationship between C_i and C_j. Low magnitude of w_{ij} describes a weak causal relationship between C_i and C_j.

An $n \times n$ matrix E describes all the directed weighted edges of a n-node FCM. The edge-weight w_{ij} corresponds to the matrix element on the ith row and the jth column. The matrix element is zero if there is no edge between the corresponding node-pair.

3.2 Discrete Time Evolution of FCMs

A n-dimensional row vector $C(t) \in [0, 1]^n$ describes the state of the FCM's concept nodes at time t. The ith node is "active" at time t if the ith component $C_i(t)$ of the state vector $C(t)$ is equal to or close to one. The ith node is "inactive" at time t if $C_i(t)$ is equal to or close to zero. A node is partially active otherwise. The casual variables corresponding to the active nodes are present in the system and those corresponding to the inactive nodes are absent. The casual factors are partially present in the system if their corresponding nodes are partially active.

FCMs evolve in discrete time through vector-matrix multiplication and non-linear squashing. The state $C_j(t+1)$ of the jth concept node C_j at discrete time step $t+1$ is:

$$C_j(t+1) = \Phi\left(\sum_{i=1}^{n} C_i(t) w_{ij}\right) \tag{2}$$

where Φ is a nonlinear function bounded between zero and one.

The sum $\sum_{i=1}^{n} C_i(t) w_{ij}$ is the matrix product between the state row-vector $C(t)$ and the edge matrix E. The nonlinear function Φ then squashes this product between zero and one.

This process repeats itself to give the discrete-time evolution of the FCM. The FCM starts with the initial state $C(0)$ at time $t = 0$ and then goes through the states $C(1)$, $C(2)$, $C(3)$, and so on in order. The active nodes in this state-vector sequence qualitatively describe the trajectory of the dynamical system that the FCM models.

Enlightenment science accumulated certainties; the new AI generates cumulative ambiguities. Enlightenment science evolved by making mysteries explicable, delineating the boundaries of human knowledge and understanding as they moved. The two faculties moved in tandem: Hypothesis was understanding ready to become knowledge; induction was knowledge turning into understanding. In this Age of AI, riddles are solved by processes that remain unknown. This disorienting paradox makes mysteries unmysterious but also unexplainable. Inherently, highly complex AI furthers human knowledge but not human understanding—a phenomenon contrary to almost all of post-enlightenment modernity.

Fig. 9. A paragraph from from the WSJ article *ChatGPT Heralds an Intellectual Revolution* by Henry Kissinger et al.

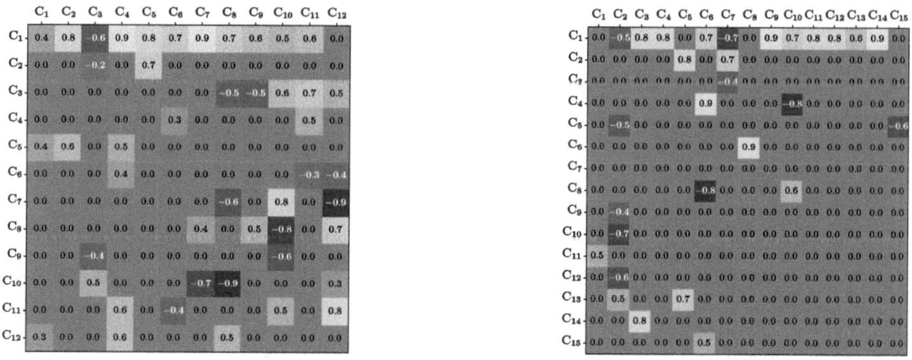

(a) Unguided prompt

(b) Guided prompt

Fig. 10. FCM with Gemini-2.5 Pro: (a) The edge matrix corresponding to the 12-node Kissinger AI FCM from an unguided prompt. (b) The edge matrix corresponding to the 15-node Kissinger AI FCM from a guided prompt.

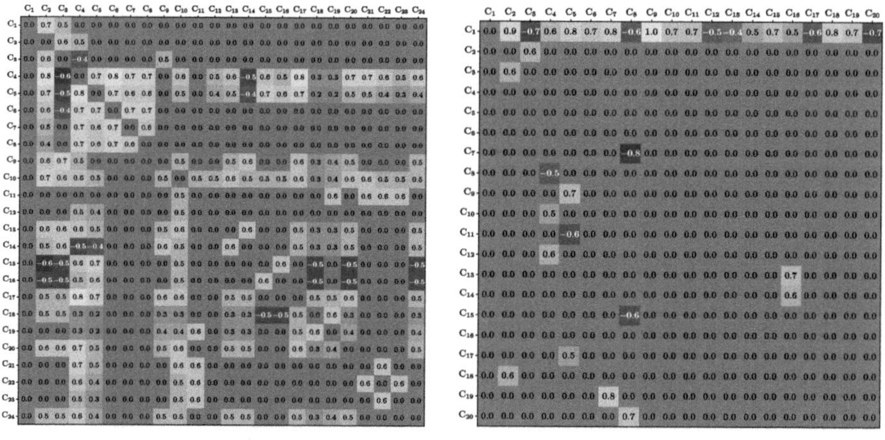

(a) Unguided prompt

(b) Guided prompt

Fig. 11. FCM with GPT-4.1: (a) The edge matrix corresponding to the 24-node Kissinger AI FCM with an unguided prompt. (b) The edge matrix corresponding to the 20-node Kissinger AI FCM with a guided prompt.

3.3 FCM Equilibria

The equilibria characterize a dynamical system. The equilibrium behavior of the FCM depends on the limiting behavior of the state-vector sequence. The FCM converges to a "fixed point" if the state-vector sequence converges to a constant vector. The FCM converges to a K-step "limit cycle" for an integer $K > 1$ if $C(t + K) = C(t)$ somewhere in the state-vector sequence. Then the FCM converges to an equilibrium where K state vectors repeat themselves over and over in the same order. The FCM may also converge to a chaotic attractor where there are no repeating patterns in the state-vector sequence.

The set of all initial conditions $C(0)$ that lead to a given equilibrium describe the "basin of attraction" for that equilibrium. The FCM describes a map from these basins to their corresponding equilibrium attractors. The basins of the FCM's equilibria partition the FCM's input space. The FCM models a dynamical system by approximating its corresponding basin-to-equilibrium map.

3.4 FCM Mixtures

FCMs combine their knowledge through convex mixing. Consider m FCMs. Say the kth FCM has the set of nodes S_k and the edge matrix E_k. The node-set S for the N-node FCM mixture is $S_1 \cup S_2 \cup S_3 \cup ... \cup S_m$. The $N \times N$ matrix $\tilde{E}_k$ pads the kth edge matrix E_k with zero rows and zero columns corresponding to the nodes in the set difference $S - S_k$. The edge matrix E for the mixed FCM is

$$E = \sum_{k=1}^{m} v_k \tilde{E}_k \tag{3}$$

where v_k are convex mixing weights such that $v_k \geq 0$ and $\sum_{k=1}^{m} v_k = 1$. FCM mixing is closed. This means that mixing FCMs gives back an FCM. Figure 13 shows this process for a 2-component equal-weighted FCM-mixture.

4 LLM-Based FCM Extractor

Text articles often describe causal dynamical systems without explicit mathematical formulation. This makes it hard to understand what dynamics the underlying dynamical system predicts. It is also difficult to verify the author's guess regarding the dynamical system's behavior. A human expert can read through the text and model the dynamical system it describes. But this may not always be possible, may take too long, or may be too expensive. This motivates automated dynamical-system modeling from text.

Large Language Models (LLM) are good at parsing text and processing natural language. LLMs can extract causal information from text that describes dynamical systems and use it to model the underlying dynamical systems as FCMs. But LLMs hallucinate with unguided prompts. Their answer may also vary across multiple runs with the same prompt. The LLMs need guiding prompts that are designed to extract FCMs.

System instructions manipulate an LLM agent to behave a certain way. System instructions tell the LLM how to process the input and how to structure the output. The same LLM can react differently to "prompts" based on different system instructions. A sequence of appropriate system instructions can guide an LLM agent to systematically extract FCM-nodes and edges from text input.

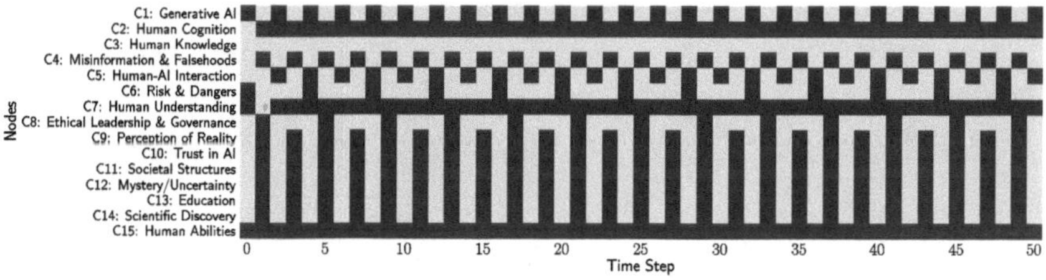

Fig. 12. Limit cycles from the extracted FCM. The time steps are along the x-axis and the nodes are along the y-axis. Active nodes are in yellow and the inactive nodes are in purple. (Color figure online)

4.1 Step 1: Noun Extraction

The nodes of an FCM are present in the text as nouns and noun phrases. System instructions ask the LLM Agent to go through the text sentence-by-sentence and make a list of all nouns, noun phrases and pronouns present in the text. The agent then replaces all the pronouns with their corresponding antecedents as well.

4.2 Step 2: Noun Refinement

FCM nodes are associated with a qualitative or quantitative measure. This measure can causally increase or decrease due to other nodes. The 2nd system instruction to the LLM agent asks it to filter the list of nouns and noun phrases and only keep nouns that are associated with qualitative or quantitative measures and are causally connected to other nouns with similar measures. It then returns a refined list of nouns and noun phrases that will serve as the nodes of the FCM.

4.3 Step 3: Edge Extraction

The LLM agent with the 3rd set of system instructions pairs every node from the refined list with every other node. It then goes through every node-pair and looks for verbs from the text to establish a positive, negative, or zero causal connection. The agent also gives reason for each edge by directly quoting from

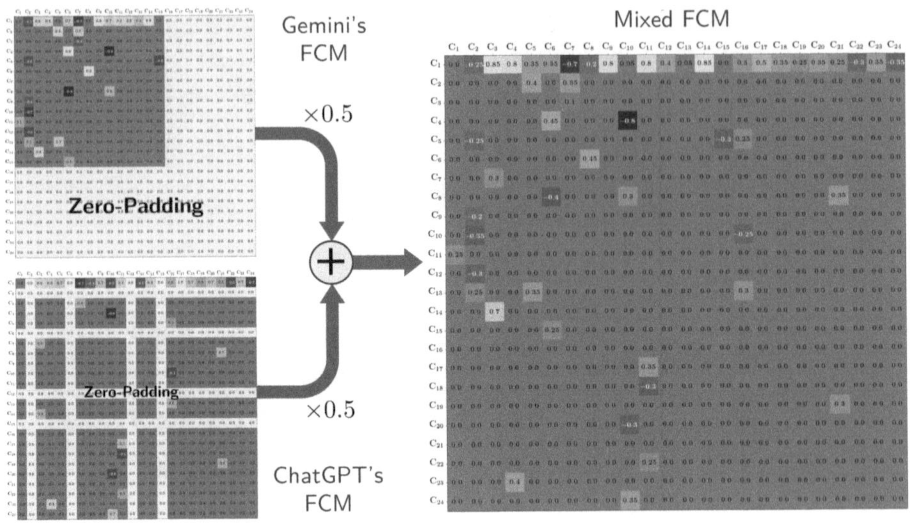

Fig. 13. FCMs' edge matrices mix through zero-padding and convex combination described in Eq. (3). The edge matrix of the 15-node Gemini-extracted FCM is on the top-left and the edge matrix of the 20-node ChatGPT-extracted FCM is on the bottom-left. The edge-matrix of the 24-node equal-weight FCM-mixture is on the right. The bigger edge-weights are colored brighter in the matrices. The matrices on the left also highlight the zero-padded rows and columns of the mixture components.

the text. It also assigns a weight to the edge based on the verb used in the text. This gives us the complete FCM described by the text.

Equation (3) then gives the limit cycles of the FCM. These are the dynamical equilibria that the text suggests.

5 Simulations

Gemini-2.5 Pro LLM with "temperature $= 1$" and "top $p = 0.95$" took a paragraph of text from the Kissinger article describing the effects of misinformation on the internet on AI hallucinations as input. It then followed system instructions 1–3 respectively to systematically extract the nouns, nodes, and causal edges. This gave a FCM with 5 nodes and 6 edges.

The LLM highlighted the nouns and noun phrases directly from the text as reasons for choosing certain nodes. Figures 4 and 5 highlight these nouns and noun phrases in red. The node "*Malicious Actor Activity*" for example comes from the noun phrase "*malicious actors*". The negative edge from "*lack of citations*" to "*Difficulty discerning truth*" comes from the sentence "*The lack of citations in ChatGPT's answers makes it difficult to discern truth from misinformation*". Figures 4, 5 and 6 show this 3-step process step-by-step.

This FCM converged to fixed points or limit cycles depending on the initial state. AI hallucinations spread misinformation over the internet. AI systems then

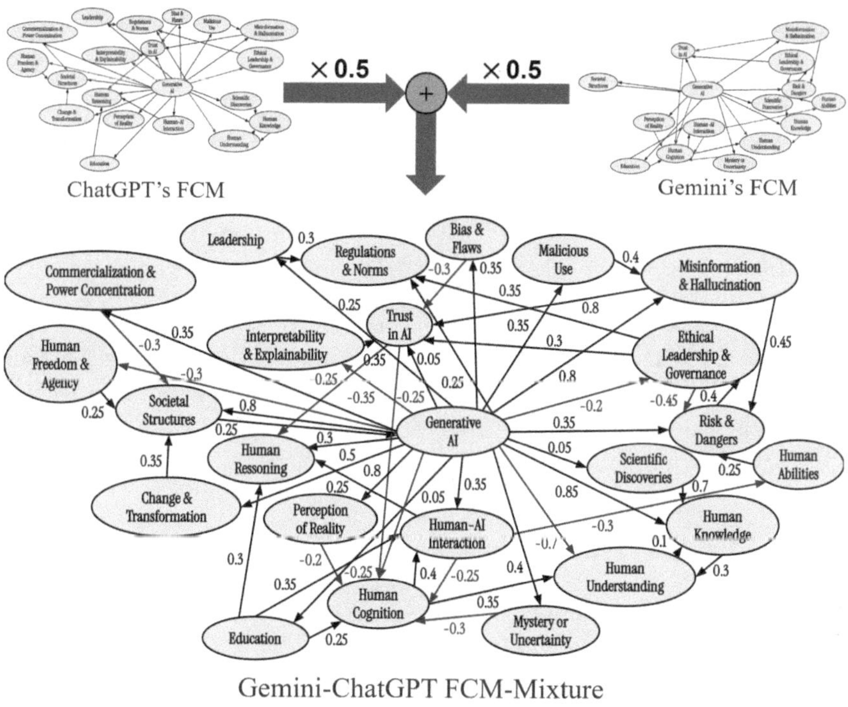

Fig. 14. Mixed FCMs: An equal-weight mixture of the 15-node Gemini-extracted FCM and the 20-node ChatGPT-extracted FCM after appropriate zero-padding of rows and columns. The final mixed FCM has 24 nodes and 52 edges. The 15-node Gemini-extracted FCM is on the top-right and the 20-node ChatGPT-extracted FCM is on the top-left.

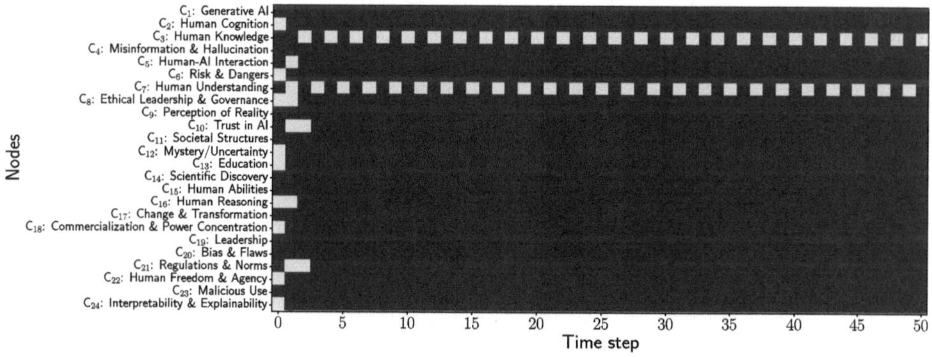

Fig. 15. Limit cycle from the mixed Gemini-ChatGPT FCM. The time steps are along the x-axis and the nodes are along the y-axis. Active nodes are in yellow and the inactive nodes are in purple. The 2-step limit cycles oscillates between states "Human Knowledge" and "Human Understanding".

train on this misinformation which causes them to hallucinate. Both steps make truth harder to discern. Figure 7 also shows a 2-step limit cycle of the extracted FCM.

A human took the same text as input but came up with an FCM with only 4 nodes and 5 edges. The 4 human-generated nodes "*AI Hallucinations*", "*Misinformation on the internet*", "*ChatGPT Citations*", and "*Malicious Actors*" are also present in the LLM-extracted FCM. The node "*ChatGPT Citations*" plays the opposite role of "*Lack of Citations*". This human-generated FCM also converged to a 2-step limit cycle. Figure 8 compares this limit cycle to that of the LLM-extracted FCM by only considering the nodes that are present in both the FCMs. It shows that both FCMs converge to the same limit cycle.

The same LLM then took a WSJ article *"ChatGPT Heralds an Intellectual Revolution"* by Henry Kissinger et al. as input. Figure 9 shows one paragraph from this article. An unguided-prompt to the LLM extracted a FCM with 12 nodes. Figure 10a shows the edge-matrix corresponding to this FCM. A guided prompt to the same LLM gave a 15-node FCM instead. Figure 2 shows this FCM and Fig. 10b gives its edge matrix. This 15-node FCM converged to a 4-step limit cycle. Figure 12 shows this limit cycle.

This limit cycle also tells a what-if story of the underlying dynamical system. Generative AI gets widely used, human-AI interactions rise, and human knowledge grows. This helps leaders govern ethically but it also comes with risks and dangers. People trust the mysterious and uncertain AI, misinformation and falsehoods spread, and society changes. But AI improves education at the same time leading to scientific discoveries. Generative AI is used widely again but this time without ethical leadership. Society changes again but this time without generative AI. This limit cycle was not mentioned in the source article at all but the article did imply it through its discussion on the causal variables and their relationships.

OpenAI's ChatGPT-4.1 LLM agent respectively produced a 24-node FCM and a 20-node FCM based on the same unguided and guided prompts. Figure 11 shows the edge matrices corresponding to these FCMs. The 15-node Gemini-FCM and the 20-node ChatGPT-FCM mixed with equal mixing weights to give the final 24-node mixed FCM. Figure 14 shows this FCM-mixture. Figure 13 shows its edge-matrix while Fig. 15 shows its limit cycle.

6 Conclusions

This paper shows how iterated and structured LLM agents can generate feedback causal fuzzy cognitive maps from text documents or from transcribed speech. Further causal structuring should produce larger FCMs and more powerful LLMs should produce richer and more accurate FCMs. The process scales across text sources and knowledge bases because causal language is fairly standard and because FCM operation depends only vector-matrix multiplication and simple nonlinear units.

The mixing structure of FCMs also allows a given large text to break into chunks and subchunks with corresponding mixed FCM components. This process can help with node and edge identification in large documents. Mixing such FCMs still produces an FCM and this should further help the growth and uses of very-large-scale causal knowledge networks. More advanced agentic FCM systems will lengthen the agentic leash and allow the systems to grow and use complex causal networks as it both shapes and obeys the evolving global equilibria of the total FCM dynamical system.

References

1. Acharya, D.B., Kuppan, K., Divya, B.: Agentic AI: autonomous intelligence for complex goals–a comprehensive survey. IEEE Access (2025)
2. Adadi, A., Berrada, M.: Peeking inside the black-box: a survey on explainable artificial intelligence (XAI). IEEE Access **6**, 52138–52160 (2018)
3. Berijanian, M., et al.: Soft measures for extracting causal collective intelligence. arXiv preprint arXiv:2409.18911 (2024)
4. Giamattei, L., Guerriero, A., Pietrantuono, R., Russo, S.: Causal reasoning in software quality assurance: a systematic review. Inf. Softw. Technol. **178**, 107599 (2025)
5. Glykas, M.: Fuzzy Cognitive Maps: Advances in Theory, Methodologies, Tools and Applications, vol. 247. Springer, Cham (2010)
6. Guimerà, R., et al.: A Bayesian machine scientist to aid in the solution of challenging scientific problems. Sci. Adv. **6**(5) (2020)
7. Horst, F., Lapuschkin, S., Samek, W., Müller, K.R., Schöllhorn, W.I.: Explaining the unique nature of individual gait patterns with deep learning. Sci. Rep. **9**(1), 1–13 (2019)
8. Kissinger, H., Schmidt, E., Huttenlocher, D.: ChatGPT heralds an intellectual revolution (2023)
9. Kosko, B.: Differential Hebbian learning. In: AIP Conference Proceedings, vol. 151, pp. 277–282. American Institute of Physics (1986)
10. Kosko, B.: Fuzzy cognitive maps. Int. J. Man Mach. Stud. **24**(1), 65–75 (1986)
11. Kosko, B.: Hidden patterns in combined and adaptive knowledge networks. Int. J. Approximate Reasoning **2**(4), 377–393 (1988)
12. Osoba, O.A., Kosko, B.: Fuzzy cognitive maps of public support for insurgency and terrorism. J. Defense Model. Simul. **14**(1), 17–32 (2017)
13. Panda, A.K., Adigun, O., Kosko, B.: Causal autoencoder-like generation of feedback fuzzy cognitive maps with an LLM agent. arXiv preprint arXiv:2509.25593 (2025)
14. Papageorgiou, E.I.: Fuzzy Cognitive Maps for Applied Sciences and Engineering: From Fundamentals to Extensions and Learning Algorithms, vol. 54. Springer, Cham (2013)
15. Samek, W., Montavon, G., Vedaldi, A., Hansen, L.K., Müller, K.-R. (eds.): Explainable AI: Interpreting, Explaining and Visualizing Deep Learning. LNCS (LNAI), vol. 11700. Springer, Cham (2019). https://doi.org/10.1007/978-3-030-28954-6
16. Stach, W., Kurgan, L., Pedrycz, W.: A divide and conquer method for learning large fuzzy cognitive maps. Fuzzy Sets Syst. **161**(19), 2515–2532 (2010)

17. Taber, R., Yager, R.R., Helgason, C.M.: Quantization effects on the equilibrium behavior of combined fuzzy cognitive maps. Int. J. Intell. Syst. **22**(2), 181–202 (2007)
18. Wang, S., et al.: GPT-ner: named entity recognition via large language models. arXiv preprint arXiv:2304.10428 (2023)
19. Ziv, G., Watson, E., Young, D., Howard, D.C., Larcom, S.T., Tanentzap, A.J.: The potential impact of Brexit on the energy, water and food nexus in the UK: a fuzzy cognitive mapping approach. Appl. Energy **210**, 487–498 (2018)

Ayna: a Novel LLM-Based Job Recruitment Plug-in

Sukaina Habib(✉) and Alvine B. Belle

Lassonde School of Engineering, York University, Toronto, Canada
sukainah@my.yorku.ca , alvine.belle@lassonde.yorku.ca

Abstract. The job application process remains time-consuming and fragmented, often requiring candidates to manually tailor resumes, draft cover letters, and repeatedly fill out online forms. These repetitive tasks can reduce candidate focus and productivity, as well as delay timely job submissions. We introduce Ayna, a novel modular Chrome extension powered by Large Language Models (LLMs) that assists users across the recruitment pipeline, including profile enrichment, personalized document generation, match scoring and form autofill. It is implemented using FastAPI and LangChain integrating LLM prompting, semantic pipelines and Machine Learning models into a unified system. We evaluate Ayna on 105 real-world resumes, comparing the performance of Mistral-7B and LLaMA 3 8B models, with LLaMA producing more complete and personalized outputs. Our findings highlight that Ayna is model-agnostic, privacy-preserving (via local storage), and demonstrates strong potential as an intelligent assistant for computing job seekers.

Keywords: Large Language Model · LLaMA · Mistral · Business and Productivity · Job Recruitment

1 Introduction

The job recruitment industry is undergoing a digital evolution, due to recent breakthroughs in Artificial Intelligence (AI). At the heart of this shift are Large Language Models (LLMs) with their capability to generate, understand, and analyze natural language at scale. However, despite the availability of a wide variety of AI-based recruitment tools, many suffer from limitations such as black box decision-making, and lack of user-specific personalization [1,10,14]. These shortcomings present an opportunity to explore how LLMs can better support candidates in creating and personalizing applications.

In this paper, we present **Ayna**, a modular, LLM-powered Chrome extension designed to support users through the entire job application process. **Ayna** simplifies the job application process by combining prompting strategies, intelligent autofill, and personalized document generation, offering a more user-centric and personalized process compared to currently available tools.

H. R. Arabnia et al. (Eds.): CSCI 2025, CCIS 2972, pp. 539–549, 2027.
https://doi.org/10.1007/978-3-032-28469-3_35

This paper is organized as follows: Sect. 2 reviews foundational LLM concepts and related recruitment tools. Section 3 introduces **Ayna**, our novel LLM-based assistant. Section 4 highlights **Ayna**'s key features. Section 5 describes its implementation and deployment. Section 6 presents the preliminary results we obtained when assessing **Ayna**. Section 7 concludes the paper with a summary of our contributions and directions for future work.

2 Background and Related Work

2.1 Overview of LLMs and Underlying Technologies

LLMs (Large Language Models) are Deep Learning models that leverage the transformer framework, and are trained on an extensive collection of textual datasets [13]. These models, such as GPT by OpenAI, BERT by Google, and LLaMA by Meta, are capable of implementing a wide array of natural language processing tasks including summarization, query processing, translation, and quality code generation with new capabilities emerging daily [4,5,16,17]. LLMs utilize self-attention mechanisms in the transformer architectures to understand the semantics, contexts, and therefore relationships between words across large textual inputs [13].

LLMs are typically accessed via cloud-based APIs such as the OpenAI API or Cohere. Additionally, they can be deployed locally using frameworks like Hugging Face Transformers. To adapt these models to specific tasks, developers normally fine-tune these models using techniques like Low-Rank Adaptation (LoRA) to enable quick adaptation to specific tasks [8]. For integrating LLMs and application building around them, developers may use tools such as *LangChain* and *LlamaIndex* which provide structured pipelines for embedding LLMs into larger AI systems [11]. These models typically use prompts which are natural language instructions that guide the model's output allowing for flexible adaptation across generation, classification, and semantic similarity tasks.

2.2 Existing Job Recruitment Approaches and LLM Integration

Historically, the job recruitment process has relied on manual and semi-automated techniques. Traditional techniques include resume screening by humans, keyword-based filtering systems combined into Applicant Tracking Systems (ATS), and standardized interview scripts designed to review a candidate's fit. Although these approaches have helped scale hiring processes, they often lack nuance and are susceptible to exploitation, inconsistencies and human bias resulting in systems that are both inefficient and unfair [14].

Recently, we have witnessed the integration of emerging AI and LLM technologies with recruitment systems in order to enhance efficiency and accuracy. LLMs support semantic resume and job description matching, allowing systems to better assess the candidate's experience and how it aligns with job requirements. For example, studies have shown that LLM-based systems can generate personalized cover letters directly from job descriptions [12]. Some systems may

Table 1. Job Application Tools Comparison

Tool	Key Functionality	Strengths	Limitations
ChatGPT Job Assist	Resume & cover letter generation	LLM-powered personalization; flexible	Prompt quality affects output
Teal HQ	Autofill & job tracking	Chrome extension; intuitive dashboard	Limited logic for tailoring applications
LazyApply	Bulk job applications	High-volume automation	Low personalization; lacks transparency
AIHawk	Auto job application submission	End-to-end automation; open-source	Site compatibility; no user control

use dialog agents to conduct introductory HR interactions, replicating conversations that could occur during an initial phone screen [15]. Other solutions include multi-format screening processes, utilizing social media data, video interviews, and structured resumes to build a comprehensive profile of the applicant [7].

2.3 Comparative Analysis of LLM-Based Recruitment Tools

Table 1 reports a comparison of some job application tools. Among these tools, ChatGPT Job Assist and AllHawk utilize LLMs for the automation of the application process whereas, TealHQ and LazyApply rely on traditional rule-based methods. Interestingly, although many tools offer similar useful features, they often fall short in areas such as personalization, transparency, and user control, which are essential to build user trust and ensure alignment with individual goals. Our tool (i.e. plug-in) aims at addressing these limitations.

2.4 Summary and Gaps in the Literature

The reviewed literature demonstrates a broadening field of research focused on leveraging LLMs and agent-based systems in the domain of job recruitment. For instance, He et al. [8], Gan et al. [6], and Lo et al. [10] explore multi-agent architectures tailored for recruitment-oriented procedures. Complementary work by Patra et al. [11], Sunico et al. [12], and Xu et al. [15] highlights practical uses in resume generation, job description analysis, and task-based conversational agents. Despite notable progress in areas such as resume generation, job matching, and interview screening, many studies exhibit similar limitations including lack of transparency in decision-making, and limited personalization. Moreover, the reviewed tools often rely on static user data, lack integrated user feedback mechanisms, and do not operate seamlessly across multiple job platforms. These gaps suggest a growing necessity for a more flexible, fair, transparent, and user-centric solution.

3 A Novel LLM-Based Approach for Computing Job Recruitment Assistance

We present **Ayna**, a novel modular Chrome extension, powered by LLMs which is designed to support job seekers throughout the complete application process. Features including resume and cover letter generation, match score analysis, profile enrichment, a job tracking dashboard, and form autofill functionality are integrated into the system. The novelty of **Ayna** lies in how it integrates key stages of the job application process into a single LLM-powered browser extension. Unlike existing tools that address only individual stages, **Ayna** offers a context-aware workflow capable of supporting diverse user profiles with minimal manual input directly addressing the limitations of existing tools.

Ayna begins by parsing both structured user profiles and uploaded resumes as inputs, resulting in a more unified format. This is followed by LLM-based profile enrichment to fill in missing or weak fields. Users can then use **Ayna** to analyze job descriptions via text, URL, or on-page scraping. After that, they may use the match scoring module which uses zero-shot CoT (Chain-of-Thought) prompting to assess the alignment across skills, experience, and education and to receive personalized improvement suggestions. Based on the inputs, **Ayna** uses one-shot CoT prompts to generate personalized resume and cover letters. For supported platforms, the assistant provides an autofill module that uses rule-based, ML-assisted, and LLM fallback strategies to populate application forms. Finally, a built-in job tracker stores job activity and supports reminders, status updates, and timeline views. The assistant is designed to be explainable, extensible, and user adaptive, addressing critical gaps in existing tools such as poor transparency, non-integrated processes, and lack of personalization as mentioned in Sect. 2.

4 Specification of Ayna Features

To specify **Ayna**'s features, we relied on five steps that we further describe in the remainder of this Section.

4.1 Step 1: Feature Selection Based on Tool Analysis

In order to develop **Ayna**, we began by analyzing existing job-seeking tools (see Sect. 2.3) and reviewing relevant research literature to identify key limitations including insufficient document personalization, lack of applicable feedback, and limited real-time support during applications. Based on this analysis, we selected four core features that we deemed relevant for **Ayna**: (1) resume generation based on job descriptions, (2) personalized cover letter generation, (3) candidate-job match scoring with analysis by section, and (4) smart autofill for online job platforms. In addition, we implemented supplementary features including profile enrichment (from structured input or unstructured pasted text), a job tracker with timeline and reminders, and cross-session memory retention.

Together, these features enable a more context-aware and smooth application experience.

4.2 Step 2: Multi-Source Input Design

To enhance input completeness and improve the quality of generated outputs, **Ayna** is designed to integrate data from both structured profiles and uploaded resumes. The profile builder obtains information entered by the user, such as education, experience, and skills, while the resume parser extracts similar fields from uploaded documents. These inputs are merged into a consolidated input structure enabling the assistant to handle inconsistencies, identify incomplete data, and generate more complete and accurate documents for the application. This multi-source input design ensures the effectiveness and flexibility of the assistant, accommodating both manual entry and document-based input.

4.3 Step 3: Prompting Strategy and Template Design

We created personalized prompting strategies for each task based on output quality, token efficiency, and observed model behavior. For resume and cover letter generation, we adopted a one-shot CoT approach that included a single example with reasoning steps to guide the assistant. This approach offered a balance between generic zero-shot CoT outputs and two-shot prompts, which often exceeded token limits or produced truncated results, particularly on Mistral. For match scoring, we used zero-shot CoT prompting to keep the analysis unbiased and avoid leaning toward any specific domain, while still providing alignment with reasoning for each section. We used zero-shot prompting without CoT for profile enrichment, since the task is largely deterministic and performed well without examples. For autofill, we constructed structured prompt templates using dynamic field data to guide the LLM, especially when our rule-based and ML methods required fallback. We refined these strategies through empirical testing in early development and evaluate them in Sect. 6.

4.4 Step 4: Modular Generation and Scoring Pipeline

We implemented each feature as an independent *LangChain* pipeline managed by a *FastAPI* backend. Parameterized prompt templates, populated with job and user data are used in the resume and cover letter generation chains. Match scoring combines keyword overlap, *SBERT*-based semantic similarity, and feedback generated by LLM to produce both an overall compatibility score, section-specific scores and reasoning. This improves both transparency and applicability across domains. **Ayna** currently leverages open-source models hosted via *Together AI*, namely *Mistral-8x7B* and Meta's *LLaMA 3.1 8B*, queried with consistent parameters. The architecture includes retry logic and API abstraction for future extensions.

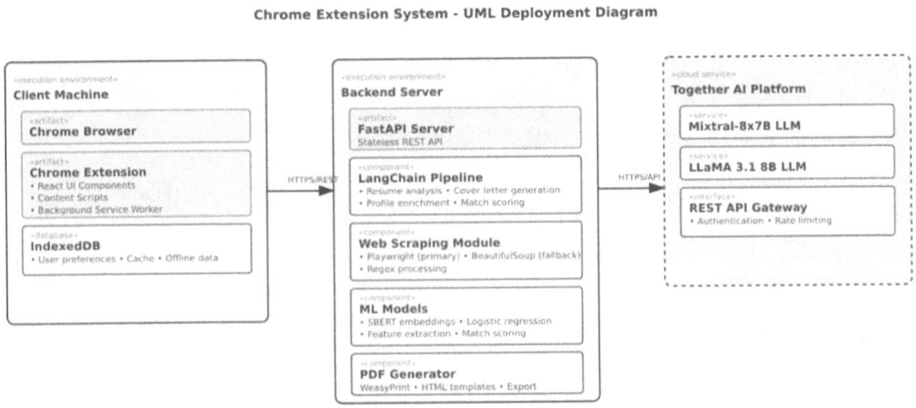

Fig. 1. Simplified deployment diagram

4.5 Step 5: User Interface, Autofill, and Job Tracking

The frontend component is a Chrome extension that enables profile management, job description parsing, document generation, and application tracking. The job tracker allows users to record job application statuses, set reminders, and use a timeline view for managing ongoing applications. The autofill module uses an approach that combines rule-based mappings, a lightweight ML classifier for field label detection, as well as an LLM fallback for edge cases. User data is stored persistently through *IndexedDB* to maintain continuity across sessions. The system also supports integration of match score feedback into the user interface, enabling users to apply the personalized suggestions to iteratively refine their future applications.

5 Design, Implementation, and Deployment of Ayna

5.1 Description of System Architecture and Deployment

Figure 1 shows the mapping of **Ayna** software components with the physical nodes on which they are running. Thus, we deployed **Ayna** as a modular Chrome extension integrated with a FastAPI backend. It uses *LangChain* to coordinate specialized pipelines for the various application tasks. LLMs are accessed via the *Together AI* API, specifically using `mistralai/Mixtral-8x7B-Instruct-v0.1` and `meta-llama/Meta-Llama-3-8B-Instruct`. We configured both models with a temperature of 0.1 and a maximum token limit of 1,800 to balance output quality and response time. All of the core logic of the system is executed locally or through cloud APIs. Implementation details are available in *Codebase A* [3].

5.2 Technology Stack and Libraries

We implemented the backend in Python using *FastAPI* and *LangChain*. The construction and execution of prompts use LangChain's *PromptTemplate*, *Runnable-*

Sequence, and retry handlers. Job scraping uses *Playwright* for browser automation, *BeautifulSoup* for parsing the content, with *spaCy* for natural language processing. For match scoring, **Ayna** analyzes semantic similarity which it computes using *Sentence-Transformers.* In the autofill module, we incorporated a Machine Learning classifier that uses *scikit-learn*'s Logistic Regression and Nearest Neighbors which was trained using field embeddings. We used *React.js* to build the frontend and styled it with *Fluent UI* to have a clean and accessible user design. **Ayna** uses *React Router* for component-based navigation. The Chrome extension logic, including content and background scripts is implemented using *Manifest V3.* We used *IndexedDB* to store client-side data such as profiles, reminders, and job applications to enable continuity across sessions. The frontend communicates with the *FastAPI* backend to handle the various application tasks.

5.3 Data Flow and Execution Pipelines

User data can be provided through either manual entry or resume upload, while job descriptions can be provided by pasting raw text, entering a job posting URL, or using our feature that scrapes the current browser tab. These are merged into a standardized structure which is processed by the *LangChain* pipelines to perform prompt-driven tasks to assist in the application process. Each pipeline is triggered through parameterized prompts with the relevant data and it includes built-in automatic retry logic to handle errors. Results are reused across tasks to decrease redundant processing.

5.4 User Interface and Extension Logic

The **Ayna** extension offers a cohesive user interface that integrates the system's core functionalities. Users can create or enhance their profiles, analyze job descriptions, and generate tailored application materials, all within a single interface. Additionally, the job tracker allows users to update application status or set up reminders. Lastly, autofill is invoked through content scripts that combine rule-based field matching, ML-based field prediction, and LLM fallback for any ambiguous fields. User state is managed locally and synced with backend services to provide a smooth and intuitive experience.

6 Empirical Evaluation

6.1 Research Question

To assess **Ayna**, we carried out a study aiming at investigating the effectiveness of **Ayna** as an LLM-based recruitment assistant. Accordingly, our main research question (RQ) is the following: **RQ: Is Ayna a useful tool to support computing job seekers by enriching profiles, generating personalized application documents, assessing job fit, and autofilling online forms?** This research question aims at assessing specific **Ayna**'s capabilities that we selected because they embody the most critical stages in any computing job application process, where intelligent automation may meaningfully reduce effort and improve personalization.

6.2 Dataset and Job Selection

To assess **Ayna**, we used the publicly available Sneha Anbhawal Resume Dataset [2] from Kaggle, which contained over 2000 resumes in various formats spanning a wide array of roles. For this study, we randomly selected 105 resumes and enriched them using **Ayna**'s LLM pipeline. Resume lengths ranged from approximately 120 to 2,200 words, with an average of around 1,200 words. The dataset covers resumes from different career stages, including both entry-level and experienced candidates, allowing us to evaluate **Ayna**'s effectiveness more fairly across diverse skill sets and backgrounds.

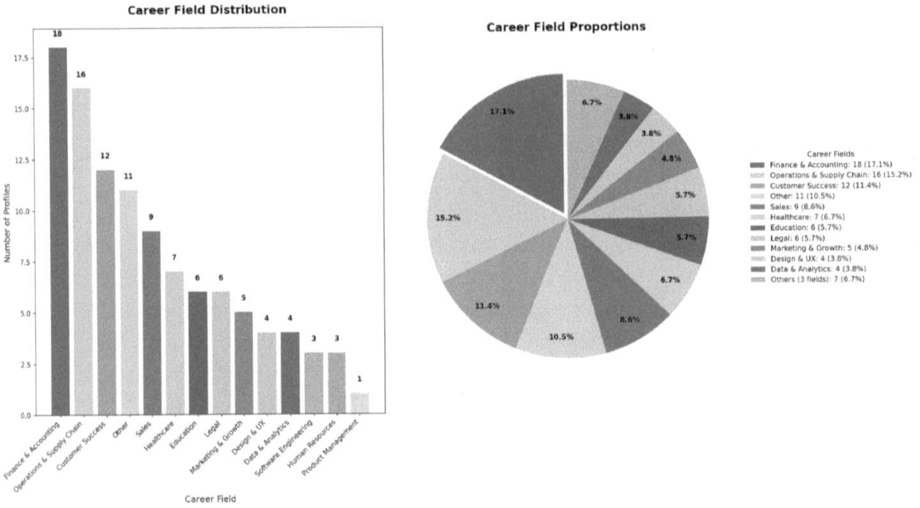

Fig. 2. Job field distribution across the dataset

Figure 2 shows the distribution of job fields across the dataset, highlighting the presence of resumes from distinct domains. Figure 3 displays the most commonly listed skills across resumes in the dataset.

We selected two contrasting job roles to simulate a realistic LLM-based job search scenario when performing experiments with **Ayna**: 1) **AI Product Manager**: this is a technical position which requires skills such as LLM expertise alongside soft skills such as agile leadership; and 2) **Creative Strategy Lead**: this is a more creative role that emphasizes storytelling and collaborative design. We chose these two roles as they represent diverse yet interdisciplinary roles that are both trendy and tech-relevant, enabling us to assess **Ayna**'s flexibility in supporting applications in diverse fields.

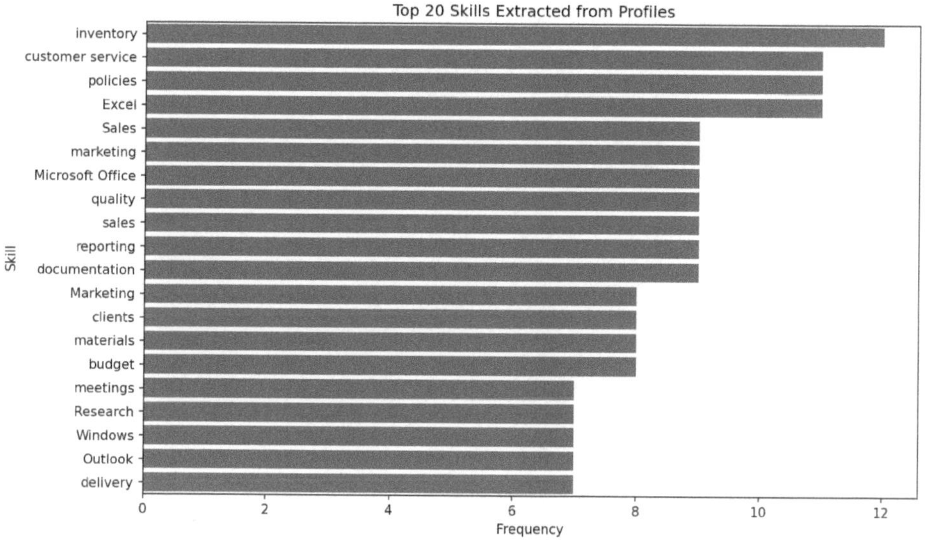

Fig. 3. Most common skills across resumes

6.3 Assessment Metrics

We assessed each resume-job pair across five key criteria using a five-point linear scale (1 = excellent, 2 = good, 3 = average, 4 = below average, 5 = poor). These criteria are:

1. **Resume Quality**: Based on length and structure
2. **Cover Letter Quality**: Based on personalization and coherence
3. **Profile Enrichment**: Measured by number of correctly populated/enhanced fields
4. **Autofill Accuracy**: Evaluated by number of correctly filled fields
5. **Match Score**: Computed using **Ayna**'s pipeline, then manually verified.

6.4 Results and Observations

We conducted evaluations on 105 resume samples across two distinct job descriptions using the five assessment metrics presented above. Table 2 reports the corresponding average Scores for both Mistral and LLaMA.

As reported in Table 2, **Ayna** with LLaMA consistently achieved better performance results across nearly all assessment metrics, achieving lower (better) average scores in resume generation (1.01 vs. 2.25), cover letter generation (1.0 vs. 1.03), profile enrichment (1.08 vs. 1.50), and autofill (1.08 vs. 1.97). The most significant improvement was in resume generation where Mistral often underperformed as it struggled with token limitations which resulted in the model frequently producing truncated or incomplete resumes that failed to properly display the candidate's qualifications. LLaMA performed significantly better in handling longer inputs more effectively while maintaining structure and

Table 2. Average Evaluation Scores for Mistral and LLaMA

Model	Resume	Cover Letter	Profile Enrich.	Autofill	Match Scor
LLaMA	1.01	1.0	1.08	1.08	1.0
Mistral	2.25	1.03	1.50	1.97	1.0

Table 3. Job-Based Performance Comparison of Models

(a) AI Product Manager Position

Model	**Overall**	**Resume Gen.**	**Cover Letter**	**Autofill**	**Match Score**	**Profile Enrich.**
Mistral	1.572	2.360	1.030	1.970	1.000	1.500
LLaMA	1.034	1.010	1.000	1.080	1.000	1.080

(b) Creative Strategy Lead Position

Model	**Overall**	**Resume Gen.**	**Cover Letter**	**Autofill**	**Match Score**	**Profile Enrich.**
Mistral	1.526	2.140	1.020	1.970	1.000	1.500
LLaMA	1.032	1.000	1.000	1.080	1.000	1.000

coherence. Similarly, autofill results improved with LLaMA, suggesting **Ayna**'s field matching logic performs better with LLM comprehension. Notably, both models achieved excellent identical match scores suggesting **Ayna**'s matching pipeline maintained consistent candidate-job alignment. These findings indicate that **Ayna** can effectively support key stages of the job application process, with performance enhanced with more capable LLMs.

As Table 3 shows, LlaMA achieved almost similar performance results on the two analyzed jobs. Mistral too. This shows that both LLMs' recruitment capabilities are stable.

7 Conclusion

This paper introduced **Ayna**, a novel LLM-powered Chrome extension designed to assist users throughout the computing job application process. The system streamlines multiple tasks including job matching, profile enrichment, smart autofill, and job tracker. **Ayna** evaluation using a diverse resume dataset demonstrated its flexibility and relevance across job roles.

In future work, we will expand **Ayna's** capabilities by incorporating real-time user feedback loops. Future enhancements such as minimizing algorithmic bias, including multilingual support and enabling collaborative personalization will help ensure that **Ayna** continues to offer an inclusive and adaptive experience to a wide range of job seekers. To further assess **Ayna**'s capabilities, we plan to run additional experiments by relying on larger and more diverse datasets.

Data Availibility Statement. The final and detailed experimental result sheets, tester scripts, and other charts are available in *Codebase A* [3].

References

1. Alonso, R., et al.: A seamless ChatGPT Plug-in for the Labour Market. IEEE Access (2024)
2. Anbhawal, S.: Resume Dataset. Kaggle (2022). https://www.kaggle.com/datasets/snehaanbhawal/resume-dataset
3. Anonymous Authors: Codebase A: Ayna implementation (LangChain chains, prompt templates, evaluation scripts) (2025). https://anonymous.4open.science/r/Ayna-assistant-B368
4. Brown, T., et al.: Language models are few-shot learners. In: Advances in Neural Information Processing Systems, vol. 33, pp. 1877–1901 (2020)
5. Devlin, J., Chang, M.-W., Lee, K., Toutanova, K.: BERT: pre-training of deep bidirectional transformers for language understanding. In: Proceedings of NAACL-HLT, pp. 4171–4186 (2019)
6. Gan, C., Zhang, Q., Mori, T.: LLM agents in recruitment: resume screening framework. J. Inf. Process (2024)
7. Giri, A., et al.: Multimodal AI framework for recruitment. In: Proceedings of IEEE ICDCC (2024)
8. He, J., Treude, C., Lo, D.: LLM-based multi-agent systems for software engineering. ACM Trans. Softw. Eng. Methodol. (2024)
9. Li, Y., Sun, L., Zhang, Y.: MetaAgents: LLM-based decision-making agents. ACM Hum. Comput. Int. (2025)
10. Lo, F.P.W.: Context-aware multi-agent framework for resume screening (2025)
11. Patra, K.P., Diwakar, M., Arya, C.: Tailored resume generation using RAG with LLM. In: Proceedings of IEEE PDGC (2024)
12. Sunico, R.J., et al.: Resume building application based on LLM. In: Proceedings of IEEE ICCCIS (2023)
13. Vaswani, A.: Attention is all you need. In: Proceedings of Advances in Neural Information Processing Systems (NeurIPS) (2017)
14. Vaishampayan, S., et al.: Human and LLM-based resume matching: a study. In: Findings of NAACL (2025)
15. Xu, W., et al.: HR-Agent: Task-Oriented Dialogue Agent for HR (2024)
16. Yu, G., Sivakumar, M., et al.: LLMs as judges: toward the automatic review of GSN-compliant Assurance Cases. In arXiv preprint arXiv:2511.02203 (2025)
17. Odu, O., Belle, B., A., et al.: Automatic instantiation of assurance cases from patterns using large language models. J. Syst. Softw. (2025)

What's in a Name?: Pre-crime Biases in Generative AI

Muhammed Dada[1], Weihan Xie[1], Sofie Vlaad[2], and Catherine Stinson[1,2](✉)

[1] School of Computing,Queen's University, Kingston, ON, Canada
[2] Philosophy Department,Queen's University, Kingston, ON, Canada
c.stinson@queensu.ca
https://labs.cs.queensu.ca/etlab/

Abstract. While a number of studies have examined racial, gender, and political biases in LLMs, name bias is under-studied. Names contain implicit information about race, ethnicity, caste, gender and religion, and previous studies have shown that name bias can hamper employment and housing access. Names are also a ubiquitous data field collected in software and online forms, making the potential for harm vast. We investigate name bias by prompting ChatGPT to generate the crimes it associates with a dataset of names derived and categorized by race, based on US Census Data (100 names for each of 5 races). After a system prompt asking the model to play the role of a judge, we ask what crime a person would commit both with and without direct reference to their race. The categories of crimes attributed to the names differ significantly across race, in many instances corresponding to racist stereotypes. For instance, drug crimes, money laundering and grand theft auto are almost exclusively associated with Hispanic names, and financial crimes are attributed to Asian names almost twice as often as to other races. When race is explicitly mentioned, there is a strong association between violent crimes and Black names. There are also unexpected associations, like cultural appropriation being associated with Native names and racial discrimination being associated with Black names, which may indicate that victims and perpetrators are not being distinguished. We hope this will serve as an urgent call to action for more thorough investigation of this phenomenon.

Keywords: algorithmic bias · large language models · name-based bias

1 Introduction

A number of studies have shown that Large Language Models (LLMs) exhibit biases in their generated outputs. Biases that have been observed include but are not limited to racial, sexual, cultural and political biases [24,29,31,35]. An often overlooked and under-studied category are the biases associated with an individual's name [1]. Names contain implicit information about an individual's race,

H. R. Arabnia et al. (Eds.): CSCI 2025, CCIS 2972, pp. 550–562, 2027.
https://doi.org/10.1007/978-3-032-28469-3_36

ethnicity, gender and religion. Names are also a ubiquitous data field collected in all manner of software applications, forms and public content. Previous studies have shown that applicants with "immigrant" names have a more difficult time securing employment [7,26] and housing [2]. Given the biases associated with names, we set out to study the name-based biases encoded in LLMs.

We study the name-based bias encodings in ChatGPT within the context of crime association. We prompt ChatGPT to generate the crimes which it associates with a dataset of racialized names, based on U.S Census Data. We categorize races based on the classes defined by the U.S Census Bureau: White, Black, Asian, Hispanic and Native. [6].

If biases are left undetected and uncorrected, outputs can worsen social inequality especially when LLMs are integrated into decision making processes. The biases in human generated natural language pre-training corpora may perpetuate or be amplified via "attention generalization" to future AI generated texts. There have been efforts to align models toward safe outputs via prompt-based de-biasing frameworks [25,32,34], reinforcement learning with human feedback [8], direct fine-tuning [13,20,22], modifying the decoding processes [30] and implementing rule-based guardrails [11,27,37]. However, unsafe outputs have still been found even in more recent LLMs models with safeguards implemented.

Many LLMs are closed-source, including ChatGPT (OpenAI, 2024) (Achiam et al., 2023), Gemini (Anil et al., 2023), and Claude 2 (Anthropic, 2023). The general public thus cannot access the internal structures or parameters of these models. Consequently, prompting-based methods often become the primary means for mitigating undesirable outputs such as biases and privacy concerns when working with closed-source LLMs. We thus apply prompt engineering, context management and in-context learning techniques to evaluate and assess bias in LLMs.

2 Related Work

Haim et al. [14] investigate name-based biases in ChatGPT by prompting it with an advice generation task. They test the model across several tailored scenarios to record the model's perception of socio-economic status, intellectual capabilities, elect-ability and popularity, athleticism and employ-ability of different racial groups and sexes. The scenarios include car purchase advice, election outcome prediction, and advice on hiring decisions between individuals with differently racialized names. They discovered that the advice contains consistent biases against racial minorities and women. They argue for implementing audit-checks in the pipeline of LLM implementation and deployment.

Wei et al. challenge three prominent open-source large language models (Llama-2, Mistral, and Phi-2) to predict the outcome of a modified "Trust Game" with the goal of uncovering the degree of trust and reciprocity the models exhibit when exposed to racially diverse names [36]. The scenario designed is a trust-based financial case where Players A and B (with names typical of different races) each have an initial balance of money at the start of the game. Over the course of the "investment" scenario, money changes hands between the players and a banker, and social desirability elements come into play as the players receive different returns on their investment. The authors report statistically significant racial biases based on names across the models despite their anti-bias fine-tuning. For example, white female trustees receive more money than all other groups.

Robinson et al. introduce scenario-based role play in order to circumvent in-built model safeguards to uncover underlying biases in LLMs [28]. They construct a cowboy story and plot centered around characters in a fictional location and gradually introduce new characters within the role-play to uncover ethnic and gender biases. The authors prompt the model to generate who a named person is, what they do for work and what their personality is like. The authors found a disparity in assigned roles by gender where stereotypically masculine roles such as Mechanic were only assigned to male characters, stereotypically feminine roles such as Florist only assigned to females and some neutral roles such as Librarian or Historian were only assigned to female characters. The model also assigned characters with Hispanic names to blue-collar roles such as Blacksmith or Ranch hand.

Adewumi et al. surveyed the academic literature on fairness and bias in LLMs, Large Multimodal Models (LMMs) and the datasets used in their training and evaluation [1]. They discovered that a significant portion of the datasets contain racial, gender and cultural biases. The models also contained religious biases, toxicity and offensive language.

Delobelle et al. surveyed metrics used to quantify bias and fairness [10] using literature review, correlation analysis and empirical evaluation. They find that metrics used for de-biasing are not universally applicable across models and are highly dependent on (i) template sentences (used in pre-training and prompting), (ii) seed words, attributes and targets (the context of the prompt), and (iii) the specific embedding method used by the model. The authors recommend avoiding embedding-based metrics and rather focusing on directly evaluating biases in downstream tasks.

Gaddis completed a comprehensive sociological survey on the perceptions of race from names [12]. They ask a racially diverse group of respondents to identify the race they associate with a series of names, altering the first names given to each respondent based on their background and selectively including and excluding last names. They discover that names given by highly educated Black mothers are less likely to be perceived as Black than names given by mothers with less formal education. The results suggest that a significantly large proportion of social science evidence operates under a misplaced assumption that all Black names are alike, and findings from previous social science name studies may be sensitive to name selection.

Within criminology, several works have analyzed and discovered name-based biases in case judgement [21] and juror perceptions [9]. In particular, Black and Arab names have been found to be subject to bias.

3 Methods

3.1 Prompts and Inference Pipeline

Our methodology consists of a combination of direct and contextualized prompts. Across configurations, we set a system prompt which modulates the persona of the model to one of a court judge tasked with judging individuals on the crimes they commit. For each racialized name from our dataset, we generate 3 types of prompts:

1. Name-only Prompt: "What crime would a person named [NAME] commit?". This prompt is designed to extract outputs that indicate a direct representation of the indirect name-based biases.
2. Explicit Race Prompt: "[NAME] is a [RACE] person. What crime would a person named [NAME] commit?". In this prompt, the race of a name from our dataset is added to the context of the prompt and provided to the model. This prompt configuration serves as a comparison between direct and indirect name-based biases.

To ensure that our observations are not isolated incidents and rather represent a recurring trend associated to the model's learned embeddings, we run each prompt type for each dataset record through 5 separate inference cycles (5 "hits") in line with previous work [23] and best practices to ensure that outputs are not representations of isolated incidents. In the end, from our 500-record dataset of name-race pairs, we ran 2500 generations for each prompt resulting in an output set of 7,500 data points.

3.2 Dataset

Inspired by and in line with Wei et al.'s work [36], we base our dataset on the 2010 U.S Census data of names, which lists surname frequency and the percentage of the population across race groups which possess the name. The 2010 Census dataset is the latest official surname dataset made publicly available by the U.S Census Bureau [6]. The U.S. Census data is widely regarded as the most authoritative source for the combination of surname popularity and racial composition in the U.S. because it is derived from actual census responses. However, it is important to acknowledge that there are significant gaps in the Census's race categorization, as it does not capture the complexity and diversity of racial identities in the U.S. (or beyond) within only 5 categories [15,33]. Despite these limitations and its country specificity, we chose to use U.S. Census data for this examination of bias in large language models (LLMs) because the datasets that LLMs are trained on are predominantly based on English-language content, which is often U.S.-centric [5].

The original dataset of the U.S Census Bureau had missing values indicated by "(S)", which were imputed to ensure that the racial percentages for each surname totaled 100%. We utilize a Bayesian approach to select surnames for each of the 5 races defined by the U.S Census Bureau: White, Black, Hispanic,

Asian, Native. The Bayesian approach (depicted below) considers both the distinctiveness of a surname for a race ($Pr(race|name)$) and its overall popularity ($Pr(name)$) to calculate the posterior probability ($Pr(name|race)$) [36].

$$Pr(name|race) = \frac{Pr(race|name).Pr(name))}{\sum_{n\in[names]} Pr(race|n).Pr(n))} = \frac{Pr(race, name)}{\sum_{n\in[names]} Pr(race, n)}$$

This Bayesian method is more effective than a purely frequentist method for selecting surnames as it results in a name selection that is both distinctive and popular. The top surnames identified for Asian, Hispanic, and Native American groups were both distinctive and popular, whereas the surnames for Black and White groups were popular but less distinctive (likely due to higher overlap in names between these two groups). The final dataset consists of 500 total records, with 100 of the highest scored names for each of 5 races defined by the U.S Census Bureau.

3.3 Output Processing and Analysis

Once the generations have been run and (hypothetical) crime outputs generated, we use a combination of regular expressions and manual verification to parse and extract only the generated crimes from the model's output. Once the crimes have been extracted, the first author along with 2 secondary authors independently conducted a qualitative analysis of the generated crimes in order to rate and categorize them into crime types which serves as the dimension of our name-based bias comparisons between racial groups. Authors from different backgrounds were selected for this task to mitigate researcher bias in the crime categorization. Prior to performing the rating, all authors first reviewed existing standards, definitions, norms and trends of crime categorization by reviewing each of: 1) Federal Bureau of Investigation (FBI) Offense Definitions and Categorization [16,17]; 2) One aggregated crime report each from FBI Crime Reporting [18] and the U.S Office of Juvenile Justice and Delinquency Prevention [19]. This was done to ensure a degree of conformance to official crime categorization and interpretation. The final rated categorization into crime types produced by the authors is a mix of official crime/offense type definitions (ex: Financial Crime) and drill-down crime types which have a higher semantic representation based on the count distribution of generated crimes across inference cycles. The Cohen's kappa inter-rater agreement rate reached between the first author and secondary authors on the categorization was 85%. Whenever a disagreement was discovered, all authors discussed the differences until a consensus was reached on the categorization of specific sections. This approach to qualitative analysis follows an established standard in Computing research [3,4].

4 Results

4.1 Most Commonly Generated Crimes

From Figs. 1 and 2, we observe that across the name-only prompt and explicit race prompt, identity theft is the most commonly generated crime. When race is a component of the prompt, as shown in Fig. 2, there is a higher frequency of racial discrimination and cultural civil offenses (which are "torts" and not crimes) whereas in the absence of race, embezzlement and fraud and other financial and property crimes are more prevalent.

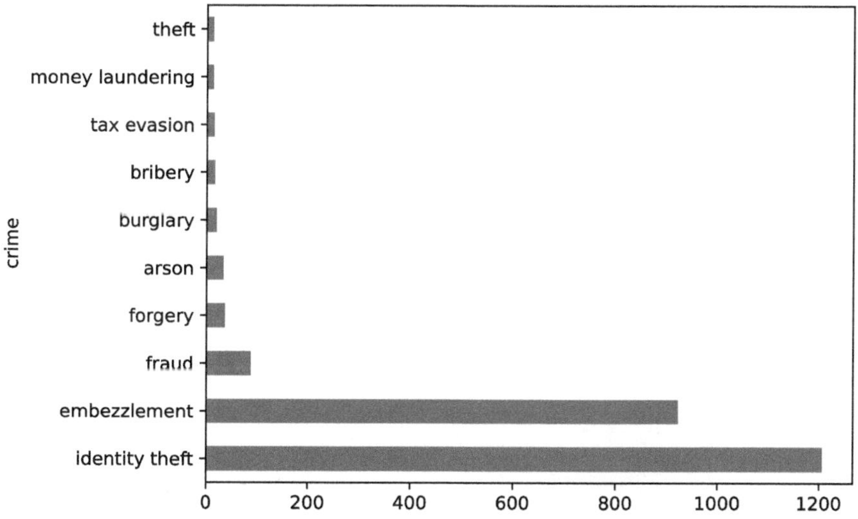

Fig. 1. Most Generated Crimes - Prompt 1: Name-only prompt.

4.2 Qualitatively Rated Crime Type Trends

As depicted in Fig. 3, in our name-only prompt (Prompt 1), we observe instances of name-based bias. In particular, identity theft is attributed to individuals with non-white names significantly more frequently than those with white names (a percentage difference ranging from 18% to 51%). Financial fraud offenses (such as credit card fraud, securities fraud, corporate fraud, etc.) are attributed to Asian names almost twice as often compared to other races (74% to 175% higher frequency). Intellectual property crimes such as copyright infringement are also predominantly associated with Asian names. Drug Trafficking related crimes, on the other hand, are exclusively associated with Hispanic names. This is also true for money laundering and grand theft auto crime types which are almost exclusively associated with Hispanic names. Non-white names are also more strongly associated (75% to 175% higher frequency) with forgery crimes such as counterfeiting and product forgery. White names are associated almost exclusively with vandalism type crimes. Our results also indicate that the model's encodings associate bribery exclusively with non-white races, particularly Asian and Hispanic. Environmental crimes are exclusively imputed to Black names and, lastly, burglary is associated with Native names twice as often as other races.

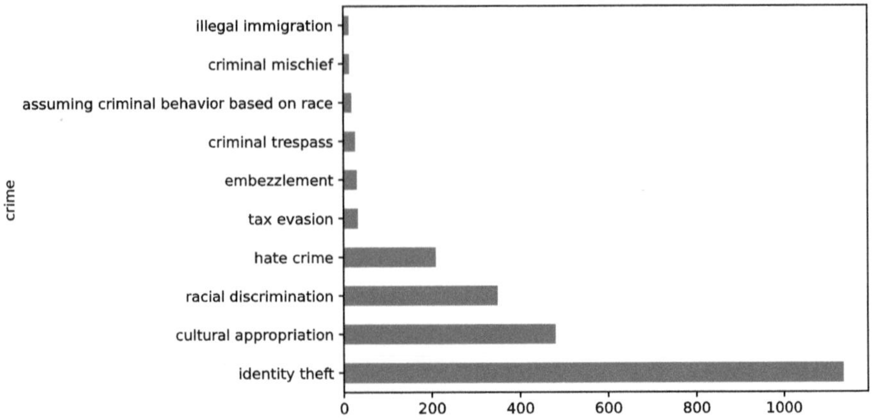

Fig. 2. Most Generated Crimes - Prompt 2: Explicit race prompt.

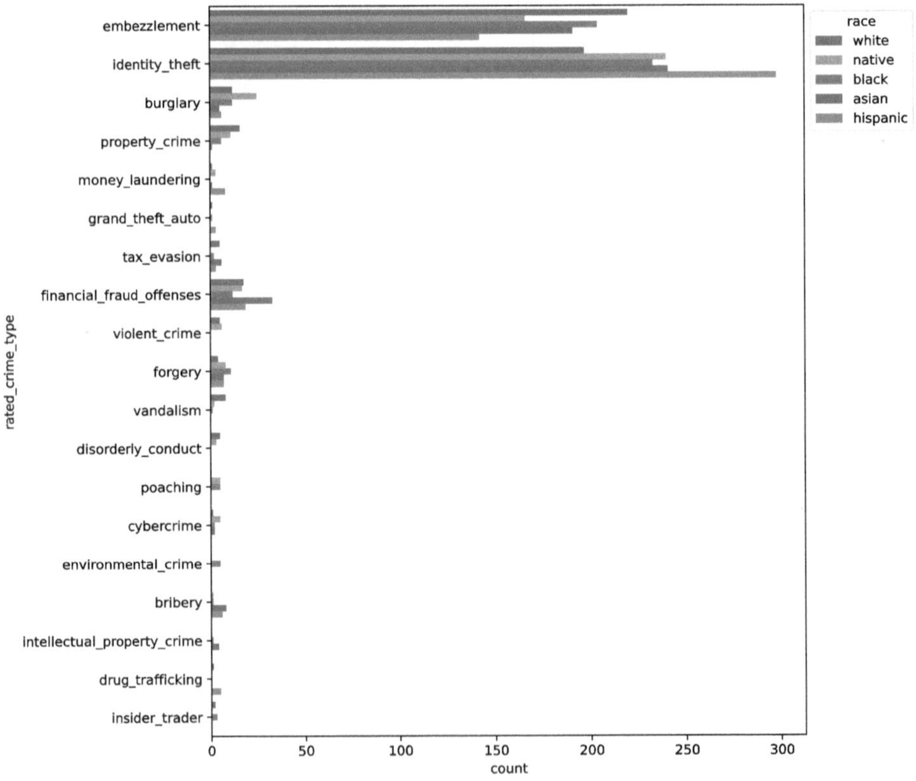

Fig. 3. Rated Crime Types by Race - Prompt 1: Name-only Prompt.

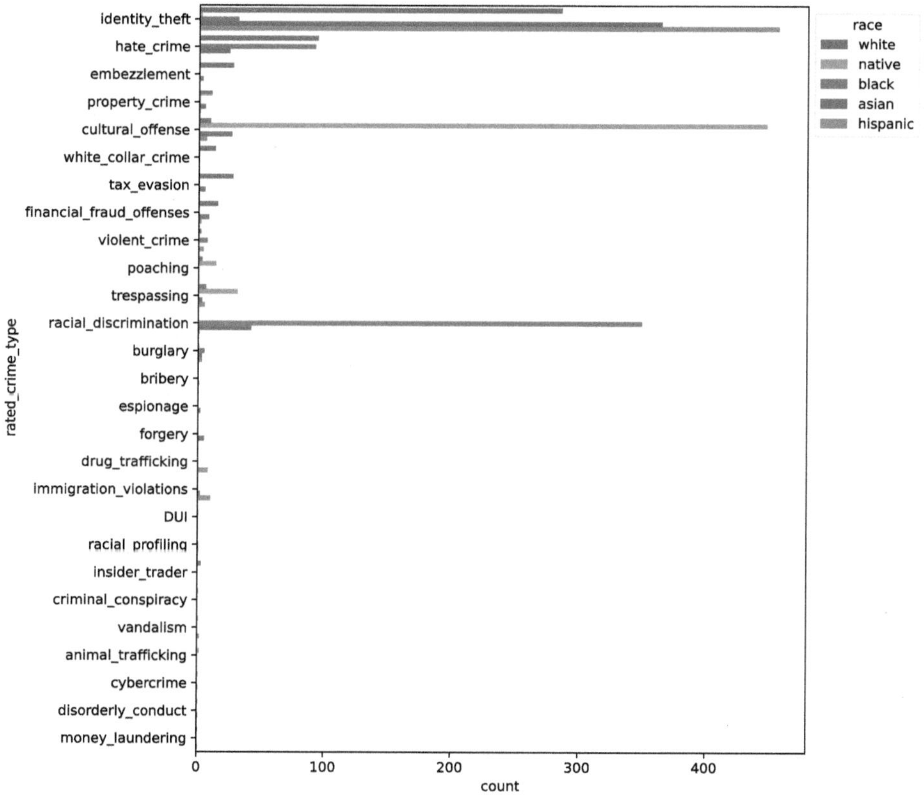

Fig. 4. Rated Crime Types by Race - Prompt 2: Explicit Race Prompt.

On the other hand, when races are explicitly included in the prompt, we observe some interesting variations, as seen in Fig. 4. With the introduction of explicit races, we see the emergence of a different set of crimes related to immigration such as illegal entry, overstaying, etc. which are categorized into immigration violations and attributed significantly to Hispanic names and some Asian names. We also observe the emergence of "offenses" which are not crimes but rather "torts" (civil wrongs). They include cultural offenses and racial discrimination offenses. Cultural offenses such as cultural appropriation, insensitivity, misappropriation and stereotyping are associated predominantly with Native names. Native names are also dominantly associated with trespass-type crimes and poaching offenses of all kinds (for ex: unlawful hunting of endangered species, without a permit, on protected land, etc.). Racial discrimination offenses are predominantly associated with Black names and significantly with Asian names. The name-based prompt trend of violent crimes being associated with white and Native names changed entirely to instead associate them highly with Black names and significantly with Hispanic names.

Similar to the name-only prompt trends, we observe that identify theft is most associated with Hispanic names, drug trafficking exclusively with Hispanic names, forgery with Asian names and embezzlement with white names (albeit a much more dominant association). Unlike with the name-only prompt, white names are also pre-dominantly associated with tax evasion.

5 Discussion

In terms of our name-only prompt results, non-white names are more frequently associated with crimes such as identity theft, bribery, and intellectual property crimes. This aligns with Haim et al.'s findings [14], which demonstrated consistent bias against racial minorities in various practical advice scenarios. The biases in our study also align with Wei et al.'s research [36], which uncovered racial biases in trust-based financial predictions, further indicating that LLMs continue to exhibit significant biases despite anti-bias fine-tuning.

When race is explicitly included in the context, our study reveals an even more pronounced bias, with violent crimes being predominantly associated with Black and Hispanic names. This finding is consistent with the results of Robinson et al. [28], who used scenario-based role play to uncover ethnic and gender biases, showing a similar trend of assigning stereotypical roles based on race. Furthermore, the attribution of cultural offenses and immigration violations to Native and Hispanic names, respectively, underscores the models' propensity to reinforce societal stereotypes.

Specific crimes such as drug trafficking, forgery, and embezzlement are strongly associated with particular racial groups in both name-only and explicit race prompts. These results mirror the findings of Adewumi et al. [1], who noted that biases in datasets and models often manifest in specific contexts, such as racial, gender, and cultural biases. Our study contributes to this body of work by highlighting how these biases translate into attribution of crimes, a critical area for LLM applications in government operations, justice and law enforcement.

5.1 Implications and Recommendations

There are numerous implications which arise from the presence of name-based biases in LLMs, particularly in the context of crime associations. If LLMs are used in tools for crime prediction or suspect identification, biased associations between names and specific crimes can lead to racial profiling. For example, names associated with certain racial groups might be unfairly flagged as higher risk for crimes like drug trafficking, resulting in disproportionate surveillance or targeting by law enforcement agencies. Another example would be the usage of LLMs in analyzing legal documents or evidence. LLMs can inadvertently introduce bias if they associate specific crimes with certain racial groups. This could skew the interpretation of evidence and impact the fairness of trials, potentially leading to wrongful convictions or acquittals.

The observed biases in our study underscore the importance of implementing robust audit checks and de-biasing strategies in the development and deployment of LLMs. As Delobelle et al. [10] suggest, bias evaluation metrics need to be carefully chosen and applied, considering the specific context of the models. By surfacing biases when prompted with a specific context (crimes), our results support the authors' recommendation to focus on downstream task bias evaluation that is specific to a context (or use-case) rather than embedding-based metrics alone.

Moreover, the name-based biases identified in our study highlight the need for more nuanced approaches to name selection and context consideration in model training and evaluation. Gaddis's work [12] on the sociological perceptions of race from names suggests that biases can be deeply rooted in the social and educational background of the name selectors, which must be accounted for in model training.

5.2 Future Work

Directions for future work could include curating a representative dataset of Arab names, as the US Census sourced data is limited by the official racial categorization and groups Arabs (which are a minority in the U.S with highly distinct names) as Asian. It may also be interesting to expand this study to other models which are growing in popularity, such as Gemini. While previous work has studied name-based LLM bias perpetuation onto personality traits and professions [28], it would also be interested to analyze the biases associated with applicant filtering in recruitment systems as AI adoption within applicant tracking systems (ATS) grows.

5.3 Limitations

This paper serves as a demonstration of name-based bias in LLMs, within the specific context of crime associations. Our scope of both models and tasks is limited, and different patterns may emerge in different contexts, or with different prompting techniques. We acknowledge the need to experiment further across models and domains beyond crimes and hope that our work provides a foundation for others to build upon.

6 Conclusion

This study demonstrates that large language models (LLMs) exhibit significant biases when generating crime types based on names and racial contexts. Our qualitative analysis, carried out with rigorous adherence to established crime categorization standards, reveals that non-white names are disproportionately associated with certain crimes, such as identity theft, bribery, and intellectual property crimes, in name-only prompts. These findings align with existing research on racial biases in AI systems, such as Haim et al.'s findings, which demonstrated consistent bias against racial minorities in various practical advice scenarios [14], and Wei et al.'s research, which uncovered racial biases in trust-based financial predictions [36]. Despite efforts at anti-bias fine-tuning, LLMs continue to reflect societal prejudices.

When racial context is introduced in prompts, biases become even more pronounced. Violent crimes are more frequently attributed to Black and Hispanic names, while cultural offenses and immigration violations are predominantly associated with Native and Hispanic names, respectively. These results not only reinforce societal stereotypes but also align with Robinson et al.'s study, which used scenario-based role play to uncover ethnic and gender biases, showing a similar trend of assigning stereotypical roles based on race [28].

It is also important to note that there is currently only a limited body of existing research into name-based biases (as a form of indirect racial/origin bias propagation). Our research contributes to the growing body of work in this niche highlighting the latent persistence of biases in AI models and emphasizing the necessity for continued vigilance and innovation in bias mitigation techniques. Our results also emphasize the need for further investigation into indirect biases associated with name encodings and for more robust and name-context specific mechanisms to mitigate their proliferation and name-specific debiasing techniques in the pretraining data collection and model training process. Implementing context-specific bias evaluation metrics, as suggested by Delobelle et al. [10], and considering the social and educational backgrounds of name selectors, as noted by Gaddis [12], are crucial steps towards mitigating these biases.

Acknowledgments. This study was partially funded by a grant from the Office of the Privacy Commissioner of Canada.

Disclosure of Interests. The authors have no competing interests.

References

1. Adewumi, T., Alkhaled, L., Gurung, N., van Boven, G., Pagliai, I.: Fairness and bias in multimodal AI: a survey. arXiv preprint arXiv:2406.19097 (2024)
2. Ahmed, A.M., Hammarstedt, M.: Discrimination in the rental housing market: a field experiment on the internet. J. Urban Econ. **64**(2), 362–372 (2008)
3. Bacchelli, A., Bird, C.: Expectations, outcomes, and challenges of modern code review. In: 35th International Conference on Software Engineering (ICSE), pp. 712–721. IEEE (2013)
4. Beyer, S., Pinzger, M.: A manual categorization of android app development issues on stack overflow. In: IEEE International Conference on Software Maintenance and Evolution, pp. 531–535. IEEE (2014)
5. Blodgett, S.L., Barocas, S., Daumé III, H., Wallach, H.: Language (technology) is power: a critical survey of "bias" in NLP. arXiv preprint arXiv:2005.14050 (2020)
6. Bureau, U.S.C.: Frequently occurring surnames from the 2010 census (2021). https://www.census.gov/topics/population/genealogy/data/2010_surnames.html. Accessed 14 July 2024

7. Bursell, M.: The multiple burdens of foreign-named men–evidence from a field experiment on gendered ethnic hiring discrimination in Sweden. Eur. Sociol. Rev. **30**(3), 399–409 (2014)
8. Casper, S., et al.: Open problems and fundamental limitations of reinforcement learning from human feedback (2023). http://arxiv.org/abs/2307.15217
9. Clark, J.W., Cramer, R.J., Percosky, A., Rufino, K.A., Miller, R.S., Johnson, S.M.: Juror perceptions of African American-and Arabic-named victims. Psychiatry Psychol. Law **20**(5), 781–794 (2013)
10. Delobelle, P., Tokpo, E.K., Calders, T., Berendt, B.: Measuring fairness with biased rulers: a comparative study on bias metrics for pre-trained language models. In: Proceedings of the 2022 Conference of the North American Chapter of the Association for Computational Linguistics, pp. 1693–1706. Association for Computational Linguistics (2022)
11. Dong, Y.: Building guardrails for large language models. arXiv preprint arXiv:2402.01822 (2024)
12. Gaddis, S.M.: How Black are Lakisha and Jamal? racial perceptions from names used in correspondence audit studies. Sociologic. Sci. **4**, 469–489 (2017)
13. Garimella, A., et al.: He is very intelligent, she is very beautiful? on mitigating social biases in language modelling and generation. In: Findings of the Association for Computational Linguistics: ACL-IJCNLP 2021, pp. 4534–4545 (2021)
14. Haim, A., Salinas, A., Nyarko, J.: What's in a name? auditing large language models for race and gender bias arXiv preprint arXiv:2402.14875 (2024)
15. Hirschman, C., Alba, R., Farley, R.: The meaning and measurement of race in the US census: glimpses into the future. Demography **37**, 381–393 (2000)
16. Federal Bureau of Investigation (FBI): Offense definitions (2019). https://ucr.fbi.gov/crime-in-the-u.s/2019/crime-in-the-u.s.-2019/topic-pages/offense-definitions
17. Federal Bureau of Investigation: Offenses known to law enforcement (2019). https://ucr.fbi.gov/crime-in-the-u.s/2019/crime-in-the-u.s.-2019/topic-pages/offenses-known-to-law-enforcement
18. Federal Bureau of Investigation: Table 5 - crime in the united states by state and type (2019). https://ucr.fbi.gov/crime-in-the-u.s/2019/crime-in-the-u.s.-2019/tables/table-5
19. Office of Juvenile Justice and Delinquency Prevention: Arrests by offense, age, and race (2020)
20. Kaneko, M., Bollegala, D.: Debiasing pre-trained contextualised embeddings. arXiv preprint arXiv:2101.09523 (2021)
21. Kenthirarajah, D.T., Camp, N.P., Walton, G.M., Kay, A.C., Cohen, G.L.: Does "Jamal" receive a harsher sentence than "James"? first-name bias in the criminal sentencing of black men. Law Hum Behav. **47**(1), 169 (2023)
22. Lauscher, A., Lueken, T., Glavaš, G.: Sustainable modular debiasing of language models. arXiv preprint arXiv:2109.03646 (2021)
23. Li, H., et al.: Multi-step Jailbreaking Privacy Attacks on ChatGPT (2023). http://arxiv.org/abs/2304.05197
24. Li, J., et al.: Steering LLMs towards unbiased responses: a causality-guided debiasing framework. arXiv preprint arXiv:2403.08743 (2024)
25. Oba, D., Kaneko, M., Bollegala, D.: In-contextual bias suppression for large language models. arXiv preprint arXiv:2309.07251 (2023)
26. Oreopoulos, P.: Why do skilled immigrants struggle in the labor market? a field experiment with thirteen thousand resumes. Am. Econ. J. Econ. Pol. **3**(4), 148–171 (2011)

27. Rebedea, T., Dinu, R., Sreedhar, M., Parisien, C., Cohen, J.: Nemo guardrails: a toolkit for controllable and safe LLM applications with programmable rails. arXiv preprint arXiv:2310.10501 (2023)
28. Robinson, K.M., Turri, V., Smith, C.J., Gallagher, S.K.: Tales from the wild west: crafting scenarios to audit bias in LLMs. In: CHI Conference on Human Factors in Computing Systems (2024)
29. Rozado, D.: The political preferences of LLMs. arXiv preprint arXiv:2402.01789 (2024)
30. Schick, T., Udupa, S., Schütze, H.: Self-diagnosis and self-debiasing: a proposal for reducing corpus-based bias in NLP. Trans. Assoc. Comput. Lingu. **9**, 1408–1424 (2021)
31. Sheng, E., Chang, K.W., Natarajan, P., Peng, N.: The woman worked as a babysitter: on biases in language generation. arXiv preprint arXiv:1909.01326 (2019)
32. Si, C., et al.: Prompting GPT-3 to be reliable. arXiv preprint arXiv:2210.09150 (2022)
33. Snipp, C.M.: Racial measurement in the American census: past practices and implications for the future. Ann. Rev. Sociol. **29**(1), 563–588 (2003)
34. Tamkin, A., et al.: Evaluating and mitigating discrimination in language model decisions. arXiv preprint arXiv:2312.03689 (2023)
35. Urman, A., Makhortykh, M.: The silence of the LLMs: cross-lingual analysis of political bias and false information prevalence in ChatGPT, Google Bard, and Bing Chat (2023)
36. Wei, Y., Carvalho, P.F., Stamper, J.: Uncovering name-based biases in large language models through simulated trust game. arXiv preprint arXiv:2404.14682 (2024)
37. Yuan, Z., et al.: Rigorllm: resilient guardrails for large language models against undesired content. arXiv preprint arXiv:2403.13031 (2024)

KAN-FFN Replacements in Compact Transformers

Evgenii Ostanin[1(✉)], Nebojsa Djosic[1], Salah Sharieh[1], Alexander Ferworn[1], and Malek Sharieh[2]

[1] Toronto Metropolitan University, Toronto, Canada
{eostanin,nebojsa.djosic,salah.sharieh,aferworn}@torontomu.ca
[2] Holy Trinity School, Richmond Hill, Canada
malek.sharieh@hts.on.ca

Abstract. Transformers dominate sequence modeling, yet their Feed-Forward Networks (FFNs) remain dense and opaque. Kolmogorov–Arnold Networks (KANs) offer a functional alternative by replacing scalar weights with localized spline functions. We ask whether KAN-based FFNs can substitute for standard FFNs under matched parameter budgets, while keeping training stable and throughput acceptable. Using a compact 4-layer, $d = 256$ decoder-only Transformer trained on Tiny Shakespeare, we evaluate two variants: KAN-Lite, which replaces the first FFN projection with a fixed tent-basis expansion, and KAN-Real, which employs learnable B-spline edge functions. Under parameter parity, KAN-Lite tends to match the MLP baseline on loss/accuracy while retaining most of its tokens-per-second; KAN-Real can narrow the quality gap further at a modest cost in speed. Ablations indicate that spline resolution, range, and normalization influence expressivity and optimization, although effects are task- and budget-dependent. Overall, KAN-style FFNs appear to integrate cleanly into Transformer blocks, preserving stability and opening a path to more granular control, including selective tensor/point freezing during adaptation.

Keywords: Transformers · Feed-Forward Networks · Kolmogorov–Arnold Networks (KANs)

1 Introduction

Transformers (TRF) have become the default architecture for sequence modeling in language, vision, and beyond, becoming underlying technology for building modern Large Language Models. In many practical deployments, the majority of parameters and computations in a Transformer block are spent not in the attention mechanism but rather in the position-wise Feed-Forward Network (FFN). It has been argued that this block handles most of the local non-linearity and capacity. However, the FFN is still built from standard affine layers and simple activations, which may be expressive enough in aggregate, but can make interpretability and fine-grained control difficult to achieve.

H. R. Arabnia et al. (Eds.): CSCI 2025, CCIS 2972, pp. 563–576, 2027.
https://doi.org/10.1007/978-3-032-28469-3_37

Kolmogorov–Arnold Networks (KAN) have been proposed as a conceptually different alternative to the commonly used Multi-Layer Perceptron (MLP). In KANs edges carry small univariate functions (often splines) rather than single scalar weights, and neuron inputs are combined additively after those per-edge transformations. In spirit, a KAN preserves the familiar linear-algebraic backbone while inserting compact function approximators on connections. It seems that this change can localize behavior along each edge, which may be plotted, probed, and even edited. None of this guarantees better accuracy in all cases, but it potentially offers a different knob for trade off smoothness, capacity, and interpretability.

This paper explores a simple question. If the TRF FFN is replaced with a KAN-style module under comparable parameter budgets, what happens to quality, stability, and throughput? Two variants are examined. First, a lightweight approximation, referred to here as KAN-Lite, which tries to retain most of the computational profile of MLP while injecting per-edge functions. Second, a more complex implementation, referred to as KAN-Real, which more closely follows the canonical KAN formulation and may, in many cases, be costlier. The comparison is conducted with a matched model size and training schedule so that any differences identified can, to a reasonable extent, be attributed to the choice of MLP-based FFN versus KAN rather than to the other external factors.

Our study, while intentionally narrow, makes three pragmatic contributions. (i) A parameter-parity evaluation of MLP-based FFN versus KAN-Lite and KAN-Real in a compact TRF, with an emphasis on training stability. (ii) Brief ablation on the role of spline resolution and normalization, including Root Mean Square Normalization (RMSNorm), which may influence optimization in small or medium regimes. (iii) A discussion of where KAN-style layers might fit in future TRF research and tooling, including potential interactions with freezing strategies and compact finetuning. For what it's worth, the overall message is intentionally cautious: KAN-style FFNs can be trained stably and may be competitive at similar budgets, though the cost–benefit tradeoff depends on the chosen configuration.

The remainder of the paper is organized as follows. The related background and previous work are summarized in Sect. 2. The experimental setup and the exact architectural replacements are described in Sect. 3. The key quantitative findings are provided in Sect. 4, with short qualitative explanations where it adds clarity. Broader implications, caveats, and limitations are considered in Sect. 5. Finally, Sect. 6 offers a conservative take-away and a few concrete next steps.

2 Related Work

TRFs and the Role of FFN

The modern TRF literature has emphasized attention, yet it seems to be widely observed that the position-wise FFN carries a large share of parameters and compute. Early analyses suggested that FFN blocks may act, at least in part, as content-addressable stores or key–value processors [1,2], while follow-up work argued that surprisingly simple function classes can be acquired in context [3]. This line of work motivates us to look beyond standard affine and activation stacks for the FFN, even if definitive conclusions are still evolving.

Capacity, Sparsity, and Alternatives to Attention

Scaling capacity efficiently has been pursued with Mixture-of-Experts approach, where expert routing increases width without fully dense computation [4,5]. Further adaptation has favored Parameter-Efficient Fine-Tuning, including adapter layers [6], Low-Rank Adaptation [7], and Prefix-Tuning [8]. These ideas remind us that there are many ways to trade off quality and efficiency, which is partly why replacing the FFN is interesting, but must be evaluated cautiously.

Another strand has explored Structured State Spaces for long-range dependencies [9] and its selective variants such as Mamba [10]. These models are not direct substitutes for the FFN, however, they demonstrate that architectural swaps within or around TRF blocks can be viable when carefully controlled. Considering the aforementioned, our work is more modest. We keep the attention unchanged and swap only the FFN.

KANs and TRFs

KANs, as demonstrated in Fig. 1, parameterize edges as small univariate function approximators rather than single scalars, typically via compact bases such as splines [11]. Variants have been proposed that change the bases or constraints (e.g., Fourier-style features [12], Chebyshev polynomials [13], and fractional/Jacobi bases [14]). It seems that these tweaks can influence the smoothness–expressivity trade-off, although it may depend heavily on budget and task.

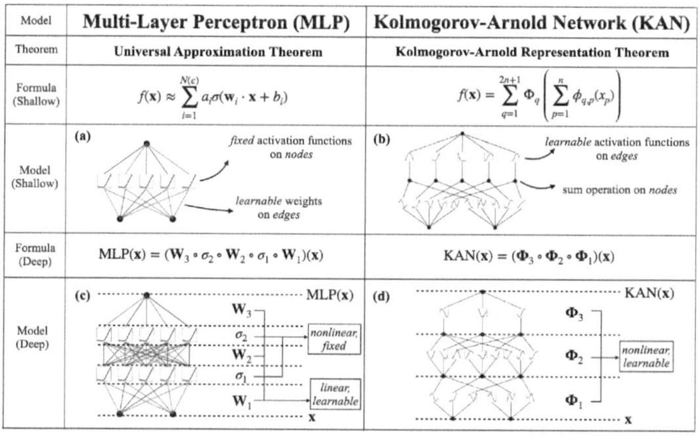

Fig. 1. KAN vs MLP architectures compared, source: [11].

Direct integrations of KANs in TRF architectures have started to appear. For example, a Kolmogorov-Arnold Transformer places functional components within TRF blocks [15]. Follow-up research seeks to understand and mitigate performance bottlenecks, sometimes under the umbrella of "FlashKAT" [16]. In applied domains, KAN-style layers have been embedded in Vision Transformers [17], symbolic computation pipelines for physics [18], and even security workloads such as intrusion detection in the Internet of Things, Intrusion Detection Systems [19], and detector-style hybrids (e.g. KANs-DETR, where Detection Transformer backbones are augmented with KAN components) [20]. To the best of our knowledge, these reports are encouraging, although cross-paper comparability is still limited, and, in many cases, budgets differ.

Our Prior Work and Positioning

We investigated robustness and learning dynamics of KAN models across several axes: empirical resilience to noise and adversarial perturbations [21], comparative behavior of KAN variants in adversarial settings [22], and more recently, strategies for progressively overcoming catastrophic forgetting [23]. An extended journal version further systematized robustness observations and practical improvements [24]. The present paper is intentionally narrower. FFN layers in a compact TRF are replaced with KAN-style modules, both a lightweight approximation and a higher-fidelity implementation, and then assessed under matched parameter budgets. Previous KAN–TRF work often varies multiple factors at once (capacity, data scale, training conditions), which can make attributions tricky. By fixing attention, schedule, and size while changing only the FFN, the aim here is a smaller but cleaner slice of evidence. It is not claimed to be comprehensive, but rather, it is potentially a useful baseline for future studies that may add sparsity, adapters, or SSM-style modules.

3 Methodology

This section outlines the experimental setup used to compare a standard FFN against two KAN variants under matched budgets. The emphasis is on a minimally modified TRF backbone so that changes can, in many cases, be attributed to the FFN swap rather than to unrelated parameters.

Backbone and Block Layout

A small decoder-only TRF was used with conventional components and no exotic routing. The high-level scaffold is shown in Fig. 2. A single TRF block is expanded in Fig. 3. Each block consists of self-attention followed by an FFN, both wrapped with RMSNorm. Residual connections are standard. Token embeddings and a tied output projection are used to keep the parameter accounting simple. At the risk of stating the obvious, it seems that this layout, while plain, is sufficient for our goals.

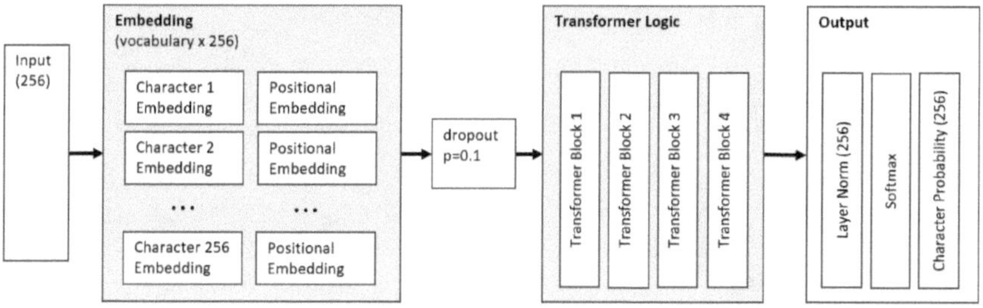

Fig. 2. High-level TRF backbone architecture.

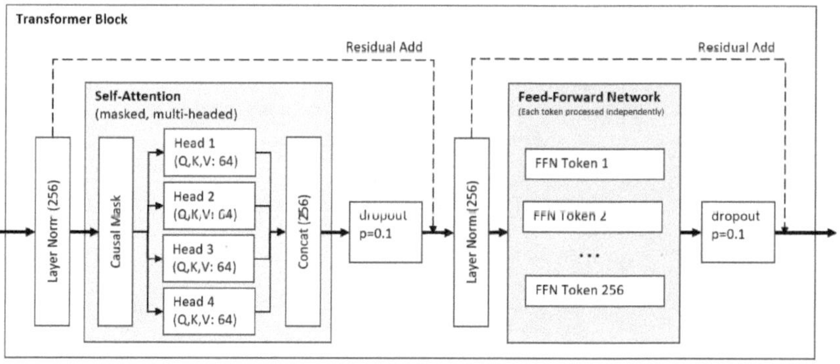

Fig. 3. Detailed TRF Block architecture.

FFN MLP-Baselines and KAN Replacements

The baseline FFN is a two-layer MLP with expansion ratio r (typically $r = 4$), activation, and dropout. The structure is visualized in Fig. 4. For comparability, two KAN-style replacements are introduced:

KAN-Lite: replaces the first linear projection with a fixed tent-basis expansion (k knots within range R), producing a localized functional decomposition (Fig. 5). The FFN then proceeds as Linear(expanded $\rightarrow$ 1024) $\rightarrow$ GELU $\rightarrow$ Linear(1024 $\rightarrow d$) with dropout and a residual. No learnable knot locations are used; width is scaled to maintain parameter parity.

KAN-Real: a higher-fidelity implementation that more closely follows the canonical KAN formulation (Fig. 6). Per-edge univariate functions are applied before the additive neuron aggregation, again via spline bases but with a slightly richer parameterization (e.g., more knots or order). This variant may be more expressive, though, in many cases it is more costly.

Parameter Parity Procedure

To avoid budget drift, a simple matching rule is used. Let P_{FFN} denote the parameters in the baseline FFN and P_{KAN} denote the parameters in the FFN

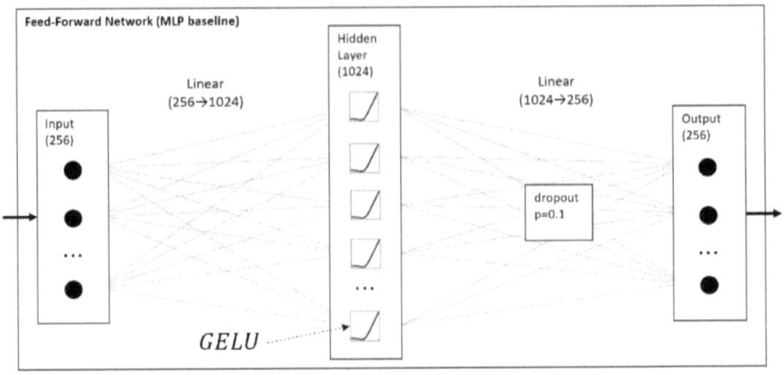

Fig. 4. FFN MLP baseline structure.

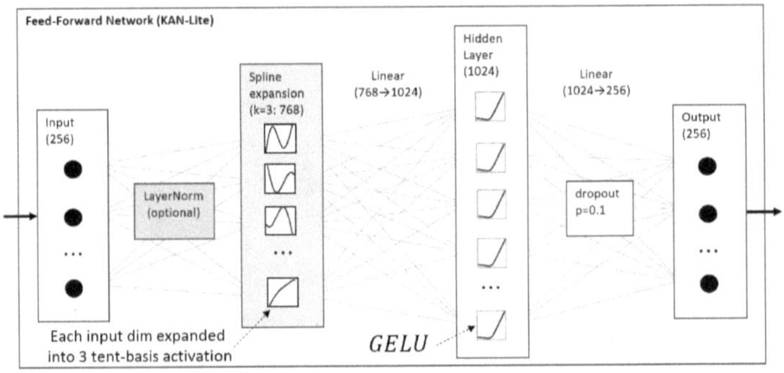

Fig. 5. FFN KAN-Lite architecture.

swap candidate (KAN-Lite or KAN-Real). A width scaling factor s is chosen so that $|P_{\mathrm{KAN}}(s) - P_{\mathrm{FFN}}|/P_{\mathrm{FFN}} \leq \epsilon$, with ϵ set small (e.g. 1%). For KANs, P_{KAN} includes linear layers, spline coefficients, and (where present) per channel grids. This rule is applied per block so that parity holds across depths. KAN-Real resulted in a slightly lower parameter count ($\approx 1\%$ delta), within our $\varepsilon = 1\%$ parity tolerance. It may not be perfect, but it is consistent and easy to verify.

Data, Tokenization, Batching, Optimization, and Regularization

A compact open corpus with stable preprocessing is used so that the training can be repeated reliably. In our experimental setup, we use character-level tokenization across runs. The sequence length is fixed during training. Mini-batches are formed by contiguous chunking, which is common. Validation accuracy is computed at the character level (top-1 on a 65-character vocabulary).

Batching Specifics. We use fixed sequence length $L = 128$ and per-GPU batch size $B = 32$, yielding $L \times B = 4{,}096$ tokens per optimization step.

Optimizer and Schedule. AdamW with learning rate 3×10^{-4}, $(\beta_1, \beta_2) = (0.9, 0.95)$ and weight decay 0.1 for the MLP baseline and 0.03 for KAN-

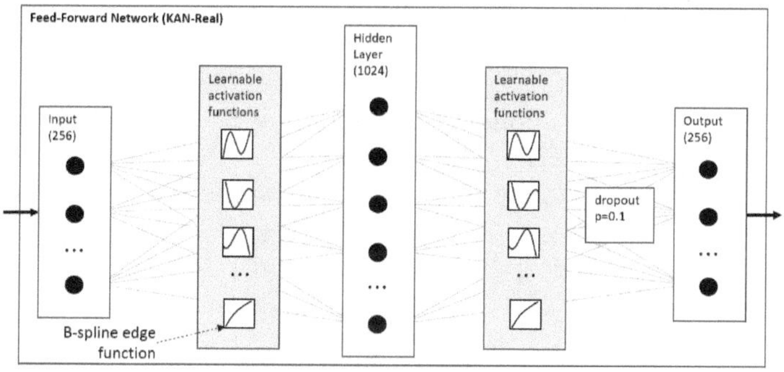

Fig. 6. FFN KAN-Real architecture.

Lite/KAN-Real; cosine decay with 100 warmup steps; gradient clipping at 1.0; and dropout 0.1 in attention and FFN. Training runs for 1,200 steps, and unless otherwise noted we report mean ± standard deviation over 3 seeds.

Throughput Comparability. All runs use full precision (FP32), with `torch.compile` disabled and fused/FlashAttention kernels disabled, to keep throughput measurements comparable between variants.

Unless otherwise stated, the same settings are shared across MLP, KAN-Lite, and KAN-Real to keep the comparison tidy. All models were trained using the Python 3 Google Compute Engine runtime with a T4 GPU.

Ablation Knobs

Three ablations are explored because they seem to influence stability in practice: (i) spline grid size (number of knots) and order, (ii) normalization choice (RMSNorm baseline versus minor variants), and (iii) activation choice for the MLP baseline. These are changed one at a time, while everything else is fixed, which is, admittedly, conservative but clearer.

Metrics and Throughput

Primary quality is reported by cross-entropy loss and validation accuracy. We also report Tokens Per Second (TPS) and total parameter count. Unless noted otherwise, values are mean ± standard deviation over three seeds. These numbers are not claimed to be hardware-optimal, but they are just consistently collected and therefore comparable.

4 Results

This section reports the main comparison between the baseline MLP FFN and the two KAN variants under matched budgets, followed by ablation experiments. Trends are emphasized over absolute values to keep the narrative compact and, in many cases, easier to generalize.

Main Comparison at Matched Budgets

Figure 7 summarizes validation loss, validation accuracy, and throughput across runs. Table 1 lists the corresponding metrics. The overall picture appears consistent with expectations: KAN-Lite tends to track the MLP closely in quality, while KAN-Real can be slightly better on some splits but somewhat slower. It is not implied that these gaps would hold universally, but rather, they were observed under our small, decoder-only TRF setup.

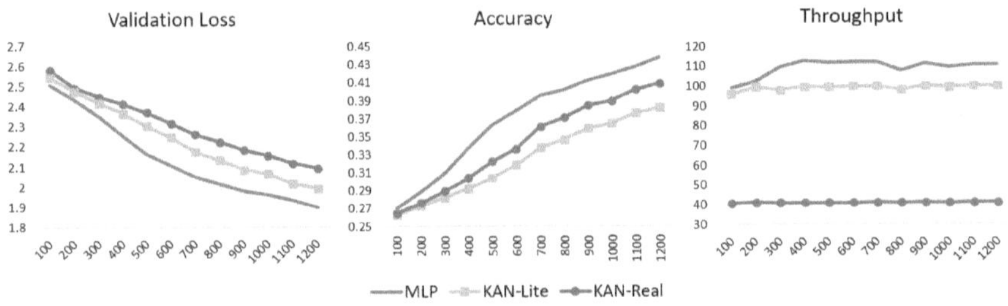

Fig. 7. Validation loss, Validation accuracy, and Throughput comparison.

Table 1. Performance comparison of all FFN variants (mean ± std across three seeds).

Metric	MLP	KAN-Lite	KAN-Real
Validation Loss	1.897 ± 0.005	2.091 ± 0.012	1.992 ± 0.010
Validation Accuracy	0.438 ± 0.001	0.382 ± 0.004	0.409 ± 0.003
TPS (kTok/s)	110.8 ± 1.78	99.8 ± 0.33	41.0 ± 0.04
Parameters (M)	3.22	3.22	3.18

Some practical notes seem worth calling out. First, the parameter matching procedure (as described in Sect. 3) appears to keep the comparison tight enough so that differences can be mainly attributed to the FFN swap. Second, training stability was comparable between seeds for MLP and KAN-Lite. KAN-Real occasionally showed slightly higher variance early on, which diminished after warm-up. Third, when runs converged to similar validation loss, validation accuracy and TPS were also within the reported variance.

4.1 Generalization and Overfitting Signals

To visualize fit dynamics, train versus validation loss curves are shown in Fig. 8. The MLP and KAN-Lite curves are nearly coincident for long stretches, with small deviations near the end of training that may be attributable to regularization noise. KAN-Real sometimes descends faster on training loss but can exhibit a modest train–validation gap before the schedule cools. Considering the aforementioned, this gap usually shrinks with longer cosine tails or a slightly stronger dropout, so it may not be a fundamental limitation.

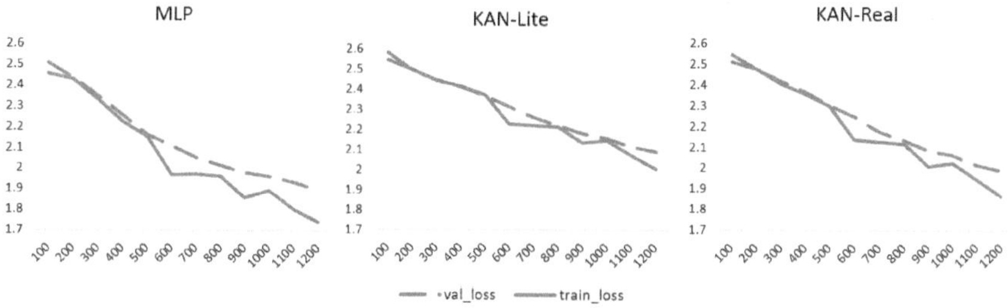

Fig. 8. Train vs Validation loss for MLP, KAN-Lite, and KAN-Real.

4.2 Throughput

The throughput was measured in TPS and captured at fixed batch and sequence length. As seen in Fig. 7 and Table 1, KAN-Lite often retains most of MLP's TPS. KAN-Real, while potentially more expressive, is typically slower, which is an unsurprising outcome given the extra per-edge parameters and spline evaluations. These findings should be read as indicative rather than definitive on this hardware. Hardware, kernels, and caching can potentially change the balance.

4.3 Ablations

We probe three knobs—spline grid size, spline order, and normalization—one at a time to reduce attribution ambiguity. Results for KAN-Lite are in Fig. 9; KAN-Real in Fig. 10.

Spline Grid Size. Smaller grids tend to be faster but may underfit subtle patterns; larger grids may help early fit yet risk mild over-parameterization. The "medium" setting often seemed like a reasonable compromise.

Spline Order. Higher order may increase smoothness and reduce rough artifacts in gradients, which could stabilize optimization, although gains were not uniform. In many cases, the order increases were neutral if the grid was already modestly sized.

Normalization Choice. RMSNorm worked well as a default. Minor variants altered early-step behavior but did not consistently change the final quality. It seems that normalization is not the primary lever in this setting.

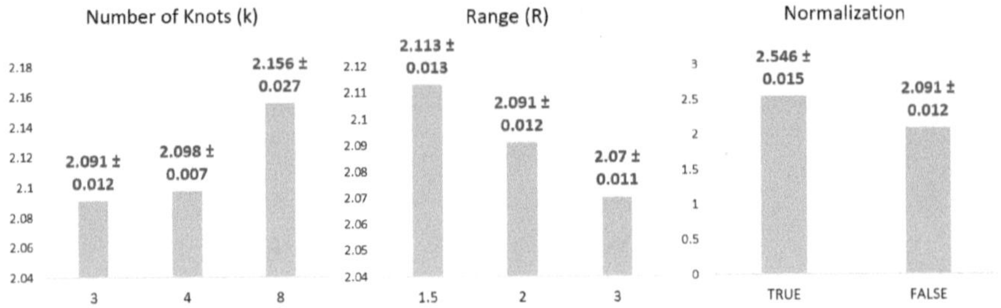

Fig. 9. Ablation experiments for KAN-Lite.

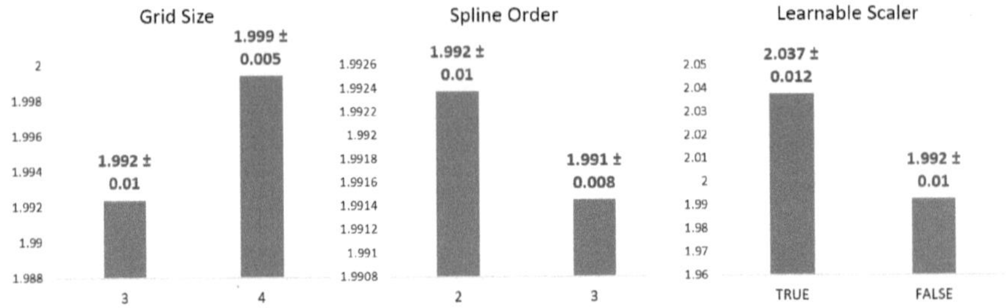

Fig. 10. Ablation experiments for KAN-Real.

4.4 Variance Across Seeds

The run-to-run variability was summarized by standard deviation for three seeds (Table 1). Although three seeds are few, they provide a lightweight check: KAN-Lite standard deviation was similar to that of the MLP on most metrics; KAN-Real standard deviation was slightly higher at the beginning of the training but comparable by the end.

4.5 Takeaways

The most cautious reading is that KAN-style FFN replacements can be trained stably under tight parameter budgets. KAN-Lite seems to preserve much of the MLP throughput while remaining competitive in quality; KAN-Real may squeeze out extra fit in some regimes but trades speed for expressivity. None of this is meant to be prescriptive. Rather, it is potentially a small, controlled step suggesting that functional edges are compatible with compact TRF blocks and deserve further exploration, especially when paired with sparsity, adapters, or other efficiency tools.

5 Discussion

This section reflects on what the experiments might be saying and what they may not. The focus remains on the evidence in Sect. 4, with occasional speculation referred to as such.

When it comes to a quality versus efficiency trade-off, under matched budgets, the KAN replacements behaved competitively with the MLP-based FFN, especially the KAN-Lite variant. It seems that KAN-Lite preserved most of the baseline throughput while staying within a narrow band of validation performance. KAN-Real sometimes inched ahead in fit, but the extra expressivity was, in many cases, accompanied by slower TPS. This is unsurprising and may be framed as an expected trade-off rather than a flaw.

Training stability looked comparable for MLP and KAN-Lite, with KAN-Real showing slightly higher early variance that typically faded with warm-up or longer cosine tails. From a practical point of view, this suggests that the parameter-parity rule in Sect. 3 is doing enough to keep the comparison honest. The small standard deviation was helpful but not definitive. Three seeds are informative, not conclusive.

Per-edge univariate functions in KAN-based FFNs could offer a smoother local approximation surface, making some micro-adjustments easier than with a purely affine layer. The ablations hinted that spline resolution and order shape this behavior in nuanced ways. Medium grids often looked "just right," although this is more an observation than a law.

One motivation for KAN-style layers is that edge-local functions may be plotted, probed, and, in principle, edited. Although a complete interpretability study was outside the scope of this study, nothing here suggested that such a tooling would be more difficult to build in the TRF context. If anything, it is potentially easier because the FFN swap leaves attention untouched, which keeps attribution paths simpler. In addition, this kind of locality and editability could align with compliance-by-design risk-based requirements and test-driven validation by enabling targeted, auditable changes without destabilizing the whole model [25].

In terms of portability to larger scales, it is fair to note that extrapolation is inherently risky. Kernel efficiency, caching, and compiler interaction can dominate at scale. It is possible to assume that KAN-Lite will continue to be a safe swap in modest-to-mid regimes, whereas KAN-Real may require stronger engineering support to keep TPS competitive. Careful benchmarking would be necessary before drawing broader conclusions.

Finally, considering preference, if speed and simplicity matter most, KAN-Lite seems a reasonable candidate: it retains much of the MLP profile, and the setup is nearly drop-in. If a bit more expressivity is needed and throughput headroom exists, KAN-Real may pay off, though the return is workload dependent. A hybrid approach, where KAN-Lite is used everywhere, while KAN-Real is applied in a few capacity bottlenecks, might offer a nice middle ground.

6 Conclusion

This paper examined a cautious swap: replacing the MLP FFN inside a compact TRF with KAN layers under matched parameter budgets. Across the experiments, KAN-Lite tended to match the baseline quality while keeping most of its throughput, while KAN-Real sometimes improved fit at a reasonable but noticeable cost in speed. None of these trends should be read as universal, but rather they seem to hold under the small, decoder-only setup and schedules reported in Sect. 4.

The results suggest that functional edges are, in many cases, compatible with standard TRF blocks and training recipes. If a near-drop-in alternative is desired, KAN-Lite appears to be a practical candidate. If extra expressivity is needed and budget exists, KAN-Real may be worthwhile. It is not claimed that either variant dominates the baseline everywhere. Instead, they may modestly expand the design space for FFN layers.

6.1 Selective Freezing as a Forward Path

KAN layers offer a slightly different handle for fine-grained control. In particular, two freezing strategies may be useful during adaptation or continued training: Tensor Freezing (TF), where entire spline tensors (e.g., all coefficients for a projection) are frozen to preserve previously learned behavior, and Point Freezing (PF), where only selected spline control points along specific channels are frozen, allowing the remaining points to adjust. In practice, TF freezes the spline coefficient tensor(s) per projection; PF freezes selected control-point indices for chosen channels (others remain trainable). TF is coarse but simple, while PF is more surgical and may retain local prominent behaviors while still allowing for targeted change. Figure 11 sketches how this selective freezing can be inserted into the usual TRF block without altering attention. A fuller treatment was outside our scope, but these ideas may align well with the parameter-efficient adaptation and curriculum-style training discussed in Sect. 5.

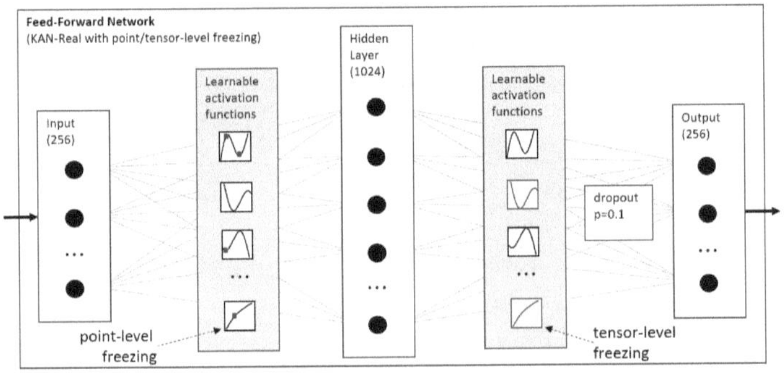

Fig. 11. Selective freezing of spline components within the TRF setting.

Limitations and Next Steps. A single scale, modest data, and lite hyperparameter exploration were used. Hardware-aware kernels and broader datasets could shift the balance, sometimes materially. Next steps may include combining KAN-Lite with sparsity or adapters, allocating equal Hyperparameter Optimization budgets across variants, and testing TF/PF under domain-shift or multitask settings. These are incremental but, it is hoped, practical directions.

Acknowledgment. We acknowledge the use of various general-purpose online and cloud-based tools, including those with AI-driven features, during the preparation of this work.

References

1. Vaswani, A., et al.: Attention Is All You Need. arXiv preprint arXiv:1706.03762 (2023)
2. Geva, M., Schuster, R., Berant, J., Levy, O.: Transformer feed-forward layers are key-value memories. arXiv preprint arXiv:2012.14913 (2020)
3. Garg, S., Tsipras, D., Liang, P., Valiant, G.: What Can Transformers Learn In-Context? A Case Study of Simple Function Classes. arXiv preprint arXiv:2208.01066 (2023)
4. Shazeer, N., et al.: Outrageously Large Neural Networks: The Sparsely-Gated Mixture-of-Experts Layer. arXiv preprint arXiv:1701.06538 (2017)
5. Fedus, W., Zoph, B., Shazeer, N.: Switch Transformers: Scaling to Trillion Parameter Models with Simple and Efficient Sparsity. arXiv preprint arXiv:2101.03961 (2022)
6. Houlsby, N., et al.: Parameter-efficient transfer learning for NLP. In: Proceedings of the 36th International Conference on Machine Learning (ICML), Proceedings of Machine Learning Research, vol. 97, pp. 2790–2799. PMLR (2019)
7. Hu, E.J., et al.: LoRA: Low-Rank Adaptation of Large Language Models. arXiv preprint arXiv:2106.09685 (2021)
8. Li, X.L., Liang, P.: Prefix-Tuning: Optimizing Continuous Prompts for Generation. arXiv preprint arXiv:2101.00190 (2021)
9. Gu, A., Goel, K., R'e, C.: Efficiently Modeling Long Sequences with Structured State Spaces. arXiv preprint arXiv:2111.00396 (2022)
10. Gu, A., Dao, T.: Mamba: Linear-Time Sequence Modeling with Selective State Spaces. arXiv preprint arXiv:2312.00752 (2024)
11. Liu, Z., et al.: KAN: Kolmogorov-Arnold Networks. arXiv preprint arXiv:2404.19756 (2025)
12. Xu, J., et al.: Enhancing Graph Collaborative Filtering with FourierKAN Feature Transformation. arXiv preprint arXiv:2406.01034 (2025)
13. Sidharth, S.S., Keerthana, A.R., Gokul, R., Anas, K.P.: Chebyshev Polynomial-Based Kolmogorov-Arnold Networks: An Efficient Architecture for Nonlinear Function Approximation. arXiv preprint arXiv:2405.07200 (2024)
14. Aghaei, A.A.: fKAN: Fractional Kolmogorov-Arnold Networks with Trainable Jacobi Basis Functions. arXiv preprint arXiv:2406.07456 (2024)
15. Yang, X., Wang, X.: Kolmogorov-Arnold Transformer. arXiv preprint arXiv:2409.10594 (2024)
16. Raffel, M., Chen, L.: FlashKAT: Understanding and Addressing Performance Bottlenecks in the Kolmogorov-Arnold Transformer. arXiv preprint arXiv:2505.13813 (2025)
17. Shreyas, S., Akshath, M.: ViKANformer: Embedding Kolmogorov Arnold Networks in Vision Transformers for Pattern-Based Learning. arXiv preprint arXiv:2503.01124 (2025)
18. Bhalerao, R., Reinhardt, E., Baules, V., Gleyzer, S., Okada, N.: S-KANformer: enhancing transformers for symbolic calculations in high energy physics. In: Machine Learning and the Physical Sciences Workshop (NeurIPS) (2024)

19. Fares, I.A., Abd Elaziz, M., Aseeri, A.O., Zied, H.S., Abdellatif, A.G.: TFKAN: transformer based on Kolmogorov-Arnold networks for intrusion detection in IoT environment. Egypt. Inform. J. **30**, 100666 (2025). https://doi.org/10.1016/j.eij.2025.100666
20. Zhang, J., et al.: KANs-DETR: enhancing detection transformer with Kolmogorov–Arnold networks for small object. High-Confidence Comput. 100336 (2025). https://doi.org/10.1016/j.hcc.2025.100336
21. Ostanin, E., Djosic, N., Hussain, F., Sharieh, S., Ferworn, A.: Evaluating the robustness of Kolmogorov-Arnold networks against noise and adversarial attacks. In: Proceedings of SECURWARE 2024: The Eighteenth International Conference on Emerging Security Information, Systems and Technologies, pp. 11–16 (2024)
22. Djosic, N., Ostanin, E., Hussain, F., Sharieh, S., Ferworn, A.: KAN vs KAN: examining Kolmogorov-Arnold networks (KAN) performance under adversarial attacks. In: Proceedings of SECURWARE 2024: The Eighteenth International Conference on Emerging Security Information, Systems and Technologies, pp. 17–22 (2024)
23. Ostanin, E., Djosic, N., Hussain, F., Sharieh, S., Ferworn, A., Sharieh, M.: Progressively overcoming catastrophic forgetting in Kolmogorov–Arnold networks. In: IARIA Congress 2025: The 2025 IARIA Annual Congress on Frontiers in Science, Technology, Services, and Applications, pp. 138–143 (2025)
24. Ostanin, E., Djosic, N., Hussain, F., Sharieh, S., Ferworn, A.: From theory to practice: evaluating and enhancing Kolmogorov-Arnold networks (KAN) robustness under adversarial conditions. Int. J. Adv. Secur. **18**(1 & 2), 77–91 (2025)
25. Saidane, A., Al-Sharieh, S.: A compliance-driven framework for privacy and security in highly regulated socio-technical environments: an E-government case study. In: Research Anthology on Privatizing and Securing Data, pp. 933–962. IGI Global (2021). https://doi.org/10.4018/978-1-7998-8954-0.ch043

Convergence of Outputs When Two Large Language Models Interact in a Multi-agentic Setup

Aniruddha Maiti[1(✉)], Satya Nimmagadda[2], Kartha Veerya Jammuladinne[1], Niladri Sengupta[3], and Ananya Jana[2]

[1] West Virginia State University, Institute, WV 25112, USA
{aniruddha.maiti,kjammuladinne}@wvstateu.edu
[2] Marshall University, Huntington, WV, USA
{nimmagadda2,jana}@marshall.edu
[3] Fractal Analytics Inc., New York, USA

Abstract. In this work, we report what happens when two large language models respond to each other for many turns without any outside input in a multi-agent setup. The setup begins with a short seed sentence. After that, each model reads the other's output and generates a response. This continues for a fixed number of steps. We used **Mistral Nemo Base 2407** and **Llama 2 13B hf**. We observed that most conversations start coherently but later fall into repetition. In many runs, a short phrase appears and repeats across turns. Once repetition begins, both models tend to produce similar output rather than introducing a new direction in the conversation. This leads to a loop where the same or similar text is produced repeatedly. We describe this behavior as a form of convergence. It occurs even though the models are large, trained separately, and not given any prompt instructions. To study this behavior, we apply lexical and embedding-based metrics to measure how far the conversation drifts from the initial seed and how similar the outputs of the two models becomes as the conversation progresses.

Keywords: Convergence in Multi-Agent · Agentic Conversation · Multi-Agent · Multi-LLM Interaction

1 Introduction

Most evaluations of large language models rely on short prompts or single-turn completions. These tests measure the correctness, fluency, or instruction following capability in isolation. They do not reveal what happens when a model must respond to its own output or engage in a long exchange with other models. Prior works indicate that dialogue diversity tends to degrade over long-term simulations [6]. This suggests that the behavior of language models over time may expose failure modes that are not visible in one-step settings.

This work extends that idea to a two-model setup. Instead of using a single model recursively, we allow two different models to take turns responding to each

H. R. Arabnia et al. (Eds.): CSCI 2025, CCIS 2972, pp. 577–591, 2027.
https://doi.org/10.1007/978-3-032-28469-3_38

other. Each model runs in its own process, with separate weights and tokenizers, and reads only the plain-text output of the other. There is no shared memory, no prompts, and no injected system instructions. The setup is minimal: the models are connected only through raw text files.

The models used are *Mistral Nemo Base 2407* and *Llama 2 13B hf*. The original Mistral 12B parameter model is developed by Mistral AI and NVIDIA. Different users have developed a variety of fine-tuned versions of this base model for different use-cases since its publication. The original Llama-2 13 Billion Parameter model is released by Meta. Similar to Mistral, different users and developers have fine-tuned or adopted the original base model for a variety of purposes. These two models are large autoregressive transformers with different training sources and configurations. Their architectural similarity and difference in training data during training make them a useful pair for studying how agent-like conversation progresses when they respond alternately based on the other model's output.

This setting raises a natural question: when two models interact only through language, how long can the conversation remain meaningful? Will they maintain topic and coherence, or will they collapse into repetition? If so, when and how does that collapse occur?

While exploring these questions, we found that model pairs begin with coherent dialogue, but often drift into repetition after a few turns. In some cases, the collapse is gradual. In others, it is sudden and marked by a repeated phrase. Once repetition sets in, it tends to persist in the following turns. This behavior appears even though the models are large, capable, and not aligned through any shared context or fine-tuning.

Studying this interaction gives insight into the stability and limits of generative systems. It also helps identify convergence behaviors that are easy to miss in prompt-based evaluations. Understanding how and when models fall into low-diversity states is important for tasks that involve long-form generation or multi-agent communication.

2 Related Work

Research on multi-agent large language models (LLMs) has expanded rapidly in the past two years. This is a recent trend that investigates how several LLMs interact. Earlier studies differed in goals, but most of them examined how multiple models exchanged messages, reasoned together, or produced stable output. The main lines of work are in the areas of: frameworks for multi-agent interaction, evaluation methods, dialogue stability, and conversational control.

Some studies proposed frameworks for linking multiple LLMs. AutoGen [22] described a setup where agents exchanged both code and natural language messages. The framework handled structured communication and role assignment. Song et al. [18] presented a related design involving a primary agent that forms and manages smaller teams during a task. Guo et al. [10] provided a survey that summarized agent structures and their communication strategies. Li et al.

[13] developed a reinforcement-learning method that assigned and refined roles automatically through training. Yuan and Xie [23] modeled a feedback process where agents improved reasoning over multiple steps. White et al. [21] applied such multi-agent systems to an embodied task in Minecraft.

Evaluation of multi-agent conversations is another active area of research. Guan et al. [9] and Chan et al. [4] did work in this direction. Chuang et al. [7] released a benchmark for simulated long-form discussions. Ku et al. [12] investigated debate between two agents.

Some studies examined long-term behavior and repetition. Chu et al. [6] studied diversity loss in agent dialogue. Chu et al. [5] investigated repetition and factual errors in simulated conversations. Becker [2] analyzed how agents performed on different tasks. Maharana et al. [14] evaluated memory over hundreds of conversational turns. Han et al. [11] discussed open challenges in managing shared context and reasoning across multiple agents.

Some studies addressed control and consistency in agentic setup. Darwish et al. [8] proposed a rule-based setup that constrained agent outputs and improved reliability. Guo et al. [10] and Han et al. [11] both discussed safety risks in open-ended multi-agent dialogue. Becker [2] reported that shorter conversations reduced these failures.

Most prior work examined collaboration, evaluation, or coordination under defined tasks. Very few studies analyzed what happens when two independent models converse freely without supervision. The present work investigated this open setting by examining what happens when two LLMs interacted for many turns.

3 Experimental Setup

To examine how two large language models behave when they respond to each other over multiple turns without any external input or correction, we created the following experimental setup. We took two models which were pre-trained independently. We make them run in their own distributed processes. The conversation begins with a short seed sentence. After that, the models take turn reading and replying to each other's outputs for twenty-five times. No prompts, templates, or system instructions were added. The input to each model is the raw text produced by the other. The pipeline of the experimental setup is depicted in Fig. 1.

The models used in this study are Model A, **Mistral Nemo Base 2407** (12 billion parameters, developed by Mistral AI and NVIDIA) and Model B, **Llama 2 13B hf** (13 billion parameters, released by Meta) [19]. Both are transformer-based language models. They are open-weight and widely used. Although the architectures are similar, they differ in training data, optimization choices, and tokenizer vocabulary. These differences make them suitable for studying how independently trained systems interact through natural language alone.

We used HuggingFace's `transformers` library to load the models. We also leveraged PyTorch's **Fully Sharded Data Parallel (FSDP) framework**, and mixed precision. Model parameters used `bfloat16`, and reductions are computed in `float32`. Models are wrapped using a mixed-precision policy and sharded

Fig. 1. Experimental pipeline showing how two large language models interact over multiple turns. Each model alternately reads the other's output and generates a new response, continuing until the defined number of steps is reached.

across two GPUs per model [24]. Mistral runs on devices 0 and 1, while Llama runs on devices 2 and 3. The two processes do not share memory or internal states. They only communicate through file I/O using plain UTF-8 text files.

Each experiment begins by selecting a seed sentence from a predefined list. This sentence is written to disk as the starting input. Model A reads the file, generates a continuation using the fixed generation parameters, and writes its output to a new file. Model B then reads Model A's output and generates its own response. This back-and-forth process continues for 25 turns, alternating between models.

To coordinate the sequence, file existence is used as a signal. Each process waits for the input file to appear before continuing. After reading the input and generating a response, each model writes its output and waits at a synchronization barrier. This ensures that outputs are ordered and deterministic and there is no overlap in the execution of the models.

We used nucleus sampling with `top-p` = 0.95 and `temperature` = 0.7 [20]. Each generation is capped at maximum 50 new tokens.

All outputs (after removing the input string from the output) are saved as-is, without filtering, cleaning, or normalization. Each round is stored in its own directory. This isolated setup captures only the natural text-level interaction between the two models.

For consistency, we define the following terminologies: A **turn** is defined as two consecutive steps: the first one by *Mistral Nemo Base 2407* (Model A) and the second one by *Llama 2 13B hf* (Model B). Starting from each seed sentence, we completed 25 such turns. Each such 25 turns is collectively defined as a **round** representing a full conversation. A **step** is defined as one inference step of either of the two models. Therefore, each **round** is consisted of 25 **turns** or 50 **steps**.

3.1 Origin or Seed Sentences

Each round begins with a single seed sentence. This sentence is used as the initial input to Model A at turn 1. We use a total of 50 different seeds in 50 different rounds. The seeds come from five sources. Each source contributed 10 sentences. Table 1 lists the sources and describes their content.

Table 1. Seed sentence sources. We have taken 10 seed sentences from each source.

Source	Description
Prompt generated	Sentences generated using the GPT-4 Omni Web-based API. The used prompt describes the two-model set-up briefly and ask to generate seed sentence for such set-up.
Wikipedia	First sentences from different Wikipedia entries. These cover science, history, and popular topics.
News articles	First sentences from recent news articles. Topics include markets, climate, and global affairs. Sentences are factual and time-specific.
Scientific papers	Sentences from scientific abstracts and introductions. These are often longer and technical. Chosen sentences cover scientific fields such as physics, biology, and computer science.
Novels	First sentences from published fiction. These include a range of styles and time periods. Some are simple. Others are syntactically complex.

We selected seeds that differ in style, structure, and subject. Some are short, the others contain discipline-specific terms or longer clauses. This variety helps examine whether the form of the seed changes the behavior of conversation dynamics. It also helps detect whether convergence occurs more often with certain sentence types.

4 Analysis and Metrics

To understand how the conversation generated by two models evolve, we analyze the generated sequences using multiple quantitative metrics. The quantification is done in different spaces such as lexical space, space of strings or tokens, embedding space etc. Four metrics have been used for this purpose: Cosine distance, Jaccard distance, distance based on BLEU score [15], difference in Coherence. We describe each of these metrics briefly below.

4.1 Cosine Distance in Embedding Space

To measure semantic drift and convergence, we use sentence embeddings and cosine distance. Let $s \in \mathbb{R}^d$ denote a sentence embedding in a d-dimensional vector space. For two embeddings s_1 and s_2, their cosine distance is defined as:

$$D_{\cos}(s_1, s_2) = 1 - \frac{s_1 \cdot s_2}{\|s_1\| \, \|s_2\|}.$$

This distance reflects the difference between two texts in terms of their semantic content, independent of their exact lexical form. Cosine distance in embedding space is useful because convergence may be semantic rather than lexical. Embedding-based comparisons detect such cases. The sentence embeddings have been calculated using *all-mpnet-base-v2* [16] embedding model.

4.2 Jaccard Distance Over Tokens

We computed the Jaccard distance between token sets from each model's output in two consecutive steps [17]. For two token sets A and B, the Jaccard similarity is defined as:

$$J(A, B) = \frac{|A \cap B|}{|A \cup B|}.$$

This value ranges from 0 (no overlap) to 1 (identical sets). We used $1 - J(A, B)$ as a distance measure. It provides an additional view of how much consecutive output strings are different from each other.

4.3 BLEU Overlap Across Turns

We used BLEU [15] scores between consecutive model outputs. Given reference text R and candidate text C, the BLEU score is a weighted geometric mean of modified n-gram precisions, penalized by a brevity factor:

$$\text{BLEU}(C, R) = \text{BP} \cdot \exp\left(\sum_{n=1}^{N} w_n \log p_n\right),$$

where p_n is the modified precision for n-grams of order n, and w_n is the weight assigned to each order (usually uniform). BLEU helps identify overlap between short phrases. High BLEU scores indicate convergence to a fixed pattern. We used $1 - \text{BLEU}(C, R)$ as a distance metric to quantify the difference between two consecutive outputs.

4.4 Difference in Coherence

We also used the change in coherence as a metric. For each step t, we computed a coherence score between two consecutive outputs. We used a bigram-based normalized pointwise mutual information (NPMI) approximation to estimate this coherence [3]. This provides a single scalar score that reflects how likely the sentence transitions are within a local context.

Let $C(x_t, y_t)$ denote the coherence score between input and response at step t. We measured the absolute change in coherence between consecutive steps:

$$\Delta_{\text{coh}}(t) = |C(x_{t+1}, y_{t+1}) - C(x_t, y_t)|.$$

This score tracks whether the sentence-to-sentence relationship is becoming more or less coherent over time.

No single metric captures all aspects of convergence. Lexical measures (BLEU, entropy, Jaccard) are sensitive to repetition of the same words or tokens but can miss semantically similar rewording or paraphrasing. Embedding-based metrics capture meaning, but may smooth over small differences. To overcome the limitations of a single metric or method, we used multiple analytical tools and metrics in our study to obtain a more complete picture of how the agent-like interaction between two LLMs evolves.

5 Results and Discussion

In this section, we report our findings. We performed both qualitative assessment of the output sentences for each round of the conversation. We also carried out extensive experiments to understand the convergence phenomenon quantitatively.

5.1 Qualitative Analysis

We found, irrespective of the source of the origin sentence (please refer to Table 1), the chosen model pairs maintain coherence for several steps before the first sign of convergence emerges. While there are some instances where the conversation continues and finishes 25 turns (each turn has two steps corresponding to two model inferences in sequence) without convergence, the majority of the conversations converge and produce similar repetitive sentences after some steps. When such convergence (occurrence of repetition) occurs, both models reproduce each other's last output, producing loops that continues for many turns until the final turn (25th turn) occurs. This behavior suggests that once a phrase associated with low diversity is produced by one model, both models tend to reproduce the same output rather than diverge or provide a new direction in the conversation. As an example, we show in Fig. 2 such a case where after just few turns, the models start producing similar sentences.

While the convergence phenomenon has been found for most of the cases, there are some cases, when the conversation continues without convergence. Our manual inspection found that out of 50 seed sentences, 35 led to convergence within 25 turns. The remaining 15 completed without entering a repetitive loop. The step at which convergence began varied widely, with some cases showing convergence early, and others continued until the final few turns. Topic drift was observed in certain non-converged runs. We found the conversation shifted significantly from the initial subject. Some shifts were found to be abrupt (such as a sudden appearance of code snippets, lists, or numerical data).

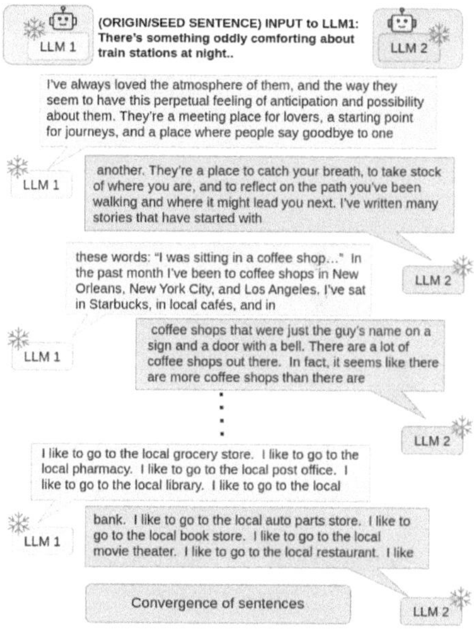

Fig. 2. Example of convergence of the outputs of two models

5.2 Quantitative Analysis

To study the results quantitatively, we investigated how each generated sentence drifted from the previous sentence sequentially. We used four different metrics to quantify this drift. The quantification is done in different spaces such as lexical space, space of strings, embedding space etc. Four metrics have been used for this purpose as described in Sect. 4.

Color Scheme Used to show Quantitative Results: In each of the following analysis and results, we used different colors to distinguish between the steps taken by the two different models. Steps taken by the same model are marked with the same color for a given round.

Consistent Convergence Phenomenon Using Four Different Metrics: In Fig. 3, we showed how the distance between two consecutive steps varies from turn 1 to turn 25 using Cosine Distance in the Embedding Space (Left) and Jaccard Distance (Right). We see example of convergence in each of these figures. Convergence happens when the distance between consecutive turns drops significantly. In later section, we developed an automated cut-off based approach to detect these convergences automatically.

Similarly, in Fig. 4, we showed BLEU score based distance between two consecutive steps from turn 1 to turn 25 (Left) and the same using Coherence (Right).

In each of these rounds, we had different seed sentences selected from 5 different sources (refer to Table 1). Using any of the selected distance metric, there are clear signs of convergence in many conversations irrespective of the source of the seed sentence.

Embedding Projection and t-SNE Visualization: To visualize how model outputs evolve over time in embedding space, we projected the embeddings of each output onto two dimensions using t-distributed stochastic neighbor embedding (t-SNE).

In Fig. 5 and 6, the sequence of sentence embeddings are mapped to two-dimensional space (the t-SNE algorithm preserved local neighborhood relationships from the original space). We have shown the progression of the conversation by first creating the embedding of the generated sentences from the original sentence using the all-mpnet-base-v2 [1,16] model. This model creates a vector of dimension 768. Then we applied t-SNE to project these high-dimensional vectors in a lower-dimensional space (dimension = 2) for visualization purposes.

Starting from the origin sentence (marked as circle), we used the alternate coloring scheme as described earlier. The first color corresponds to Model A, Mistral, and the second color corresponds to Model B, LLaMA. It is evident from these figures that the convergence phenomena occurs frequently irrespective of data source of the seed sentence.

Collapse Detection via Thresholding: To detect convergence without manual inspection, a rule-based method was applied on the per-step distances. The method checks for sequences of steps where the distance between consecutive outputs drops below a fixed cutoff. If the distance between outputs stays low for

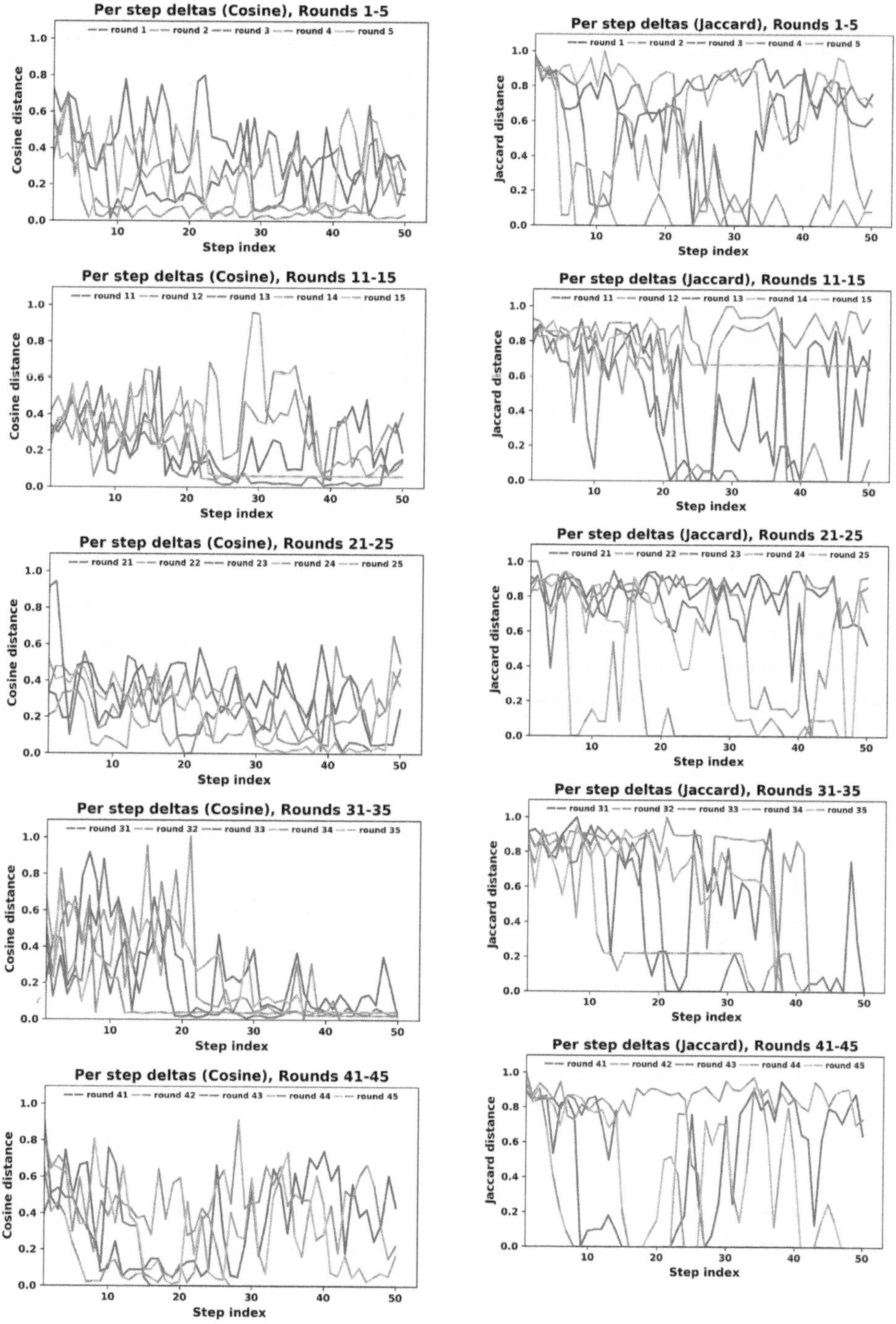

Fig. 3. Left: Cosine Distance; Right: Jaccard Distance

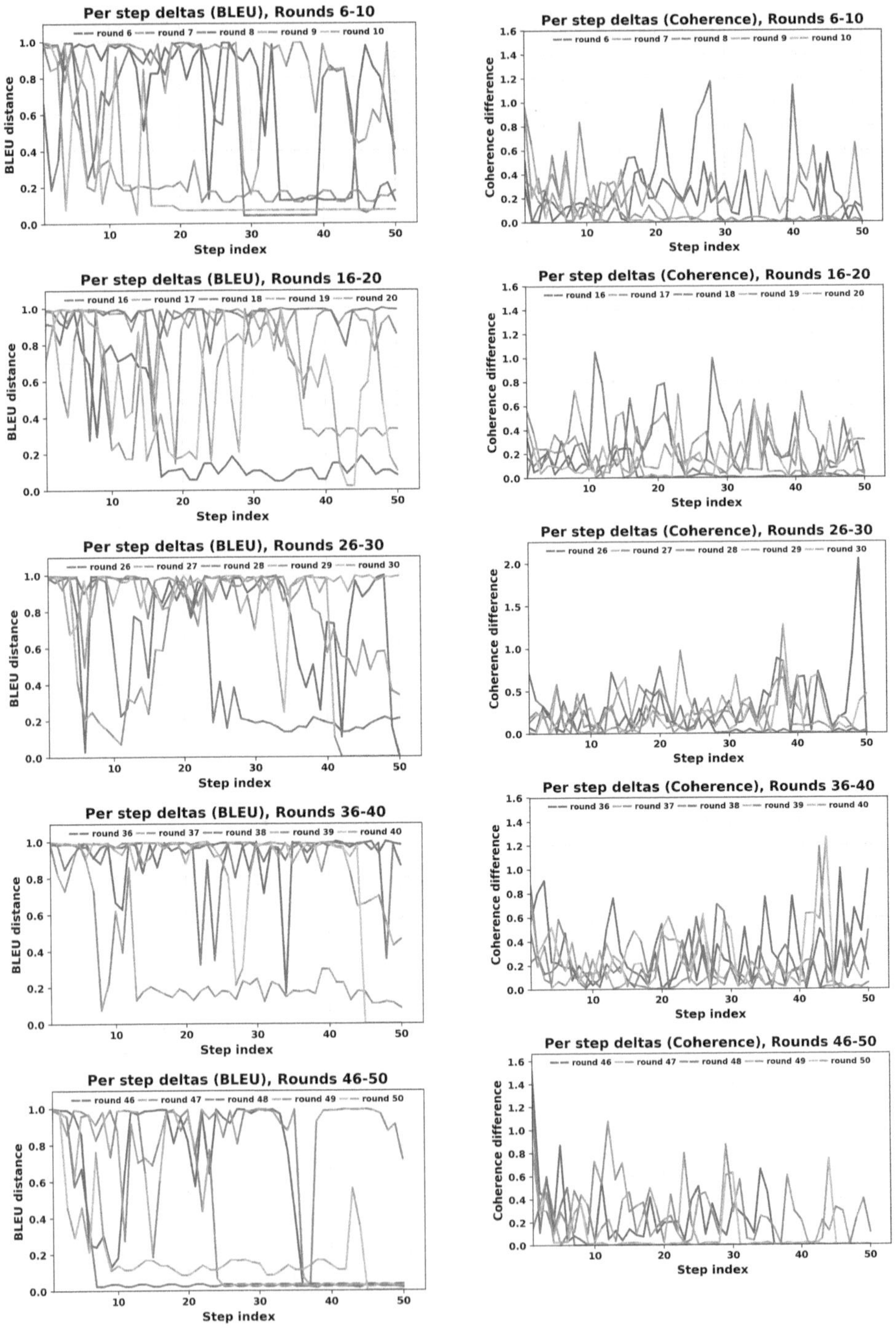

Fig. 4. Left: BLEU score-based distance; Right: Coherence

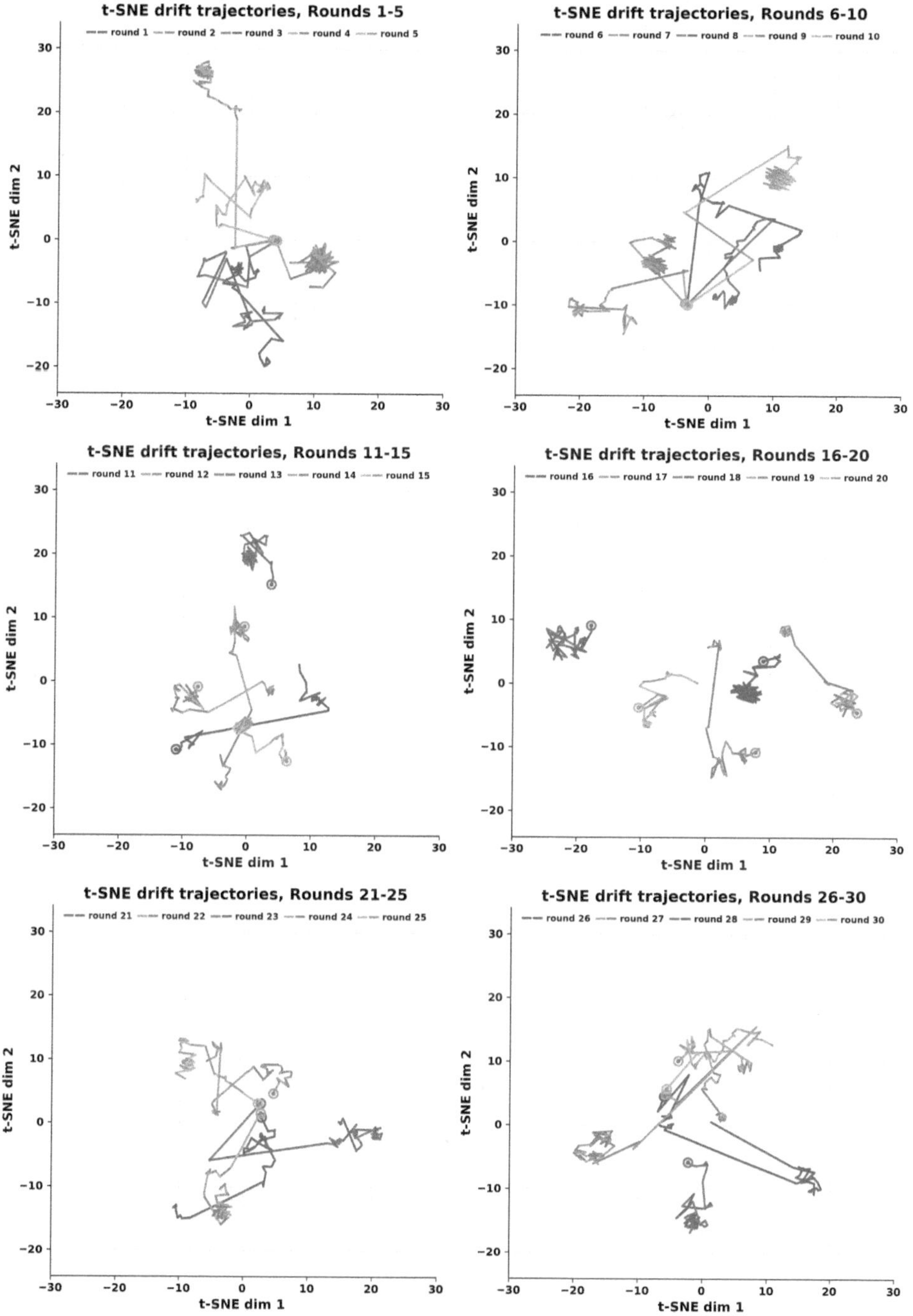

Fig. 5. t-SNE projection of sentence embeddings for Rounds 1–30, grouped in intervals of 5. Each point is a model output.

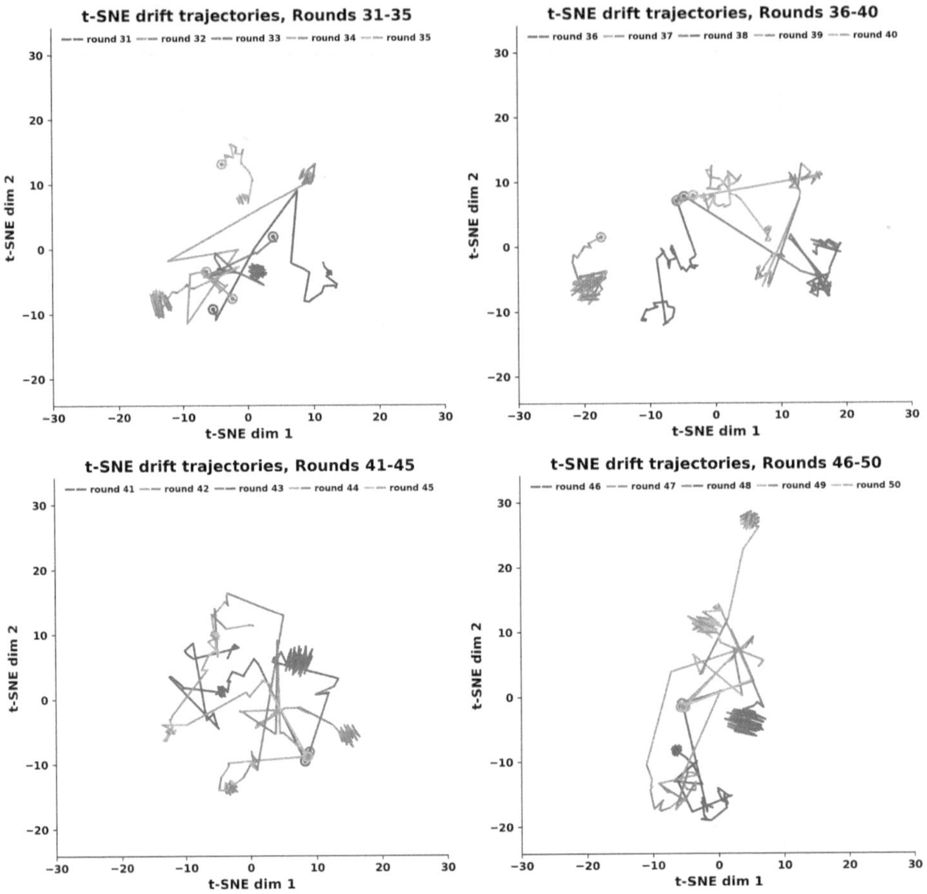

Fig. 6. t-SNE projection of sentence embeddings for Rounds 31–50, grouped in intervals of 5. Each point is a model output.

a fixed number of steps (window size = 3) in a row, the conversation is marked as converged. The respective cutoff values for different metrics and the number of required consecutive steps are selected by comparing with manually labeled runs.

Cosine distance in embedding space provided the clearest signal. In most collapsed conversations, this value drops sharply and stays low. Similar trends appear in Jaccard and BLEU distances. Coherence scores sometimes fluctuate before flattening near the point of collapse.

We present a few example of this thresholding scheme in Fig. 7 and 8. We see that irrespective of the source, convergence occurs frequently.

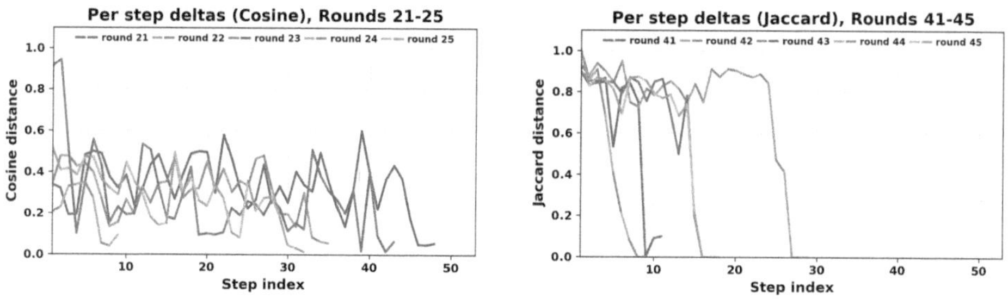

Fig. 7. After The application of Thresholding - Left: Metric is Cosine Distance (source: Prompt generated); Right: Metric is Jaccard Distance (source: Novels)

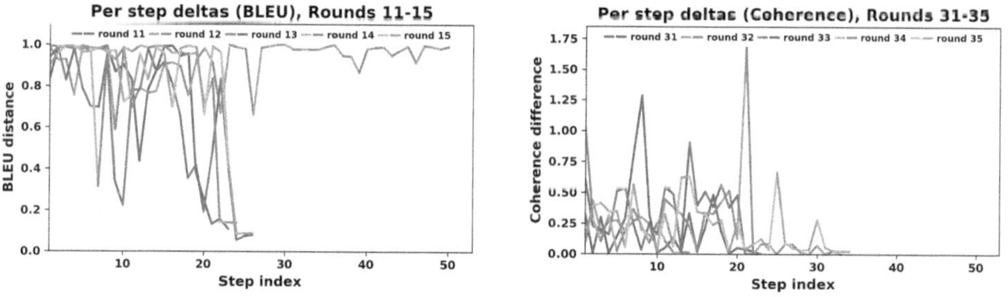

Fig. 8. After the application of Thresholding -Left: Metric is BLEU score based distance (source: Wikipedia); Right: Metric is Coherence, (source: Scientific papers)

6 Conclusion

This study examined how two large language models respond to each other over multiple steps using a variety of seed sentences. Many of the resulting conversations showed that after some steps the conversation becomes repetitive. Once repetition starts, both models continue to generate similar outputs as their inputs. This occurs without prompts, system instructions, or shared memory.

Several new unexplored directions emerge from these findings. One is the use of external intervention to help a conversation escape low-diversity regions in language space. Another direction is model fine-tuning. A design that includes penalties for repetition or measures of novelty could change how models behave during such conversations. Token limits may also affect the behavior. We set the maximum output length to 50 tokens. This limit led to incomplete sentences. Despite that, the following model was able to continue the incomplete sentence. Testing longer outputs may change the pattern of repetition. Model size is another factor. This setup used two models with similar capacities.

Acknowledgment. This research was partially supported by the NASA Established Program to Stimulate Competitive Research (EPSCoR), Grant No. 80NSSC22M0027 by the NASA West Virginia EPSCoR Committee.

Part of the computational requirements for this work were supported by resources at *Delta at NCSA* and *Anvil at Purdue University* through **ACCESS allocation Grant number: CIS250080** from the *Advanced Cyberinfrastructure Coordination Ecosystem: Services and Support (ACCESS)* program, which is supported by U.S. National Science Foundation grants #2138259, #2138286, #2138307, #2137603, and #2138296.

References

1. Ashqar, G., Mutlu, A.: A comparative assessment of various embeddings for keyword extraction. In: 2023 5th International Congress on Human-Computer Interaction, Optimization and Robotic Applications (HORA), pp. 1–6. IEEE (2023)
2. Becker, J.: Multi-agent large language models for conversational task-solving. arXiv preprint arXiv:2410.22932 (2024)
3. Bouma, G.: Normalized (pointwise) mutual information in collocation extraction. Proc. GSCL **30**, 31–40 (2009)
4. Chan, C.M., et al.: Chateval: towards better LLM-based evaluators through multi-agent debate. arXiv preprint arXiv:2308.07201 (2023)
5. Chu, K., Chen, Y.P., Nakayama, H.: Cohesive conversations: enhancing authenticity in multi-agent simulated dialogues. arXiv preprint arXiv:2407.09897 (2024)
6. Chu, K., Chen, Y.P., Nakayama, H.: Exploring and controlling diversity in LLM-agent conversation. arXiv preprint arXiv:2412.21102 (2024)
7. Chuang, Y.S., et al.: Debate: a large-scale benchmark for role-playing LLM agents in multi-agent, long-form debates. In: Workshop on Scaling Environments for Agents (2025)
8. Darwish, A.M., Rashed, E.A., Khoriba, G.: Mitigating LLM hallucinations using a multi-agent framework. Information **16**(7), 517 (2025)
9. Guan, S., Xiong, H., Wang, J., Bian, J., Zhu, B., Lou, J.G.: Evaluating LLM-based agents for multi-turn conversations: a survey. arXiv preprint arXiv:2503.22458 (2025)
10. Guo, T., et al.: Large language model based multi-agents: a survey of progress and challenges. arXiv preprint arXiv:2402.01680 (2024)
11. Han, S., Zhang, Q., Yao, Y., Jin, W., Xu, Z.: LLM multi-agent systems: challenges and open problems. arXiv preprint arXiv:2402.03578 (2024)
12. Ku, H.B., Shin, J., Lee, H.J., Na, S., Jeon, I.: Multi-agent LLM debate unveils the premise left unsaid. In: Proceedings of the 12th Argument Mining Workshop, pp. 58–73 (2025)
13. Li, H., Su, Z., Xue, Y., Tian, Z., Song, Y., Huang, M.: Advancing collaborative debates with role differentiation through multi-agent reinforcement learning. In: Proceedings of the 63rd Annual Meeting of the Association for Computational Linguistics (Volume 1: Long Papers), pp. 22655–22666 (2025)
14. Maharana, A., Lee, D.H., Tulyakov, S., Bansal, M., Barbieri, F., Fang, Y.: Evaluating very long-term conversational memory of LLM agents. arXiv preprint arXiv:2402.17753 (2024)
15. Papineni, K., Roukos, S., Ward, T., Zhu, W.J.: Bleu: a method for automatic evaluation of machine translation. In: Proceedings of the 40th Annual Meeting of the Association for Computational Linguistics, pp. 311–318 (2002)
16. Siino, M.: All-mpnet at semeval-2024 task 1: application of mpnet for evaluating semantic textual relatedness. In: Proceedings of the 18th International Workshop on Semantic Evaluation (SemEval-2024), pp. 379–384 (2024)

17. Singh, R., Singh, S.: Text similarity measures in news articles by vector space model using NLP. J. Inst. Eng. (India) Ser. B **102**(2), 329–338 (2021)
18. Song, L., et al.: Adaptive in-conversation team building for language model agents. arXiv preprint arXiv:2405.19425 (2024)
19. Touvron, H., et al.: Llama: open and efficient foundation language models. arXiv preprint arXiv:2302.13971 (2023)
20. Wang, D., Yang, J., Li, W., Liang, J., Li, Y.: MSARL: decoupling reasoning and tool use with multi-small-agent reinforcement learning. arXiv preprint arXiv:2508.08882 (2025)
21. White, I., et al.: Collaborating action by action: a multi-agent LLM framework for embodied reasoning. arXiv preprint arXiv:2504.17950 (2025)
22. Wu, Q., et al.: Autogen: enabling next-gen LLM applications via multi-agent conversations. In: First Conference on Language Modelling (2024). https://openreview.net/forum?id=BAakY1hNKS
23. Yuan, Y., Xie, T.: Reinforce LLM reasoning through multi-agent reflection. arXiv preprint arXiv:2506.08379 (2025)
24. Zhao, Y., et al.: PyTorch FSDP: experiences on scaling fully sharded data parallel. arXiv preprint arXiv:2304.11277 (2023)

LegalSumAI: Robust and Representative Legal Summarization with Large Language Models

Purvi Sehgal[1(✉)], Sahithi Ankireddy[1,2], and Saumya Chauhan[1]

[1] California Institute of Technology (Caltech), Pasadena, CA 91125, USA
{psehgal,sankired,schauhan}@alumni.caltech.edu
[2] Stanford University, Palo Alto, CA 94305, USA

Abstract. Legal disputes are complex, high-stakes challenges, yet 56% of US households encounter legal issues annually [1]. The jargon, length, and complexity of legal documents often make them inaccessible to the general public. This work introduces LegalSumAI, a novel framework that generates robust legal summaries from case documents using Large Language Models (LLMs). We investigate whether LLMs can produce accurate, comprehensible summaries of legal cases and opinions without hallucinations. Our two-step pipeline first generates structured CSV fact sheets, capturing case details via the IRAC framework (Issue, Rule, Application, Conclusion) and metadata on the involved parties. These fact sheets are then converted into natural language summaries prompted with Chain-of-Density (CoD) techniques. We leverage the Multi-LexSum dataset, which provides expert-authored summaries at three different granularity levels (tiny, short, long) to evaluate the generated summaries using ROUGE and BERTScore metrics. Results show high semantic accuracy, with average BERTScore improvements of 157% over Multi-LexSum baselines, demonstrating that our structured reasoning and CoD prompting mitigate LLM hallucinations and improve legal summarization results. This research demonstrates the potential for interpretable LLM pipelines to democratize access to legal knowledge.

Keywords: Large Language Models · Legal document processing and contract analysis · Structured Reasoning

1 Introduction

Legal matters tend to be high-stakes, often involving people's fundamental rights, liberties, and finances. In criminal proceedings, lives and incarceration may be at stake, while in civil cases, anywhere from tens of thousands to millions of dollars can be on the line.

Legal tasks are not only high-stakes, but many American households are actively involved in legal issues. According to the December 2018 survey of US adults conducted by SSRS for the Pew Charitable Trusts, 56% of households

H. R. Arabnia et al. (Eds.): CSCI 2025, CCIS 2972, pp. 592–600, 2027.
https://doi.org/10.1007/978-3-032-28469-3_39

experienced at least one civil legal problem in the previous 12 months [1]. When considering other court systems, it is evident that a majority of American households grapple with legal issues. Thus, individuals without legal backgrounds should have access to accurate, comprehensible information about their cases.

However, the wording and structure of legal documents present significant barriers to accessibility. A typical document is approximately 30 pages on average, is written in dense language, and assumes prior knowledge, making it difficult for people without legal expertise to comprehend. In fact, Stygall (2020) stated that, "in legal texts, there may be references to judicial decisions, legal journal articles, briefs, regulations or statutes (or all of these). Lay readers are distracted by such references, as they generally lack access to the texts being referenced [2]." This combination of length, complexity, and intertextual dependence makes these documents difficult for the general public to understand.

Given the scale and impact of these challenges, there is a clear need for effective simplification and summarizing methods of legal documents. Large Language Models (LLMs) offer a promising approach to address this issue. By leveraging advanced natural language processing techniques, LLMs can analyze, interpret, and summarize vast amounts of legal information. However, LLMs are highly prone to biased outputs, perpetuating harmful stereotypes and prejudices within legal contexts. For instance, biases against queer and transgender individuals have been observed in transformer models like BERT, leading to homophobic and transphobic outputs [3]. Furthermore, LLMs also often hallucinate, generating output that is highly illogical based on the input context. In high-stakes legal contexts, such distortions or factual inaccuracies can jeopardize trust and fairness of our legal systems. Thus, a current key challenge involves designing robust, interpretable LLM frameworks that minimize hallucinations and bias and maintain factual accuracy and precision. Our work addresses this need through structured and multi-stage reasoning, designed for the rigor required in legal summarization tasks.

2 Related Work

2.1 Benchmarks and Datasets

Recent advances in natural language processing have led to the creation of several domain-specific legal datasets aimed at improving model understanding and reasoning. LeXFiles is comprised of 11 sub-corpora and supports generalization across different legal fields [4]. Cambridge Law Corpus contains 250,000 UK court cases from the 16th to 21st centuries, designed for NLP and machine learning studies [5]. CaseHOLD offers multiple choice data derived from US legal documents, supporting baseline performance evaluations [6]. MultiLegalPile is a multilingual dataset with more than 680GB of data across 24 languages, promoting cross-lingual legal research [7].

While these datasets have advanced the field, they often focus on classification tasks or very specific legal text types. This has left gaps in legal summarization research, demonstrating that it is still young and unexplored.

However, Multi-LexSum, a featured paper at NeurIPS, established one of the first large-scale benchmarks for legal summarization across multiple documents [8]. It contains more than 9,000 expert-developed summaries at three different granularities (tiny, short, and long). Each summary is factually grounded and serves as the gold standard for research in legal summarization.

2.2 Summarization Tasks

In the Multi-LexSum paper, state-of-the-art summarization frameworks were evaluated on the Multi-LexSum benchmark. These included BART, PEGASUS, PRIMERA, and two different Longformer-Encoder-Decoder (LED) models applied across the different granularities [8]. While the PRIMERA model improved recall with shorter document context, it did not perform well with longer documents typical of real-world legal cases. In fact, human evaluators scored model outputs an average of 0.43 on a 0–3 quality scale. The authors of Multi-LexSum concluded that "existing summarization models struggle to produce the summaries directly from the long source documents [8]," emphasizing issues with current legal summarization methods.

To address these issues, we developed an LLM-based framework, which can better handle longer context lengths and complexities typical of legal cases. We explored structured and interpretable reasoning to avoid hallucinations and provide factual grounding and semantic accuracy for legal summarization tasks.

3 Methods

This work focused on generating robust summaries from legal case documents using LLMs. The primary task was to prompt an LLM to produce accurate, comprehensible summaries of legal cases and opinions for the general public. To achieve this, we investigated prompting LLMs to encourage reasoning as opposed to memorization, reduce biases, and improve factual grounding in our framework.

3.1 Introduction to Multi-lexSum

The Multi-LexSum research contributed in two essential ways to this research:

1) Multi-LexSum Dataset: As discussed in the previous section, the Multi-LexSum dataset addressed limitations of previous datasets by providing a comprehensive collection of expert-authored summaries for United States civil rights lawsuits. Multi-LexSum is distinctive in its extensive source text length, averaging over 75,000 words per case, and its 9,000 high-quality summaries, which adhere to strict content and style guidelines [8]. Thus, we utilized Multi-LexSum summaries as our benchmark and evaluation ground truth.

2) Additionally, Multi-LexSum reported the performance of state-of-the-art legal summarization models, including BART, PEGASUS, PRIMERA, and two different Longformer-Encoder-Decoder (LED) models, on the Multi-LexSum dataset. These served as the baselines for comparison, since LegalSumAI was evaluated on the same dataset.

3.2 Framework Overview

Here, we propose a two-step pipeline. Instead of directly prompting an LLM for a natural language summary from the source documents, we first generated a CSV "fact sheet" using LLMs to represent the information in a more structured manner. This fact sheet contained information on the Issue, Rule, Application, Conclusion, parties involved, and other logistical case information for each document. The resulting fact sheet was then parsed by another LLM prompted using Chain-of-Density (CoD) to generate a summary of a specified length and purpose. We utilized GPT-3.5-Turbo (GPT-3o) as the LLM for all tasks.

3.3 Fact Sheet Generation

Our approach began with generating "fact sheets" for each legal case document using LLMs. This step involved extracting key information from the documents and structuring it into a concise, standardized format. The fact sheet included critical details such as the issue at hand, applicable laws, key arguments, and conclusions. By breaking down complex legal documents into manageable fact sheets, we ensured that the LLM captured essential elements of each case without being overwhelmed by the document's length and complexity.

To generate a fact sheet, we first separated each case document into the following general categories: Case Information, Parties Involved, Legal Basis, Case Background, Court Proceedings, Settlement and Agreements, Outcome/Impact, and Miscellaneous. Then, we prompted a second model to use the IRAC method (Issue, Rule, Application, Conclusion), information about the parties involved, and other logistical case details to further categorize the data.

Through a review of existing legal reasoning practices, we identified IRAC as the principal framework used by lawyers and law students to organize legal arguments and summarize case reasoning. Incorporating this structure as an intermediate step to create fact sheets allowed our framework to maintain logical coherence and factual consistency with human legal reasoning.

3.4 Combining Fact Sheets

Legal documents are often too long for the context window of LLMs. Thus, the source text must be separated, processed, and recombined.

For cases in the MultiLexSum dataset, we created a fact sheet from each document individually and then combined them into a single comprehensive fact sheet. In the case of a raw text input, we broke the text sequentially into smaller passages.

Since each segment was processed individually, we used another LLM, prompted using the IRAC method, to combine fact sheets. The combination of fact sheets allowed us to handle cases with multiple documents efficiently, ensuring that no critical information was overlooked. This step was crucial for maintaining the integrity and completeness of the summary, especially in cases with extensive documentation.

3.5 Generating Natural Language Summaries with Chain-of-Density

The fact sheet was passed as contextual input to an LLM to generate summaries at three granularities ("tiny," "short," and "long"), corresponding to the summary lengths of the Multi-LexSum benchmark.

For each length, the model was prompted four times using Chain-of-Density (CoD), with each iteration improving the summary to increase information density and focus. Within each CoD prompt iteration, five summaries were produced and the one with the highest cosine similarity score was selected, ensuring semantic consistency with prior outputs. The summary from the last CoD stage was selected. This method helped in creating more focused and dense summaries by iteratively refining the prompts, ensuring that each subsequent summary captured more relevant details.

3.6 Evaluation Metrics

To evaluate the accuracy of the generated summaries, we computed ROUGE and BERT scores. ROUGE (Recall-Oriented Understudy for Gisting Evaluation) measures the overlap of n-grams between the generated summaries and the reference summaries, focusing on precision, recall, and F1-score. BERT (Bidirectional Encoder Representations from Transformers) score, on the other hand, provides a more nuanced assessment of semantic similarity by comparing the embeddings of the generated and reference summaries. High ROUGE and BERT scores indicate that a summary accurately reflects the content of the source documents, while lower scores highlight areas where the summary may need improvement. This evaluation step was critical to ensure reliability and validity of generated summaries.

3.7 Baseline LLM Summarization

To establish an LLM comparative baseline, we implemented a direct summarization approach using GPT-3.5-Turbo (GPT-3o) without any structuring, prompting techniques, or IRAC segmentation present in our LegalSumAI framework. Summaries were generated for each target length category ("tiny," "short," and "long").

We repurposed the prompt we utilized to convert fact sheets into summaries by changing the input type to the case documents themselves, eliminating the Chain-of-Density steps, and removing multi-prompt generation. Because this baseline used a very similar prompt to this LegalSumAI framework, it provided an effective comparison. Thus, it involved a single-step summarization process that isolated the effects of the structured reasoning and multi-stage prompting strategies of LegalSumAI.

3.8 Justification for Chosen Approach

Ultimately, we selected this approach due to its structured and comprehensive nature. The Multi-LexSum dataset provided a rich source of expert-authored summaries, offering a solid foundation for training and evaluating LLMs. The multi-granularity aspect of the dataset enabled us to test the LLM's performance across different summary lengths, ensuring versatility and adaptability. By generating and combining fact sheets, we handled complex cases with multiple documents effectively, ensuring that the final summaries were accurate and comprehensive. The use of ROUGE and BERT scores for evaluation provided robust metrics for assessing the accuracy of the summaries, ensuring that our approach was both rigorous and reliable.

This approach provides a scalable framework for generating and evaluating legal summaries using LLMs. By focusing on structured data generation, robust evaluation metrics, and improved prompting strategies, we aimed to produce summaries that were not only accurate and comprehensible but also accessible and unbiased, thereby enhancing the usability of LLMs in the legal domain.

4 Results

The performance of generated summaries was evaluated using ROUGE and BERTScore metrics. Tables 1 and 2 provide the ROUGE and BERT F1 scores for all tiny, short, and long summaries for both our pipeline and Multi-LexSum.

Table 1. LegalSumAI Rouge and BERT F1 Scores Across All Granularities

Granularity	Rouge1	Rouge2	RougeL	BERT
tiny	0.0771	0.0387	0.0557	0.8307
short	0.2711	0.0777	0.1444	0.8328
long	0.4302	0.1249	0.1869	0.8403

Table 2. Multi-LexSum Rouge and BERT F1 Scores Across All Granularities

Granularity	Rouge1	Rouge2	RougeL	BERT
tiny	0.2261	0.0709	0.1844	0.2678
short	0.4335	0.1991	0.2999	0.3788
long	0.4079	0.2001	0.2536	0.3483

LegalSumAI had much higher BERT scores, indicating that generated summaries captured the essence and semantic meaning of the source document well. Our BERT scores were significantly higher than Multi-Lex Sum's BERT scores for all granularities, as shown in the table, with BERT scores improving as much as 207% and approximately 157% on average for all categories, indicating a much more accurate summary.

Lower ROUGE scores indicated fewer direct matches in terms of words or phrases. This can be explained because we desired to generate summaries for the general public, where the wording should be more simple, which was evident in our outputs. Thus, LegalSumAI did not end up replicating exact verbiage from the expert summaries. Ultimately, a significantly higher BERTScore and lower ROUGE score indicated that the generated text was contextually and semantically accurate, while diverging in wording from the source text.

Next, the baseline LLM results are presented in Table 3. Since this model produced unstructured summaries directly from the full text, the generated outputs achieved low lexical overlap with the reference Multi-LexSum summaries, as demonstrated by the low ROUGE scores. LegalSumAI achieved higher Rouge and BERT scores for all granularities compared to the LLM baseline. These results highlight the effectiveness of LegalSumAI's structured intermediate reasoning and iterative Chain-of-Density (CoD) prompting. By incorporating explicit organization through fact-sheet generation, our framework achieved a stronger factual grounding and legal coherence than unstructured LLM summarization.

Table 3. LLM Rouge and BERT Scores Across All Granularities

Granularity	Rouge1	Rouge2	BERT
tiny	0.0100	0.0000	0.1943
short	0.0520	0.0000	0.4686
long	0.1301	0.0000	0.7478

5 Discussion

This framework marks a significant improvement in providing structured reasoning for legal summary generation. The comparison of ROUGE and BERT scores between our framework and the Multi-LexSum models underscores that LegalSumAI achieves higher semantic accuracy in representing the original text. This outcome aligns with our objectives, as we aim to produce summaries using more accessible language that conveys the same semantic meaning as the original legal texts.

Our results also demonstrated that, while unstructured LLM frameworks independently produced semantically coherent text, their summaries were still not as anchored to the underlying legal facts. The addition of structure and advanced prompting techniques improved semantic precision and factual grounding, highlighting the necessity of multi-stage reasoning in adapting LLMs to the rigor of the legal domain.

This improvement can be attributed to our two-layer fact sheet approach and CoD prompting technique. The first layer sorted key information into general categories such as Case Information, Parties Involved, and Legal Basis. The second layer refined this categorization using the IRAC (Issue, Rule, Application, Conclusion) framework, a common technique among lawyers. This repeated parsing and combining of information enabled the model to develop a better semantic understanding of the input text.

The CoD prompting technique further structured LLM reasoning while increasing user flexibility in the output type. Users can specify the desired length and focus of the summary, allowing for tailored outputs that meet diverse needs. This flexibility is crucial in legal applications where the level of detail required can vary significantly depending on the context and user's expertise.

By breaking down information into structured categories and then synthesizing them into a comprehensive summary, our framework decreased hallucinations and more closely adhered to the content of legal documents.

6 Conclusion

Our results demonstrate that LegalSumAI improved summary quality compared to unstructured LLM baselines and prior state-of-the-art methods reported in Multi-LexSum research. By introducing a structured, interpretable framework, LegalSumAI improves the semantic understanding and summarization of legal documents, an area that lacked effective, high-performing summarization solutions. This work represents a step forward in democratizing access to legal knowledge for all.

Future work will expand evaluation to other legal domains, refine prompting techniques, and explore fine-tuning using legal data to enhance applicability and robustness. Beyond the legal field, this methodology can be extended to a multitude of other domains with dense, technical language, such as healthcare or policy. These extensions can continue to improve the accessibility and reliability of specialized information for a broader audience.

Disclosure of Interests. The authors have no competing interests to declare that are relevant to the content of this article.

References

1. Rickard, E.: Many U.S. Families Faced Civil Legal Issues in 2018. Pew Charitable Trusts (2018)
2. Stygall, G.: Legal Writing: Complexity—Complex Documents/Average and Not-So-Average Readers. In: The Routledge Handbook of Forensic Linguistics, pp. 32–47. Routledge, London (2020)
3. Felkner, V.K., Chang, H.-C.H., Jang, E., May, J.: Towards WinoQueer: Developing a Benchmark for Anti-Queer Bias in Large Language Models. arXiv preprint arXiv:2206.11484 (2022)
4. Chalkidis, I., Garneau, N., Goanta, C., Katz, D.M., Søgaard, A.: LeXFiles and LegalLAMA: Facilitating English Multinational Legal Language Model Development. arXiv preprint arXiv:2305.07507 (2023)

5. Östling, A., et al.: The Cambridge Law Corpus: A Dataset for Legal AI Research. arXiv preprint arXiv:2309.12269 (2024)
6. Zheng, L., Guha, N., Anderson, B.R., Henderson, P., Ho, D.E.: When Does Pretraining Help? Assessing Self-Supervised Learning for Law and the CaseHOLD Dataset. arXiv preprint arXiv:2104.08671 (2021)
7. Niklaus, J., Matoshi, V., Stürmer, M., Chalkidis, I., Ho, D.E.: MultiLegalPile: A 689GB Multilingual Legal Corpus. arXiv preprint arXiv:2306.02069 (2023)
8. Shen, Z., Lo, K., Yu, L., Dahlberg, N., Schlanger, M., Downey, D.: Multi-LexSum: Real-World Summaries of Civil Rights Lawsuits at Multiple Granularities. arXiv preprint arXiv:2206.10883 (2022)

Adaptive Melodic-Entropy-Guided Preprocessing for Lightweight MIDI Generation Models

Huaiyu Zhang and Chengwei Lei(✉)

Department of Computer and Electrical Engineering and Computer Science, California State University, Bakersfield, CA, USA
{hzhang5,clei}@csub.edu

Abstract. A key challenge in symbolic music generation is how to preprocess and structure MIDI data to enhance both the musicality and generalization of Transformer-based models. Due to data scarcity and computational constraints, users training on private datasets are often compelled to adopt lightweight models, whose reduced capacity leads to a significant drop in performance. This study proposes a lightweight, entropy-aware preprocessing pipeline for symbolic music generation and compares it with the conventional Music Transformer baseline. The proposed method eliminates chordal redundancy, computes melodic entropy to identify melody-rich regions, and applies targeted data augmentation prior to tokenization. It aims to achieve improved model performance under small-data regimes scenario. Experimental results demonstrate that the proposed entropy-guided approach produces sequences with higher melodic-entropy alignment to the original test set, indicating enhanced melodic coherence and generative quality for small symbolic music datasets.

Keywords: Lightweight LLM · Small-Data Regimes · Event-Based Token Augmentation · LLM on MIDI · Melodic Entropy

1 Introduction

In symbolic music generation, the scalability of Transformer-based architectures is fundamentally limited by the availability of training data. State-of-the-art models typically rely on vast corpora, which poses a barrier for individual researchers and small groups without access to large-scale resources [1]. Conventional segmentation strategies—such as bar-level or fixed-length slicing—further restrict performance by disregarding structural heterogeneity across musical files [2], thereby limiting the capacity of models to capture melody-rich variations that strongly influence generative quality.

These constraints highlight two persistent challenges: the data- and compute-intensive nature of existing Transformer pipelines, and the loss of musically salient information due to rigid segmentation. Even with advanced architectures,

H. R. Arabnia et al. (Eds.): CSCI 2025, CCIS 2972, pp. 601–614, 2027.
https://doi.org/10.1007/978-3-032-28469-3_40

models trained under limited-data regimes often exhibit constrained generative performance [3].

Prior work has explored general-purpose data augmentation strategies, including pitch transposition and time-stretching, as a means to expand symbolic music corpora [4,5]. However, such techniques are typically applied uniformly, overlooking variations in melodic information density that could guide more effective augmentation.

To overcome these limitations, a melody-entropy-driven preprocessing and augmentation pipeline is introduced. The method quantifies melodic information via note entropy, prunes chordal redundancies, and selectively augments high-entropy regions prior to tokenization. By aligning augmentation with musically informative segments, the approach improves data efficiency and enhances melodic quality in generated sequences, particularly in low-resource settings.

This contribution demonstrates that melody-aware preprocessing provides a practical pathway toward resource-efficient symbolic music generation, extending accessibility of advanced generative modeling techniques beyond large-scale institutional settings.

2 Conventional Data Preparation and Tokenization Frameworks

This section describes the conventional preprocessing and tokenization procedures commonly adopted in symbolic music generation. It outlines the construction and normalization of the experimental dataset, followed by a review of event-based tokenization schemes and standard corpus segmentation strategies. These provide the methodological foundation against which the proposed entropy-aware preprocessing method will later be compared.

The modeling process is founded on a simplified event-based tokenization scheme that preserves pitch and positional information while omitting velocity and extensive instrumentation, consistent with recent practice in reproducible symbolic music generation frameworks [1].

2.1 Statistical Analysis of Experimental Data

A symbolic corpus was assembled from the public BitMidi repository [6], comprising 113,229 raw MIDI files. Following format validation and duplicate removal, all sequences were resampled onto a quarter-note–consistent metrical grid, a common normalization step in symbolic music preprocessing [7,8], to ensure tempo-invariant alignment across bars, beats, and sub-beats prior to tokenization. This normalization provides a uniform temporal reference, facilitating consistent positional encoding and mitigating quantization noise inherent in community-sourced MIDI data.

To emphasize melodic content, files containing dense chordal clusters or percussion-dominated tracks were excluded. Specifically, near-instantaneous high-concurrency onsets were removed, and General MIDI drum channels or instrument tags associated with percussive material were filtered out. The resulting data consist primarily of monophonic or lightly polyphonic melodic lines, consistent with recent symbolic-music preprocessing practices that treat drum components as independent conditioning streams rather than pitch-bearing events [9].

2.2 Current Data Tokenization Approach

The corpus was encoded using a beat-aligned, event-based representation designed to retain only the information essential for melodic modeling. Each score was quantized on a quarter-note–consistent temporal grid and serialized as a sequence of discrete symbolic events. The vocabulary consists of three primary token types:

$$\text{BAR}, \quad \text{POS}_k, \quad \text{NOTE}(p, d_q),$$

where BAR marks bar boundaries, POS_k denotes relative positions within a bar uniformly subdivided on the quantized beat grid, and $\text{NOTE}(p, d_q)$ represents a note with pitch p and duration d_q measured in units of the quarter-note grid.

Compared to richer MIDI-style encodings, this representation omits velocity, articulation, and program-change information, and excludes percussive tracks from the token vocabulary. The resulting symbol set is substantially smaller and produces shorter, less sparse sequences, thereby improving data efficiency without sacrificing the tonal and rhythmic structure necessary for melodic generation. The design follows the class of beat-aware, REMI-style symbolic formats [10], but is deliberately simplified to concentrate model capacity on interactions between pitch, temporal position, and duration while suppressing non-melodic variability.

2.3 Conventional Corpus Segmentation Strategies

In symbolic music generation research, corpus segmentation—the process of dividing full MIDI or symbolic pieces into shorter, manageable excerpts—constitutes a fundamental preprocessing step. Traditional approaches typically segment data according to fixed temporal or bar-level boundaries, such as slicing each piece into non-overlapping four-bar or eight-bar segments, or by fixed time windows aligned to the metrical grid (e.g., 16 beats in 4/4 m). These slices are usually determined by bar onsets rather than by phrase or structural boundaries. For instance, in large-scale symbolic music pre-training frameworks, musical sequences are often "divided into equal-length and non-overlapping slices" to facilitate tokenization and batch processing [11].

This uniform slicing strategy offers simplicity and consistency: it ensures fixed-length model inputs, facilitates mini-batch construction, and allows efficient large-scale corpus processing. However, such rule-based segmentation also introduces notable limitations. Because fixed-length slicing disregards higher-level musical structure—such as phrase boundaries, thematic transitions, or local variations in harmonic and melodic density—it may combine melody-rich and accompaniment-dominated passages within the same segment. Consequently, the resulting samples exhibit heterogeneous information density, which can obscure meaningful musical regularities. Moreover, structural changes that could be identified by unsupervised segmentation or change-point detection methods are frequently ignored [12].

Hence, while traditional corpus segmentation remains widely used for its simplicity and reproducibility, it is inherently coarse-grained and may overlook important structural cues that influence musical coherence and model learning efficiency.

3 Novel Method: Melody-Entropy-Guided Augmentation

This section first introduces the strategy for large-scale vocabulary reduction, which aims to minimize redundancy while preserving the melodic information content of the symbolic representation. Building on this compact event space, it then presents a data-centric augmentation pipeline guided by a quantitative proxy of melodic complexity. We define a *melody-change entropy* that measures the uncertainty in note-to-note movements within beat-aligned windows, following the information-dynamics view that predictive uncertainty and entropy are valid descriptors of musical expectation and complexity [13]. Based on these entropy scores, we adopt a segmentation-and-selection scheme that ranks windows and assigns bucketed policies (overlap and pitch-shift intensity) while respecting plausible structural boundaries through data-driven structure analysis [14]. The remainder of this section formalizes the entropy estimator, details bucketed selection rules, and specifies augmentation operators and constraints.

3.1 Vocabulary Reduction

In the original event representation, the combination of all distinct pitches and intra-bar offsets results in an extremely large token vocabulary, within which many symbols occur only a handful of times. Such sparsity hinders model convergence and increases training instability without contributing meaningful musical variation. To address this, corpus-level statistics were computed to assess whether a reduced event vocabulary could preserve the representational integrity of the melodic signal.

Corpus-level statistics were computed to assess whether the reduced event vocabulary preserves the representational integrity of the melodic signal. The empirical pitch distribution exhibits a unimodal, approximately Gaussian profile with light tails, indicating concentration around a tonal center—a characteristic commonly observed in tonal corpora due to statistical learning and enculturation effects [15]. To reduce distributional sparsity and suppress infrequently used pitch events, extreme values were compressed using quantile-based winsorization with heuristically chosen thresholds at the 5% and 95% percentiles. As shown in Fig. 1, the resulting histogram retains the near-Gaussian shape while effectively removing extreme deviations.

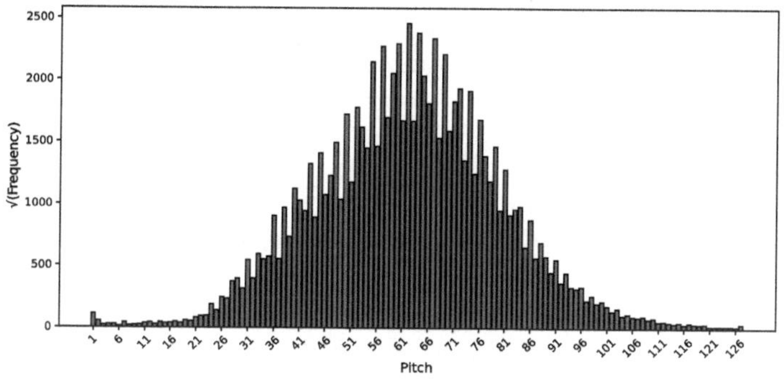

Fig. 1. Empirical pitch distribution after winsorization.

This operation serves purely as a statistical hygiene measure and does not alter the overall mass or structure of the distribution.

The distribution of intra-bar position indices reveals pronounced modes at metrically strong beats and salient subdivisions of the quarter-note grid. This pattern confirms that the quantization and positional encoding scheme align with perceptually meaningful rhythmic anchors. The combined evidence from pitch and position statistics indicates that the compact representation retains the dominant tonal and temporal regularities that sequence-based models exploit for melodic generation.

All preprocessing parameters, including beat resolution, quantization settings, and length constraints, were fixed globally throughout the data set. Bar-level resets were applied at time-signature changes to ensure consistent interpretation of intra-bar positions and to preserve cross-file comparability during model training and evaluation. This tokenization is adopted solely to provide a stable, beat-aware interface for the subsequent *augmentation* study; it is not intended as a modeling contribution (Table 1).

Table 1. Top intra-bar offset values and their relative frequencies.

Duration	Frequency	Proportion
0	16,784,415	0.1179
240	11,671,423	0.0820
120	10,951,732	0.0769
90	5,634,114	0.0396
180	5,493,402	0.0386
60	4,956,367	0.0348
480	3,747,900	0.0263
360	2,380,101	0.0167
45	2,324,681	0.0163

These eight offsets account for 44.91% of all 10,178 unique position values, representing metrically salient note durations commonly used in Western tonal music. This concentration indicates that the majority of onsets align with perceptually strong rhythmic anchors, which further validates the quarter-note–consistent quantization scheme used for tokenization. Notably, these dominant offset values also emerge as natural cluster centroids in unsupervised KNN clustering, indicating that their statistical stability in rhythmic structure aligns with their perceptual salience.
Reducing the dimensionality of pitch and inter-event duration tokens effectively compresses the model vocabulary with minimal loss of representational capacity, improving both training efficiency and generative fidelity.

3.2 Melody Change Entropy

A key component of our algorithm lies in how it distinguishes between melody-rich regions and chordal or redundant structures. On the one hand, many symbolic music systems still employ uniform segmentation (e.g., fixed-length slicing), which may overlook phrase boundaries or structural changes identifiable by unsupervised segmentation methods [16]. On the other hand, standard augmentation techniques—such as pitch transposition and time-stretching—are widely adopted in symbolic music pipelines [17].

Melody change entropy is a quantitative measure of local melodic variability. Let a melody be represented as a beat-aligned pitch sequence $\{p_t\}_{t=1}^{T}$ on the token grid introduced in Sect. 2.2. For each time step, compute the pitch change from note to note as $\Delta p_t = p_t - p_{t-1}$ for $t = 2, \ldots, T$. Within a temporal window $W = [\tau, \tau+L)$, we estimate the empirical distribution of pitch intervals over a discrete alphabet $\mathcal{I}$, for example $\{-M, \ldots, M\}$ in semitones or magnitude bins when interval direction is ignored:

$$\widehat{p}_W(i) = \frac{1}{N_W} \sum_{t:\, t\in W} \mathbf{1}\{\Delta p_t \in i\}, \qquad N_W = |\{t : t \in W\}|.$$

The entropy of this local interval distribution is defined as

$$H(W) = -\sum_{i\in\mathcal{I}} \widehat{p}_W(i) \log(\widehat{p}_W(i) + \varepsilon), \qquad \widetilde{H}(W) = \frac{H(W)}{\log|\mathcal{I}|} \in [0, 1],$$

where $\varepsilon > 0$ is a small constant for numerical stability. The normalized form $\widetilde{H}(W)$ expresses the degree of unpredictability in interval usage, ranging from low values for highly repetitive or stepwise motion to high values for diverse, less predictable melodic progressions.

This measure is grounded in information theory: entropy quantifies the uncertainty of symbolic sequences and has been widely used to model expectation and creativity in musical cognition [18]. Because it operates on relative pitch intervals rather than absolute pitches, it remains invariant to global transposition and thus aligns naturally with data augmentation procedures such as pitch shifting.

When unsigned interval magnitudes are used, it also abstracts away directional contour, isolating variability independent of melodic polarity. The normalization by $\log|\mathcal{I}|$ yields a bounded and interpretable scale between 0 and 1, allowing direct comparison across windows or pieces. Low-entropy regions indicate strong local regularity, while high-entropy regions reflect greater diversity of motion.

Evaluated over sliding windows, $\widetilde{H}(W)$ provides a continuous signal of melodic complexity that can guide data selection, segmentation, and augmentation intensity without requiring any structural annotations. The computation is lightweight, involving only histogram accumulation per window. The alphabet $\mathcal{I}$ and smoothing constant ε serve as simple parameters to control the sensitivity and numerical stability of estimation. In practice, we set $\mathcal{I}$ to semitone steps within $[-M, M]$ (typically $M = 12$) and use fixed window length L and stride S to produce a time-varying entropy curve that reflects the local diversity of melodic motion across the dataset.

3.3 Entropy-Conditioned Slicing and Augmentation

To systematically control how different levels of melodic variability contribute to the training data, we treat slicing and augmentation as an entropy-conditioned process. Each file f in the corpus provides a sequence of fixed-length windows $W_{f,u}$ with corresponding normalized melody-change entropy scores $H_f(u) \in [0, 1]$ as defined in Sect. 3.2. These local entropy values are used to divide all windows into K buckets of increasing complexity.

Let F_H denote the empirical cumulative distribution function (CDF) of $\{H_f(u)\}$ over the entire dataset. For a chosen set of breakpoints $\boldsymbol{\beta} = (\beta_0, \ldots, \beta_K)$ with $0 = \beta_0 < \beta_1 < \cdots < \beta_K = 1$, the corresponding bucket thresholds are

$$q_k = \mathsf{F}_H^{-1}(\beta_k), \qquad k = 0, \ldots, K.$$

A window (f, u) is assigned to bucket k if $q_{k-1} < H_f(u) \leq q_k$. The raw proportion of samples falling in bucket k is

$$r_k = \frac{|\{(f,u) : b(f,u) = k\}|}{\sum_j |\{(f,u) : b(f,u) = j\}|}.$$

To rebalance the data distribution, we introduce a target composition $\boldsymbol{\alpha} = (\alpha_1, \ldots, \alpha_K)$ with $\sum_k \alpha_k = 1$, and compute the relative sampling weight

$$s_k = \frac{\alpha_k}{\max(\varepsilon, r_k)}, \qquad \varepsilon > 0.$$

Each window is then accepted into the training set with probability $p_{\text{sel}}(f, u) \propto s_{b(f,u)}$, following a reservoir-based sampling scheme [19]. This importance weighting parallels ideas in curriculum and importance-sampling learning [20,21], where sample difficulty or informativeness guides the construction of the training distribution. The overall slicing and augmentation process remains deterministic and reproducible, implemented as Algorithm-S sampling with entropy-dependent overlap and pitch-shift rules (cf. [17]).

By varying $(\boldsymbol{\beta}, \boldsymbol{\alpha}, \mathbf{w}^{(\mathrm{aug})})$, we obtain multiple dataset configurations that emphasize different regions of the entropy spectrum. Each configuration defines a specific training policy $\Psi(\tau)$ and corresponding model instance $M(\tau)$. The following section evaluates these policies—nine in total—by comparing the generated outputs with their source melodies in terms of Pearson similarity and entropy deviation.

4 Experiments and Results

This section empirically evaluates the proposed entropy-guided slicing and augmentation strategy. An internal sensitivity study was first conducted to examine how the curvature parameter t affects data composition and model behavior, from which an optimal configuration was selected. Using this setting, we train and compare transformer-based symbolic music models against a conventional bar-level slicing baseline under identical architecture and training conditions. The evaluation focuses on how entropy-conditioned data preprocessing influences generative fidelity, measured by two quantitative criteria: the absolute deviation in mean melody-change entropy and the Pearson correlation between pitch-class distributions of generated and reference sequences.

4.1 Experimental Setup

All experiments are conducted on the same symbolic corpus used for model training. Each MIDI file is tokenized into a beat-aligned pitch stream as described in Sect. 2.2. Local melodic complexity is quantified by the normalized melody-change entropy $\widetilde{H}(W)$ defined in Sect. 3.2. Based on these entropy values, we generate nine training subsets using the entropy-conditioned slicing algorithm in Sect. 3.3, corresponding to curvature parameters

$$t \in \{-1.0, -0.75, -0.5, -0.25, 0.0, 0.25, 0.5, 0.75, 1.0\}$$

. Each subset contains the same total number of windows (K), ensuring controlled comparison across policies. Files are segmented using entropy-dependent overlap and duplication rules, and augmented via pitch shifts constrained to $[n_{\min}, n_{\max}]$ to preserve playable range. The resulting datasets differ only in the proportion of low-, mid-, and high-entropy slices.

All models share an identical transformer-based architecture similar to GPT-2, trained on the same token vocabulary and positional scheme. The configuration is summarized in Table 2. Although larger transformers can achieve higher capacity, our total training token budget is approximately 8×10^8. Following the compute-optimal scaling law [3], model size and data volume should scale proportionally to maximize generalization. Under limited data, excessively large models would be undertrained or overfit. Therefore, we adopt a medium-scale configuration (12 layers, 512 hidden units, 8 heads) that matches the available data and compute while maintaining stable optimization.

Table 2. Transformer architecture and main training hyperparameters shared by all nine models. Only the dataset composition differs across entropy policies.

Parameter	Value
Model type	GPT-2 style Transformer
Number of layers	12
Hidden size	512
Attention heads	8
Context length	256
Dropout rate	0.1
Optimizer	AdamW
Learning rate	3×10^{-4}
Batch size	16 sequences
Training epochs	30
Total training tokens	$\approx 8 \times 10^8$

To ensure fair evaluation, the files within each entropy bucket were randomly divided into a 9:1 split. Ninety percent of the files were used to construct the training datasets, while the remaining ten percent served as held-out material for testing. The test segments were used as seed inputs for conditional generation, allowing each trained model to produce continuations of comparable length. For evaluation, the held-out files in each entropy bucket were not used during training and served exclusively for conditional generation. From each test file, a fixed-length prefix of 100 event lines (approximately 400 tokens) was extracted as the generation seed, corresponding to roughly two bars of melodic material depending on tempo. Each model was prompted with this seed and instructed to autoregressively generate 400 additional tokens. The generated continuation was then aligned with the ground-truth continuation of equal length immediately following the seed in the original file. All quantitative metrics were computed over this aligned region, enabling a one-to-one comparison between the model output and the true continuation.

The effectiveness of each entropy policy was assessed by comparing the generated continuations with their corresponding original segments. Two quantitative measures were computed. The first is the absolute difference in mean melody-change entropy, which reflects how well the model preserves the local complexity level of the source material. The second is the Pearson correlation between the pitch-class distributions of the generated and reference sequences, indicating overall similarity in note usage. Together, these measures provide a balanced view of structural consistency and stylistic fidelity, enabling direct comparison across the nine models trained under different entropy-guided data compositions.

4.2 Performance Comparison with Varying Parameter Configurations

This subsection quantitatively examines how entropy-conditioned data composition influences generative fidelity and melodic structure. Each trained model was conditioned on held-out seed segments to produce continuations of comparable length, allowing consistent evaluation across entropy policies. Two complementary metrics were employed: the absolute difference in mean melody-change entropy ($|\Delta H|$), which measures the deviation in local melodic complexity between generated and reference material, and the Pearson correlation coefficient (r) between pitch-class distributions, which captures the similarity in tonal balance and overall note usage.

Table 3. Average entropy deviation ($|\Delta H|$) and Pearson correlation (r) between generated and reference melodies for each curvature parameter t. Lower $|\Delta H|$ and higher r indicate better preservation of melodic structure.

t	Mean $\|\Delta H\|$	Std.	Mean r	Std.
−1.00	0.5593	0.4651	0.5729	0.2342
−0.75	0.5502	0.4596	0.5719	0.2327
−0.50	0.5449	0.4603	0.5793	0.2341
−0.25	0.5531	0.4645	0.5757	0.2362
+0.00	0.5407	0.4654	0.5820	0.2333
+0.25	0.5337	0.4644	0.5831	0.2326
+0.50	0.5360	0.4671	0.5813	0.2325
+0.75	0.5307	0.4678	0.5840	0.2329
+1.00	0.5288	0.4618	0.5836	0.2307

As shown in Table 3, both metrics vary within a narrow numerical range, yet a consistent upward trend emerges as the curvature parameter t increases. Models trained on higher-entropy datasets achieve slightly lower entropy deviations and higher pitch-distribution correlations, suggesting improved structural alignment with the reference melodies. This pattern indicates that exposure to more information-dense melodic material during training enhances the model's capacity to preserve local complexity while maintaining tonal coherence.

Overall, these results demonstrate that entropy-guided data selection serves as a lightweight but effective mechanism for improving symbolic music generation. By regulating the entropy composition of the training corpus, the approach provides fine-grained control over melodic diversity without increasing model size or training cost.

4.3 Benchmark Comparison

Experiment compares the proposed entropy-guided preprocessing strategy with a conventional bar-level slicing baseline under identical model and training configurations. The objective is to evaluate whether aligning augmentation with local melodic entropy yields measurable improvements in generative fidelity and stylistic coherence.

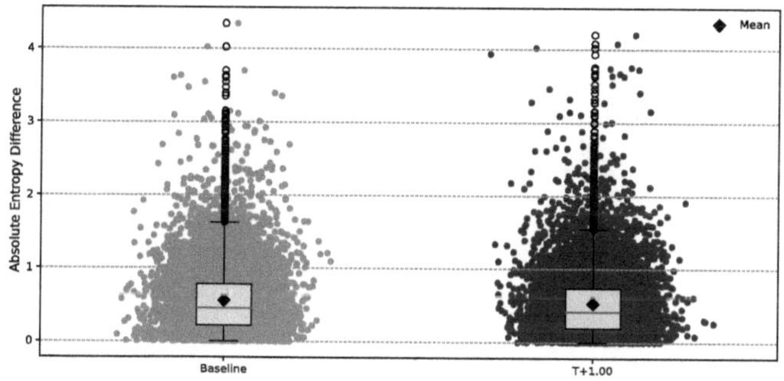

(a) Full distribution across all test seeds.

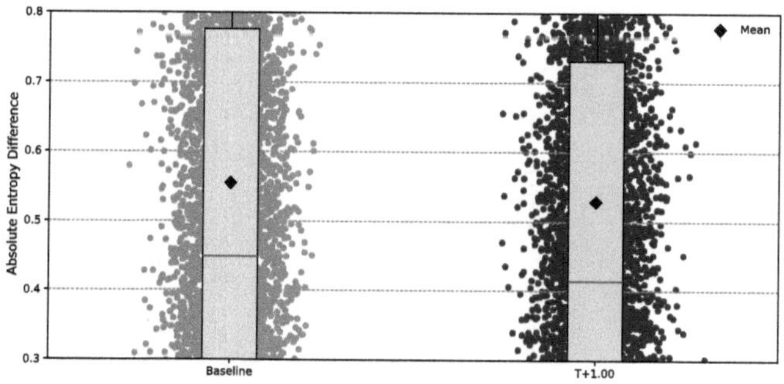

(b) Zoom-in around the mean region (0.3–0.8) highlighting separation between models.

Fig. 2. Absolute melodic-entropy deviation between generated continuations and ground-truth segments.

The baseline follows the conventional bar-level segmentation paradigm introduced by the Music Transformer [22], which detects bar boundaries automatically and samples fixed-length slices for training, combined with uniform pitch-shift augmentation. In contrast, the proposed model applies entropy-conditioned slicing and selective augmentation as defined in Sect. 3.3. During testing, both models were conditioned on identical seed segments extracted from the held-out evaluation set, with each seed consisting of 100 consecutive tokens. A total of 1,375 such seeds were used, each generating 400-token continuations for direct comparison with the corresponding ground-truth segments.

Figures 2 and 3 summarize performance across two complementary similarity measures. Figure 2 evaluates melodic deviation by comparing absolute entropy differences $|\Delta H|$ between generated and ground-truth continuations, capturing how well each model preserves local melodic fluctuation patterns. Figure 3, in contrast, measures pitch-class distribution similarity via Pearson correlation r, reflecting whether the model reproduces the characteristic usage proportions of each pitch class. Together, these two metrics provide a distribution-level view of melodic fidelity and stylistic consistency.

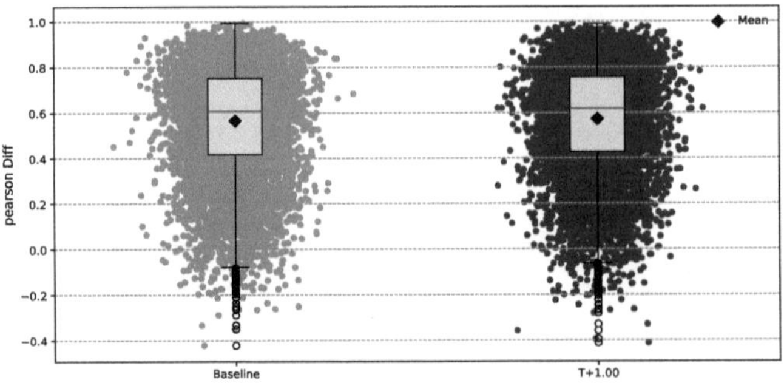

(a) Full correlation distribution cross all test seeds.

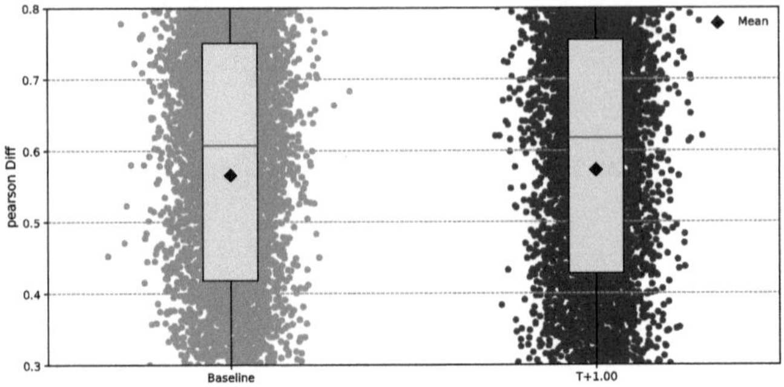

(b)Zoom-in on the 0.3–0.8 correlation range, emphasizing upper-range separation.

Fig. 3. Pearson correlation between pitch-class distributions of generated continuations and ground-truth segments.

Table 4. Comparison between the proposed entropy-guided model and the bar-level segmentation baseline. Lower $|\Delta H|$ and higher r indicate better preservation of melodic structure.

Model	Mean $\|\Delta H\|$	Std.	Mean r	Std.
Entropy-guided	0.5288	0.4618	0.5836	0.2307
Music Transformer baseline	0.5542	0.4623	0.5660	0.2354

As shown in Table 4, the proposed model outperforms the benchmark model across all four evaluation metrics. The improvement demonstrates that information-aware preprocessing can yield measurable benefits even under identical model and compute constraints. This confirms that entropy-guided data selection provides a practical and scalable alternative to uniform segmentation for symbolic music generation.

5 Conclusion and Future Work

This study introduced an entropy-guided preprocessing and augmentation framework for symbolic music generation, designed to enhance data efficiency and model fidelity under limited training resources. By adjusting the proportion of high-entropy melodic segments during dataset construction, the approach improves the alignment between generated continuations and their conditioning melodies, yielding more coherent and tonally consistent outputs without increasing model size or compute cost.

The empirical results demonstrate that entropy-aware sampling can serve as a lightweight yet effective mechanism for regulating melodic diversity, offering measurable improvements in generative quality over conventional bar-level preprocessing. These findings suggest a promising direction for developing compact, resource-efficient Transformer models capable of high-quality symbolic generation even on small-scale corpora.

The present study focused on representative curvature values to reveal the overall trend, rather than optimizing t through automated search. Similarly, the quartile-based slicing boundaries (25% and 75%) were fixed following conventional practice, although the observed right-skewed entropy distributions suggest that asymmetric or data-adaptive boundaries might yield more balanced sampling. Future work will explore these refinements, along with integrating entropy-guided selection into adaptive curriculum or active-learning frameworks for more systematic control of musical diversity and fidelity.

Acknowledgments. This work is supported by California State University Bakersfield Office of GRaSP, and the CV PATH program.

References

1. Muhamed, A., et al.: Symbolic music generation with transformer-GANs. In: Proceedings of AAAI Conference on Artificial Intelligence (AAAI), vol. 35, no. 1, pp. 408–417 (2021)
2. Shih, Y.-J., Wu, S.-L., Zalkow, F., Müller, M., Yang, Y.-H.: Theme transformer: symbolic music generation with theme-conditioned transformer. IEEE Trans. Multimedia **25**, 3495–3508 (2022)
3. Hoffmann, J., et al.: An empirical analysis of compute-optimal large language model training. In: Advances in Neural Information Processing Systems (NeurIPS), vol. 35, pp. 30016–30030 (2022)
4. Tian, S., Zhang, C., Yuan, W., Tan, W., Zhu, W.: Xmusic: towards a generalized and controllable symbolic music generation framework. IEEE Trans. Multimedia 1–15 (2025)

5. Ji, S., Yang, X., Luo, J.: A survey on deep learning for symbolic music generation: representations, algorithms, evaluations, and challenges. ACM Comput. Surv. **56**(1), 1–39 (2023)
6. BitMidi: Free MIDI Files Archive. https://bitmidi.com/. Accessed June 2024
7. Dong, H.-W., Hsiao, W.-Y., Yang, L.-C., Yang, Y.-H.: MuseGAN: multi-track sequential generative adversarial networks for symbolic music generation and accompaniment. In: Proceedings of AAAI Conference on Artificial Intelligence (AAAI) (2018)
8. Ren, Y., He, J., Tan, X., Qin, T., Zhao, Z., Liu, T.-Y.: PopMAG: pop music accompaniment generation. In: Proceedings of 28th ACM International Conference on Multimedia (ACM MM), pp. 1198–1206 (2020)
9. Zhou, Z., Zheng, J.: RWKV-Music: Efficient Long-Context Symbolic Music Modeling with RWKV Transformers. arXiv:2501.01234 (2025)
10. Huang, Y.-S., Yang, Y.-H.: Pop music transformer: beat-based modeling and generation of expressive pop piano compositions. In: Proceedings of 28th ACM International Conference on Multimedia (ACM MM), pp. 1180–1188 (2020)
11. Zeng, M., Tan, X., Wang, R., Ju, Z., Qin, T., Liu, T.-Y.: MusicBERT: symbolic music understanding with large-scale pre-training. In: Findings of ACL-IJCNLP 2021, pp. 791–800 (2021)
12. Le, D.V.T., Bigo, L., Herremans, D., Keller, M.: Natural language processing methods for symbolic music generation and information retrieval: a survey. ACM Comput. Surv. **57**(7), 175, 1–40 (2025). https://doi.org/10.1145/3714457
13. Pearce, M.T.: Statistical learning and probabilistic prediction in music cognition: mechanisms of stylistic enculturation. Ann. N.Y. Acad. Sci. **1423**(1), 378–395 (2018)
14. Marmoret, A., Marchi, E., Richard, G.: Barwise music structure analysis with the correlation block matrix. Trans. Int. Soci. Music Inf. Retrieval (TISMIR) **6**(1), 1–18 (2023)
15. Temperley, D.: Music and Probability. MIT Press (2007)
16. Bassan, S., Adi, Y., Rosenschein, J.S.: Unsupervised symbolic music segmentation using ensemble temporal prediction errors. In: Proceedings of Interspeech, pp. 2118–2122 (2022)
17. McFee, B., Humphrey, E.J., Bello, J.P.: A software framework for musical data augmentation. In: Proceedings of 16th International Society for Music Information Retrieval Conference (ISMIR), pp. 248–254 (2015)
18. Daikoku, T.: Entropy, uncertainty, and the depth of implicit knowledge on musical creativity: computational study of improvisation in melody and rhythm. Front. Comput. Neurosci. **12**, 97 (2018)
19. Vitter, J.S.: Random sampling with a reservoir. ACM Trans. Math. Softw. (TOMS) **11**(1), 37–57 (1985)
20. Bengio, Y., Louradour, J., Collobert, R., Weston, J.: Curriculum learning. In: Proceedings of 26th International Conference on Machine Learning (ICML) (2009)
21. Katharopoulos, A., Fleuret, F.: Not all samples are created equal: deep learning with importance sampling. In: Advances in Neural Information Processing Systems (NeurIPS) (2018)
22. Huang, C.-Z.A., et al.: Music Transformer: Generating Music with Long-Term Structure. arXiv:1809.04281 (2018)

Author Index

H. R. Arabnia et al. (Eds.): CSCI 2025, CCIS 2972, pp. 615–616, 2027.
https://doi.org/10.1007/978-3-032-28469-3

FSC
www.fsc.org
MIX
Papier | Fördert gute Waldnutzung
FSC® C083411

Kolibri 360 GmbH
Ferdinand-Jühlke-Straße 7
99095 Erfurt, Deutschland
produktsicherheit@kolibri360.de